AF775149

# Characterization and Control of Interfaces for High Quality Advanced Materials III

# Characterization and Control of Interfaces for High Quality Advanced Materials III

Ceramic Transactions, Volume 219

*Proceedings of the Third International Conference on Characterization and Control of Interfaces for High Quality Advanced Materials*
*Kurashiki, Japan (2009)*

Edited by
Kevin Ewsuk
Makio Naito
Tomoyuki Kakeshita
Soshu Kirihara
Keizo Uematsu
Hiroya Abe

The American Ceramic Society

WILEY

A John Wiley & Sons, Inc., Publication

Copyright © 2010 by The American Ceramic Society. All rights reserved.

Published by John Wiley & Sons, Inc., Hoboken, New Jersey.
Published simultaneously in Canada.

No part of this publication may be reproduced, stored in a retrieval system, or transmitted in any form or by any means, electronic, mechanical, photocopying, recording, scanning, or otherwise, except as permitted under Section 107 or 108 of the 1976 United States Copyright Act, without either the prior written permission of the Publisher, or authorization through payment of the appropriate per-copy fee to the Copyright Clearance Center, Inc., 222 Rosewood Drive, Danvers, MA 01923, (978) 750-8400, fax (978) 750-4470, or on the web at www.copyright.com. Requests to the Publisher for permission should be addressed to the Permissions Department, John Wiley & Sons, Inc., 111 River Street, Hoboken, NJ 07030, (201) 748-6011, fax (201) 748-6008, or online at http://www.wiley.com/go/permission.

Limit of Liability/Disclaimer of Warranty: While the publisher and author have used their best efforts in preparing this book, they make no representations or warranties with respect to the accuracy or completeness of the contents of this book and specifically disclaim any implied warranties of merchantability or fitness for a particular purpose. No warranty may be created or extended by sales representatives or written sales materials. The advice and strategies contained herein may not be suitable for your situation. You should consult with a professional where appropriate. Neither the publisher nor author shall be liable for any loss of profit or any other commercial damages, including but not limited to special, incidental, consequential, or other damages.

For general information on our other products and services or for technical support, please contact our Customer Care Department within the United States at (800) 762-2974, outside the United States at (317) 572-3993 or fax (317) 572-4002.

Wiley also publishes its books in a variety of electronic formats. Some content that appears in print may not be available in electronic format. For information about Wiley products, visit our web site at www.wiley.com.

***Library of Congress Cataloging-in-Publication Data is available.***

ISBN 978-0-470-90917-1

10 9 8 7 6 5 4 3 2 1

# Contents

## MATERIALS DESIGN

# Preface

Interfaces are critically important to a broad spectrum of materials and technologies. In 2003, the First International Conference on Characterization and Control of Interfaces for High Quality Advanced Materials (ICCCI2003) established an international forum to discuss interface science and technology. Interest and participation doubled in 2006, at the Second International Conference (ICCCI2006). The discussion continued in 2009 at the Third International Conference (ICCCI2009), where interface characterization and control technology for nano- to micro-scale materials synthesis, powder processing, composite processing, joining, and the control of airborne particulates were addressed by scientists and engineers from academia, industry, and national laboratories. Additionally, joining technology for new metallic glasses and inorganic materials, and interface phenomena at high temperature were featured as special topics at ICCCI2009.

ICCCI2009 was supported by the Global COE Program "Center of Excellence for Advanced Structural and Functional Materials Design" lead by Professor Tomoyuki Kakeshita at Osaka University, the Project on Joining Technology for New Metallic Glasses and Inorganic Materials, the Institute of Materials Research (IMR) of Tohoku University, the Materials and Structures Laboratory (MSL) of the Tokyo Institute of Technology, Kobe Gakuin University, Hosokawa Powder Technology Foundation, the Japan JSPS 124th Committee, and the Joining and Welding Research Institute (JWRI) of Osaka University.

Over 160 scientists and engineers from academia and industry from 18 different countries attended ICCCI2009 to see and discuss 140 invited and contributed presentations and posters on the state-of-the-art of interface characterization and control for particulate materials, joining, and nanotechnology. This proceedings volume features 57 peer-reviewed papers from ICCCI2009 on interface characterization and control technology, powder and composite processing, joining, the control of airborne particulates, new metallic glasses, and interface phenomena at high temperature.

KEVIN EWSUK
MAKIO NAITO
TOMOYUKI KAKESHITA
SOSHU KIRIHARA
KEIZO UEMATSU
HIROYA ABE

# Joining Technology for New Metallic Glasses and Inorganic Materials

# DEVITRIFICATION BEHAVIOR AND CRYSTAL-GLASSY MIXED-PHASE STRUCTURES OBSERVED IN PARTIALLY CRYSTALLIZED Cu-BASED GLASSY ALLOYS

D. V. Louzguine-Luzgin[1,2], G. Xie[2], Q. Zhang[1] and A. Inoue[1,2]
[1] *WPI Advanced Institute for Materials Research, Tohoku University, Sendai 980-8577, Japan*
[2] *Institute for Materials Research, Tohoku University, Aoba-Ku, Sendai 980-8577, Japan*

## ABSTRACT

We studied Cu-Zr-based alloys having exceptionally high glass-forming ability (GFA), and investigated the influence of Ag and Al on the GFA and crystallization behavior. Most of the bulk glassy alloys studied do not contain any crystals in the as-solidified state. The crystallization kinetics of $Cu_{55}Zr_{45}$, $Cu_{50}Zr_{50}$, $Cu_{55-x}Zr_{45}Ag_x$ (x = 0, 10, 20), $Cu_{45}Zr_{45}Al_5Ag_5$, $Cu_{55}Zr_{30}Ti_{10}Ag_5$, $Cu_{55}Hf_{25}Ti_{15}Ag_5$, $Cu_{44}Ag_{15}Zr_{36}Ti_5$ and $Cu_{36}Zr_{48}Al_8Ag_8$ glassy alloys was analyzed. $Cu_{35}Zr_{45}Ag_{20}$ alloy exhibits possible phase separation upon heating within the supercooled liquid region prior to crystallization. The depletion of Ag in the residual glassy phase results in the higher thermal stability of $Cu_{35}Zr_{45}Ag_{10}$ versus $Cu_{35}Zr_{45}Ag_{20}$. $Cu_{50}Zr_{50}$ undergoes primary crystallization, forming a metastable monoclinic CuZr phase, while $Cu_{55}Zr_{45}$, $Cu_{45}Zr_{45}Ag_{10}$, $Cu_{35}Zr_{45}Ag_{20}$ and $Cu_{45}Zr_{45}Al_5Ag_5$ undergo eutectic crystallization, forming mostly stable phases. $Cu_{50}Zr_{50}$, $Cu_{55}Zr_{30}Ti_{10}Ag_5$ and $Cu_{55}Hf_{25}Ti_{15}Ag_5$ undergo nanocrystallization. Although some differences in the crystallization behavior between $Cu_{55-x}Zr_{45}Ag_x$ (x = 0, 10, 20) and $Cu_{45}Zr_{45}Al_5Ag_5$ were found, this may not be the only reason Al and Ag additions improve the GFA of the Cu-Zr alloy. The difficulty in nucleating the tP4 $AgZr^{ss}$ phase may be responsible for the high glass-forming ability of the $Cu_{36}Zr_{48}Al_8Ag_8$ and $Cu_{44}Ag_{15}Zr_{36}Ti_5$ alloys.

## INTRODUCTION

The structure and properties of bulk glassy alloys have been widely studied since the late 1980's [1,2,3]. Cu-Zr-based alloys exhibit good potential for industrial applications. However, the glass-forming ability (GFA) of Cu-rich alloys has been limited [4]. Several attempts have also been made to increase the GFA of these alloys based on different criteria. These include a reduced glass transition temperature, $T_{rg} = T_g/T_l$ [5], where $T_g$ is the glass-transition temperature and $T_l$ is the liquidus temperature; the width of the supercooled liquid region ($\Delta T_x$) defined as $T_x$-$T_g$, where $T_x$ is the crystallization onset temperature, and $\gamma=T_x/(T_g+T_l)$ [6]. Binary Cu-Zr and Cu-Hf bulk glassy alloys were reported to form, though only in narrow composition ranges [7]. It was also reported that some of these binary glasses contained nanocrystals in the as-quenched condition [8]. The addition of a third element like Ti [9] or Al [10], for example, produced dense packing [11], and enhanced the GFA [12] of the binary alloys [13]. Thus, $Cu_{36}Zr_{48}Al_8Ag_8$ could be produced in a fully glassy state with a maximum diameter of 25 mm [14]. The structure and crystallization behavior of some binary Cu-Zr [15], ternary Cu-Zr-Ti [16, 17, 18, 19], and some quaternary Cu-based [20] glassy alloys were studied in detail. The nanocrystal-glassy Cu-Zr [8] and Cu-Zr-Al [21, 22] composites showed a large room-temperature compressive plastic deformation of up to 50%. Ag drastically improved the GFA of Cu-Zr-Ti alloys [23]. $Cu_{45}Zr_{45}Ag_{10}$ exhibited a relatively high GFA and a high mechanical strength[24].

Following the above findings, the role of Al and Ag to improve the GFA of different Cu-based alloys, their thermal stability, and their crystallization behavior were studied in the present work. Another purpose is to find a possible link between the crystallization behavior and GFA of the alloys.

## EXPERIMENTAL PROCEDURE

The ingots of the $Cu_{55}Zr_{45}$, $Cu_{50}Zr_{50}$, $Cu_{55-x}Zr_{45}Ag_x$ (x = 10, 20), $Cu_{45}Zr_{45}Al_5Ag_5$, $Cu_{55}Zr_{30}Ti_{10}Ag_5$, $Cu_{55}Hf_{25}Ti_{15}Ag_5$, $Cu_{44}Ag_{15}Zr_{36}Ti_5$ and $Cu_{36}Zr_{48}Al_8Ag_8$ glassy alloys (compositions are given in nominal atomic percentages) were prepared by arc-melting mixtures of (99.9 wt.% purity) under an argon atmosphere. From these ingots, ribbon samples of about 20 μm thickness and 1 mm width were prepared by melt spinning on to a single copper roller at a roller tangential velocity of about 40 m/s. Bulk rod samples of more than 1 mm diameter were prepared by the copper-mold casting technique. The structure of both the ribbon and rod samples was examined by X-ray diffractometry (XRD) with monochromatic $CuK_{\alpha}$ radiation. A high-resolution field-emission scanning electron microscope (SEM Hitachi S-4800) equipped with an energy dispersive X-ray spectrometer (EDS) was also used for structural characterization of the polished cross-sections. Phase transformations were studied by differential scanning calorimetry (DSC) at a heating rate of 0.67 K/s and differential isothermal calorimetry (DIC). Transmission electron microscopy (TEM) investigations were carried out using a JEOL JEM 2010 microscope operating at 200 kV and equipped with an energy dispersive X-ray spectrometer (EDS) of 0.1 keV resolution. Samples for TEM were prepared first mechanically and subsequently by an ion-polishing technique. In order to avoid structural damage the ion-beam energy was kept below 2 keV. The oxygen content in the studied alloys was kept below 500 ppm (by weight).

## RESULTS AND DISCUSSION

Most of the bulk glassy alloys studied do not contain any crystals in the as-solidified state while some samples like $Cu_{44}Ag_{15}Zr_{36}Ti_5$ studied by TEM were found to contain well developed medium-range order zones and even nanoparticles in a bulk form. For example, the structure of the $Cu_{36}Zr_{48}Al_8Ag_8$ metallic bulk glass-forming alloy studied by high-resolution transmission electron microscopy (HRTEM) is shown in Fig. 1. On prolonged observation in the TEM, the glassy phase partially crystallized. Nanocrystals were detected after about 120 s of observation under focused beam which is typical for some Cu-based glassy alloys[25].

Following our Fig. 1 and our recent XRD studies one can suggest that the medium-range order in the studied Cu-Zr-based alloys maintains up to about 2 nm distance [26]. It is also shown that the interatomic distances in the first coordination shell correspond to those of some crystalline compounds used as glassy-phase approximants [27].

The GFA and the crystallization behavior of glassy alloys were analyzed and are summarized in Table I. Natural logarithm of the critical diameter as an indicator of the GFA is plotted as a function of $T_{rg}$ and $\gamma$ parameter in Fig. 2. Although, some other indicators of the GFA give a better correspondence than $T_{rg}$ [28,29], $T_{rg}$ correlates quite well with the critical diameter of Cu-Zr-based alloys belonging to the similar systems studied in the present work.

Both Ag-bearing and Ag-free alloys exhibit devitrification forming of a supercooled liquid region and subsequent formation of the equilibrium crystalline phases. The addition of Ti causes nanocrystallization of Cu-Zr-Ag as well as Cu-Zr and Cu-Hf alloys (see Fig. 3 illustrating HRTEM image and nanobeam electron diffraction (NBD) pattern) though the XRD pattern of the annealed sample still does not show any sharp peaks (Fig. 4). The $Cu_{50}Zr_{50}$ glassy alloy exhibited a primary crystallization behavior forming a metastable monoclinic mP4 CuZr phase while $Cu_{55}Zr_{45}$, $Cu_{45}Zr_{45}Ag_{10}$, $Cu_{35}Zr_{45}Ag_{20}$, $Cu_{50}Zr_{45}Al_5$ and $Cu_{45}Zr_{45}Al_5Ag_5$ glassy alloys showed rather eutectic crystallization forming mostly equilibrium phases (Fig. 5). For example, the simultaneous formation of oC68 $(Cu,Ag)_{10}Zr_7$ and tP4 (Ag,Cu)Zr solid solution phases in the Cu-Zr-Ag alloys by nucleation and 3-dimensional interface-controlled growth of nuclei indicates rather eutectic crystallization. Although, some difference in the crystallization behavior of $Cu_{55-x}Zr_{45}Ag_x$, $Cu_{50}Zr_{45}Al_5$ and $Cu_{45}Zr_{45}Al_5Ag_5$ glassy alloys was found as shown in Table I it may be not the only reason for

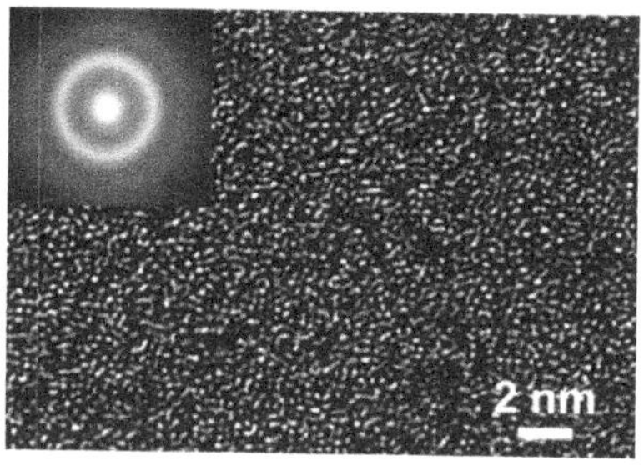


Figure 1. Typical HRTEM image of the $Cu_{36}Zr_{48}Al_8Ag_8$ glassy alloy. The insert is selected-area diffraction pattern.

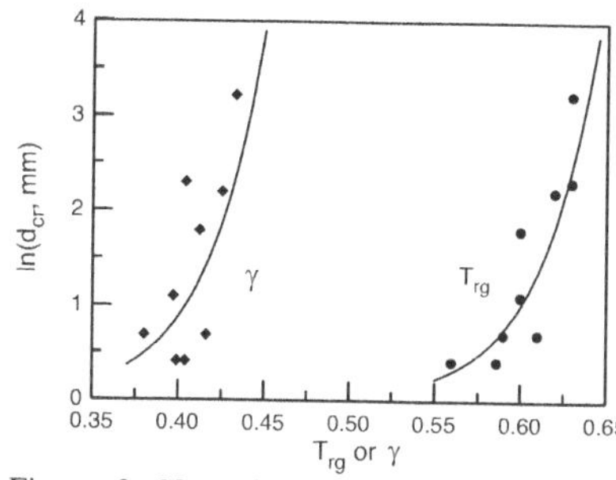


Figure 2. Natural logarithm of the critical diameter as a function of $T_{rg}$ and $\gamma$ parameter given in Table I.

Table I. GFA, natural logarithm of critical diameter ($d_{cr}$) and crystallization behavior of the Cu-Zr-based alloys studied. IR-initial crystallization reaction.

| Alloy | ln($d_{cr}$, mm) | $T_g/T_l$ | $\gamma$ | IR | Phase composition* |
|---|---|---|---|---|---|
| $Cu_{55}Zr_{45}$ | 0.41 | 0.59 | 0.399 | Eutectic | oC68 $Cu_{10}Zr_7$+ cP2 CuZr |
| $Cu_{50}Zr_{50}$ | 0.69 | 0.59 | 0.380 | Primary | mP4 CuZr |
| $Cu_{45}Zr_{45}Ag_{10}$ | 1.79 | 0.60 | 0.412[30] | Eutectic | oC68 $Cu_{10}Zr_7$ and tP4 AgZr |
| $Cu_{35}Zr_{45}Ag_{20}$ | 0.41 | 0.56 | 0.404 | Eutectic | oC68 $Cu_{10}Zr_7$ and tP4 AgZr |
| $Cu_{55}Zr_{30}Ti_{10}Ag_5$ | 1.1 | 0.60 | 0.397 | Primary | nanoscale, unidentified |
| $Cu_{50}Zr_{45}Al_5$ | 0.69 | 0.61[31] | 0.416 | Eutectic | oC68 $Cu_{10}Zr_7$ + unidentified |
| $Cu_{45}Zr_{45}Al_5Ag_5$ | 2.20 | 0.62[32] | 0.425 | Eutectic | oC68 $Cu_{10}Zr_7$, cF16 $Cu_2AlZr$ and tP4 AgZr |
| $Cu_{44}Ag_{15}Zr_{36}Ti_5$ | 2.30 | 0.63[25] | 0.404 | Primary | tP4 AgZr |
| $Cu_{36}Zr_{48}Al_8Ag_8$ | 3.22 | 0.63 | 0.433 | Primary | tP4 AgZr |

* a certain solid solubility of other alloying elements is found in some of the binary phases.

the improvement of the GFA of the Cu-Zr alloys by Al and Ag addition.

Contrary to the $Cu_{45}Zr_{45}Al_5Ag_5$ alloy in which a mixture of oC68 $Cu_{10}Zr_7^{ss}$, cF16 $Cu_2AlZr$ and tP4 $AgZr^{ss}$ (ss denotes solid solution) phases was found to form directly by a single-stage eutectic-type reaction and other similar eutectic alloys, the $Cu_{36}Zr_{48}Al_8Ag_8$ alloy exhibited a primary precipitation of tP4 $AgZr^{ss}$ phase upon crystallization on heating. $Cu_{36}Zr_{48}Al_8Ag_8$ and $Cu_{44}Ag_{15}Zr_{36}Ti_5$ alloys have a higher glass-forming ability than the $Cu_{45}Zr_{45}Al_5Ag_5$ alloy but exhibited a primary crystallization. The first DSC exothermic peak in the $Cu_{36}Zr_{48}Al_8Ag_8$ alloy exhibits two shoulders while a single-stage transformation takes place in the $Cu_{45}Zr_{45}Al_5Ag_5$ alloy suggesting that this is close to the eutectic composition. The values of the Avrami exponent close to 4 observed upon the isothermal crystallization of the eutectic alloys supports this conclusion.

$Cu_{35}Zr_{45}Ag_{20}$ alloy exhibited possible phase separation upon heating within the supercooled liquid region prior to crystallization of the liquid. Fig. 6 represents a globular residual amorphous phase surrounded by crystalline (Ag,Cu)Zr and $(Cu,Ag)_{10}Zr_7$ phases. The depletion of the residual glassy phase in Ag as tested by EDX in TEM argues for the higher thermal stability of the $Cu_{35}Zr_{45}Ag_{10}$ glassy phase versus $Cu_{35}Zr_{45}Ag_{20}$.

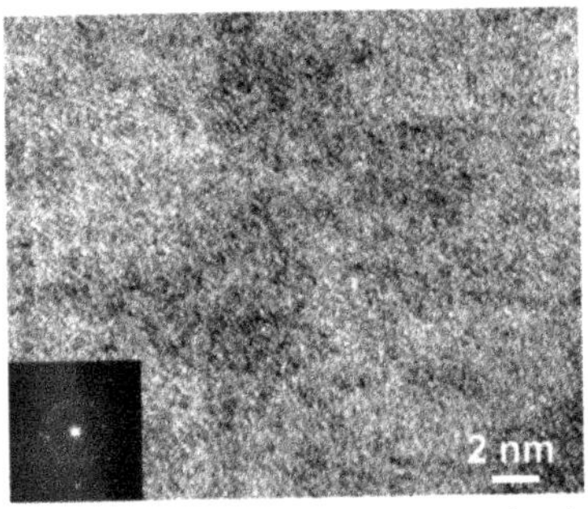


Figure 3. HRTEM image of the $Cu_{55}Hf_{25}Ti_{15}Ag_5$ alloy annealed at 770 K for 300 s. The insert is NBD pattern.

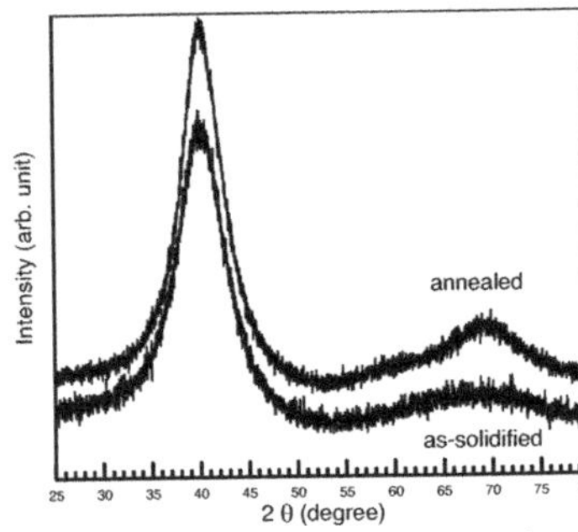


Figure 4. XRD patterns of the $Cu_{55}Hf_{25}Ti_{15}Ag_5$ alloy in as-solidified and annealed at 770 K for 300 s state.

The structure of the nanoscale phases in $Cu_{55}Zr_{30}Ti_{10}Ag_5$ and $Cu_{55}Hf_{25}Ti_{15}Ag_5$ alloys remained unknown owing to their small size. One should also admit a very high density of the precipitates (Fig. 3). Contrary to Cu-Zr-Ti-(Pd,Au) alloys containing noble metals in which formation of the nanoscale icosahedral phase was observed[33] the Ag-containing glassy alloys studied in the present work did not form the icosahedral phase on crystallization.

No significant difference is found in the glass-transition temperature of the melt-spun ribbons and bulk glassy rods of $Cu_{36}Zr_{48}Al_8Ag_8$. On annealing the melt-spun ribbon sample at 744 K for 480 s, primary crystallization led to the formation of tP4 $AgZr^{ss}$ phase. An unidentified phase was also present. Additionally, the oC68 $Cu_{10}Zr_7^{ss}$ phase formed at higher temperatures. On annealing the ribbon sample at a higher temperature of 900 K for 300 s, crystallization was complete resulting in the formation of the tI6 $Zr_2Cu$ crystalline phase.

In the case of the bulk (10-mm-diameter rod) samples formation of the $AgZr^{ss}$ phase was also observed in the initial stage of crystallization. The $Cu_{10}Zr_7$ phase also formed in the bulk sample on heat treating at 900 K, and a well-defined fraction of the $CuZr_2$ phase is present after long term annealing. However, even at such a high temperature the structure is not in equilibrium after annealing for 84.6 ks. Only on annealing at 1000 K the ribbon and bulk samples attained the equilibrium structure which indicates a very high stability of the intermediate phases.

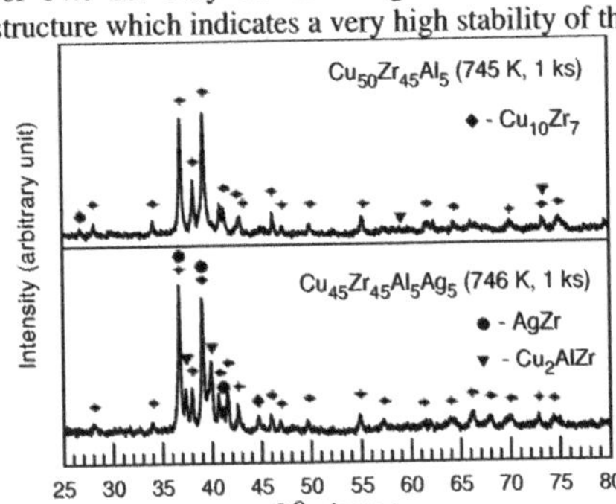


Figure 5. XRD patterns of some alloys in a heat-treated state.

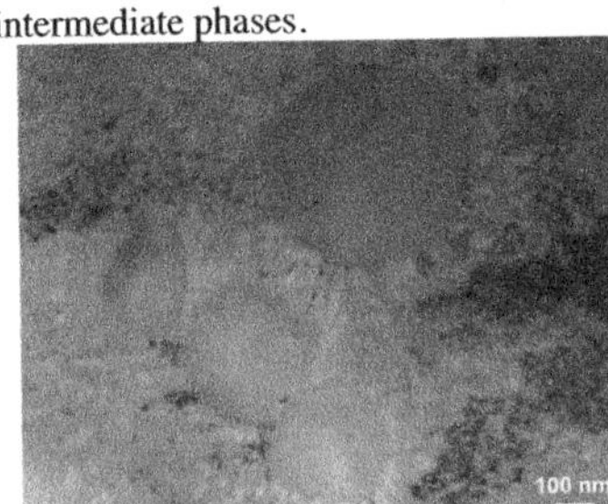


Figure 6. $Cu_{35}Zr_{45}Ag_{20}$ alloy annealed at 722 K for 1 ks, bright-field image, TEM.

Difficulty of nucleation of tP4 $AgZr^{ss}$ phase may be the reason for the high glass-forming ability of $Cu_{36}Zr_{48}Al_8Ag_8$ and $Cu_{44}Ag_{15}Zr_{36}Ti_5$ alloys. Although this phase has quite a simple structure it contains a large fraction of Ag which is rather a minor alloying element in the alloys, and thus, diffusion redistribution of the alloying elements is required upon its formation, which may slow down the crystallization kinetics.

## CONCLUSIONS

The crystallization behavior of $Cu_{55}Zr_{45}$, $Cu_{50}Zr_{50}$, $Cu_{55-x}Zr_{45}Ag_x$ (x = 0, 10, 20), $Cu_{45}Zr_{45}Al_5Ag_5$, $Cu_{55}Zr_{30}Ti_{10}Ag_5$, $Cu_{55}Hf_{25}Ti_{15}Ag_5$, $Cu_{44}Ag_{15}Zr_{36}Ti_5$ and $Cu_{36}Zr_{48}Al_8Ag_8$ glassy alloys was analyzed. $Cu_{35}Zr_{45}Ag_{20}$ alloy exhibited possible phase separation upon heating within the supercooled liquid region prior to crystallization. The depletion of Ag in the residual glassy phase results in the higher thermal stability of the $Cu_{35}Zr_{45}Ag_{10}$ versus $Cu_{35}Zr_{45}Ag_{20}$. $Cu_{55}Zr_{45}$, $Cu_{45}Zr_{45}Ag_{10}$, $Cu_{35}Zr_{45}Ag_{20}$ and $Cu_{45}Zr_{45}Al_5Ag_5$ glassy alloys show a eutectic-type crystallization, forming mostly stable phases. The addition of Ti causes nanocrystallization of Cu-Zr-Ag as well as Cu-Zr and Cu-Hf alloys. Although some difference in the crystallization behavior of $Cu_{55-x}Zr_{45}Ag_x$ (x = 0, 10, 20) and $Cu_{45}Zr_{45}Al_5Ag_5$ glassy alloys was found, this may not be the only reason for the improvement in the GFA of the Cu-Zr alloys with the addition of Al and Ag. The difficulty in nucleating tP4 $AgZr^{ss}$ may be responsible for the high glass-forming ability of $Cu_{36}Zr_{48}Al_8Ag_8$ and $Cu_{44}Ag_{15}Zr_{36}Ti_5$ alloys.

## REFERENCES

[1] A. Inoue, "High Strength Bulk Amorphous Alloys with Low Critical Cooling Rates", *Mater. Trans. JIM* 36, 866 (1995).
[2] W. L. Johnson, "Bulk glass-forming metallic alloys: science and technology", *MRS Bull.* 24(10), 42 (1999).
[3] A. Inoue, "Stabilization of metallic supercooled liquid and bulk amorphous alloys," *Acta Mater.* 48, 279 (2000).
[4] D. H. Xu, G. Duan and W. L. Johnson, "Unusual Glass-Forming Ability of Bulk Amorphous Alloys Based on Ordinary Metal Copper", *Phys. Rev. Lett.* 92, 245 (2004).
[5] D. Turnbull and M. H. Cohen, "Free-Volume Model of the Amorphous Phase: Glass Transition", *J. Chem. Phys.* 34, 120 (1961).
[6] Z.P. Lu, C.T. Liu, "A New Glass-Forming Ability Criterion for Bulk Metallic Glasses", *Acta Mater.* 50, 3501 (2002).
[7] A. Inoue and W. Zhang, "Formation, Thermal Stability and Mechanical Properties of Cu-Zr and Cu-Hf Binary Glassy Alloy Rods", *Mater. Trans.* 45, 584 (2004).
[8] A. Inoue, W. Zhang, T. Tsurui, A. R. Yavari and A. L. Greer, "Unusual Room-Temperature Compressive Plasticity in Nanocrystal-Toughened Bulk Copper-Zirconium Glass", *Phil. Mag. Lett.* 85, 221 (2005).
[9] A. Inoue, W. Zhang, T. Zhang and K. Kurosaka, "High-Strength Cu-Based Bulk Glassy Alloys in Cu-Zr-Ti and Cu-Hf-Ti Ternary Systems", *Acta Mater.* 49, 2645 (2001).
[10] A. Inoue and W. Zhang, "Formation, Thermal Stability and Mechanical Properties of Cu-Zr-Al Bulk Glassy Alloys", *Mater. Trans.* 43, 2921 (2002).
[11] D. B. Miracle, W. S. Sanders, "The Influence of Efficient Atomic Packing on the Constitution of Metallic Glasses", *Phil. Mag.* 83, 2409–2428, (2003).
[12] H.-J. Fecht, W.L. Johnson, "Thermodynamic Properties and Metastability of Bulk Metallic Glasses", *Mater. Sci. Eng. A* 375–377, 2–8 (2004).
[13] A. L. Greer, "Metallic Glasses", *Science* 267, 1947 (1995).
[14] Q. Zhang, W. Zhang and A. Inoue, "Preparation of $Cu_{36}Zr_{48}Ag_8Al_8$ Bulk Metallic Glass with a Diameter of 25 mm by Copper Mold Casting", *Mater. Trans.* 48, 629 (2007).

[15] N. Mattern, A. Schops, U. Kuhn, J. Acker, O. Khvostikova and J. Eckert, "Structural Behavior of $Cu_xZr_{100-x}$ Metallic Glass (x = 35-70)", *J. Non-Crystalline Solids* 354, 1054 (2008).
[16] J. Z. Jiang, J. Saida, H. Kato, T. Ohsuna and A. Inoue, "Is $Cu_{60}Ti_{10}Zr_{30}$ a Bulk Glass-Forming Alloy?", *Appl. Phys. Lett.* 82, 4041 (2003).
[17] D. V. Louzguine and A. Inoue, "Nanocrystallization of Cu-(Zr or Hf)-Ti Metallic Glasses", *J. Mater. Res.* 17, 2112 (2002).
[18] D. V. Louzguine and A. Inoue, "Evaluation of the Thermal Stability of a $Cu_{60}Hf_{25}Ti_{15}$ Metallic Glass", *Appl. Phys. Lett.* 81, 2561 (2002).
[19] M. Kasai, J. Saida, M. Matsushita, T. Osuna, E. Matsubara and A Inoue, "Structure and Crystallization of Rapidly Quenched Cu-(Zr or Hf)-Ti Alloys Containing Nanocrystalline Particles", *J. Phys. Condensed Matter* 14, 13867 (2002).
[20] D. V. Louzguine and A. Inoue, "Influence of Ni and Co Additions on Supercooled Liquid Region, Devitrification Behavior and Mechanical Properties of Cu-Zr-Ti Bulk Metallic Glass", *J. Metastable & Nanocrystalline Mater.* 15-16, 31 (2003).
[21] J. Das, M. B. Tang, K. B. Kim, R. Theissmann, F. Baier, W. H. Wang and J. Eckert, ""Work-Hardenable" Ductile Bulk Metallic Glass", *Phys. Rev. Lett.* 94, 205501 (2005).
[22] S. Pauly, J. Das, C. Duhamel and J. Eckert, Martensite Formation in a Ductile $Cu_{47.5}Zr_{47.5}Al_5$ Bulk Metallic Glass Composite", *Adv. Eng. Mater.* 9, 487 (2007)
[23] C.-L. Dai, H. Guo, Y. Shen, Y. Li, E. Ma and J. Xu, "A New Centimeter–Diameter Cu-Based Bulk Metallic Glass", *Scripta Mater.* 54, 1403 (2006).
[24] F. Jia, W. Zhang and A. Inoue, "Effects of Additional Hf on the Thermal Stability and Mechanical Properties of Cu-Zr-Ag Bulk Glassy Alloys", *Mater. Trans.* 47, 1922 (2006).
[25] G. Xie, Q. Zhang, D. V. Louzguine, W. Zhang and A. Inoue, "Stability of Nanocrystallites Dispersed in $Cu_{50}Zr_{45}Ti_5$ Metallic Glass Under Electron Irradiation", *J. Nanosci. and Nanotech.* 7, 3286 (2007).
[26] D. V. Louzguine-Luzgin, K. Georgarakis, A. R. Yavari, G. Vaughan, G. Xie and A. Inoue "Effect of Ag Addition on Local Structure of Cu-Zr Glassy Alloy", *Journal of Materials Research*, 24, 274 (2009).
[27] D. V. Louzguine-Luzgin, A. R. Yavari, G. Vaughan and A. Inoue, "Clustered Crystalline Structures as Glassy Phase Approximants", *Intermetallics*, 17, 477 (2009).
[28] E. S. Park, D. H. Kima and W. T. Kim, Parameter for glass forming ability of ternary alloy systems, *Appl. Phys. Lett.* 86, 061907 (2005).
[29] Z.P. Lu, Y. Li, S.C. Ng, "Reduced Glass Transition Temperature and Glass Forming Ability of Bulk Glass Forming Alloys", *Journal of Non-Crystalline Solids* 270, 103 (2000).
[30] W. Zhang, Q. Zhang, C. Qin and A. Inoue, "Synthesis and Properties of Cu-Zr-Ag-Al Glassy Alloys with High Glass-Forming Ability", *Materials Science and Engineering: B* 148, 92, (2008).
[31] W. Zhang and A. Inoue, "Formation and Mechanical Strength of New Cu-Based Bulk Glassy Alloys", *Materials Transactions*, 45, 1210 (2004).
[32] W. Zhang, F. Jia, Q. S. Zhang and A. Inoue, "Formation and Properties of Cu-Zr Binary Glassy Alloys", *Mater. Sci. Eng. A* 459, 330 (2007).
[33] D. V. Louzguine-Luzgin and A. Inoue, "Formation and Properties of Quasicrystals", *Annual Review of Materials Research*, 38, 403 (2008).

# DIRECT BONDING OF COPPER TO ALUMINA AND ITS CHARACTERIZATION

Wei-Hsing Tuan and Shao-Kuan Lee
Dept. of Materials Science and Engineering, National Taiwan University, Taipei, Taiwan 106

## ABSTRACT

In the present study, $Al_2O_3/Cu/Al_2O_3$ sandwich substrates were prepared by a direct bonding process. Various amounts of oxygen are first introduced into the copper by pre-oxidizing treatments. A very small amount of $Cu_2O$ is formed on the surface of the pre-oxidized Cu foil. As the joining takes place above 1065°C, a strong bond between Cu and $Al_2O_3$ is formed. The energy release rate of the $Cu/Al_2O_3$ interface can reach a value of 40 $J/m^2$. The strength of the sandwich also shows a strong dependence on the oxygen content in the Cu foil.

## INTRODUCTION

The characteristics of the metal-ceramic interface are important in many applications such as electronic packaging, multilayer ceramic capacitors (MLCCs), wear resistant coatings, and micro-electromechanical systems (MEMS). Among the many metal-ceramic systems, the Cu-$Al_2O_3$ system has received wide attention[1-3]. Metallic copper is superior in electrical conductivity, thermal conductivity, and migration resistance. Ceramic alumina is widely used for its high electrical resistance and availability. By combining Cu and $Al_2O_3$ together, the Cu-$Al_2O_3$ substrate can be used for both thermal dissipation and electrical insulation applications.

Among all the bonding techniques, direct bonding through the help of a eutectic liquid is an important one. The process is carried out below the melting point of copper, which is 1083°C. Furthermore, no expensive brazing material is needed. The eutectic temperature for the Cu-O system is 1065°C. The eutectic bonding is carried out between 1065°C and 1083°C. The direct bonding process is thus also described as eutectic bonding[4]. The solubility of oxygen in copper is very small, ~0.03at%[5]. Above the solubility limit, a cuprite oxide ($Cu_2O$) is formed at low oxygen concentration. The amount of oxygen content in copper should affect the bonding characteristics; however, this issue is seldom addressed in previous studies. In the present study, the effect of oxygen content in copper on the bonding of Cu and $Al_2O_3$ is investigated.

## EXPERIMENTAL PROCEDURES

The $Al_2O_3/Cu/Al_2O_3$ sandwich substrates were prepared in the present study. The alumina substrates with the dimensions of $30 \times 7 \times 0.64$ mm were used. The purity of the substrate was 96% $Al_2O_3$. The copper foils with purity > 99% were used. The thickness of Cu foil was 50 μm. The Cu foil was pre-oxidized at a temperature varied from 300°C to 600°C for various times. The oxygen content in Cu after pre-oxidation treatments was determined by applying an IR spectroscopy technique

(Nitrogen/Oxygen Analyzer, TC-136, LECO Co., USA). Apart from the IR technique, X-ray diffraction (XRD, Philips Co., NJ, USA) method was also used to determine the phases on the surface of the pre-oxidized Cu foil. The bonding of alumina to the pre-oxidized copper was carried out at 1075°C for 10 minutes. Flowing nitrogen with the rate of 50 cc/min was used during joining.

The four-point bending technique was used to determine the flexural strength and interface strength of the $Al_2O_3/Cu/Al_2O_3$ sandwich specimen. The configuration for the bending test is illustrated in Fig. 1. The loading rate was 0.001mm/sec. The fracture surface was observed with scanning electron microscopy (SEM). The electron probe microanalysis (EPMA, JXA-8600, JEOL Co., Tokyo, Japan) was also used to detect the composition profile.

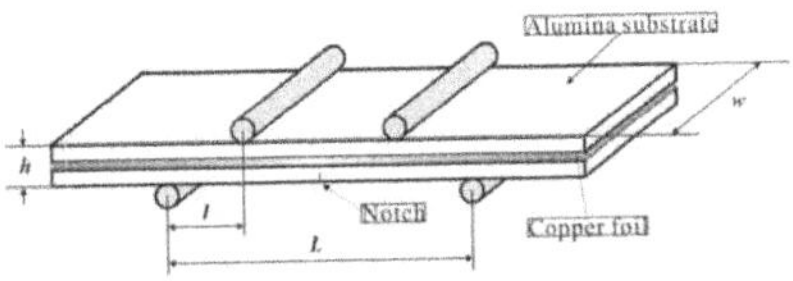


Figure 1. Schematic of the specimen and test configuration.

RESULTS

By using the IR spectroscopy technique, the oxygen content in the pre-oxidized Cu foil varies from 0.11 wt% to 1.8 wt%. To be demonstrated later, the concentration is much higher near the surface region. The oxygen content should thus be treated as the average value in the Cu foil after the pre-oxidation treatments. Figure 2 shows the XRD patterns of the pre-oxidized Cu foil. As the oxygen content is low, the amount of copper oxide, if any, is below the detection limit of the present XRD technique used. By increasing the oxygen content, the $Cu_2O$ is formed. The amount of $Cu_2O$ increases with the increase of oxygen content. As the amount of oxygen is 1.8 wt%, copper oxide (CuO) is formed together with the Cu-rich $Cu_2O$.

Figure 3 shows the typical micrographs of the $Al_2O_3/Cu$ interfaces after eutectic bonding. As the starting oxygen content in Cu is lower than 0.66 wt%, no reaction phase at interface is observed. As the starting oxygen content in Cu foil is higher than 1 wt%, a reaction phase is observed. The load-displacement curves for the $Al_2O_3/Cu/Al_2O_3$ sandwich specimen are shown in Figure 4. Two peaks are observed in the load-displacement curves. From the in-situ observation, the first and second peaks correspond respectively to the breaking of the lower and upper alumina substrates. By using the value at the first peak, the flexural strength of the $Al_2O_3/Cu/Al_2O_3$ sandwich specimen can be calculated. The flexural strength of the $Al_2O_3/Cu/Al_2O_3$ sandwich specimen is shown as a function of oxygen content in Figure 5. The lowest value between the two peaks corresponds to the debonding at the interface, as demonstrated in Fig. 6. This value is used to calculate the interface energy release rate. Several theoretical relationships had been proposed to estimate the interface energy release rate[7-10]. However, some relationships are suitable for the bilayered system[7-8]; some for thin-film systems[9-10].

Hsueh et al[11] has developed recently a closed-form solution to calculate the interface energy release rate for multilayered beams. Their relationship can be applied to a multilayered system with any number of layers. The relationship is thus adopted for the trilayered system investigated in the present study. The interface energy release rate as a function oxygen content is also shown in Figure 5. Figure 7 shows the oxygen profile near the $Cu/Al_2O_3$ interface. The oxygen content at the surface is higher than 50 at%, then drops to a very low value within a distance of 4 μm.

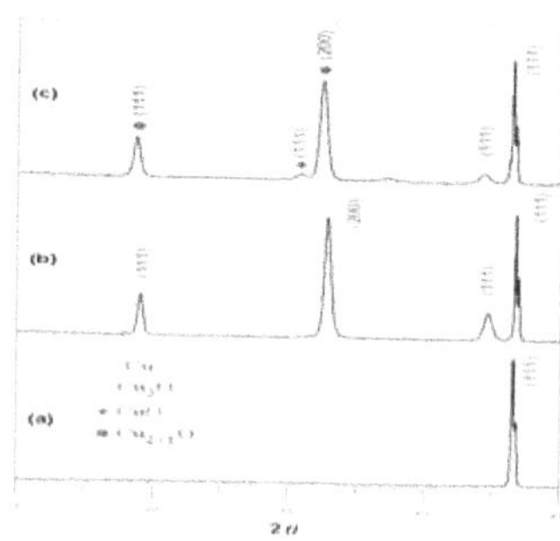


Figure 2. XRD patterns of the surface of pre-oxidized Cu foil. The oxygen content as determined by IR analysis is (a) 0.11, (b) 0.66 and (c) 1.8 wt%.

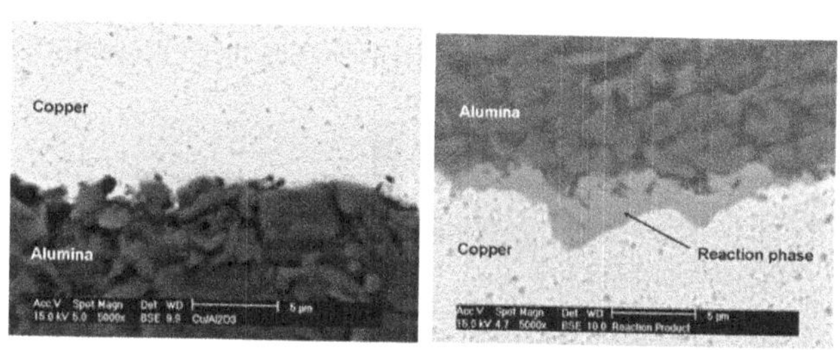


(a) (b)

Figure 3. Microstructures at the $Cu/Al_2O_3$ interface. The oxygen content in Cu before eutectic bonding was (a) 0.11 wt% and (b) 1.8 wt%.

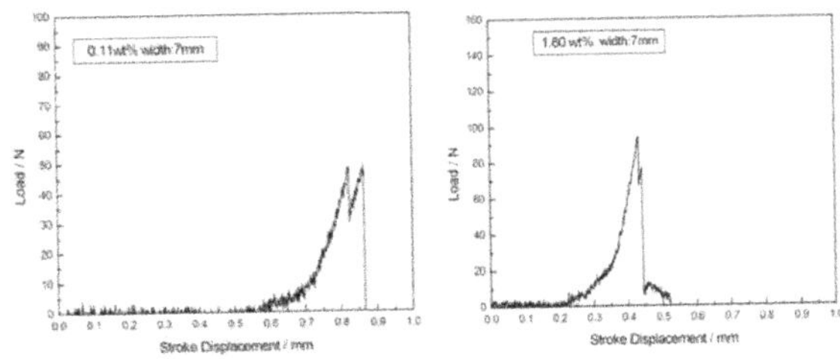


(a) (b)

Figure 4. Load-displacement curves for the $Al_2O_3/Cu/Al_2O_3$ sandwich specimens. The starting oxygen content in Cu is (a) 0.11wt% and (b) 1.8wt%.

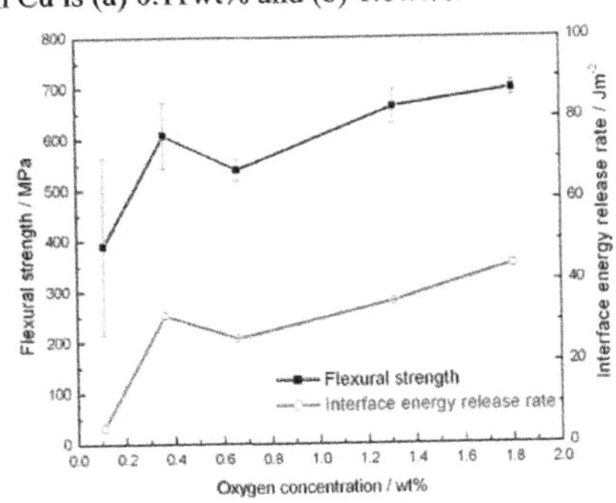


Figure 5. Flexural strength and interface energy release rate for the $Al_2O_3/Cu/Al_2O_3$ sandwich specimens as function of oxygen content.

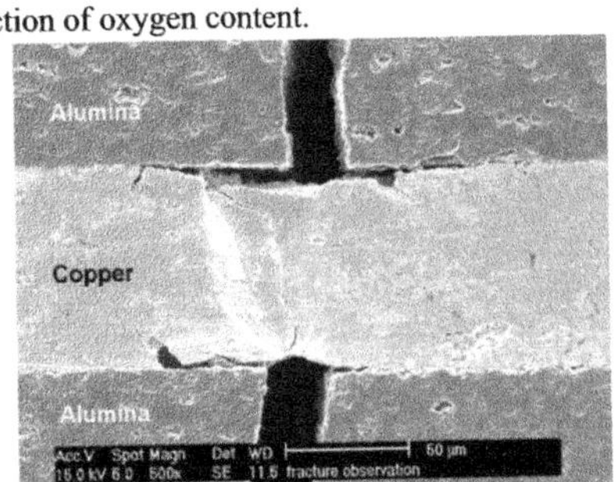


Figure 6. Typical SEM micrograph for the $Al_2O_3/Cu/Al_2O_3$ sandwich specimen after bending.

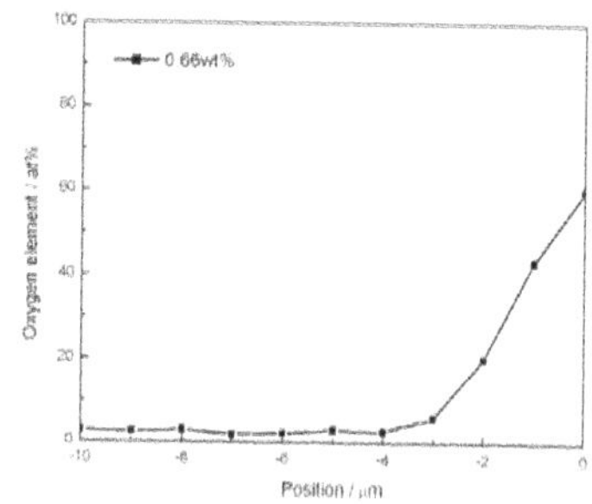


Figure 7. Oxygen profile in Cu as determined by the EPMA technique.

## DISCUSSION

Both the flexure strength and interface energy release rate depend strongly on the starting oxygen content in copper. The strength of the monolithic alumina substrate (with notch as well) has also been determined with the same 4-point bending technique. The flexural strength of the notched alumina substrate is only 190±5 MPa. To insert a copper foil into two alumina substrates increases the strength considerably.

The maximum solubility of oxygen in Cu is as low as 0.03 at%[5]. As the oxygen content is higher than the solubility limit, cuprite oxide ($Cu_2O$) or copper oxide (CuO) may be formed, depending on the amount of oxygen in the surrounding environment. Contradicted experimental results had been reported in the previous studies on the Cu-$Al_2O_3$ system[12]. According to the thermodynamic analysis conducted by Trumble[12], the formation of $CuAlO_2$ interfacial phase depends on the amount of oxygen solute in Cu and temperature. He had suggested that the $CuAlO_2$ phase should be formed at the Cu-$Al_2O_3$ interface as the oxygen solute in Cu is higher than 10~20 at. ppm. The $CuAl_2O_4$ spinel phase is not a stable phase at the interface as the oxygen content is low.[12] In the present study, various amounts of oxygen are introduced into copper foil before bonding. The oxygen content, especially at the interface, is much higher than the solubility of oxygen in copper. The reaction phase, $CuAlO_2$ interphase should be formed. However, such reaction phase is found only for the system containing huge amount of oxygen, > 1 wt%. In the present study, the FESEM technique was used to observe the interface. The reaction phase is hardly identified at the interface as the oxygen content is low. Detailed observation at interface by TEM is being conducted in our laboratory.

From Figure 6, a considerable plastic deformation in copper has observed after debonding. The plastic deformation may consume partly the fracture energy. The interface energy release rate may be consequently affected.

## CONCLUSIONS

A pre-oxidation treatment was applied to introduce oxygen into the surface region of a copper foil. A strong bond is then formed at the Cu-$Al_2O_3$ interface by eutectic bonding. The flexural strength and interface energy release rate of the $Al_2O_3$-Cu-$Al_2O_3$ sandwich specimen increases with the increasing of oxygen content. A reaction phase, $CuAlO_2$, is found at the interface when the oxygen content is high.

## REFERENCES

[1] C. Beraud, M. Courbiere, C. Esnouf, D. Juve, and D. Treheux, "Study of Copper-Alumina Bonding", *J. Mater. Sci.*, 24 4545-4554 (1989).

[2] Y. Yoshino and T. Shibata, "Structure and Bond Strength of a Copper-Alumina Interface", *J. Am. Ceram. Soc.*, 75 [10] 2756-2760 (1992).

[3] C. Scheu, M. Gao, S. H. Oh, G. Dehm, S. Klein, A. P. Tomsia and M. Rühle, "Bonding at Copper-Alumina Interfaces Established by Different Surface Treatments: a Review", *J. Mater. Sci.*, 41 5161-5168 (2006).

[4] J. F. Burgess, C. A. Neugebauer and G. Flanagan, "The Direct Bonding of Metals to Ceramics by the Gas-Metal Eutectic Method", *J. Electrochem. Soc.*, 122 [5] 688-690 (1975).

[5] R. Schmid, "A Thermodynamic Analysis of the Cu-O System with an Associated Solution Model", *Metall. Trans.* 14B 473-481 (1983).

[6] Y. Yoshino, "Role of Oxygen in Bonding Copper to Alumina", *J. Am. Ceram. Soc.*, 72 [8] 1322-1327 (1989).

[7] P. G. Charalambides, J. Lund, A. G. Evans and R. M. McMeeking, "A Test Specimen for Determining the Fracture Resistance of Bimaterial Interfaces", *J. Appl. Mech.*, 56 77-81 (1989).

[8] Z. Suo and J. W. Hutchinson, "Interface Crack Between Two Elastic Layers", *Inter. J. of Fracture*, 43 1-18 (1990).

[9] N. W. Klingbeil and J. L. Beuth, "Interfacial Fracture Testing of Deposited Metal Layers under Four-Point Bending", *Engineering Fracture Mechanics*, 56 [1] 113-126 (1997).

[10] A. A. Volinsky, N. R. Moody, W. W. Gerberich, "Interfacial Toughness Measurements for Thin Films on Substrates", *Acta Materialia*, 50 441-446 (2002).

[11] C. H. Hsueh, W. H. Tuan and W. C. J. Wei, "Analyses of Steady-State Interface Fracture of Elastic Multilayered Beams under Four-Point Bending", *Scripta Mater.*, 60 721-724 (2009).

[12] K.P. Trumble, "Thermodynamic Analysis of Aluminate Formation at Fe/$Al_2O_3$ and Cu/$Al_2O_3$ interfaces", *Acta metall. mater.*, **40**, Suppl., S105-S110 (1992).

# SYNTHESIS AND CHARACTERIZATION OF MATERIALS HARMONIZED WITH THE ENVIRONMENT - PROTON CONDUCTIVE SILICA-BASED HYBRID MEMBRANES FOR FUEL CELLS AT INTERMEDIATE TEMPERATURES

Masaki Kato, Junji Umeda*, Masashi Suzuki, Makoto Moriya*, Wataru Sakamoto*, Toshinobu Yogo*
Research and Development Division, Aichi Industrial Technology Institute,
Ondacho 1-107-1, Kariya 448-0013, Japan
*Divisison of Nanomaterials Science, EcoTopia Science Institute, Nagoya University
Furo-cho, Chikusa, Nagoya 464-8603, Japan

## ABSTRACT

Novel proton conductive silica-based hybrid membranes were synthesized from styryl-substituted alkoxysilanes containing different numbers of methoxy groups and phenylvinylphosphonic acid (PhVPA) by radical copolymerization and sol-gel processing. The proton conductivity of the hybrid membranes increased with decreasing inorganic Si-O crosslinking. The membrane from difunctional methyl(dimethoxy)silylmethylstyrene/PhVPA=1/4 showed conductivities of $4.0 \times 10^{-2}$ S $cm^{-1}$ under 100%R.H. at 130°C, and $3.4 \times 10^{-3}$ S $cm^{-1}$ under about 19.2%R.H. at 130°C.

## INTRODUCTION

Polymer electrolyte fuel cells (PEFCs) attract special attention for application in various energy generation systems including electrical devices because of their energy efficiency and clean exhaust gas[1]. Perfluorosulfonic polymers are widely used as the electrolyte in PEFCs because of their mechanical strength, good chemical stability and high proton conductivity under high relative humidity conditions at low temperatures (<100°C)[2]. However, these membranes have the problem of the Pt anode catalysts being poisoned by CO impurities in the fuel gas due to low operation temperatures below 100°C[3].

In contrast, fuel cell operation at intermediate temperatures under low humidity conditions has several advantages, such as an increase in the rate of reaction within the fuel cells, a larger temperature difference relative to the ambient for heat radiation from a fuel cell stack, and less CO poisoning of the Pt catalyst. Therefore, operation at elevated temperatures is effective in simplifying the heat management in fuel cells and the reformer system for hydrogen generation[4]. Thus, proton-conducting materials of high conductivity in the intermediate temperature range from 100°C to 200°C are required.

Among various alternative proton-conducting materials, the organosiloxane-based inorganic/organic hybrid is one of the promising materials for the proton-conducting membranes[5-8]. This is due to their marked thermal stability, the flexibility of the organosiloxane networks, and their excellent processing versatility. Phosphoric acid has been incorporated into organosiloxane-based matrices[8], producing materials with good proton conductivities. However, the proton conductivity decreases with time due to the leaching of the phosphoric acid from the membranes because of their insufficient interaction with the matrices[9].

One of the possible solutions to increase the chemical stability is to anchor acid groups onto stable inorganic/organic hybrid matrices via chemical bonds. The copolymerzaition of an organosiloxy compound with unsaturated phosphonic acid was selected in the present work. Organosilyl compounds are known to undergo hydrolysis depending upon the nature of the chemical bonds. Si-$OCH_3$ undergoes acidic hydrolysis, yielding Si-O-Si linkages, while Si-$CH_3$ is intact under the same conditions. Thus, the organosilyl groups of -$Si(OCH_3)_3$ and -$Si(CH_3)(OCH_3)_2$ are trifunctional and difunctional, respectively, to the hydrolysis reaction. Hence, the three-dimensional structure of the Si-O-Si linkages is controlled by the number of methoxy groups on the Si atoms.

This paper describes the synthesis of proton-conductive membranes from styryl-substituted alkoxysilanes containing different numbers of functional groups and phenylvinylphosphonic acid via copolymerization and sol-gel processes. The effects of structural variation of hybrid membranes on their conductive properties were studied. The proton conductivities of each hybrid membrane at various temperatures and relative humidities were investigated.

## EXPERIMENTAL

### Materials

The molecular structures used in this study are shown in Figure 1. (Methyldimethoxysilylmethyl)styrene (MDMSMS) and trimethoxysilylstyrene (TMSMS) were prepared according to the literature[10].

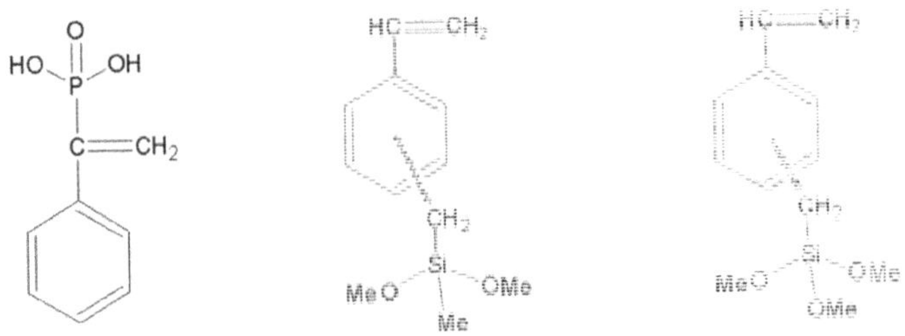


PhVPA | MDSMS(D) | TMSMS(T)

Figure 1. Molecular structure of starting compounds.

The compounds are abbreviated as MDMSMS (D, disubstituted) and TMSMS (T, trisubstituted) based on the number of methoxy groups.

### Synthesis of hybrid membranes from styryl-substituted alkoxysilane and PhPVA

First, the copolymerization was carried out in dimethylformamide (DMF) solution with various monomer ratios of organoalkoxysilane to PhVPA at 85°C for 8h with AIBN as an initiator. The samples were labeled as x/y= MDMSMS(D)/PhVPA or TMSMS(T)/PhVPA, more specifically, Si/P, according to their molar ratios. After the sealed glass capsule was heated at 85°C for 8h, a viscous solid copolymer was separated by centrifuging. The organophosphonic copolymer obtained was dissolved in anhydrous DMF again immediately after separation. Then, 1N hydrochloric acid solution was added dropwise to the copolymer solution to hydrolyze silicon alkoxide. The solution was stirred at room temperature for 24h, yielding a sol, and the sol obtained was casted onto a Teflon plate. The casted film was heated in a dry oven at 100°C for 12h, then at 120°C for 2h, and finally at 140°C for 2h..

Characterization of membranes

Thermal behavior was investigated by thermogravimetry-differential thermal analysis (TG-DTA) (DTG-50H, Shimadzu) at a heating rate of 10°C/min from ambient temperature to 800°C in an air flow. Solid-state $^{13}C$ and $^{29}Si$ cross-polarization (CP)/magic angle spinning (MAS) NMR spectra of these hybrid membranes were recorded with an Avance 300 spectrometer (Bruker Biospin) operating at 75.477 and 59.624 MHz, respectively.

Proton conductivity and fuel cell studies

The proton conductivity of the hybrid membrane was measured by an AC impedance method using a frequency response analyzer including a potentiostat/galvanostat (PAR 2263, Princeton Applied Research) from 0.1Hz to 1MHz. The thickness of the hybrid membranes for the measurement was from 250 to 950μm. The methods are the same as reported by the authors[11]. A saturated aqueous solution of $MgCl_2$, a saturated aqueous solution of $NaNO_3$ and water vapor were used to control the relative humidity. The resistances of the membranes ($R_{dc}$) were estimated from the values of the real part of the impedance at the intersection in the complex AC impedance plots (Nyquist plots). The conductivity ($\sigma$) of the membrane in the transverse direction was calculated using the relation $\sigma = t/R_{dc}S$, where t and S are the thickness and the electrode area of the membrane, respectively. The electrodes for the test cell were prepared from MDMSMS(D)/PhVPA=1/4 and Pt/C powder. The membrane electrode assemblies (MEA) were tested using a potentio/galvanostat (PAR 2263).

## RESULTS AND DISCUSSION

NMR analysis of the hybrid membranes

Figure 2 shows the $^{13}C$ NMR spectra of hybrid membranes from MDMSMS(D)-PhVPA and TMSMS(T)-PhVPA with a Si/P ratio of 1/2. The membrane from MDMSMS(D)-PhVPA (Fig. 2(a)) shows almost the same spectrum as that from TMSMS(T)-PhVPA (Fig. 2(b)) except for the signal at around 0.0 ppm. This peak is assigned to the methyl carbon bound to the Si atom in MDMSMS(D)-PhVPA hybrid. TMSMS(T) has no $Si\text{-}CH_3$ group in its structure. Other peaks observed at around 27.8, 35.1, 49.8, from 127.3 to 137.0 ppm are assigned to silylmethylene carbon, α-carbons of the organic polymer backbone, β-carbons of organic polymer backbone, and carbons of benzene ring[12], respectively, as shown in Figure 2. The results of $^{13}C$ CP-MAS NMR support that the copolymerization proceeds successfully in both hybrid membranes.

Figure 3 shows the $^{29}Si$ CP-MAS NMR spectra of these hybrid membranes. Figure 3(a) reveals a broad signal appearing from -25 to -14 ppm for the hydrolyzed MDMSMS(D)-PhVPA. The peak

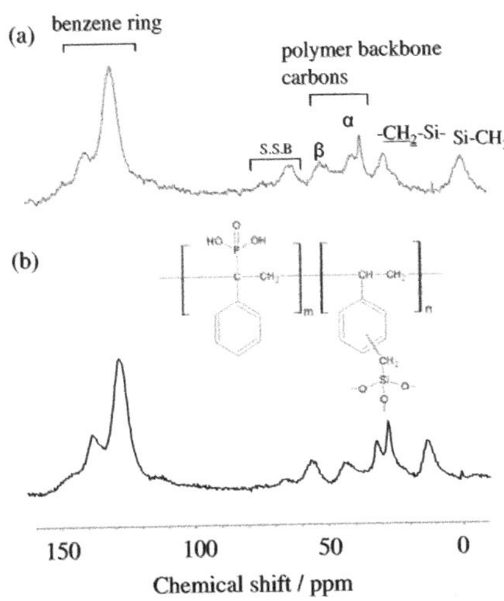


Figure 2. [13]C NMR spectra of hybrid membranes
(a) hybrid from MDMSMS(D) /PhVPA=1/2,
(b) hybrid from TMSMS(T) /PhVPA=1/2.

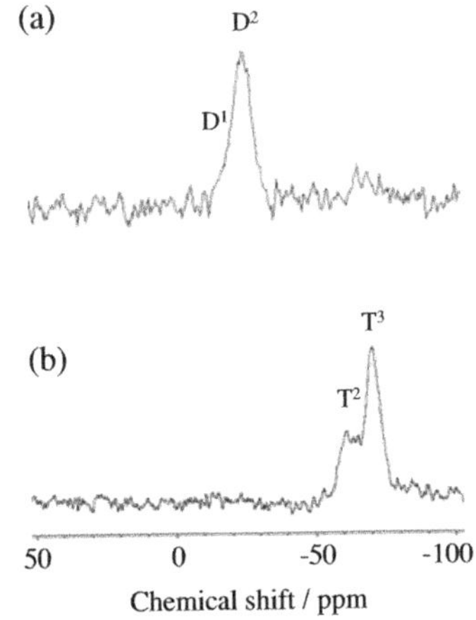


Figure 3. [29]Si NMR spectra of hybrid membranes
(a) hybrid from MDMSMS(D) /PhVPA=1/2,
(b) hybrid from TMSMS(T) /PhVPA=1/2.

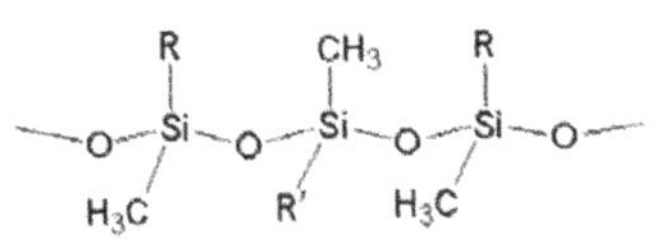

Figure 4. (a) Two-dimensional Si-O linkage from MDMSMS(D)/PhVPA

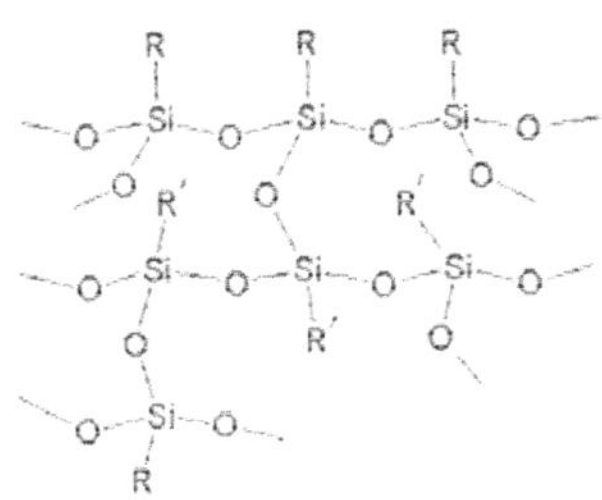

(b) Three-dimensional Si-O linkage from TMSMS(T)-PhVPA

seems to consist of $D^1$ (-15.0 ppm, Si-O-**Si** linkage) and $D^2$ (-24.6ppm, Si-O-**Si**-O-Si linkage) units, since the chemical shift corresponds to that of the D unit[13]. When trifunctional TMSMS(T)-PhVPA was hydrolyzed, two signals are observed at -48 and -76 ppm in the T region (Fig. 3(b)). The former is $T^2$ (Si-O-**Si**-O-Si linkage), the latter a $T^3$ (Si-O-**Si**(-OSi)-O-Si) unit[14]. The presence of $T^3$ units suggests that the product comprises a mixture of multi-chain, polycyclic and branched structures. The possible linkages for membranes from the three starting materials are shown in Figure 4. Compared to the TMSMS(T)-PhVPA hybrid that consists of dense and three dimensional Si-O-Si cross-linking (Fig. 4(b)),

MDMSMS(D)-PhVPA mainly consist of chain structure (Fig. 4 (a)) because the number of methoxy groups in the alkoxysilane is less than 2. Therefore, it is considered that the Si-$OCH_3$ bond undergoes hydrolysis and condensation, yielding Si-O-Si linkages as reported in sol-gel process.

Thermal analysis of the hybrid membranes

Figure 5 shows TG–DTA curves of membranes MDMSMS(D)/PhVPA (Fig. 5(a)) and TMSMS(T)/PhVPA (Fig. 5(b)) with a ratio of 1/4 measured in dry air from room temperature to 800°C. From room temperature to 180°C, no significant endothermic peak corresponding to the melting point of PhVPA (113°C) is observed for both hybrid membranes, because PhVPA is chemically bound to the matrices of the copolymers. Small weight losses measured up to 180°C by TGA are attributed to the desorption of physically absorbed water during the storage.

From 180°C to 310°C, small weight losses of about 8.4% (Fig. 5(a)) and 7.2% (Fig. 5(b)) in the TGA are observed for the hybrid membranes. These are attributed to the desorption of water generated by the condensation of residual silanol groups in the hybrid membranes and the evaporation of water as a by-product of the condensation of P-OH of PhVPA. Between 310°C and 450°C, the differences in the systems are clearly observed. The weight losses are due to both the condensation of phosphonic acid groups and the thermal decomposition of the organic polymer network and aromatic rings. The thermal decomposition begins at a higher temperature in the hybrid prepared from the alkoxysilane copolymer with larger numbers of methoxy groups. The thermally stable Si-O-Si bonds are considered to suppress the thermal degradation of the organic portion of the hybrid membrane. Thus, the thermal stabilities of

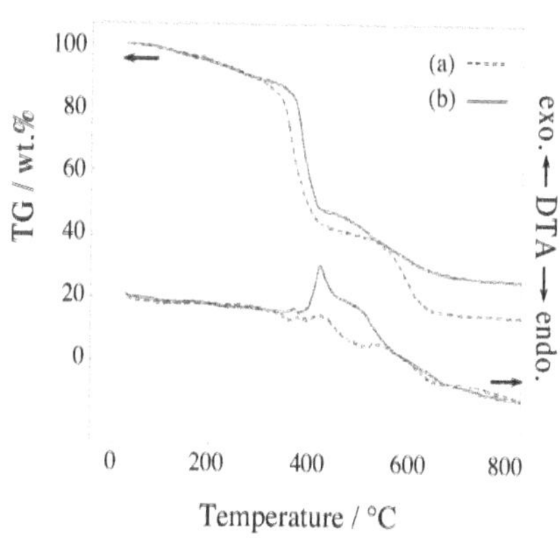


Figure 5. DTA-TG curves for hybrid membranes
(a) membrane from MDMSMS(D)/PhVPA=1/4,
(b) membrane from TMSMS(T)/PhVPA=1/4

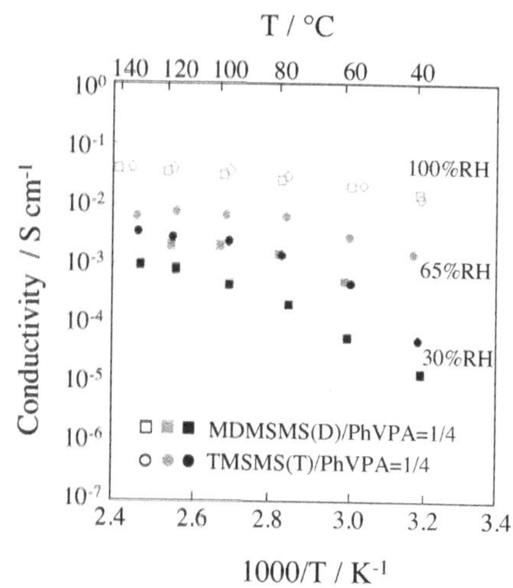


Figure 6. Temperature dependences of the proton conductivities of hybrid membranes under different relative humidities.

the hybrids are influenced by the structure of the inorganic cross-linking remarkably. Hence, all the hybrid membranes are thermally stable up to 180°C in air.

Proton conductivity of the hybrid membranes

Figure 6 shows the proton conductivities of MDMSMS(D)/PhVPA and TMSMS(T)/PhVPA hybrid membranes with ratios of 1/4 at various humidities from 35.9%R.H to 100 R.H. Both membranes show an increase in conductivity with increasing temperature up to 130°C at each humidity. The proton conductivities also increase with an increase of relative humidity from about 35.9% R.H. to 100% R.H. Under saturated humidity conditions, the conductivities of MDMSMS(D)/PhVPA=1/4 and TMSMS(T)/PhVPA=1/4 are comparable. The proton conductivities of MDMSMS(D)/PhVPA=1/4 and TMSMS(T)/PhVPA=1/4 are $4.0 \times 10^{-2}$ S $cm^{-1}$ and $3.7 \times 10^{-2}$ S $cm^{-1}$, respectively, at 130°C under 100%R.H. Also, under fully hydrated conditions, the slope of the plot decreases in comparison with that under low humidity conditions.

On the other hand, under low and medium humidities, the proton conductivity of the hybrid membrane of MDMSMS(D)/PhVPA=1/4 is obviously higher by about a half order as compared to TMSMS(T)/PhVPA=1/4. The proton conductivities of MDMSMS(D)/PhVPA=1/4 and TMSMS(T)/PhVPA=1/4 hybrid membranes were $3.4 \times 10^{-3}$ S $cm^{-1}$ and $9.5 \times 10^{-4}$ S $cm^{-1}$, respectively, at 130°C under 19.2% R.H.

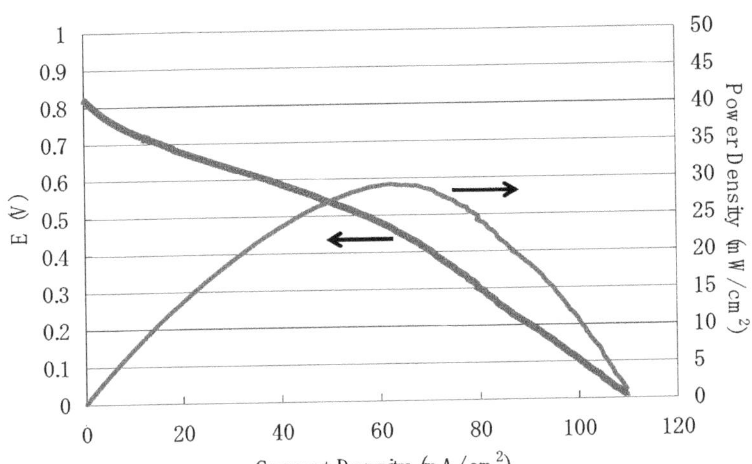


Figure 7. Voltage-current curve of MEA made of a TMSMS(T)/PhVPA=1/4 electrolyte operated at 140°C and 30%RH.

Water is introduced into the hybrid membranes from a humidified atmosphere because of the hygroscopic nature of the phosphonic groups, resulting in high proton mobility directly. The P-OH groups act as the proton conducting moieties which form hydrogen bonds with water. The proton mobility in hydrated polymers reflects the diffusion of hydrated protons[15]. The hybrid consists of an organic polymer and an inorganic Si-O-Si linkage. The proton mobility as $H_3O^+$ was found to depend largely upon the type of Si-O-Si linkage. In comparison with TMSMS(T)/PhVPA, the three-dimensional Si-O-Si cross-linking is not formed in MDMSMS(D)/PhVPA. Instead, the linear region of the siloxane structure constitutes the organosiloxane backbone. Therefore, the membrane becomes more flexible and increases both in water capacity and proton mobility. The diffusion of hydrated proton is considered to be promoted by segmental relaxation of the polymer network. The restriction of segmental motion of the polymer chains decreases with decreasing Si-O-Si cross-linkages in the membrane.

Figure 7 shows the voltage-current curve for TMSMS(T)/PhVPA=1/4 hybrid membranes at 140°C under 30% R.H. The open circuit voltage was about 0.82 V. The maximum current density and power were 110 mA/cm$^2$ and 29 mW/cm$^2$, respectively. The MEA was found to exhibit a cell performance at the intermediate temperature and low humidity conditions.

## CONCLUSION

Proton conductive silica-based hybrid membranes were synthesized from styryl-substituted alkoxysilanes containing different numbers of methoxy groups and phenylvinylphosphonic acid (PhVPA). The silica-based hybrid membranes had sufficient thermal stability up to 180°C. The proton conductivity of the membranes depends on the structure of the Si-O linkages. The membrane from difunctional methy(dimethoxy)silylmethylstyrene/PhVPA=1/4 revealed higher conductivities than those from trifunctional (trimethoxysilyl)methylstyrene/PhVPA=1/4. These membranes can be incorporated into novel membrane-electrode assemblies for PEFCs, because the membranes exhibit sufficient conductivity at medium temperatures under low humidity conditions.

## REFERENCES


[1]G. Cacciola, V. Antonucci, and S. Freni, "Technology Up Date and New Strategies on Fuel Cells", *J. Power Sources* **100** 67-79 (2001).

[2]F. Damay and L.C. Klein, "Transport properties of Nafion Composite Membranes for Proton-Exchange Membranes Fuel Cells", *Solid State Ionics* **162-163** 261-267 (2003).

[3]M. Aparicio and A. Duran, "Hybrid Organic/Inorganic Sol-gel Materials for Proton Conducting Membranes", *J. Sol-Gel Sci. Tech.* **31** 103-107 (2004).

[4]M.F.H. Schuster, W.H. Meyer, M. Schuster, and K.D. Kreuer, "Toward a New Type of Anhydrous Organic Proton Conductor Based on Immobilized Imidazole", *Chem. Mater.* **16** 329-337 (2004).

[5]T. Xu, "Ion Exchange Membranes: State of Their Development and Perspective", *J. Membr. Sci.*, **263**, 1-29 (2005).

[6]X.J. Cui, S.L. Zhong and H.Y. Wang, "Organic–Inorganic Hybrid Proton Exchange Membranes Based on Silicon-Containing Polyacrylate Nanoparticles with Phosphotungstic Acid", *J. Power Sources* **173**

28-35 (2007).
[7]Y. Jin, S. Qiao, J.C.D. da Costa, B.J. Wood, B.R. Ladewig and G.Q. Lu, "Hydrolytically Stable Phosphorylated Hybrid Silicas for Proton Conduction", *Adv. Func. Mater.* **17** 3304-3311 (2007).
[8]K. Tadanaga, H. Yoshida, A. Mastuda, T. Minami, M. Tatsumisago, "Preparation of Proton Conductive Inorganic-Organic Hybrid Films Using Epoxycyclohexylethyltrimethoxysilane and Orthophosphoric acid" *J. Sol-Gel Sci. Tech.* **31** 365-368 (2004).
[9]S.-P. Szu, L.C. Klein, M. Greenblatt, "Effect of Precursors on the Structure of Phosphosilicate Gels: 29Si and 31P MAS-NMR study", *J. Non-Cryst. Solids* **143** 21-30 (1992).
[10]M. Kokubo, H. Inagawa, M.Kawahara, D. Terunuma, H. Nohira, "A Study of Some New Types of Silane Coupling Agents for Mica Flake or Glass Flake/Plastics Composites", *Kobunshi Ronbunshu,* **38** 201-207 (1981) [JAPANESE].
[11]M. Kato, W. Sakamoto, T. Yogo, "Synthesis of Organosiloxane-Based Inorganic/Organic Hybrid Membranes with Chemically Bound Phosphonic Acid for Proton-Conductors", *Electrochim. Acta* **52** 5924-5931 (2007).
[12]R.M. Silverstein, F.X. Webster, D.J. Kiemle, *Spectrometric Identification of Organic Compounds*, 7th ed., John Wiley & Sons, New York, 2005, p. 72, 127, 204.
[13]S.E. Rankin, C.W. Macosko, A.V. Mccormick, "Copolymerization Kinetics of a Modelsiloxane System". *J. Polym. Sci. A, Polym. Chem.* **35** 1293-1302 (1997).
[14]Y. Sugahara, S. Okada, S. Sato, K. Kuroda, C. Kato, "$^{29}$Si NMR Study of Hydrolysis and Initial Polycondensation Processes of Organoalkoxysilane". *J. Non-Cryst. Solid* **167** 21-2 (1994).
[15]K.D. Kreuer, "On the Development of Proton Conducting Polymer Membranes for Hydrogen and Methanol Fuel Cells, *J. Membrane Sci.* **185** 29-39 (2001).

# FORMATION OF BONE-LIKE HYDROXYAPATITE ON SURFACE-MODIFIED BULK METALLIC GLASS USING A HYDROTHERMAL-ELECTROCHEMICAL METHOD

Nobuhiro Matsushita[a], Naota Sugiyama[a], Sayaka Maruyama[a], Rieko Ito[a], Kenichi Katsumata[a] Masahiro Yoshimura[a], Masahiro Tsukamoto[b], Xinmin Wang[c], Takeshi Wada[c], Akihisa Inoue[c], and Kiyoshi Okada[a]

a) Materials and Strucutres Laboratory, Tokyo Institute of Technology, 4259 Nagatsuta, Midori-ku, Yokohama 226-8503, Japan

b) Joining and Welding Research Institute, Osaka University, 11-1 Mihogaoka. Ibaraki, Osaka 567-0047, Japan

c) Institute for Materials Research, Tohoku University, 2-1-1 Katahira, Aoba, Sendai 980-8577, Japan

## ABSTRACT

Bioactive nano-structured titanate layers were fabricated on the surface of Ti-based bulk metallic glass (BMG) using a hydrothermal-electrochemical method. A constant electric current of 0.5 mA/cm$^2$ was applied to the $Ti_{40}Zr_{10}Cu_{36}Pd_{14}$ substrate using sodium hydroxide aqueous solutions of 1.0 mol/L as an electrolyte. The morphology of the nano-structured layers depended on the temperature during the hydrothermal-electrochemical treatments. The surface area of a sample 1 cm$^2$ in area increased from 185 to 325 cm$^2$ with the increase in temperature from 90 to 150°C. The as-prepared samples with amorphous sodium titanate surface layers having a nanomesh structure were immersed in simulated body fluid (SBF). The growth of hydroxyapatite crystallite was observed only for the sample treated at 150°C having a surface area larger than 300 cm$^2$.

## INTRODUCTION

Bulk metallic glass (BMG) exhibits many properties that are superior to those of conventional crystalline metallic materials[1-4]. BMG families based on Fe[5-6], Ni[7-8], and Ti[9-12] have been investigated for their various applications as spring material, transport or high speed machine parts, hydrogen storage, and as soft magnetic materials. Since Ti-based BMGs such as Ti-Cu-Ni-Co, Ti-Cu-Ni-Zr, and Ti-Cu-Ni-Zr-Be have a high elastic limit, low Young's modulus, and high corrosion resistance[11,13-16], they have been investigated for biomedical applications. However, they contain toxic elements such as Ni or Be for good glass-forming ability, and this restricts their use in applications such as dental, load bearing, and orthopedic implants. Zhu et al. succeeded in preparing Ti-based BMG($Ti_{40}Zr_{10}Cu_{36}Pd_{14}$) which does not contain Ni and Be[17]. However, this BMG cannot be joined directly to human bone because of its high chemical stability[15, 18]. Therefore, the a surface modification process that enables the bone to bond to BMG is highly desired. One promising method to obtain bioactivitiy is to fabricate a ceramic layer on it. Various methods for joining ceramics to metallic materials have been investigated, and techniques such as sputtering[19-20], CVD[21], and thermal decomposition[22] have been tested as a way to enhance the surface biocompatibility of Ti metal and Ti alloys. These processes require high temperature, and mechanical and thermal stresses during cooling to room temperature result in surface or body cracks. Since the excellent mechanical properties of BMG degrade over the glass transition point, around 350-400°C[17], a low temperature coating process is required for ceramics on BMG.

Our group found that a hydrothermal method, which is a gentle, low temperature process, was useful for converting the bioinert surface of Ti and Ti alloys to a bioactive one[23-24]. However, with BMG, it is difficult to coat or modify the surface, even by hydrothermal methods, due to its high chemical stability and corrosion resistance.

In previous studies, we succeeded in fabricating crack-free and homogenous titanate films, such as $BaTiO_3$[25], $SrTiO_3$[26-28], and $CaTiO_3$[29] on Ti metals and alloys using a hydrothermal-electrochemical (H-E) method at temperatures below 200°C. The H-E treated ceramic enabled good adhesion at the ceramic/metal interfaces.

In this study, titanate nano-structured layers with various types of morphology were fabricated on a Ti-based BMG ($Ti_{40}Zr_{10}Cu_{36}Pd_{14}$) by changing the temperature during the H-E treatments, and the ability to form hydroxyapatite on them was investigated.

## EXPERIMENTS

### Fabrication of titanate nano-structured layers

The hydrothermal-electrochemical experimental equipment is shown in Figure1. A ribbon-shaped titanium-based BMG ($Ti_{40}Zr_{10}Cu_{36}Pd_{14}$), fabricated by arc-melting a mixture of pure metal (>99.9%) in an argon atmosphere, was cut into a 10×40×0.1 $mm^3$ in size and used as the anode. A platinum substrate 10 × 50 × 0.5 $mm^3$ in size was used as the cathode. The substrates were hung in a 1.0 mol/L NaOH aqueous solution (pH 14.0) with an active surface of about 3 $cm^2$, keeping the distance between electrodes at 6 cm. A constant current of 0.5 $mA/cm^2$ was applied between the electrodes for 2 hours at different temperatures from 90-150°C during the H-E treatments. After the H-E treatments, the samples obtained were removed from the electrolyte, washed in distilled water to remove the alkali, and dried at 80°C for 2 hours in air.

### Evaluation of the nano-structured layers

The morphology of the layers obtained was observed by a scanning electron microscope (Hitachi SP 4500), and the surface area of a 1 $cm^2$ sample was determined by the BET method with nitrogen adsorption[30] (SHIMADZU Micromeritics FlowSorb III). Raman spectroscopic studies (Jobin-Yvon T64000 with an Ar laser (wavelength=514.5 nm) operated at 50 mW were carried out on the as-prepared samples. Their ability to form hydroxyapatite on the nano-structured layers was examined by immersing them in simulated body fluid (SBF)[31]. This electrolyte, with an ion concentration almost equal to that of human blood plasma, was prepared by dissolving reagent grade NaCl, $NaHCO_3$, KCl, $K_2HPO_4 \cdot 3H_2O$, $MgCl_2 \cdot 6H_2O$, $CaCl_2$ and $Na_2SO_4$ in distilled water, and buffering it at pH 7.40 with tris-aminomethane and hydrochloric acid. The ion concentration of the SBF was: (mmol/L); $Na^+$ 142.0, $K^+$ 5.0, $Mg^{2+}$ 1.5, $Ca^{2+}$ 2.5, $Cl^-$ 147.8,

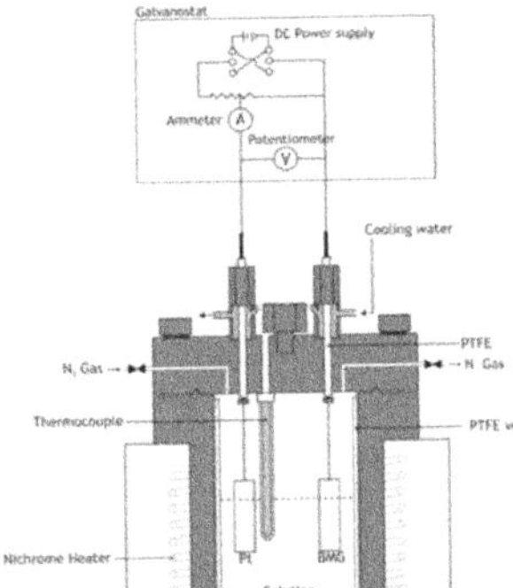


Figure 1. Hydrothermal-electrochemical equipment used in this study.

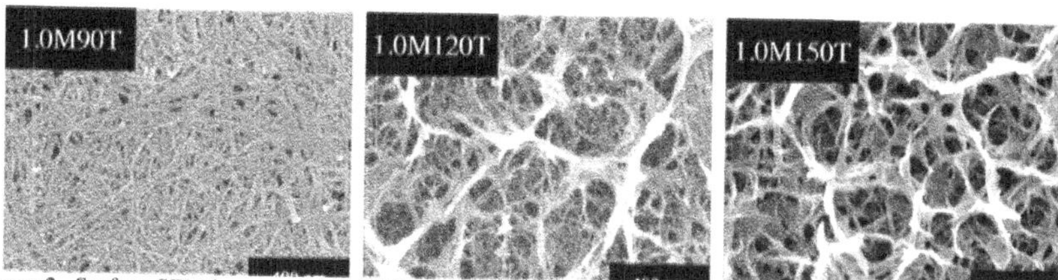


Figure 2. Surface SEM images of 1.0M90T, 1.0M120T and 1.0M150T samples hydrothermal-electrochemical treated for 2 hours with constant current density of 0.5 mA/cm$^2$ in NaOH of 1.0 mol/L.

$HCO^{3-}$ 4.2, $HPO_4^{2-}$ 1.0, and $SO_4^{2-}$ 0.5. The samples were soaked in 30 mL of SBF in a polypropylene vial and stored in an incubator at 37.0°C for 12 days. The specimen was rinsed with distilled water after SBF immersion and dried in air prior to the SEM observation.

## RESULTS and DISCUSSION

Figure 2 shows the surface SEM images of the samples prepared at 90, 120, and 150°C for 2 hours with a current density of 0.5 mA/cm$^2$ in 1.0 mol/L NaOH solutions. Hereafter these samples are called as 1.0M90T, 1.0M120T and 1.0M150T, respectively. The growth of nano wires several tens of nm in width was observed for the 1.0M90T sample. It formed a dense net-like reticular structure. However, 1.0M120T exhibited a more complicated surface structure that consisted of not only nano wires but also a nano sheet like wire. The 1.0M150T sample had a larger spacing in the nano-structured layer than the 1.0M120T sample. The samples treated hydrothermal-electrochemically at different temperature possessed wide variety of morphologies. The mechanical properties of these nano structures including the adhesion to BMG surface are now under investigation. The surface area increased from 185, 205, and 325 cm$^2$ with increasing reaction temperature as shown in Figure 3.

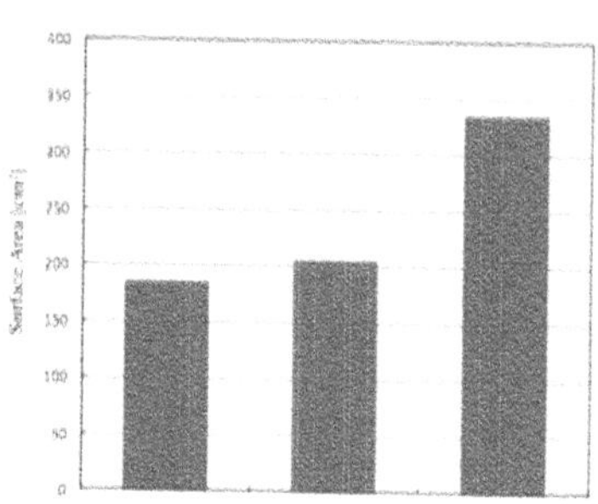

Figure 3. Surface area of the sample 1 cm$^2$ in area measured by BET method using a nitrogen.

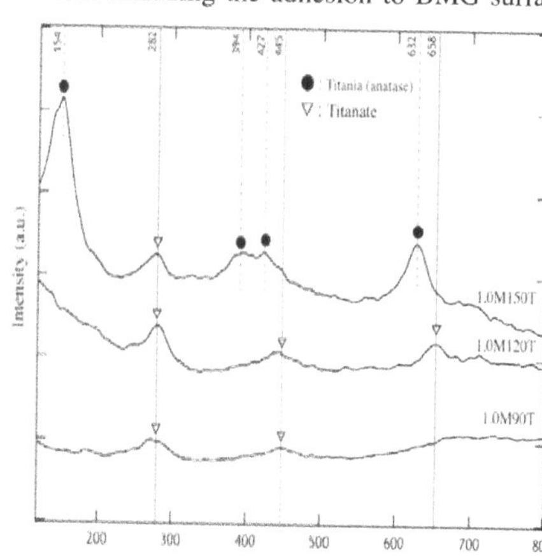


Figure 4. Visible Raman spectra of samples treated hydrothermal-electrochemically at 90, 120, and 150°C.

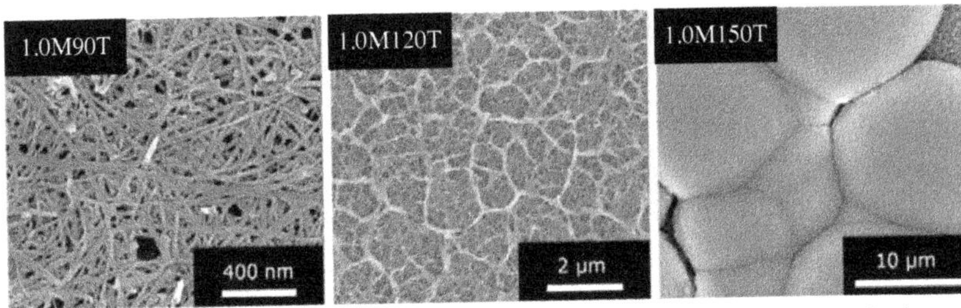


Figure5 Surface SEM images of 1.0M90T, 1.0M120T, and 1.0M150T samples after immersing in SBF for 12 days.

Figure 4 shows the visible Raman spectra of the as-prepared 1.0M90T, 1.0M120T, and 1.0M150T samples. Peaks at 282, 445, and 658 $cm^{-1}$ were detected in the 1.0M90T and 1.0M120T sample, indicating the Na-O-Ti stretching vibration mode in the titanate[32-33]. This is similar to our previous result[34]. Peaks at 154, 394, 427, and 632 $cm^{-2}$ indicate the anatase phase of $TiO_2$[32] in the 1.0M150T sample.

Figure 5 shows the surface SEM images of the samples after immersing in SBF for 12 days. The formation of hydroxyapatite was clearly observed for the 1.0M150T sample, which had large surface of 325 $cm^2$, while hydroxyapatite formation was hardly observed on the surface of 1.0M90T and 1.0M120T samples having a surface area of about 185 and 205 $cm^2$, respectively. In a previous study, we also obtained better hydroxyapatite forming ability for the sample having a surface area larger than 300 $cm^2$, in which a higher NaOH concentration (5.0 mol/L) was used during the H-E treatment[34]. The data obtained were insufficient to explain the relationship between the surface area of sample and their hydroxyapatite forming ability. However, higher surface areas of samples correspond to higher Na concentrations on their surfaces, and changing from Na to Ca ions on the surface of the titanate layer promote the nucleation of the hydroxyapatite crystallite. We think this is the reason that samples having a higher surface area exhibited better hydroxyapatite forming ability, and the surface area of 300 $cm^2$ was critical for obtaining the growth of hydroxyapatite crystallite in 12 days.

The hydroxyapatite crystallite formed on 1.0M150T sample was several μm in height. The adhesion to BMG is now under investigation and will be presented in another paper.

## CONCLUSIONS

Bioactive titanate nano-structured layers were fabricated on Ti-based bulk metallic glass ($Ti_{40}Zr_{10}Cu_{36}Pd_{14}$) using a hydrothermal-electrochemical process. Larger surface area 1 $cm^2$ in area samples were obtained with a higher temperature hydrothermal-electrochemical treatment. A SBF immersion test showed that sodium titanate nano-mesh layers with a large surface area enhanced hydroxyapatite-forming ability compared with a bare BMG, which is rather bioinert. A porous structure with a large surface area proved to be essential for hydroxyapatite formation. The successful fabrication of bioactive ceramic layers on BMG using the H-E method opens new possibilities for BMG, enabling their use in practical biomedical applications.

ACKNOWLEDGEMENTS

This research was supported by the MEXT project, "Ceramics/Bulk Metallic Glass Joining and Bonding" and by three COE Laboratories: Institute of Materials Research (IMR), Tohoku University, Materials and Structures Laboratory (MSL), Tokyo Institute of Technology, Joining and Welding Research Institute (JWRI), Osaka University. The authors would like to acknowledge the contribution of Prof. M. Aizawa (Meiji University) for assisting with BET measurements.

REFERENCES

[1] A. L. Greer, METALLIC GLASSES, *Science*, 1995, **267**, 1947-1953.

[2] W. L. Johnson, "Bulk glass-forming metallic alloys", *Mrs. Bull.*, 1999, **24**, 42-56.

[3] A. Inoue, "Stabilization of metallic supercooled liquid and bulk amorphous alloys", *Acta. Mater*, 2000, **48**, 279-306.

[4] Z. P. Lu and C. T. Liu, "A new glass-forming ability criterion for bulk metallic glasses", *Acta. Mater.*, 2002, **50**, 3501-3512.

[5] A. Inoue, B. L. Shen and C. T. Chang, "Super-high strength of over 4000 MPa for Fe-based bulk glassy alloys in [(Fe1-x,COx)(0.75)B0.2Si0.05](96)Nb-4 system", *Acta Mater.*, 2004, **52**, 4093-4099.

[6] K. F. Yao and C. Q. Zhang, "Fe-based bulk metallic glass with high plasticity", *Appl. Phys. Lett.*, 2007, **90**, 3.

[7] R. Akatsuka, T. Zhang, H. Koshiba and A. Inoue, "Preparation of new Ni-based amorphous alloys with a large supercooled liquid region", *Mater. Trans. Jim.*, 1999, **40**, 258-261.

[8] H. Choi-Yim, D. H. Xu and W. L. Johnson, "Ni-based bulk metallic glass formation in the Ni-Nb-Sn and Ni-Nb-Sn-X (X = B,Fe,Cu) alloy systems", *Appl. Phys. Lett.*, 2003, **82**, 1030-1032.

[9] L. E. Tanner and R. Ray, "Physical Properties of Ti50Be40Zr10 Glass", *Scr. Metall.*, 1977, **11**, 783-789.

[10] D. E. Polk, A. Calka and B. C. Giessen, "Preparation an Thermal and Mechanical Properties of New Titanium Rich Metallic Glasses", *Acta. Metall.*, 1978, **26**, 1097-1103.

[11] K. Amiya, N. Nishiyama, A. Inoue and T. Masumoto, "Mechanical Strength and Thermal Stability of Ti-Based Amorphous Alloys with Large Glass Forming Ability", *Mater. Sci. Eng. A.*, 1994, **179**, 692-696.

[12] G. He, W. Loser, J. Eckert and L. Schultz, "Enhanced plasticity in a Ti-based bulk metallic glass-forming alloy by in situ formation of a composite microstructure", *J. Mater. Res.*, 2002, **17**, 3015-3018.

[13] Y. C. Kim, D. H. Bae, W. T. Kim and D. H. Kim, "Glass forming ability and crystallization behavior of Ti-based amorphous alloys with high specific strength", *J. Non-Cryst Solids.*, 2003, **325**, 242-250.

[14] J. M. Park, Y. C. Kim, W. T. Kim and D. H. Kim, "Ti-based bulk metallic glasses with high specific strength. Materials Transactions", *Mater. Trans*, 2004, **45**, 595-598.

[15] M. L. Morrison, R. A. Buchanan, A. Peker, P. K. Liaw and J. A. Horton, "Electrochemical behavior of a Ti-based bulk metallic glass", *J. Non-Cryst Solids*, 2007, **353**, 2115-2124.

[16] A. Inoue, N. Nishiyama, K. Amiya, T. Zhang and T. Masumoto, "Ti-Based Amorphous Alloys with a Wide Supercooled Liquid Region", *Mater. Lett.*, 1994, **19**, 131-135.

[17] S. L. Zhu, X. M. Wang, F. X. Qin and A. Inoue : A new Ti-based bulk glassy alloy with potential for biomedical application, *Mater. Sci. Eng. A.*, 2007, **459**, 233-237.

[18] J. J. Oak, D. V. Louzguine-Luzgin and A. Inoue, "Fabrication of Ni-free Ti-based bulk-metallic glassy alloy having potential for application as biomaterial, and investigation of its

mechanical properties, corrosion, and crystallization behavior", *J. Mater. Res.*, 2007, **22**, 1346-1353.

[19] N. Ohtsu, K. Sato, A. Yanagawa, K. Saito, Y. Imai, T. Kohgo, A. Yokoyama, K. Asami, T. Hanawa, "CaTiO3 coating on titanium for biomaterial application - Optimum thickness and tissue response", *J. Biomed. Mater. Res.*, 2007, **82A**, 304-315.

[20] K. Ueda, T. Narushima, T. Goto, M. Taira and T. Katsube, "Fabrication of calcium phosphate films for coating on titanium substrates heated up to 773 K by RF magnetron sputtering and their evaluations", *Biomed Mater*, 2007, **2**, S160-S166.

[21] M. Sato, R. Tu, T. Goto, K. Ueda and T. Narushima, "Hydroxyapatite formation on CaTiO3 film prepared by metal-organic chemical vapor deposition", *Mater. Trans*, 2007, **48**, 1505-1510.

[22] P. Zhou and M. Akao, "Preparation and characterization of double layered coating composed of hydroxyapatite and perovskite by thermal decomposition", *Bio-Med. Mater. Eng.*, 1997, **7**, 67-81.

[23] H. M. Kim, F. Miyaji, T. Kokubo and T. Nakamura, "Effect of heat treatment on apatite-forming ability of Ti metal induced by alkali treatment", *J. Mater. Sci-Mater. Med.*, 1997, **8**, 341-347.

[24] S. Nishiguchi, H. Kato, H. Fujita, M. Oka, H. M. Kim, T. Kokubo and T. Nakamura, "Titanium metals form direct bonding to bone after alkali and heat treatments", *Biomaterials*, 2001, **22**, 2525-2533.

[25] K. Kajiyoshi, M. Yoshimura, Y. Hamaji, K. Tomono and T. Kasanami, "Growth of (Ba,Sr)TiO3 thin films by the hydrothermal-electrochemical method and effect of oxygen evolution on their microstructure", *J. Mater. Res.*, 1996, **11**, 169-183.

[26] S. E. Yoo, M. Hayashi, N. Ishizawa and M. Yoshimura, "Preparation of Strontium Titanate Thin Film on Titanium Metal Substrate by Hydrothermal-Electromechanical Method," *J. Am. Ceram. Soc.*, 1990, **73**, 2561-2563.

[27] K. Kajiyoshi, K. Tomono, Y. Hamaji, T. Kasanami and M. Yoshimura, "Growth of Strontium Titanate Thin Films of Controlled Thickness by the Hydrothermal-Electrochemical Method", *J. Am. Ceram. Soc.*, 1994, **77**, 2889-2897.

[28] K. Kajiyoshi and M. Yoshimura, "Growth mechanism of ATiO(3) (A=Ba, Sr) thin film by the hydrothermal-electrochemical method", Eur. *J. Solid. State. Inorg. Chem.*, 1996, **33**, 623-635.

[29] M. Yoshimura, W. Urushihara, M. Yashima and M. Kakihana, "$CaTiO_3$ Coating on TiAl by Hydrothermal Electrochemical Technique, *Intermetallics*, 1995, **3**, 125-128.

[30] S. Brunauer, P. H. Emmett and E. Teller, "Adsorption of Gases in Multimolecular Layers", *J. Am. Chem. Soc.*, 1938, **60**, 309.

[31] T. Kokubo, H. Kushitani, S. Sakka, T. Kitsugi and T. Yamamuro, "Solutions able to Reproduce Invivo Surface Structure Chantes in Bioactive Glass-Ceramic A-W3", *J. Biomed. Mater. Res.*, 1990, **24**, 721-734.

[32] X. D. Meng, D. Z. Wang, J. H. Liu and S. Y. Zhang, "Preparation and characterization of sodium titanate nanowires from brookite nanocrystallites", *Mater. Res. Bull.*, 2004, **39**, 2163-2170.

[33] T. Kasuga, M. Hiramatsu, A. Hoson, T. Sekino and K. Niihara, "Titania nanotubes prepared by chemical processing", *Adv. Mater.*, 1999, **11**, 1307.

[34] N. Sugiyama et al., "Bioactive titanate nanomesh layer on the Ti-based bulk metallic glass by hydrothermal-electrochemical technique", *Acta Biomaterialia*, 2009, **5**, 1367-1373.

[35] N. Sugiyama et al. : Novel Growing Integration Layer (GIL) method for joining/bonding of metallic and ceramic materials, and its applications for bulk metallic glasses with high bioactivities, *Mater Sci Eng B*, 2009, **161**, 31-35.

[36] BD. Yao, YF. Chan, XY. Zhang, WF. Zhang, ZY Yang, N. Wang : Formation mechanism of TiO2 nanotubes, *Applied Physics Letters*, 2003, **82**, 281.

# LOW TEMPERATURE BONDING OF BULK METALLIC GLASS USING AN ULTRASONIC PROCESS

Hiroshi Nishikawa[1], Krit WongPiromsarn[1], Hiroya Abe[1], Tadashi Takemoto[1], Masao Kubo[2], Yoshiharu Sanagawa[2], Takamasa Sakai[2], Mikio Fukuhara[3], and Akihisa Inoue[3]

[1] Joining and Welding Research Institute, Osaka University, 11-1 Mihogaoka, Ibaraki, Osaka 567-0047, Japan

[2] Micro Fabrication Process Development Center, Panasonic Electric Works Co., Ltd., 1048, Kadoma, Osaka 571-8686, Japan

[3] Institute for Materials Research, Tohoku University, 2-1-1, Katahira, Aoba-Ku, Sendai 980-8677, Japan

## ABSTRACT

For bulk metallic glasses (BMGs) to be adopted over a broader range of engineering applications, it is imperative to develop appropriate bonding processes for BMG/BMG and BMG/crystalline metal. The glass transition temperature and crystallization temperature of BMGs are relatively low. Therefore, to avoid the recrystallization of the glassy phase during bonding, low-temperature processes such as common soldering and ultrasonic wedge bonding process have been studied. The purpose of this study is to determine the feasibility of ultrasonic Al wire wedge bonding on Cu-, Ni-, and Zr-based BMGs. We achieved the bonding of Al wire to BMGs at ambient temperatures, and it was found that the interfaces are straight and sharp without any visible intermetallic compound layers under conventional magnifications of a scanning electron microscope.

## INTRODUCTION

Bulk metallic glasses (BMGs) exhibit remarkable combinations of mechanical, chemical and physical properties compared with the conventional crystalline metals[1-2]. Because of their excellent properties, BMGs are expected to emerge as a new type of industrial or engineering material. Therefore, significant research has been devoted to the development of BMGs, both for fundamental science, and for industrial applications. As a result, many types of BMGs such as Zr-, Ni-, Pd-, Cu-, Fe- and Pt-based BMGs have been prepared, and many unique and useful properties of BMGs have been discovered. Basically, BMG has been formed by avoiding the nucleation and growth of crystalline phases when the alloy is cooled from the molten liquid. Consequently, the stabilization has enabled the production of large and thick BMGs.

For BMGs to be adopted over a broader range of engineering applications, it is imperative to develop appropriate bonding processes for BMG/BMG and BMG/crystalline metal.

During BMG bonding, the most serious issue is the reformation of the glassy phase at high temperatures. Specifically, for melting processes such as electron beam welding and laser welding, it is very important to control the heating and cooling rates. To date, several bonding processes have been attempted for joining two BMGs and a BMG to a crystalline metal, and successful bonds have been obtained by some processes such as electron beam welding, friction welding, explosion, and laser welding[3-5].

In our group, to avoid the recrystallization of the glassy phase, the low-temperature bonding processes of common soldering and ultrasonic wedge bonding have been examined. Regarding the soldering process, in our previous work, the solderability of BMGs using lead-free solder was investigated, and the interface between the solder and a Pd-based BMG was characterized[6]. The ultrasonic wire bonding process is an important method that is usually used for interconnections in electronic packaging[7]. At ambient temperature, the metal wire and the substrate metallization can be bonded together using the ultrasonic energy that is associated with the deformation energy produced by wire plastic deformation. The purpose of the present study is to investigate the feasibility of the ultrasonic Al wire wedge bonding on BMGs. At first, the parameters for ultrasonic wedge bonding were discussed, and then the interface between the Al wire and the BMGs was observed.

EXPERIMENTAL

In this study, three types of bulk metallic glasses (BMGs) were prepared as test samples: $Cu_{60}Zr_{30}Ti_{10}$ (Cu-based BMG), $Zr_{55}Cu_{30}Ni_5Al_{10}$ (Zr-based BMG), and $Ni_{36}Nb_{24}Zr_{40}$ (Ni-based BMG). These BMGs were selected to examine whether the difference in alloy composition affects the bonding properties. Figure 1 shows a schematic diagram of the ultrasonic wedge bonding process using an Al wire. The pure Al (99.99%) wire with 200-μm diameter was bonded to the BMGs. The ultrasonic wedge bonding was implemented at an ultrasonic frequency of 110 kHz and a bonding force of 300 gf (1 gf =9.8 mN) without external heating. The ultrasonic power and the length of time the vibrations were applied were the test parameters. For the experiments, the ultrasonic power was set to 2.94 W, 1.99 W, and 1.16 W, and the vibration duration was set to 300, 150, and 100 ms, respectively.

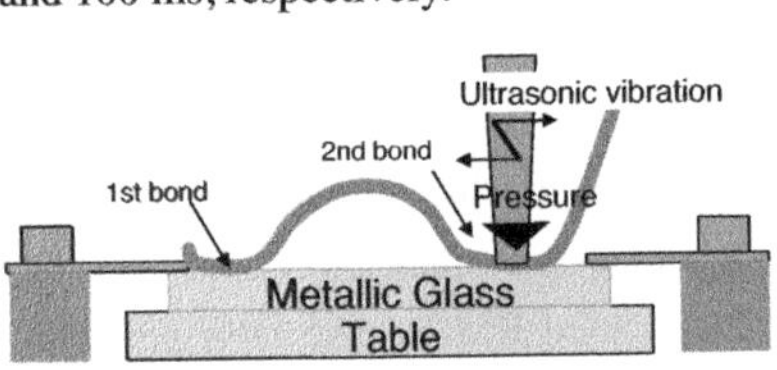


Figure 1. Ultrasonic Al wire wedge bonding diagram.

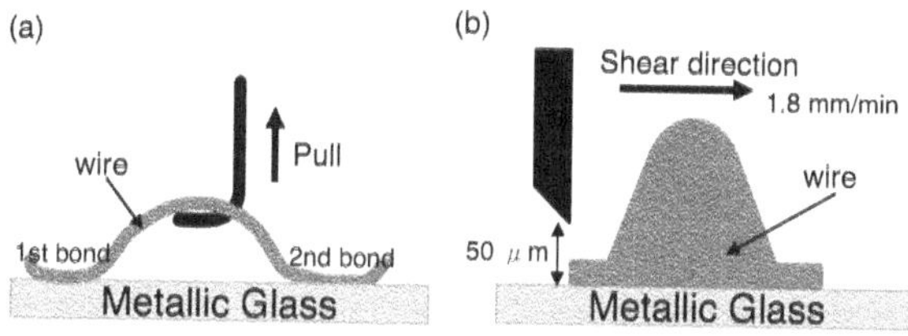


Figure 2. Diagram of pull test (a) and shear test (b).

To investigate the bonding quality, the wire bonds were subjected to a pull test and a shear test, as shown in Figure 2. At least three samples were used for each test. To observe the Al/BMG interface, a scanning electron micrograph (SEM) was used.

## RESULTS AND DISCUSSION

The joint quality and the reliability of the ultrasonic wedge bond are related to processing parameters such as ultrasonic vibration speed, ultrasonic power, bonding force, bonding time, and surface condition. In the present study, the bonding behavior of the ultrasonic Al wedge bond on BMGs was investigated. In particular, the ultrasonic power and the bonding time were varied to get good bond quality at ambient temperature. Table I summarizes the effect of the ultrasonic power and the bonding time on the bond quality. For 3.44 W of ultrasonic power

Table I Result of ultrasonic Al wire wedge bonding on bulk metallic glasses.

| BMG | Bonding time (ms) | Output power (W) | | |
|---|---|---|---|---|
| | | 2.94 | 3.44 | 4.08 |
| Ni-based | 100 | × | × | × |
| | 150 | — | × | — |
| | 300 | × | ○ | — |
| Cu-based | 100 | × | × | × |
| | 150 | — | △ | — |
| | 300 | × | ○ | — |
| Zr-based | 100 | × | × | × |
| | 150 | — | △ | — |
| | 300 | × | ○ | — |

○: No failure (0/3), △: One failure (1/3), ×:Two or three failure (2/3 or 3/3)

and a bonding time of 300 ms, all wedge bonds were sufficiently strong, and would stick to all BMGs. In other conditions, regardless of the BMG, all or some bonds were lifted off the BMGs because of weak bonding. It was found that a good bond between the Al wire and BMG can be formed using the ultrasonic wedge bonding process, and that the ultrasonic power and the bonding time significantly influence the bond quality of the ultrasonic Al wedge bonds on BMGs. The final bonding parameters were set as follows: ultrasonic power 3.44 W, bonding time 300 ms, ultrasonic frequency 110 kHz, and bonding force 300 gf.

The completed bonded sample was subsequently subjected to the pull test shown in Figure 2(a) to determine the pull strength and failure mode. The failure could be clearly classified into the following three modes: an interfacial break between the Al wire and the BMG, a neck break of Al wire, and a wire break. Wire break and neck break are the preferred mode of failure in this experiment because these breaks indicate that the bonding interface between the wire and BMG is sufficiently robust. Figure 3 shows the pull strength and failure mode for each bond sample. The average tensile strengths were 272 gf for Ni-based BMG, 292 gf for Cu-based BMG, and 290 gf for Zr-based BMG. For the failure mode, the ratio of wire break to neck break for Cu-based BMG was higher than that for Ni- and Zr-based BMGs, which indicates that the bond quality of the Cu-based BMG is superior to that of Ni- and Zr-based BMGs. However, the origin of these phenomena is not clear at present.

Figure 4 shows how the results of the pull test of the wire part are related to the shear test of the bonding part. It is clear that a good relationship exists between the pull strength and the shear strength for all BMG systems tested here. Thus, it is reasonable to estimate the bond quality by using the pull test.

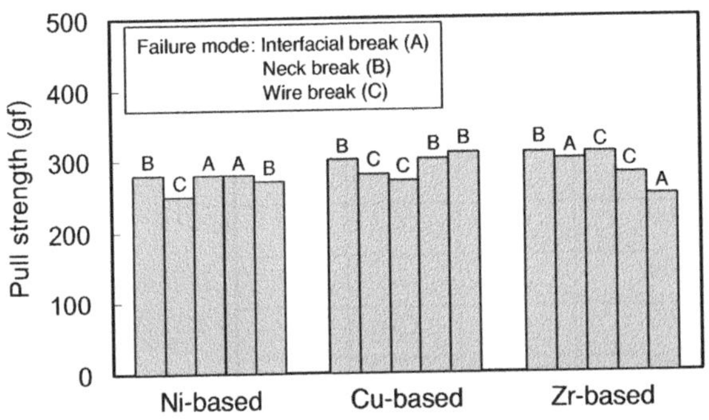


Figure 3. Pull strength and failure mode for each bond.

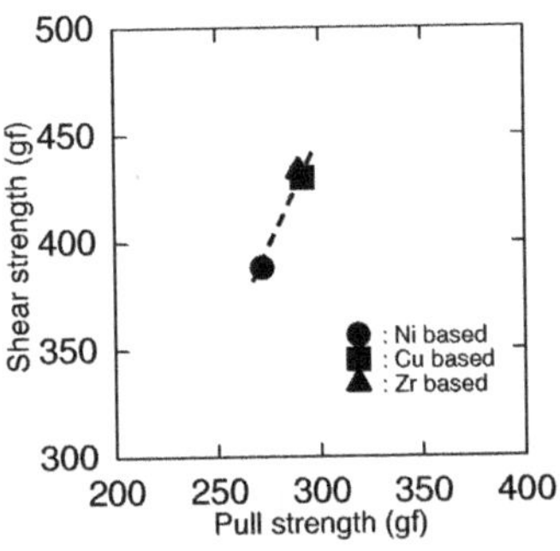


Figure 4. Relationship between shear strength and pull strength.

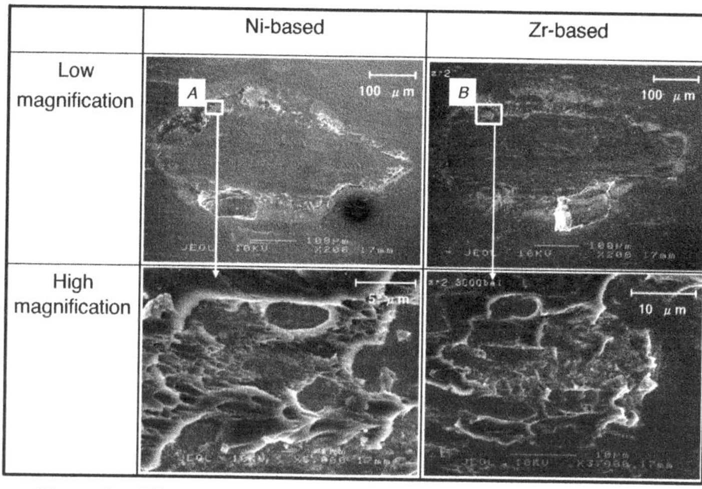


Figure 5. Ultrasonic wedge bond fracture surfaces of after pull tests.

After the pull test, the interfacial break surfaces were observed using the SEM. Figure 5 shows the interfacial break surfaces of Ni-based BMG and Zr-based BMG after the pull test. At low magnification, fragments of Al wire were found on both surfaces, suggesting that the Al wire was perhaps strongly bonded on the BMG at the periphery. At high magnification, the fracture surfaces in areas *A* and *B* are dimpled, indicating a desirable ductile fracture mode. Normally, substances inhibiting bonding must be removed to form a good joint. In this case, the relative movement would occur between the Al wire and the BMG, resulting in the removal of oxides and contaminants. Furthermore, the clean surfaces are exposed so the metallurgical bonding can be realized locally.

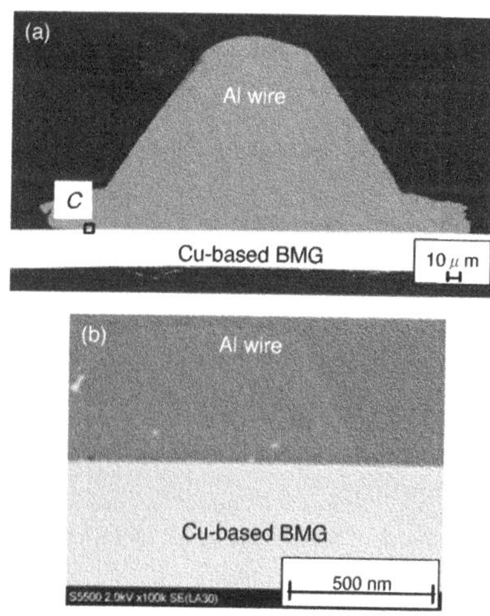


Figure 6. Interface between Al wire and Cu-based BMG after ultrasonic wedge bonding: (a) low magnification, (b) high magnification of area *C*.

Figure 6 shows SEM images of a typical interface between the Al wire and the Cu-based BMG. The direction of ultrasonic vibration in this case is perpendicular to the plane of the figure. Figure 6(a) presents the cross-section of the bond and Figure 6(b) shows a high magnification of area *C* that is indicated in Figure 6(a). It is apparent from Figure 6(a) that a large deformation in the Al wire exists, especially at the periphery of the wire. In Figure 6(b), the interface was straight and sharp without any visible intermetallic compound (IMC) layer under this magnification for structural characterization, although it is well known that the IMC formation at the interface enhances the interfacial bonding strength.

## CONCLUSIONS

Ultrasonic Al wire wedge bonding on Cu-based, Ni-based and Zr-based BMGs at ambient temperature was investigated, and the interface between the Al wire and the BMG was observed. Based on the results, we formulated the following conclusions:

(1) Ultrasonic Al wire wedge bonding on BMGs at ambient temperature is achieved regardless of which of the three BMG systems are used. The ultrasonic power and the bonding time strongly influence the bond quality.

(2) The results suggest that the Al wire may be strongly bonded to the BMG at the periphery. The fracture surfaces on the pull test samples show dimple patterns, which indicate a desirable ductile fracture mode.

(3) The interface between the Al wire and the BMG is straight and sharp without any visible intermetallic compound layer.

## ACKNOWLEDGEMENTS

This work was supported by Grant-in-Aid for Cooperative Research Project of Nationwide Joint-Use Research Institute on Development Base of Joining Technology for New Metallic Glasses and Inorganic Materials from The Ministry of Education, Culture, Sports, Science and Technology, Japan.

## REFERENCES

1) A. Inoue, "Stabilization of Metallic Supercooled Liquid and Bulk Amorphous Alloys", *Acta Mater.*, **48**, 279-306 (2000).

2) A. Inoue, "Bulk Amorphous and Nanocrystalline Alloys with High Functional Properties", *Mater. Sci. Eng. A*, **304-306**, 1-10 (2001).

3) Y. Kawamura, "Liquid Phase and Supercooled Liquid Phase Welding of Bulk Metallic Glasses", *Mater. Sci. Eng. A*, **375-377**, 112-119 (2004).

4) S. Kagao, Y. Kawamura, and Y. Ohno, "Electron-beam Welding of Zr-based Bulk Metallic Glasses, Mater". *Sci. Eng. A*, **375-377**, 312-316 (2004).

5) B. Li, Z. Y. Li, J. G. Xiong, L. Xing, D. Wang, and Y. Li, "Laser Welding of $Zr_{45}Cu_{48}Al_7$ Bulk Glassy Alloy", *J. Alloys and Compounds*, **413**, 118-121 (2006).
6) H. Nishikawa, K. WongPiromsarn, H. Abe, T. Takemoto, M. Fukuhara, T. Wada and, A. Inoue, "Wetting Characteristics of Sn-Ag-Cu Solder on Pd-based Metallic Glass", *Mater. Sci. Eng. B*, **148**, 124-127 (2008).
7) U. Geissler, S.-R. Martin, K.-D. Lang, and H. Reichl, Investigation of Microstructural "Processes during Ultrasonic Wedge/Wedge Bonding of AlSi1 Wires", *J. Electron. Mater.*, **35**, 173-180 (2006).

# MICROSTRUCTURE AND PROPERTIES OF A NI-BASED METALLIC GLASS COATING PRODUCED BY GAS TUNNEL TYPE PLASMA SPRAYING

Akira Kobayashi*, Toshio Kuroda*, Hisamichi Kimura** and Akihisa Inoue**
*Joining & Welding Res. Inst., Osaka University, Ibaraki, Osaka, Japan
**Inst. for Materials Res., Tohoku University, Sendai, Miyagi, Japan

## ABSTRACT

Metallic glass has excellent characteristics including high toughness and corrosion resistance. Developmental research studies on metallic glass have been conducted by many researchers. Thermal spraying is one method to produce metallic glass composites. The gas tunnel type plasma system, which has a high energy density and efficiency, is useful to obtain high quality ceramic coatings such as $Al_2O_3$ and $ZrO_2$ coatings. In this study, Ni-based metallic glass coatings were produced by gas tunnel type plasma spraying, and their microstructure and mechanical properties were characterized. Dense, Ni-based metallic glass coatings of more than 200 μm in thickness having a Vickers hardness of $Hv$=600-800 were formed.

## INTRODUCTION

Among various functional materials, metallic glass has excellent physical and chemical properties such as high toughness and corrosion resistance[1-3]. Therefore it is one of the most attractive advanced materials on which many researchers have conducted various developmental research studies. However, because metallic glass material is expensive, it has only found application for small size parts in limited industrial fields. In order to widen the industrial application, a lower cost composite material is preferred.

In the processing of metallic glass coating with conventional deposition techniques such as plasma sputtering, low deposition rates make it difficult to forming thick coatings. Thermal spraying is one potential method to produce metallic glass coatings on a large scale at low cost, and therefore could widen the application fields.

Gas tunnel plasma spraying is one of the most effective technologies for depositing high quality ceramic coatings[4,5] and synthesizing functional materials[6], because the plasma jet has high speed and high energy density under various operating conditions[7]. The performance of gas tunnel type plasma jets were determined in previous studies[8,9]. Because of its superior advantages compared with other conventional plasma jets[10], this plasma has great possibilities for various applications in thermal processing[7]. High quality ceramic coatings have been fabricated by the gas tunnel type plasma spraying method[11]. For example, typical alumina coatings produced have a high Vickers hardness of $Hv$ =1200-1600 [12]. Also, it is possible to produce sprayed coatings of refractory materials such as W[13]. In another application, the gas tunnel type plasma jet was applied for the surface nitridation of titanium[14].

Regarding metallic glass coatings, thick coatings of Fe-based metallic glass were easily produced by gas tunnel type plasma spraying[15]. The characteristics of the metallic glass coatings were investigated in a previous study, and new results were obtained using the plasma spraying method. The amorphous phase of metal glass coating was confirmed by x-ray diffraction (XRD). Dense Fe-base metal glass coatings of about 200 μm in thickness with a Vickers hardness of about $Hv$ =1100 were obtained with a plasma current of 300A.

In this study, a Ni-based metallic glass coating was deposited on a stainless-steel substrate by gas tunnel plasma spraying using Ni based metallic glass powder (Ni60-Nb20-Zr20) as starting material. The microstructure and the morphology of the cross section of the as-sprayed metallic glass coatings were examined. The structure of the metallic glass coatings was analyzed using XRD. The Vickers hardness was measured on the cross section of the coating.

## EXPERIMENTAL PROCEDURE

### Preparation of metal glass coatings

The gas tunnel type plasma spraying torch is shown schematically in Figure 1. The powder was atmospherically plasma-sprayed (APS) onto a flat 304 stainless steel substrate, and the metallic glass coating was formed. Ni-based metallic glass powder was provided from the outside of the gas divertor nozzle (20 mm in diameter) of the torch in this study. In this case also, Ni-based metallic glass powder was fed inside the plasma flame and melted effectively.

Table 1 shows the spraying conditions in the present investigation. The plasma torch was operated at a power level of 10-20kW and the arc current *I* was 100-350 A. The plasma jet was generated with an argon gas flow rate of 200 l/m, at a spray distance of 40 mm. The powder feed rate was about 7 g/min. The stainless steel substrate was traversed 16 times during the spraying time of 24 s. The substrate of dimension of 50 x50 x2 $mm^3$ was grit-blasted with alumina grit on one side.

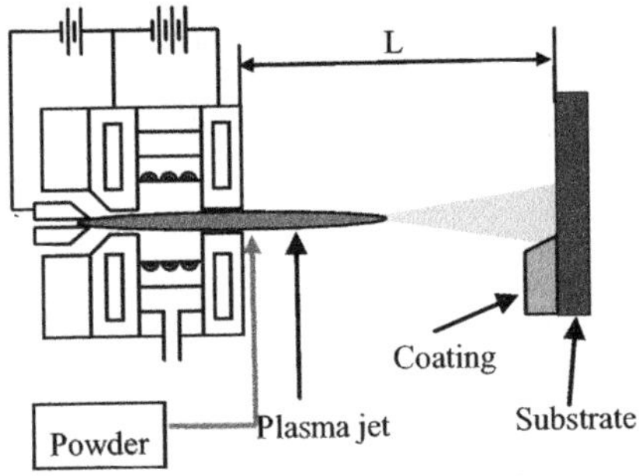


Figure 1. Schematic of the gas tunnel type plasma spraying torch.

Table I Spraying condition.

| | |
|---|---|
| Arc current | 100-350 A |
| Voltage | 40-50 V |
| Spraying distance | 40 mm |
| Working gas Ar flow rate | 200 l/min |
| Powder feed rate | 7 g/min |
| Spraying time | 24 s |

### Ni-based metallic glass powder

Ni-based metallic glass powder (Ni60-Nb20-Zr20) was used in this investigation. The morphology of the Ni-based metallic glass powder is shown in Fiure.2. This was taken with a scanning electron microscope (SEM). The particles are spherical with sizes ranging from 10 μm to 25 μm (average size: 18 μm). Figure 3 shows the XRD pattern of this Ni-based metallic glass powder. This XRD pattern of metallic glass powder showed a broad peak at about 42 degree corresponding to the amorphous phase, although it contained some small crystalline peaks of the composite as indicated by the arrows (The main content is thought as $Ni_{10}$-$Zr_{7}$.).

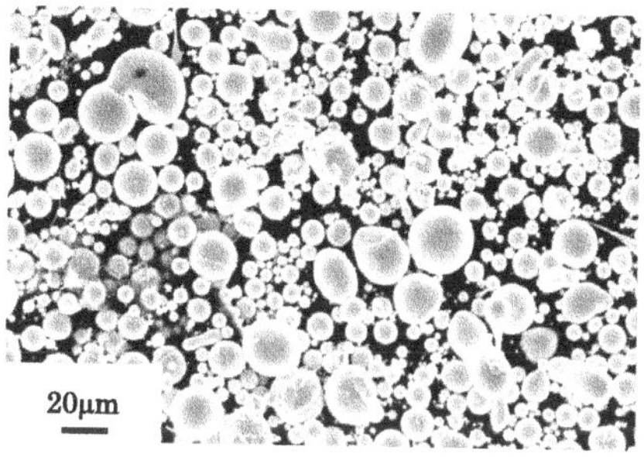


Figure 2. SEM micrographs of the Ni based metal glass powder.

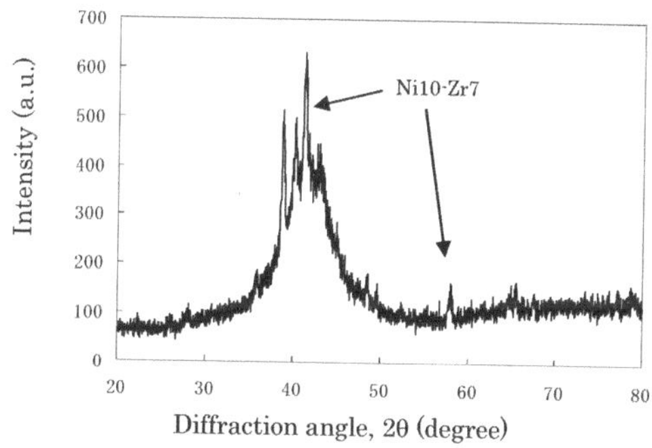


Figure 3. XRD patterns of the Ni-based metallic glass powder.

Characterization of Ni-based metallic glass coatings

Microscopic observation of the coatings was performed using an optical microscope (OM) and an ERA8800FE scanning electron microscope (SEM). The average thickness of the sprayed coatings was determined through the cross section of the coating.

The phase constituents of the metallic glass coating were identified by using a JEOL JDX- 3530M X-ray diffractometer with a CuKα radiation source at a voltage of 40 kV and a current of 40 mA. Vickers microhardness measurement was made on polished sample surfaces with a load of 50g. The indentation parameter was set at 20s loading time. An average hardness was derived from five measurements.

## RESULTS AND DISCUSSION

### Microstructure of Ni-based metallic glass coatings

Figure 4 shows the SEM micrographs of the cross-section of Ni-based metallic glass coatings sprayed at spraying distance of L=40 mm with a plasma current of 300A. As shown in Figure 4, the coating thickness was more than 250 μm. This coating had some pores in the cross section, but there was not more than 5% porosity. The bonding between the coating and the substrate was good.

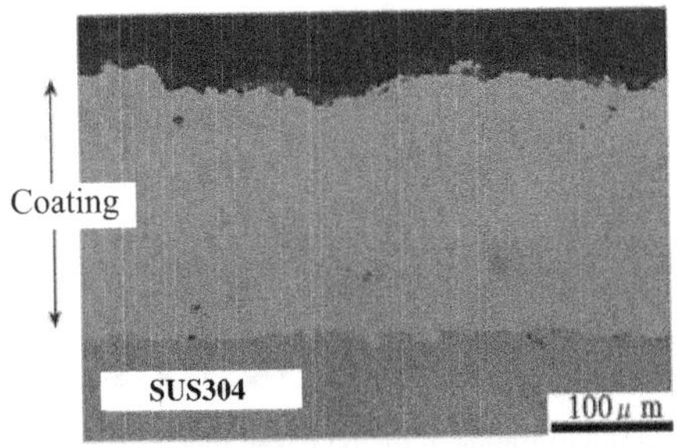


Figure 4. SEM micrographs of the cross-section of the Ni based metallic glass coating sprayed at 300A, on the 16 times traversed substrate.

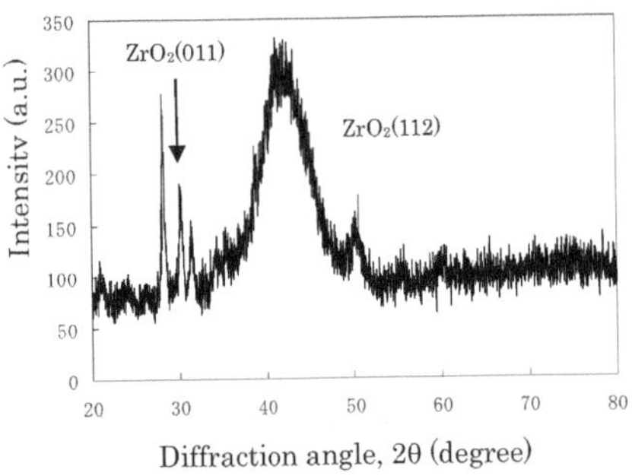


Figure 5. XRD patterns of the Ni-based metallic glass coating sprayed on stainless-steel substrate.

The XRD pattern from the surface of the coating shown in Figure 4 is shown in Fiure.5. In this case, a broad amorphous phase peak with maximum at about 42 degree was observed at the same diffraction angle as that of the as-received powder. Although a number of crystalline peaks of intermetallic compounds were observed in the same pattern of the powder (see the XRD pattern shown in Fig. 3), those peaks disappeared after coating. On the other hand, some crystalline peaks of Zr oxides appear in the XRD pattern. Also a few unknown peaks appear near the peak of $ZrO_2$(011).

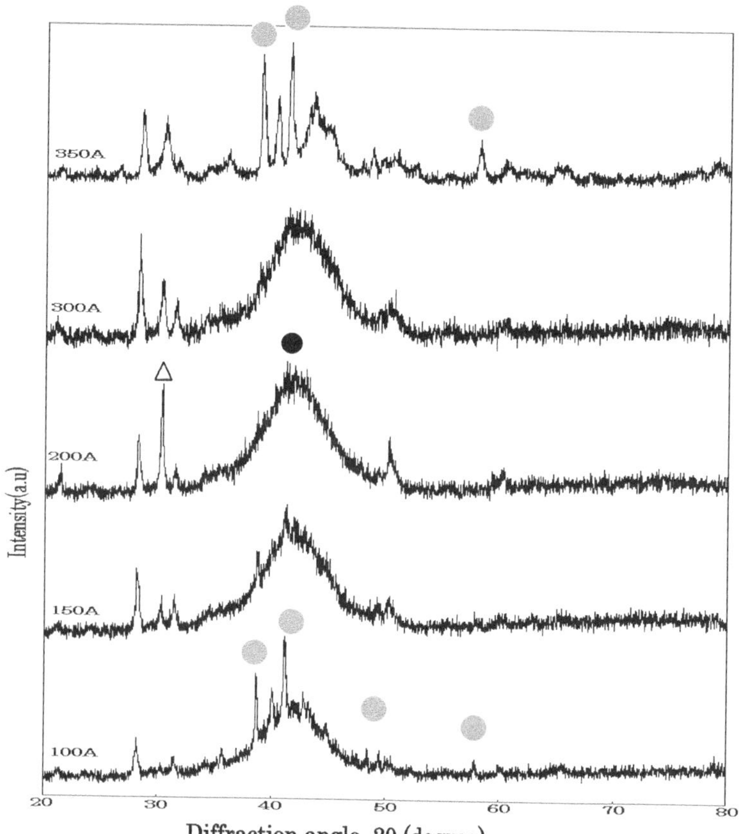


Figure 6. XRD patterns of the Ni-based metallic glass coating sprayed on stainless-steel substrate at various plasma currents of 100A to 350A.

Effect of plasma current on Ni-based metallic glass coating.

Pores in the coating were observed for a small current of below 200 A. However, for 250A and 300A, there were few pores, and dense coatings were obtained, indicating a proper condition for obtaining a good quality coating. Also the bonding between the coating and the substrate was good at the interface in these photographs.

Figure 6 shows the XRD patterns from the surface of the Ni based metallic glass coating. The amorphous phase appears at the broad peak with maximum at about 42 degree for the current, 150 A, 200 A, 250A and 300A, but the shape was destroyed at 350A. And many crystalline peaks are observed at the same time (100 A also similar to powder). Some crystalline phases of Zr oxides or intermetallic compounds were formed during the plasma spraying.

The XRD patterns at 200-300A shows a broad peak with a maximum at about 42 degrees. The oxidation peak of $ZrO_2$ also appears at about 30 degree (a triangle mark in Fig.6). This means that oxidation occurred during the plasma spraying process. It seems to be at a minimum intensity at about 250 A.

These results show that processing at 250 A will be best for obtaining a good coating, as the intensities of the crystalline peaks were a little lower. This suggests that the coating quality was improved by increasing to the proper plasma current.

## CONCLUSIONS

Ni-based metallic glass coatings were formed by gas tunnel type plasma spraying, and the results are as follows:

(1) A Ni-based metallic glass sprayed coating of thickness of more than 250 μm was obtained at 300 A, which, depended on the spraying parameters. A dense metal glass coating with low porosity was formed on a stainless-steel substrate using a plasma current of 200-300 A.

(2) From the XRD pattern of the coating, a broad amorphous peak at about 42 degree was observed, which is the same diffraction angle as the powder. Additionally, the XRD analyses showed that Zr based metallic glass particles had some decomposition or oxidation during deposition, in spite of the presence of the amorphous phases in the sprayed coatings.

## REFERENCES


1) A. Inoue, A. Takeuchi and B. Shen, "Formation and Functional Properties of Fe-Based Bulk Glassy Alloys", Mater. Trans., Vol 42 2001, p970-978

2) A. Inoue, B.L. Shen and H.M. Kimura, "Fundamental Properties and Applications of Fe-Based Bulk Glassy Alloys", J. Metastable Nanocrys. Mater. 20-21 2004, p3-12

3) A. Inoue, and B.L. Shen, "A New Fe-based Bulk Glassy Alloy with Outstanding Mechanical Properties", Adv. Mater., Vol 16 2004, p2189-2192.

4) Y. Arata, A. Kobayashi, and Y. Habara, "Ceramic coatings produced by means of a gas tunnel type plasma jet", *J.Appl.Phys.*, 62-12 (1987), 4884-4889

5) A. Kobayashi, Y. Habara, and Y. Arata, "Effects of Spraying Conditions in Gas Tunnel Type Plasma Spraying", *J.High Temp.Soc.*, 18-2 (1992), 25-32.

6) A. Kobayashi, "New Applied Technology of Plasma Heat Source", *Weld. International*, 4-4 (1990), 276-

7) M. Okada and Y. Arata, "Plasma Engineering", Pub. *Nikkan Kogyo Shinbun-sha*, Tokyo (1965).

8) Y. Arata and A. Kobayashi, "Development of Gas Tunnel Type High Power Plasma Jet", *J.High Temp.Soc.*, 11-3 (1985), 124-131.

9) Y. Arata and A. Kobayashi, "Application of gas tunnel to high-energy-density plasma beams", *J.Appl.Phys.* 59-9 (1986), 3038-3044.

10) Y. Arata, A. Kobayashi, and Y. Habara, "Basic Characteristics of Gas Tunnel Type Plasma Jet Torch", *Jpn.J.Appl.Phys.*, 25-11 (1986), 1697-1701.

11) A. Kobayashi, Y. Habara, and Y. Arata, "Effects of Spraying Conditions in Gas Tunnel Type Plasma Spraying", *J.High Temp.Soc.*, 18-2 (1992), 25-32.

12) A. Kobayashi, Property of an Alumina "Coating Sprayed with a Gas Tunnel Plasma Spraying", *Proc.of ITSC.*, (1992)57-62.

13) A. Kobayashi, S. Sharafat and N.M. Ghoniem, "Formation of Tungsten Coatings by Gas Tunnel Type Plasma Spraying", *Surface and Coating Technology*, (2006), 4630-4635.

14) A. Kobayashi, "Surface Nitridation of Titanium Metals by Means of Gas Tunnel Type Plasma Jet", J.Mater.Eng.& Performance, Vol.5, (No.3), 1996, p373-380

15) A. Kobayashi, S. Yano, H. Kimura and A. Inoue, "Formation of Fe-base Metal Glass Coating Produced by Gas Tunnel Type Plasma Spraying", JWRI, Vol 35-1 (2006) p23-27.

# TENSILE BEHAVIOR OF Zr-BASED BULK METALLIC GLASS IN HIGH-SPEED TENSILE TESTS

Yoshihiro Sakino, Toshio Kuroda and You-Chul Kim
Joining and Welding Research Institute, Osaka University
Ibaraki, Osaka, Japan

## ABSTRACT

High-speed tensile tests on Zr-based bulk metallic glass were carried out to determine the effect of strain rate on the tensile behavior of metallic glasses. In these tests, two strain rates, $1.7 \times 10^{0}$/s and $2.3 \times 10^{1}$/s, were selected, and the results were compared to those of the static test ($1.8 \times 10^{-5}$/s). $Zr_{48}Cu_{36}Al_{8}Ag_{8}$ bulk metallic glass was chosen for this study. The diameter, gauge length, and overall length of the bulk specimen were 6 mm, 30 mm, and 60 mm, respectively.

Regardless of the strain rate, all the specimens fractured in the elastic range. Sparking phenomena were observed at the fracture points. The tensile strength of the Zr-based bulk metallic glass becomes large; in particular, it is approximately four times the corresponding value for the crystallized specimen for all the strain rates. The tensile strength and fracture elongation of metallic glass tend to become small with an increase in the strain rate. Young's moduli were almost in the same range for all strain rates. Drastic changes in the tensile behavior of Zr-based metallic glass caused by an increase in the strain rate were not observed.

## INTRODUCTION

It is well known that strain rate affects the tensile properties of metals, such as tensile strength, fracture elongation, and fracture properties. For example, a decrease in the elongation and an increase in the transition temperature caused by high strain rates have been reported[1]. It can be considered that the tensile behavior of metallic glass is also affected by the strain rate[2,3]. To determine the effect of the strain rate on the tensile behavior of metallic glasses, high-speed tensile tests on Zr-based bulk metallic glass were carried out.

## OUTLINE OF EXPERIMENT

The high-speed-tensile-test unit (Figure 1) of the dynamic structural testing system can load 1,200 kN at a speed of 1,200 mm/s. This unit can support both large loads and high speeds. However, this unit cannot be used for small and brittle specimens, like those of the metallic glasses, and cannot be used to measure small loads with accuracy at high speeds. This unit is designed for large and ductile specimens, like steel and welded connections that are fractured

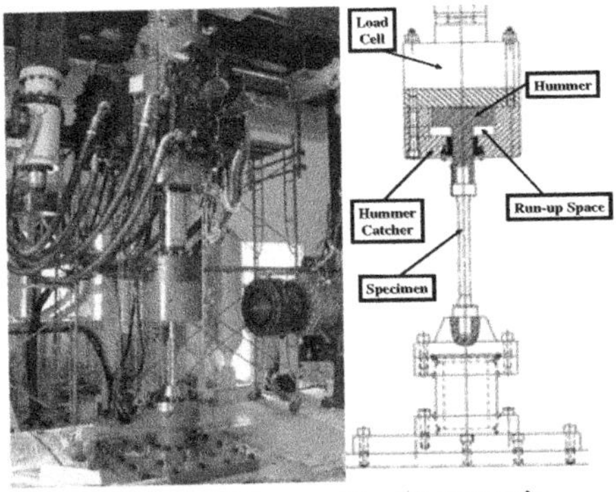


Figure1. High-speed tensile test unit of the dynamic structural testing system.

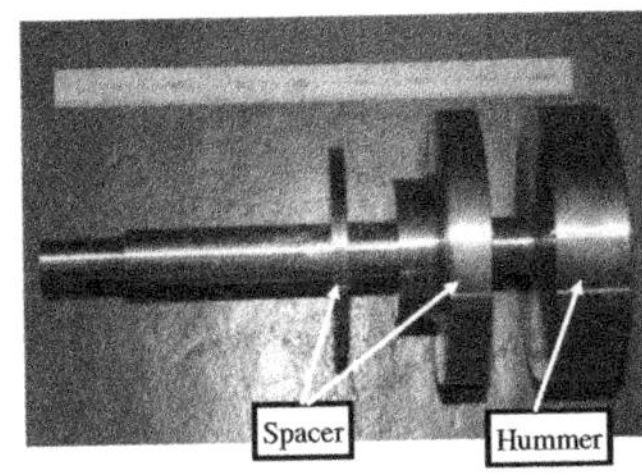


Figure 2. Light weight hummer and spacer for metallic glass

only by loads over 500 kN. Therefore, accurate methods for measuring loads by using a coupler with strain gauges and by using low-pass filters were developed, and a special lightweight hummer and spacer (Figure 2) were also developed to enable high-speed tensile tests on brittle specimens like metallic glasses to be carried out. By using these experimental methods, it has been confirmed that the high-speed tensile test on the small and brittle specimens can be carried out using the high-speed tensile test unit of the dynamic structural testing system.

$Zr_{48}Cu_{36}Al_{8}Ag_{8}$ bulk metallic glass was chosen in this study. As shown in Figure 3, the diameter, gauge length, and overall length of the specimen were 6 mm, 30 mm, and 60 mm, respectively. Visible light photographs and X-ray photographs are shown in Figure 4. All specimens were inspected to confirm that there were no defects.

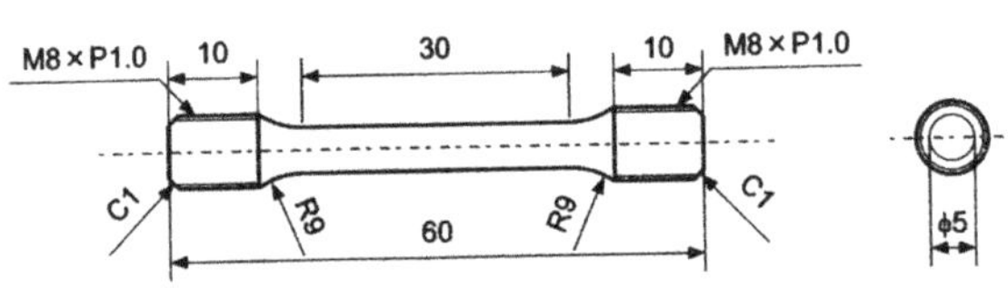


Figure 3. Specimen geometries

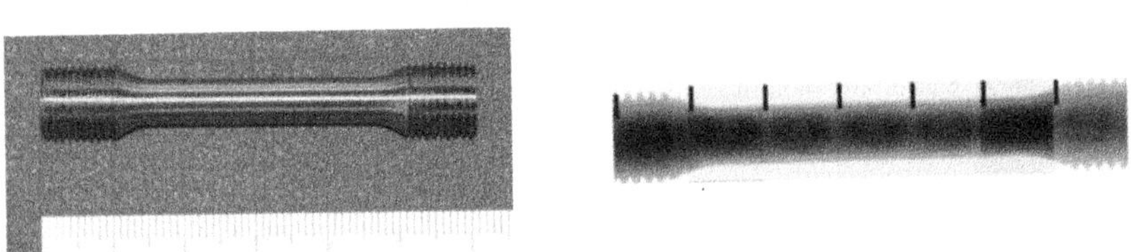

Figure 4. Visible and X-ray photographs of the specimen

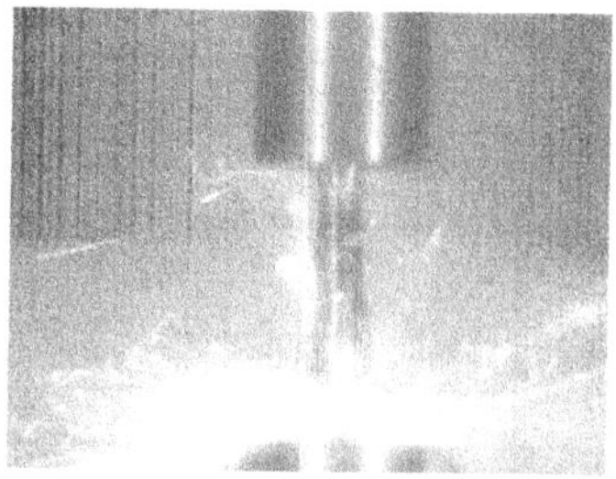

Figure 5. Sparking phenomena (Strain rate: $1.8\times10^{-5}$/s)

Figure 6. Sparking phenomena (Strain rate: $1.7\times10^{0}$/s)

In the high-speed tensile test, two strain rates, $1.7 \times 10^{0}$/s and $2.3 \times 10^{1}$/s, were selected, and the results were compared to those of the static test ($1.8 \times 10^{-5}$/s). The moments of fracture of the specimens were recorded in high-speed video.

## EXPERIMENTAL RESULTS

Figure 5 and 6 show moment of fracture of specimens. Sparking phenomena were observed at the fracture points for all strain rates. Strong sparks were simultaneously generated at several points. X-ray diffraction scans revealed the amorphous nature of the fracture surface. SEM observations show the occurrence of local melting phenomena near the surface of the specimen. No other studies have reported this sparking phenomenon in a tensile test. However, sparking was observed at fracture in a fatigue test on Zr-based metallic glass[5].

Figures 7 and 8 show examples of stress-strain relationships. Irrespective of the strain

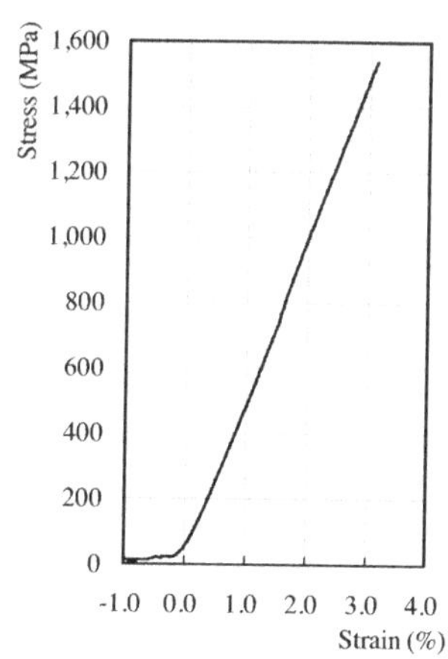


Figure 7. Stress-strain relationship (Strain rate: $1.8\times10^{-5}$/s)

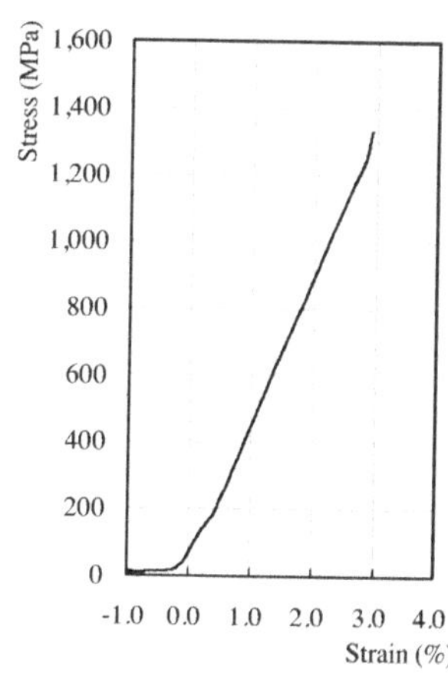


Figure 8. Stress-strain relationship (Strain rate: $1.7\times10^{0}$/s)

Table I. Test results

| Loading speed | Strain rate | Maximum load | Tensile Strength | Fracture elongation | | Young's modulus |
|---|---|---|---|---|---|---|
| (mm/sec) | (/sec) | (kN) | (MPa) | (mm) | (%) | (GPa) |
| 0.001 | $1.8\times10^{-5}$ | 30.3 | 1,525 | 0.936 | 3.12 | 48.7 |
| 100 | $1.7\times10^{0}$ | 26.2 | 1,333 | 0.877 | 2.92 | 45.0 |
| 100 | $1.7\times10^{0}$ | 22.2 | 1,077 | 0.790 | 2.63 | 44.5 |
| 1,200 | $2.3\times10^{1}$ | 26.2 | 1,335 | 0.790 | 2.63 | 50.4 |

rate, all the specimens fractured in the elastic range. Hence, a reduction in Young's modulus and the occurrence of plastic deformations with an increase in stress was not observed.

All test results are listed in Table I. The behavior of the tensile strength, fracture elongation, and Young's modulus in terms of the strain rate is shown in Figures 9–11. The tensile strength of the crystallized specimen ($Zr_{55}Cu_{30}Al_{10}Ni_5$) is also shown in Figure 9. The tensile strength of the general Zr-based crystallized alloy is in almost the same range as that of the crystallized specimen. However, the tensile strength of the Zr-based metallic glass specimen (•) becomes large; in particular, it is around four times the corresponding value for the crystallized specimen (■) for all the strain rates. The tensile strength and fracture elongation of metallic glass

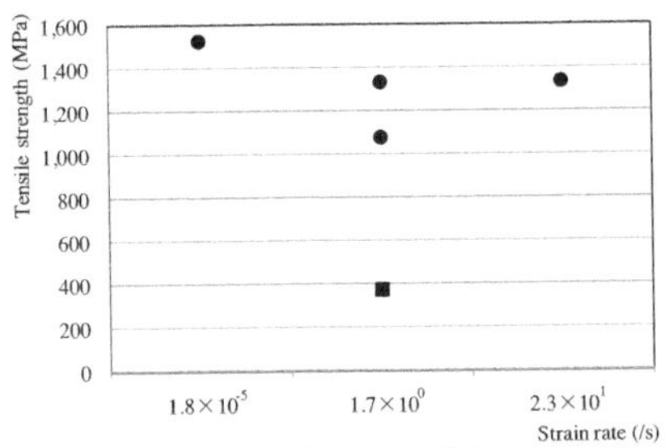


Figure 9. Comparison of tensile strength in term of strain rate

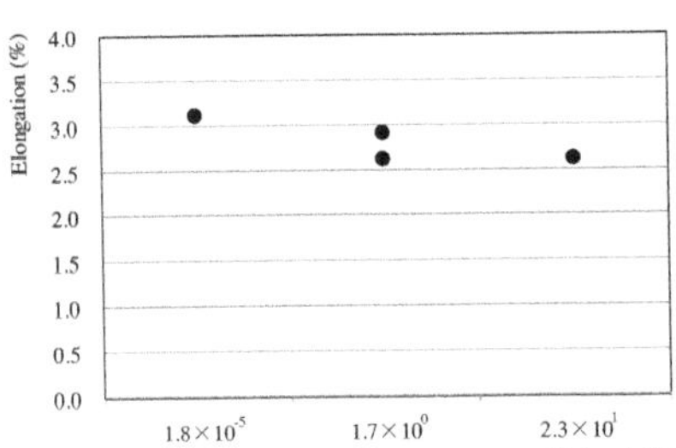


Figure 10 Comparison of elongation in term of strain rate

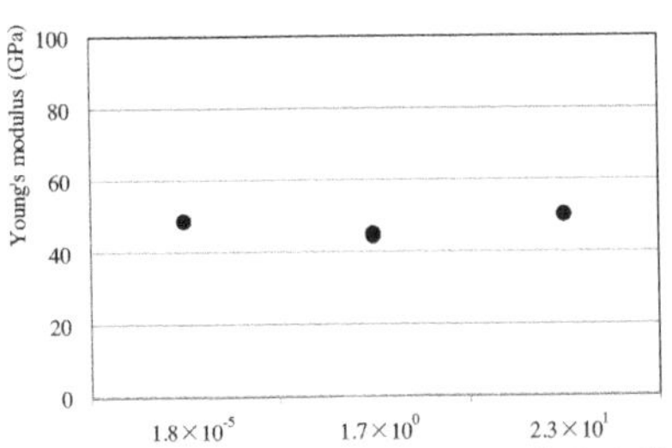


Figure 11 Comparison of Young's modulus in term of strain rate

tend to become small with an increase in the strain rate. Young's moduli were in almost the same range for all the strain rates. As mentioned above, drastic changes in the tensile behavior of Zr-based bulk metallic glass caused by an increase in the strain rate were not observed.

## CONCLUSIONS

1) Irrespective of the strain rate, all the Zr-based bulk metallic glass specimens fractured in the elastic range, and strong sparks were simultaneously generated at several fracture points.
2) The tensile strength of the Zr-based bulk metallic glass becomes large; in particular, it is around four times the corresponding value for the crystallized specimen for all strain rates.
3) Drastic changes in the tensile behavior of Zr-based bulk metallic glass caused by an increase in the strain rate were not observed.

## ACKNOWLEDGEMENTS

Authors sincerely thank Professor Akihisa Inoue and Associate Professor Hisamichi Kimura, Tohoku University for supplying the samples for this study. Authors also thank Technical specialist Mr. Nakatuji, Osaka University for his assisting the experiment. This work was supported by Grant-in-Aid for Cooperative Research Project of Nationwide Joint-Use Research Institute on Development Base of Joining Technology for New Metallic Glasses and Inorganic Materials from The Ministry of Education, Culture, Sports, Science and Technology, Japan.

## REFERENCES

[1] Y. Sakino, S. Takahashi, H. Kamura and Y.-C.KIM, "Effect of Strain Rate on Tensile Properties of HAZ of Structural Steels", *Proc. National Symposium on Welding Mechanics and Design 2006*, **No.2**, 455-458 (2006).

[2] Z. F. Zhang, J. Eckert and L. Schultz, "Difference in Compressive and Tensile Fracture Mechanisms of $Zr_{59}Cu_{20}Al_{10}Ni_8Ti_3$ Bulk Metallic glass", *Acta Materialia*, **51**, 1167-1179 (2003).

[3] L. F. Liu, L. H. Dai, Y. L. Bai and B. C. Wei, "Initiation and Propagation of Shear Bands in Zr-based Bulk Metallic Glass under Quasi-static and Dynamic Shear Loadings", *J. Non-Crystalline Solids*, **351**, 3259-3270 (2005).

[4] Y. F. Xue, H. N. Cai, L. Wang, F. C. Wang and H.F. Zhang, "Effect of Loading Rate on Failure in Zr-based Bulk Metallic Glass", *Mater. Sci. and Eng.*, A, **473**, 105-110 (2008).

[5] G. Y. Wang, P. K. Liaw, W. H. Peter, B. Yang, Y. Yokoyama, M. L. Benson, B. A. Green, M. J. Kirkham, S. A. White, T. A. Saleh, R. L. McDaniels, R. V. Steward, R. A. Buchanan, C. T. Liu and C. R. Brooks, "Fatigue Behavior and Fracture Morphology of $Zr_{50}Al_{10}Cu_{30}Ni_{10}$ Bulk-metallic Glasses", *Intermetallics*, **12**, 1219-1227 (2004).

# MICROSTRUCTURE FORMATION ON A TITANIUM-BASED BULK METALLIC GLASS PLATE BY FEMTOSECOND LASER IRRADIATION

Masahiro Tsukamoto[1], Togo Shinonaga[2], Nobuhiro Matsushita[3], Xinmin Wang[4], Takeshi Wada[4] and Nobuyuki Abe[1]

[1]*Joining and Welding Research Institute, Osaka University, Suita, Osaka, Japan*

[2]*Graduate School of Engineering, Osaka University, Suita, Osaka, Japan*

[3]*Materials and Structures Laboratory, Tokyo Institute of Technology, Nagatsuka, Midoriku, Yokohama, Japan*

[4]*Institute for Materials Research, TohokuUniversity, Sendai, Miyagi, Japan*

## ABSTRACT

Bulk metallic glasses (BMGs) are the subject of important research in materials science and processing. Ti-based BMGs are expected to be utilized as new biomaterials because they have high strength, low Young's modulus and excellent corrosion resistance. Bioactivity might depend on developing a microstructure on the surface of the BMG. A femtosecond laser is a good tool for forming a surface microstructure. Ti-Zr-Cu-Pd bulk alloys were irradiated with femtosecond lasers. After the irradiation, the BMGs were treated using a hydrothermal electrochemical method. The results of soaking BMGs in simulated body fluid suggest that the BMGs with a surface microstructure have better bioactivity than the BMGs without.

## INTRODUCTION

Ti-based bulk metallic glasses (BMGs) are expected to be used as new biomaterials because of their high strength, low Young's modulus, and excellent corrosion resistance. Recently, Zhu et al. developed Ti-based BMGs, Ti-Zr-Cu-Pd, without toxic elements, such as Ni, Be and Al[1]. However, this may not be directly joined to human bones due to their high chemical stability and bioinertness. This can be overcome by surface modification of the BMGs[2] and microstructure formation on BMGs[3]. For the surface modification of, and the microstructure formation on BMGs, low temperature processes are more desirable because such processes maintain the BMGs' excellent properties. If not, these BMGs will crystallize, resulting in degradation of their properties. Surface modification of BMGs has been developed. It is reported that bioactive titanate nanomesh layers were coated using a hydrothermal-electrochemical treatment of the BMGs [2]. This nanomesh layer can induce the growth of hydroxyapatite (HAp) in simulated body fluid (SBF), which has good biocompability. In another surface modification method, it is reported that aHAp film was created on a titanium plate by aerosol beam irradiation[4]. These surface modifications were performed at low temperature. For microstructure formation on BMGs, the femtosecond laser is a useful tool that provides considerable advantages for precision material processing, such as drilling, cutting and grooving into the metal[5] in comparison with nanosecond lasers[6,7]. The advantages are based on very rapid creation of vapor and plasma phase, negligible heat conduction, and the absence of liquid phase. It is reported that various microstructures, such as periodic nanostructures, periodic microstructures, and cone-like protrusions (CLP), are formed on the titanium plate[8,9]. However, microstructure formation on BMGs with the femtosecond laser by controlling the laser fluence and the number of laser pulses has not been investigated yet.

In this study, we report on the formation of surface microstructures on $Ti_{40}Zr_{10}Cu_{36}Pd_{14}$ BMGs treated with femtosecond laser irradiation. The influence of the laser fluence and the number of laser pulses on the microstructure was investigated. An experiment on the effects of nanosecond laser irradiation on the BMGs was also carried out. The microstructures formed with the femtosecond laser were compared with those produced with the nanosecond laser. After femtosecond laser irradiation, the BMGs were treated using a hydrothermal electrochemical method, and soaked in simulated body fluid (SBF) to evaluate their bioactivity.

## EXPERIMENTAL

A femtosecond (Ti:sapphire) laser and a nanosecond (Nd:YAG) laser were employed in these experiments. The wavelength, pulse duration, repetition rate, and beam diameter of the femtosecond laser were 775 nm, 150 fs, 1 kHz, and approximately 5 mm, respectively. The laser beam was focused onto the BMG surface by a lens with a 100 mm focal length. The Gaussian laser beam had a diameter of 60 μm (at the $1/e^2$ intensity points) on the BMG surface. The wavelength, pulse duration, repetition rate, and beam diameter of the nanosecond laser were 1064 nm, about 6 ns, 10 Hz, and approximately 3 mm, respectively. The laser beam was focused onto the BMG surface by a lens with a 150 mm focal length. The Gaussian laser beam had a diameter of 250 μm (at the $1/e^2$ intensity points) on the BMG surface. The beam spot scanned the BMG surface with XY stages connected to a computer. An attenuator to reduce the output energy of the laser was composed of polarizing filters.

In the first experiment, the BMG surface was irradiated with the femtosecond laser at the average laser fluence of 0.1, 0.5 and 1.0 J/cm$^2$, respectively. The number of laser pulses for the irradiation was 100 pulses. In the second experiment, the number of femtosecond laser pulses for the irradiation on the BMG was varied in the range of 10 to 500. The laser fluence was fixed at 0.5 J/cm$^2$. In the third experiment, the BMG surface was irradiated with the nanosecond laser at the average laser fluence of 0.6, 1.0 and 1.4 J/cm$^2$, respectively. The number of laser pulses for the irradiation was 100 pulses. In the forth experiment, the number of nanosecond laser pulses for the irradiation on the BMG was varied in the range of 10 to 500. The laser fluence was fixed at 1.4 J/cm$^2$. The BMG surfaces irradiated by femtosecond and nanosecond laser were observed with a scanning electron microscope (SEM).

For SBF test, microstructures were formed on the BMG in the area 5 mm×5 mm at 1.0 J/cm$^2$ for the scanning velocity of 1 mm/s. The BMG was treated by hydrothermal electrical chemical method for 2 hours before the SBF test. The SBF test was performed for 12 days. Deposited materials were analyzed with Raman spectroscopy.

## EXPERIMENTAL RESULTS AND DISCUSSION

SEM images of BMG surface irradiated with the femtosecond laser for 100 pulses at 0.1, 0.5 and 1.0 J/cm$^2$ in the first experiment are shown in Figures 1 (a), (b) and (c), respectively. High magnification images of Figs. 1 (a), (b) and (c), the center region of the irradiation area, are shown in Figures 1 (d), (e) and (f), respectively. As Figure 1 (d) shows, the periodic nanostructures, lying perpendicular to the laser electric field polarization vector $E$, were formed in the irradiation area at 0.1 J/cm$^2$. The period of the periodic nanostructure was about 600 nm. As Figures 1 (b) and (e) show, the periodic microstructures, which lay parallel to the laser electric field polarization vector $E$, were observed in the center region at 0.5 J/cm$^2$. The period of the periodic microstructures (parallel periodic microstructure) was about 2 μm. At 1.0 J/cm$^2$, Figure 1 (c) and (f) show, the microstructures were not observed in the center region.

In the second experiment with the femtosecond laser, SEM images at 0.5 J/cm$^2$ for 10, 50,

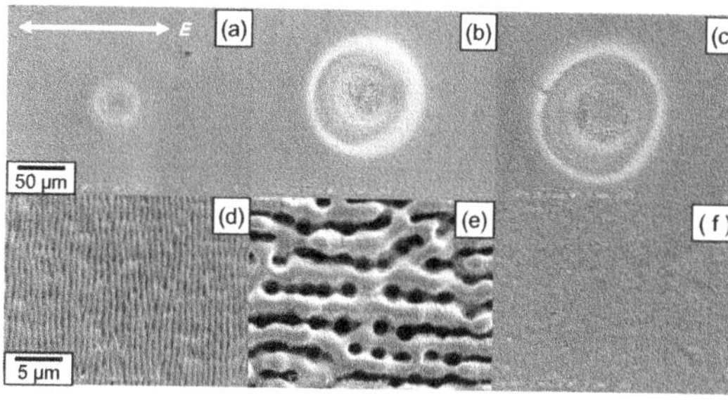


Figure 1. SEM images of the femtosecond laser irradiation area for 100 pulse at 0.1 J/cm$^2$ ((a) and (d)), 0.5J/cm$^2$ ((b) and (e)) and 1.0 J/cm$^2$ ((c) and (f)). (a), (b) and (c) at low magnification and (d), (e) and (f) at high magnification.

100, 500 pulses are shown in Figs. 2 (a), (b), (c) and (d), respectively. High magnification images of Figures 2 (a), (b), (c) and (d), the center region of the irradiation area, are shown in Figs. 2 (e), (f), (g) and (h), respectively. As Figure 2 (e) shows, periodic nanostructures were formed in the irradiation area for 10 pulses. The period of the periodic nanostructure was about 600 nm. For 50 pulses, Figures 2 (b) and (f) show, the periodic nanostructures were superimposed on the parallel periodic microstructures. The period of the parallel periodic microstructure was about 2 μm. For 100 pulses, Figures 2 (g) shows the parallel periodic microstructures which were clearly observed compared with those for 50 pulses shown in Figure 2 (f). For 500 pulses, Figures 2 (d) and (h) show the parallel periodic microstructures which were broken in the center area. Laser fluence and number of laser pulses for the periodic nanostructures (PN) and the parallel periodic microstructures (PPM) described above are shown in Table 1.

In the third experiment, SEM images of BMG surfaces irradiated with the nanosecond laser for 100 pulses at 0.6, 1.0 and 1.4 J/cm$^2$ are shown in Figures 3 (a), (b) and (c), respectively. High magnification images of Figures 3 (a), (b) and (c), the center region of the irradiation area, are shown in Figures 3 (d), (e) and (f), respectively. As Figures 3 (a), (d) and (b) show, the microstructures were not observed at 0.6 and 1.0 J/cm$^2$, respectively. It was suggested that the BMG surface was melted during the laser irradiation. As Figure 3 (e) shows, fringes, which lay parallel to the laser electric field polarization vector *E*, are observed in the center area. The period of the fringes was about 1 μm.

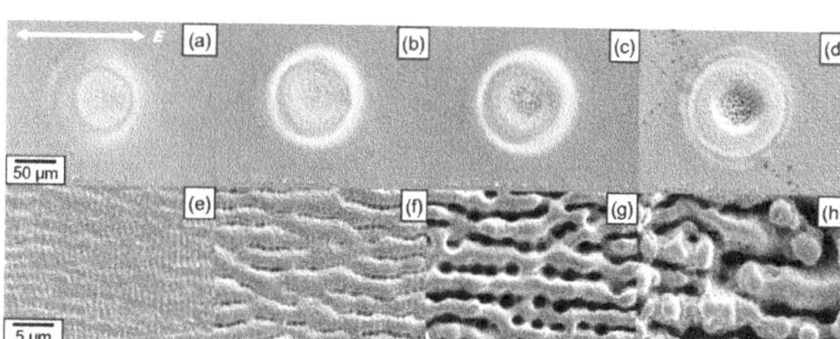


Figure 2. SEM images of the femtosecond laser irradiation area at 0.5 J/cm$^2$ for 10 ((a) and (e)), 50 ((b) and (f)), 100 ((c) and (g)) and 500 pulses ((d) and (h)). (a), (b), (c) and (d) at low magnification and (e), (f), (g) and (h) at high magnification.

Table 1. Laser fluence and number of laser pulses for the periodic nanostructures (PN) and the parallel periodic microstructures (PPM)

| Number of laser pulses | Laser fluence ($J/cm^2$) | | |
|---|---|---|---|
| | 0.1 | 0.5 | 1.0 |
| 10 | | PN | |
| 50 | | PN+PPM | |
| 100 | PN | PPM | — |
| 500 | | — | |

As Figure 3 (c) shows, the microstructures were not observed at 1.4 $J/cm^2$. It was that BMG's surface was melted during the laser irradiation. As Figure 3 (f) shows, fringes, which lay parallel to the laser electric field polarization vector *E*, are also observed in the center area. The period of the fringes was about 1 μm.

In the fourth experiment with the nanosecond laser, SEM images at 1.4 $J/cm^2$ for 10, 50, 100, and 500 pulses are shown in Figures 4 (a), (b), (c) and (d), respectively. High magnification images of Figs. 4 (a), (b), (c) and (d), the center region of the irradiation area, are shown in Figures 4 (e), (f), (g) and (h), respectively. As Figures 4 (a) and (e) show, the microstructures were not observed in the laser irradiation area. For 50 pulses, Figure 4 (b) shows the microstructures were not observed. It was suggested that BMG's surface was melted during the laser irradiation. As Figure 4 (f) shows, fringes, which lay parallel to the laser electric field polarization vector *E*, are observed in the center area. The period of the fringes was about 1 μm. For 100 pulses, Figure 4 (c) shows that the microstructures were not observed. It was suggested that BMG's surface was melted during the laser irradiation. As Figure 4 (g) shows, fringes, which lay parallel to the laser electric field polarization vector *E*, are also observed in the center area. The period of the fringes was about 1 μm. For 500 pulses, Figure 4 (d) shows the microstructures were not observed. It was suggested that BMG's surface was melted during the laser irradiation. As Figure 4 (h) shows, the fringes, which lay parallel to the laser electric field polarization vector *E*, was also observed in the center area. However, the fringes for 500 pulses were not clearly compared with the fringes for 50 and 100 pulses shown in Figures 4 (f) and (g).

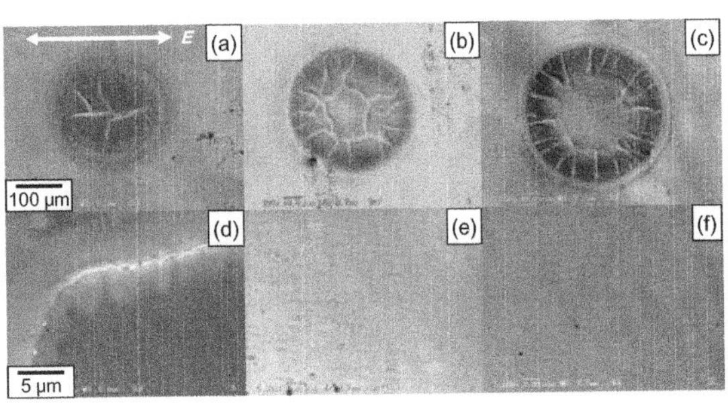


Figure 3. SEM images of the nanosecond laser irradiation area for 100 pulses at 0.6 $J/cm^2$ ((a) and (d)), 1.0 $J/cm^2$ ((b) and (e)) and 1.4 $J/cm^2$ ((c) and (f)). (a), (b) and (c) at low magnification and (d), (e) and (f) at high magnification.

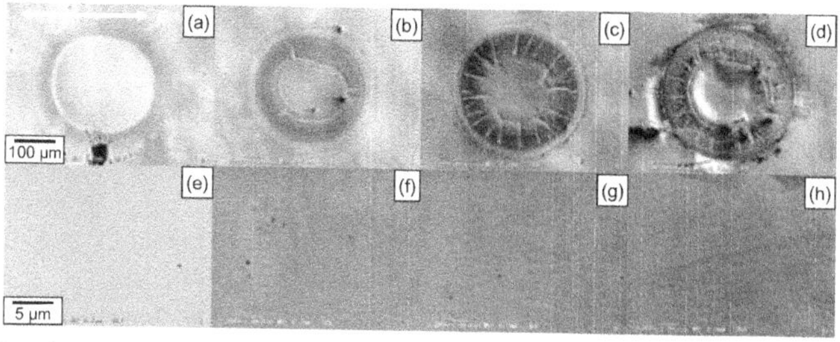


Figure 4. SEM images of the nanosecond laser irradiation area at 1.4 J/cm$^2$ for 10 ((a) and (e)), 50 ((b) and (f)), 100 ((c) and (g)) and 500 pulses ((d) and (h)). (a), (b), (c) and (d) at low magnification and (e), (f), (g), and (h) at high magnification.

The period of the fringes was about 1 μm. SEM images were unable to measure depth of the peak of the hill to the bottom of the fringes. Measurement of the peak of the hill to the bottom of the fringes is required.

No surface microstructures are shown in Figure 5 (a) since the BMG surface was not irradiated with the femtosecond laser. Figure 5 (b) shows nanomesh produced on the BMG surface shown in Figure 5 (a) by the hydrothermal electrochemical method. The nanomesh layer after the SBF test for 12 days was shown in Figure 5 (c). As shown in Figure 5 (c), spherical materials were produced on the nanomesh layer. Figure 5 (d) shows the microstructures formed with the femtosecond laser at 1.0 J/cm$^2$ for 1 mm/s, which were the broken parallel periodic microstructures. Periodic nanostructures were also produced and superimposed on the microstructures. After hydrothermal electrochemical treatment, the nanomesh layer was created on the microstructures. Then, shape of the microstructures was maintained as shown in Figure 5 (e) although the periodic nanostructures were not observed. Many deposited materials were observed on the microstructures as Figure 5 (f) shows. Quantity of the deposited materials in Figure 5 (f) was more than that in Fig. 5 (c). In Raman spectroscopy analysis for the specimens shown in Figures 5 (c) and (f), peaks of the calcium phosphate hydroxide were detected. These results indicated that the HAp might be created on the nanomesh coated BMG. The difference between quantity of the deposited materials in Figs.5 (c) and (f) suggested that bioactivity of the nanomesh coated BMG with microstructures might be stronger than that without the microstructures.

## SUMMARY

We tried to form surface microstructures on BMGs by treatment with femtosecond and nanosecond laser irradiation. With femtosecond laser irradiation, periodic nanostructures, which lay perpendicular to the laser electric polarization, were clearly formed on the BMG surface at 0.1 J/cm$^2$ for 100 pulses and at 0.5 J/cm$^2$ for 10 pulses. With 50 pulses at 0.5 J/cm$^2$, periodic nanostructures were superimposed on the parallel periodic microstructures. With 100 pulses at 0.5 J/cm$^2$, parallel periodic microstructures were observed. With 500 pulses at 0.5 J/cm$^2$, the parallel periodic microstructures were broken. The results of the SBF test suggest that the bioactivity of the nanomesh coated BMG was increased when the BMG had the femtosecond laser produced microstructures on its surface.

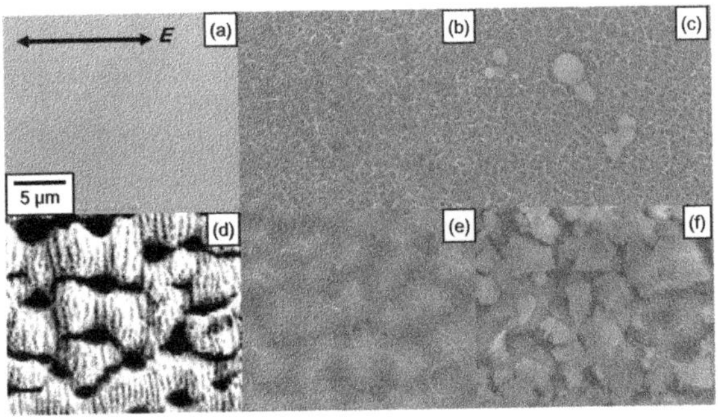


Figure 5. SEM images of the BMGs surfaces without femtosecond laser irradiation (a), after hydrothermal electrochemical treatment (b) and SBF test (c). SEM images of the femtosecond laser irradiation area at 1.0 J/cm$^2$ for 1 mm/s (d), after hydrothermal electrochemical treatment (e) and SBF test (f).

REFERENCES

[1]S.L. Zhu, X.M, Wang, F.X. Qin and A. Inoue, "A new Ti-based bulk glassy alloy with potential for biomedical application", *Mater Sci Eng A*, **459**, 233-237, (2007).

[2]N. Sugiyama, H.Y. Xu, T. Onoki, Y. Hoshikawa, T. Watanabe, N. Matsushita, X.M. Wang, F.X. Qin, M. Fukuhara, M. Tsukamoto, N. Abe, Y. Komizo, A. Inoue and M. Toshimura, "Bioactive titanate nanomesh layer on the Ti-based bulk metallic glass by hydrothermal-electrochemical technique", *Acta Bionaterialia*, **5**, 1367, (2009).

[3]X.L. Zhu, J. Chen, L. Scheideler, R. Reichl, and J. Geis-Gerstorfer, "Effects of topography and composition of titanium surface oxides on osteoblast response", *Biomaterials*, 25, 4087, (2004).

[4]M. Tsukamoto, T. Fujiwara, N. Abe, S. Miyake, M. Katto, T. Nakayama and F. Akedo, "Hydroxyapatite Coating on Titanium Plate with Ultrafine Particle Beam", *Jpn. J. Appl. Phys.*, **42**, L120-L122, (2003).

[5]P.S. Banks, B.C. Stuart, M.D. Perry, M.D. Feit, A.M. Rubenchik, J.P. Armstrong, H. Nguyen, F. Roeske, R.S. Lee, B.R. Myers and J.A. Sefcik, "Material effects in ultra-short pulse laser drilling of metals", *Technical Digest of Conf. on Lasers and Electro-Optics*, **6**, 510, (1998).

[6]C. Momma, B.N. Chichkov, S. Nolte, F.V. Alvensleben, A. Tunnermann, H. Welling and B. Wellegehausen, "Short-pulse laser ablation of solid targets", *Opt Commun*, **128**, 101, (1996).

[7]B.N. Chichkov, C. Momma, S. Nolte, F.V. Alvensleben and A. Tunnermann, "Femtosecond, picosecond and nanosecond laser ablation of solids", *Appl. Phys. A*, **63**, 109-115, (1996).

[8]M. Tsukamoto, K. Asuka, H. Nakano, M. Hashida, M. Katto, N. Abe, and M. Fujita, "Periodic microstructure produced by femtosecond laser irradiation on titanium plate", *Vacuum*, **80**, 1346-1350, (2006).

[9]M. Tsukamoto, T. Kalahari, H. Nakano, M. Hashed, M. Kato, M. Fujita, M. Tanaka and N. Abe, "Microstructures formation on titanium plate by femtoseccond laser ablation", *J. Phys. Conf. Ser.*, **59**, 666-669, (2007).

LASER BUTT WELDING OF Mg-BASED METALLIC GLASS

Takeshi Terajima[1], Kazuhiro Nakata[1], Hisamichi Kimura[2], and Akihisa Inoue[3]
[1]Joining and Welding Research Institute, Osaka University, 11-1 Mihogaoka, Ibaraki, Osaka, Japan
[2]Institute for Materials Research, Tohoku University, 2-1-1 Katahira, Aobaku, Sendaishi, Miyagi, Japan
[3]Tohoku University, 2-1-1 Katahira, Aobaku, Sendaishi, Miyagi, Japan

ABSTRACT

The metallic glass $Mg_{65}Cu_{25}Gd_{10}$ has low specific gravity and excellent mechanical properties and thus, is expected to be a useful structural material. However, welding of $Mg_{65}Cu_{25}Gd_{10}$ is difficult since it immediately crystallizes if the quench rate is lower than a critical value of approximately 1 K/s. In this study, a high-power fiber laser with an ultrahigh power density of several megawatts per square millimeter, which corresponds to the power density of an electron beam and is a well suited for this purpose, is examined for welding $Mg_{65}Cu_{25}Gd_{10}$. The laser beam was transmitted thorough an optical fiber and focused on two $Mg_{65}Cu_{25}Gd_{10}$ rods that were rotated at 3000 rpm at 0.5 MPa. Consequently, the molten alloy that forms upon laser irradiation was extruded from the joint, and the specimens were successfully welded. The microstructure observed by micro-focus X-ray diffraction showed that the crystalline phase was not present inside the joint, but was in the extruded region. The success of the proposed welding is likely attributable to the pressure applied to the specimens, which plays an important role in extruding the crystalline phase from the joint.

INTRODUCTION

Metallic glass possesses unique properties such as high mechanical strength, low elastic coefficient, small solidification shrinkage, and high resistance to corrosion. These excellent properties, which are not found in crystalline metals, have attracted the attention of many researchers, and metallic glass is therefore expected to be useful for applications in a variety of fields. Some types of metallic glass, such as Mg-[1], Zr-[2,3], Fe-[4,5], Pd-[6], and Ni-based[7] alloys, have low critical cooling rates. $Mg_{65}Cu_{25}Gd_{10}$, in particular, has one of the lowest critical cooling rates of only 1 K/s[8]. The strong glass-forming ability of $Mg_{65}Cu_{25}Gd_{10}$ allows the fabrication of amorphous alloy rods with a maximum diameter of 8 mm[8]. In addition, the specific strength of $Mg_{65}Cu_{25}Gd_{10}$ is twice that of typical AZ91 Mg alloy. Therefore, $Mg_{65}Cu_{25}Gd_{10}$ is expected to be used as a lightweight structural material such as for vehicle bodies and car appliances. However, it is necessary to develop joining techniques for metallic glass in order to broaden the scope of its industrial applications. There have been attempts to join pieces of Zr-based metallic glass using electron beam welding, explosive welding, pulse current welding, friction welding[9-11], and laser welding[12-13]. The most serious problem with welding metallic glass is crystallization in the molten and heat-affected zones. If the cooling rate from the liquidus temperature to the glass transition temperature is slower than the critical cooling rate of the glass formation, metallic glass rapidly transforms into intermetallic compounds, which are mostly brittle and low strength compared with the amorphous bulk glass. Therefore, a high energy density source, such as a laser, is well suited for welding of metallic glass. A fiber laser can focus the beam diameter to less than 100 μm. Therefore, it can produce an ultrahigh power density on the order of megawatts per square millimeter, which corresponds to the power density of a focused electron beam. Such a fiber laser has the potential for processing with low heat input, which leads to a rapid cooling rate and consequently minimizes processing risks such as crystallization and excessive enlargement of the heat-affected zone. An additional advantage of the proposed method, in comparison with electron beam welding, is that fiber lasers can operate in a vacuum-free environment. In this work, laser buttwelding of $Mg_{65}Cu_{25}Gd_{10}$ metallic glass is demonstrated.

## EXPERIMENTAL

$Mg_{65}Cu_{25}Gd_{10}$ pre-alloy was prepared by the following procedure. Cu-Gd alloy was prepared by arc-melting a mixture of Cu and Gd together under a Ti-gettered argon atmosphere in a water-cooled copper crucible. Then the Cu-Gd alloy was alloyed with Mg in a graphite crucible in Ar atmosphere by an induction-heating furnace. The resulting $Mg_{65}Cu_{25}Gd_{10}$ pre-alloy button was cast by the liquid quench method to form a metallic glass rod with dimension of $\phi 3\times 75$ mm. The $Mg_{65}Cu_{25}Gd_{10}$ rod was cut to a dimension of $\phi 3\times 15$ mm. The prepared $Mg_{65}Cu_{25}Gd_{10}$ metallic glass rod was welded using a fiber laser welding facility (IPG Co., Ltd., YLR-10000) with a maximum laser power of 10 kW and beam parameter product (BPP) of 4.5 mm·mrad. The laser beam was transmitted through optical fibers and focused on the specimen surface with a focal lens. The spot size at the focal position was $\phi$=200 μm.

Figure 1 shows an illustration for laser buttwelding of $Mg_{65}Cu_{25}Gd_{10}$ metallic glass. Two $Mg_{65}Cu_{25}Gd_{10}$ specimens were attached to the specimen chuck and butted together. The specimen chuck can simultaneously rotate at 3000 rpm and apply a maximum pressure to the specimens of 0.5 MPa. Shrinkage of the welded specimen was measured by an LED displacement sensor between the specimen chucks. Welding was performed by irradiating the interface of the $Mg_{65}Cu_{25}Gd_{10}$ rods rotating at 3000 rpm with a laser pulse of 0.5 kW for 80 ms. Applied pressures of 0.05 MPa and 0.5 MPa were tested.

The cross-sectional surface of the bead was polished. The microstructure of the joint was analyzed with an optical microscope, and the formation of a single glassy phase was confirmed by micro-focused X-ray diffractometry (XRD) using Co $K_\alpha$ radiation. In addition, preliminary differential scanning calorimetry (DSC) measurements at a heating rate of 20 K/min were conducted to evaluate the thermal stability of the prepared $Mg_{65}Cu_{25}Gd_{10}$ metallic glass.

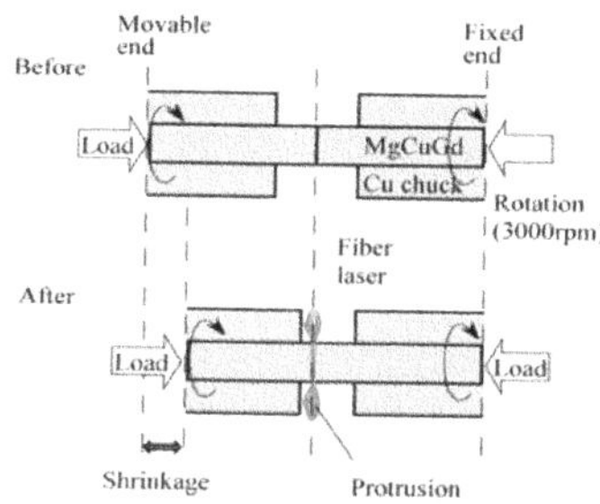


Figure 1. Illustration of laser buttwelding of $Mg_{65}Cu_{25}Gd_{10}$ metallic glass.

The glass transition temperature $T_g$, the crystallization temperature $T_x$, and the liquidus temperature $T_l$ were 423 K, 484 K, and 740 K, respectively. The strength of the joint was measured by compressive test at a strain rate of $5\times10^{-4}$ $s^{-1}$ for a specimen with dimensions of $\phi3\times6$ mm.

RESULT AND DISCUSSION

A schematic illustration of the laser buttwelding of $Mg_{65}Cu_{25}Gd_{10}$ is shown in Figure 1. A pair of $Mg_{65}Cu_{25}Gd_{10}$ specimens was rotated at a rotation speed of 3000 rpm. Pressure of 0.05 MPa or 0.5 MPa was applied to the rotating specimens, and the specimens were successfully irradiated with the fiber laser at a power of 0.5 kW for 80 ms. Figures 2(a) and (b) show the microstructures of the cross section of the $Mg_{65}Cu_{25}Gd_{10}$ joints welded at pressure of 0.05 MPa and 0.5 MPa, respectively. Clear contrast with the crystalline phase was observed along the weld line in the joint welded at 0.05 MPa. No contrast was observed in the joint welded at 0.5 MPa, indicating an amorphous material. Although most of the molten region was spattered by the

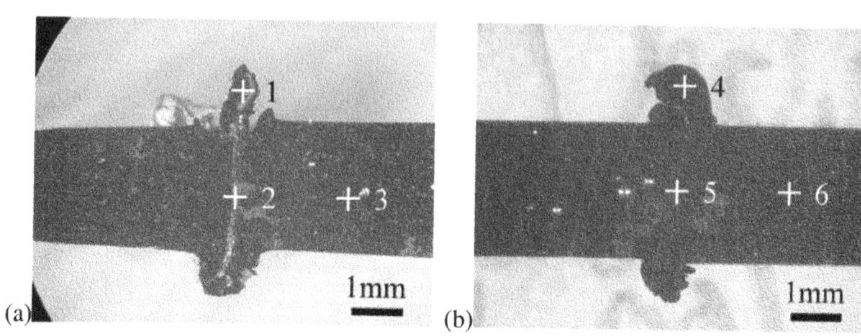


Figure 2. Microstructures of cross section of $Mg_{65}Cu_{25}Gd_{10}$ joints welded at (a) 0.05 MPa and (b) 0.5 MPa. Note that the slight contrasts beside the protrusions were caused by stain adhering to specimens during the desiccation process after polishing.

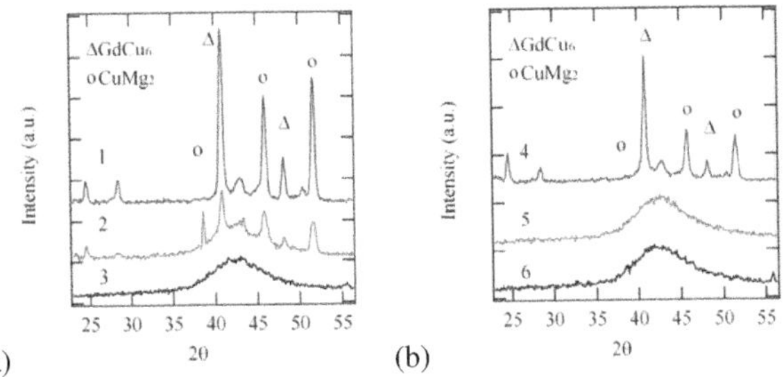


(a) (b)

Figure 3. Micro-focused XRD patterns at (a) positions 1, 2, and 3 in Figure 2 and (b) positions 4, 5, and 6 in Figure 2.

applied pressure and centrifugal force of the specimen rotation, a surface protrusion with a thickness of approximately 1 mm was observed outside the joint. Neither deformation nor swelling was observed other than the protrusion. Figures 3(a) and (b) show micro-focused XRD patterns corresponding to positions 1, 2, and 3 shown in Figure 2(a) and positions 4, 5, and 6 in Figure 2(b), respectively. The results show that the protrusion (position 1) and the center of the joint (position 2) welded at a pressure of 0.05 MPa were mainly composed of $GdCu_6$ and $CuMg_2$ intermetallic compounds. Although the protrusion (position 4) welded at a pressure of 0.5 MPa was also composed of these intermetallic compounds, the center of the joint (position 5) was amorphous, which is characterized by a halo pattern. Furthermore, the region within 3 mm of the center of the joint was also confirmed to be amorphous in each specimen (positions 3 and 6).

Figure 4(a) shows the shrinkage of the specimen during the welding. The shrinkage proceeded in two stages. In the first stage (0-80 ms), the specimens rapidly shrank during laser irradiation. In the second stage (80-500 ms), although the laser was off, the specimens continued

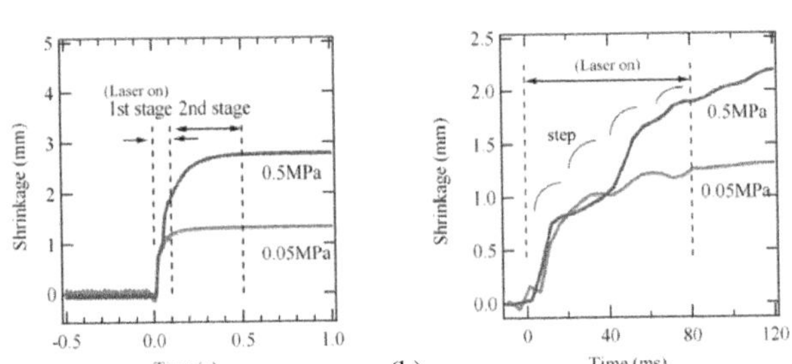


(a) (b)

Figure 4. (a) Shrinkage of specimen during laser buttwelding. (b) Expanded view from 0 to 120 ms.

to shrink slightly. The shrinkage was approximately 0.2 mm and 1 mm for the specimens welded at 0.05 MPa and 0.5 MPa, respectively. After the second stage, the shrinkage was ultimately 1.3 mm and 2.9 mm for the specimens welded at 0.05 MPa and 0.5 MPa, respectively. It appeared that the shrinkage in the first stage was caused by extrusion of the molten alloy, and the shrinkage in the second stage was caused by the superplastic deformation of the supercooled liquid alloy in the heat-affected zone (HAZ). Figure 4(b) shows an expanded view of the shrinkage during the first stage. The specimens shrank in four steps at intervals of 20 ms. The step interval corresponds to the rotation period of the specimen. Thus, the butt joint interface of $Mg_{65}Cu_{25}Gd_{10}$ was melted by the laser irradiation and rotation, and the molten alloy was extruded by the applied pressure. Undesirable crystalline phases such as $GdCu_6$ and $CuMg_2$ were removed from the joint by this extrusion. However, if the applied pressure was too small to extrude an adequate amount of molten alloy, the crystalline phases remained in the joint, as was observed at the pressure of 0.05 MPa.

Figure 5 shows the compressive stress-strain curve of the as-cast $Mg_{65}Cu_{25}Gd_{10}$ and the $Mg_{65}Cu_{25}Gd_{10}$ welded at pressures of 0.05 MPa and 0.5 MPa. As indicated, the behavior of the $Mg_{65}Cu_{25}Gd_{10}$ metallic glass is strictly elastic without any macroscopic plastic deformation. The compressive fracture strength of $Mg_{65}Cu_{25}Gd_{10}$ welded at a pressure of 0.5 MPa was 690 MPa, which was the same as that of the as-cast $Mg_{65}Cu_{25}Gd_{10}$. In contrast, the specimen welded at a pressure of 0.05 MPa had a compressive fracture strength of 390 MPa. This decrease of the strength was expected since brittle intermetallic compounds formed in the joint welded at the lower pressure.

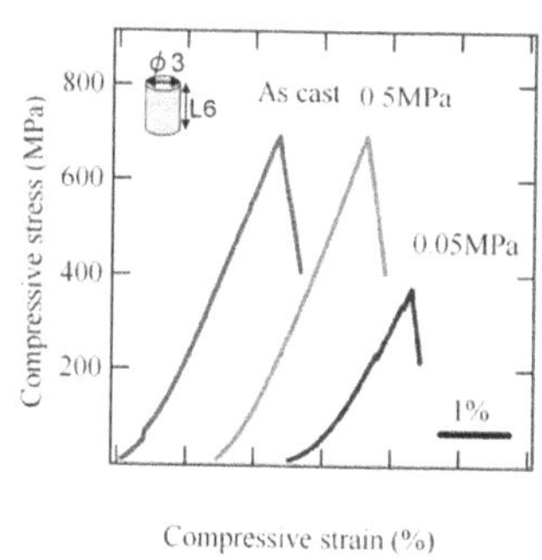


Figure 5. Compressive stress-strain curve of welded $Mg_{65}Cu_{25}Gd_{10}$.

## CONCLUSIONS

Laser buttwelding was performed for $Mg_{65}Cu_{25}Gd_{10}$ metallic glass. When 0.5 MPa was applied to the specimens, $Mg_{65}Cu_{25}Gd_{10}$ rods were successfully welded together and maintained an amorphous structure. In contrast, when 0.05 MPa was applied, the joint contained crystalline phases composed of $GdCu_6$ and $CuMg_2$ intermetallic compounds. It was found that the applied pressure plays an important role in extruding the molten alloy and undesirable crystalline phases from the joint during laser buttwelding of $Mg_{65}Cu_{25}Gd_{10}$ metallic glass.

## REFERENCES

[1] A. Inoue, K. Ohtera, K. Kita, and T. Masumoto, "New Amorphous Mg-Ce-Ni Alloy with High Strength and Good Ductility", *Japan. J. Appl. Phys.*, **27**, L2248-L2251 (1988).

[2] A. Inoue, T. Zhang, and T. Masumoto, "Zr-Al-Ni Amorphous Alloys with High Glass Transition Temperature and Significant Supercooled Liquid Region", *Mater. Trans. JIM.*, **31**, 177-183 (1990).

[3] A. Peker and W. L. Johnson, "A highly processable metallic glass: $Zr_{41.2}Ti_{13.8}Cu_{12.5}Ni_{10}Be_{22.5}$", *Appl. Phys. Lett.*, **63**, 2342-2344 (1993).

[4] A. Inoue and J. S. Gook, "Fe-Based Ferromagnetic Glassy Alloys with Wide Supercooled Liquid Region", *Mater. Trans. JIM.*, **36**, 1180-1183 (1995).

[5] A. Inoue, T. Zhang, T. Itoi, and A. Takeuchi, "New Fe-Co-Ni-Zr-B Amorphous Alloys with Wide Supercooled Liquid Regions and Good Soft Magnetic Properties", *Mater. Trans. JIM.*, **38**, 359-362 (1997).

[6] A. Inoue, N. Nishiyama, and T. Matsuda, "Preparation of Bulk Glassy $Pd_{40}Ni_{10}Cu_{30}P_{20}$ Alloy of 40 mm in Diameter by Water Quenching", *Mater. Trans. JIM.*, **37**, 181-184 (1996).

[7] R. Akatsuka, T. Zhang, M. Koshiba, and A. Inoue, "Preparation of New Ni-based Amorphous Alloys with a Large Supercooled Liquid Region", *Mater. Trans. JIM.*, **40**, 258-261 (1999).

[8] H. Men and D. H. Kim, "Fabrication of ternary Mg-Cu-Gd bulk metallic glass with high glass-forming ability under air atmosphere", *J. Mater. Res.*, **18**, 1502-1504 (2003).

[9] Y. Kawamura, T. Shoji, and Y. Ohno, "Welding technologies of bulk metallic glasses", *J. Non-cryst. Solids*, **317**, 152-157(2003).

[10] Y. Kawamura and Y. Ohno, "Superplastic bonding of bulk metallic glasses using friction", *Scr. Mater.*, **45**, 279-285 (2001).

[11] Y. Kawamura and Y. Ohno, "Super welding of $Zr_{55}Al_{10}Ni_5Cu_{30}$ bulk metallic glasses", *Scr. Mater.*, **45**, 127-132 (2001).

[12] J. Kim, D. Lee, S. Shin, and C. Lee, "Phase evolution in $Cu_{54}Ni_6Zr_{22}Ti_{18}$ bulk metallic glass Nd:YAG laser weld", *Mater. Sci. and Eng. A*, **A434**, 194-201 (2006).

[13] B. Li, Z. Y. Li, J. G. Xiong, L. Xing, D. Wang, and Y. Li, "Laser welding of $Zr_{45}Cu_{48}Al_7$ bulk glassy alloy", *J. Alloys Comp.*, **413**, 118-121 (2006).

# TERAHERTZ WAVE PROPERTIES OF MICRO PATTERNED TITANIA AND METALLIC GLASS PARTICLES IN HEXAGONAL TABLETS FABRICATED USING MICROSTEREOLITHOGRAPHY

Soshu Kirihara, Naoki Komori, Maasa Nakano, Noritoshi Ohta and Toshiki Niki
Smart Processing Research Center, Joining and Welding Research Institute, Osaka University
11-1 Mihogaoka, Ibaraki, Osaka 567-0047, Japan

## ABSTRACT

Periodic arrangements of acrylic polygon tablets including fine particles with high dielectric constant and permeability were micro patterned using micro stereolithography in order to control terahertz waves effectively. In this process, a 10 μm thick layer of photosensitive resin paste containing a dispersion of titania particles and metallic glass (Fe-14.4%B-9.6%Si-4%Nb) particles was spread on a substrate by moving a knife edge, and patterned images were exposed to UV rays by using digital micro mirror device (DMD) with 2 μm in part accuracy. After the layer stacking, micro patterns in hexagonal tablets of 120 μm in edge length and 200 μm in pattern thickness were obtained. The transmission properties of electromagnetic waves through the patterned samples were measured using a terahertz spectroscopic (TDS) device. Forbidden band formations in the transmission spectra and localization behaviors in the magnetic micro patterns were observed in the THz wave frequency range. The electric field profiles in the patterns were calculated using transmission line modeling (TLM) simulations with a finite difference time domain (FDTD) method.

## INTRODUCTION

Terahertz waves are well known to resonate with various types of saccharide and protein molecules, and expected to control the biological material syntheses through frequency excitements[1-6]. In our research group, micro stereolithography with computer aided design and manufacturing (CAD/CAM) system was newly developed to realize a spatially structural joining of ceramic or metal components on the micrometer scale to control terahertz waves effectively[7-12]. By using this system, diamond type photonic crystals with periodic arrangements of micrometer order ceramic or metal lattices were fabricated successfully to exhibit forbidden gaps through Bragg diffraction in the terahertz wave frequency rages[13-16]. In this investigation, two dimensional periodic patterns composed of acrylic resins with titania and metallic glass particles dispersions were fabricated to realize wave diffractions and resonations in the terahertz frequencies. Filtering effects of the electromagnetic waves for a perpendicular direction to the dielectric and magnetic patterns were observed through time domain spectroscopic

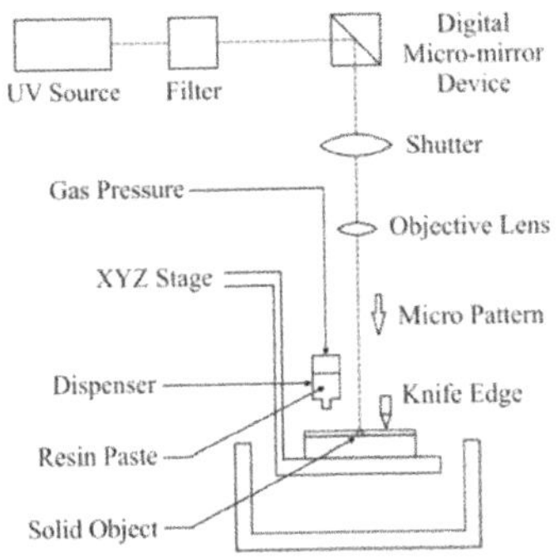


Figure 1 A schematically illustrated free forming system of a micro-stereolithography machine by using computer aided design and manufacturing (CAD/CAM) processes. (D-MEC Co. Ltd., Japan, SI-C 1000, http://www. d-mec.co.jp).

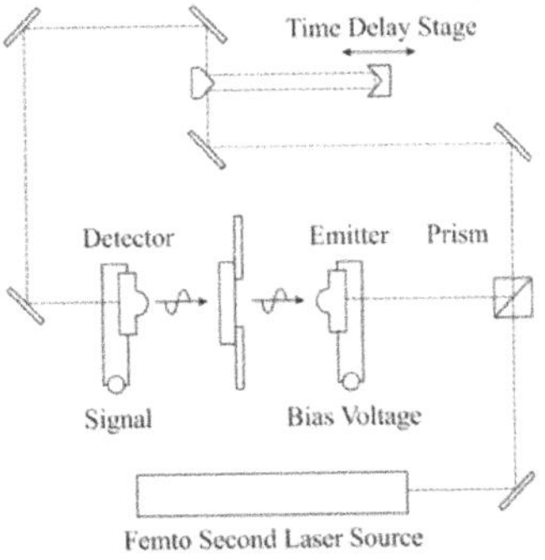


Figure 2 The schematically illustrated measuring system of a terahertz wave analyzer by using a time domain spectroscopic (TDS) detect method (Advanced Infrared Spectroscopy, Co. Ltd. Japan, J-Spec 2001, http://www. aispec.com).

measurements. These micro geometric patterns of extremely thin devices with a high dielectric constant and permeability were designed to concentrate the terahertz wave energies effectively through a theoretical simulation method.

## EXPERIMENTAL PROCEDURE

The micro dielectric pattern was designed as the periodic structure composed of micro hexagonal tablets of 120 μm in edge length at intervals of 60 μm. These micro tablets of 9×9=81 in numbers were arranged to form the extremely thin dielectric device of 200 μm in pattern thickness. The real sample was fabricated trough the micro stereolithography system. A designed graphic model was converted for stereolithography (STL) data files and sliced into a series of two dimensional layers. These numerical data were transferred into the fabrication equipment (SIC-1000, D-MEC, Japan). Figure 1 shows a schematic illustration of the fabrication system. As the raw material, nanometer sized titania particles of 270 nm in average diameter were dispersed into a photo sensitive acrylic resin at 40 volume percent. And micrometer sized metallic glass (Fe-14.4%B-9.6%Si-4%Nb) particles of 5 μm were dispersed the resin at 45 volume percent. The mixed slurry was squeezed on a working stage from a dispenser nozzle. This material paste was spread uniformly by a moving knife edge. Layer thickness was controlled to 10 μm. Ultra violet light of 405 nm in wavelength was exposed to the resin surface according to the computer operation. Two dimensional solid patterns were obtained by light induced photo polymerization. High resolutions in these micro patterns had been achieved by using a digital micro-mirror

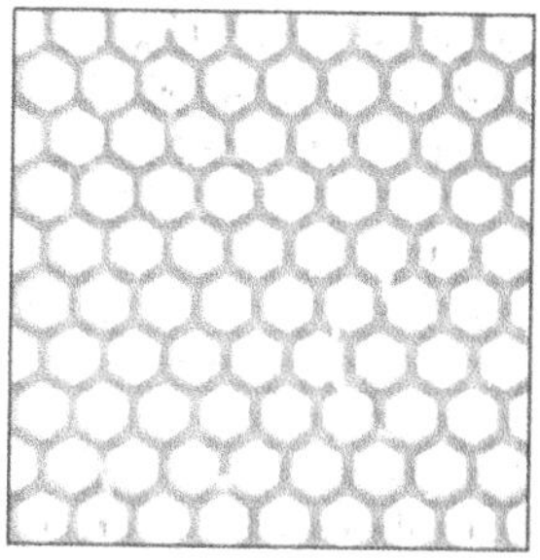


Figure 3 A micro periodic pattern with hexagonal tablets composed of titania dispersed acrylic resin.

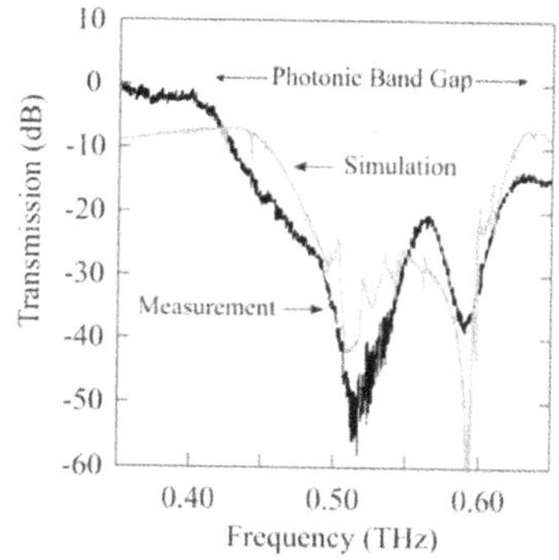


Figure 4 Measured and simulated spectra of terahrtz wave transmission through the dielectric micro pattern.

device (DMD). In this optical device, square aluminum mirrors of 14 μm in edge length were assembled with 1024×768 in numbers. Each micro mirror can be tilted independently, and cross sectional patterns were dynamically exposed through objective lenses as bitmap images of 2 μm in space resolution. After stacking and joining these layers through photo solidifications, the periodical arrangements of the micro dielectric tablets were obtained. A terahertz wave attenuation of transmission amplitudes through these micro periodic patterns were measured by using a terahertz time domain spectrometer (TDS) apparatus (Pulse-IRS 1000, AISPEC, Japan). Figure 2 shows the schematic illustration of the measurement system. Femto second laser beams were irradiated into a micro emission antenna formed on a semiconductor substrate to generate the terahertz wave pulses. The terahertz waves were transmitted trough the micro patterned samples perpendicularly. The terahertz wave diffraction and resonation behaviors in the micro patterns were calculated theoretically by using a transmission line modeling (TLM) simulator (Micro-stripes Ver. 7.5, Flomerics, UK) of a finite difference time domain (FDTD) method.

## RESULTS AND DISCUSSION

The dielectric micro pattern with the periodic arrangement of the acrylic tablets with the titania particle dispersion was fabricated successfully by using the micro-stereolithography system as shown in Figure 3. Dimensional accuracies of the fabricated micro tablets and the air gaps were approximately 0.5 percent in length. The nanometer sized titania particles were verified to disperse uniformly in the acrylic resin matrix thorough a scanning electron microscope (SEM) observation. The dielectric constant of the composite material of the titania dispersed acrylic resin was measured as 40 by using the TDS system. Figure 4 shows transmission spectra measured and simulated by using the TDS and TLM methods, respectively.

Figure 5 A micro hexagonal pattern with acrylic tablets with metallic glass particles dispersions.

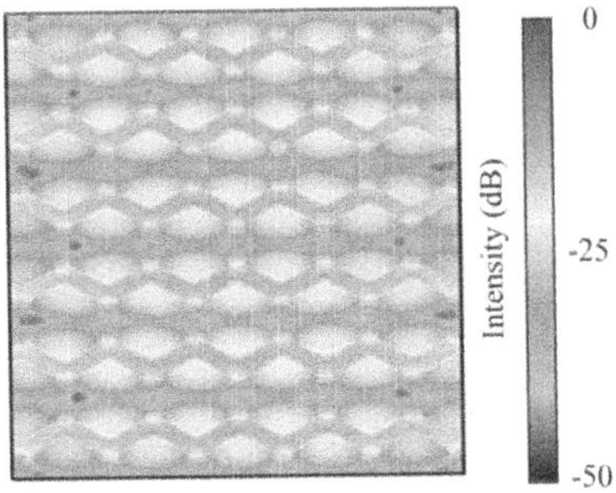


Figure 6 A simulated distribution of electric field intensity in the micro pattern at a localized frequency.

The measured forbidden gap agrees with the calculated photonic band gap. Opaque regions were formed in both spectra form 0.4 to 0.6 THz approximately. Maximum attenuation was measured as about -50 dB in transmission amplitude, and the minimum transmittance showed below 0.001 percent. The magnetic micro pattern with the periodic arrangements of hexagonal tablets was also fabricated successfully as shown in Figure 5. The homogeneous distributions of the metallic glass particles were verified through the SEM observation. The permeability of hexagonal tablets was measured and converted into the dielectric constant of 40 by using the TDS system. The forbidden gap formation was measured in the terahertz wave frequency range. The two dimensional photonic crystals with periodic arrangement with the lower dielectric contrasts and permeability were well known to open the band gaps limitedly for the parallel directions to the plane structures. However, the micro patterns with the periodically arranged hexagonal tablets with the higher dielectric constant and permeability could exhibit the clear forbidden bands in the transmission spectra toward the perpendicular direction to the plane patterns. The fabricated micro pattern is considered to totally reflect the terahertz wave at the wavelength comparable to the optical thickness. Figure 6 shows a simulated distribution of electric field intensity in the micro pattern at the localized frequency. The black area indicates that the electric field intensity is high, whereas the white area indicates it is low. The incident terahertz wave is resonated and localized along the two dimensionally arranged magnetic tablets at the specific frequency.

## CONCLUSIONS

Micro patterns composed of hexagonal acryl tablets with titania and metallic glass particle dispersions were arranged periodically in two dimensions by using a stereolithography system. These dielectric and magnetic micro patterns were verified to be able to exhibit a forbidden band in the opaque region. Additionally, terahertz wave energies were concentrated and localized in thin layers thorough the multiple resonations and standing wave formation. The fabricated micro pattern can incorporate various types of solutions in the gaps between the hexagonal tablets; therefore, it will be applied to form novel micro reactors to create useful biological materials.

ACKNOWLEDGEMENTS

This study was supported by Priority Assistance for the Formation of Worldwide Renowned Centers of Research - The Global COE Program (Project: Center of Excellence for Advanced Structural and Functional Materials Design) from the Ministry of Education, Culture, Sports, Science and Technology (MEXT), Japan.

REFERENCES

[1] Martin Van Exter, Ch. Fattinger, and D. Grischkowsky, "Terahertz Time-domain Spectroscopy of Water Vapor", Optics Letters Vol. 14, Iss. 20, pp. 1128-1130, October, 1989.

[2] D. Clery, "Brainstorming Their Way to an Imaging Revolution", Science, Vol. 297, pp. 761-763, August, 2002.

[3] K. Kawase, Y. Ogawa, Y. Watanabe, and H. Inoue, "Non-destructive Terahertz Imaging of Illicit Drugs Using Spectral Fingerprints", Optics Express, Vol. 11, Iss. 20, pp. 2549-2554, October, 2003.

[4] R. M. Woodward, V. P. Wallace, D. D. Arnone, E. H. Linfield, and M. Pepper, "Terahertz Pulsed Imaging of Skin Cancer in the Time and Frequency Domain", J. Biological Physics, Vol. 29, No. 2-3, pp. 257-259, June, 2003.

[5] V. P. Wallace, A. J. Fitzgerald, S. Shankar, N. Flanagan, "Terahertz Pulsed Imaging of Basal Cell Carcinoma ex Vivo and in Vivo", The British Journal of Dermatology, Vol. 151, No. 2, pp. 424–432, August, 2004.

[6] Y. Oyama, L. Zhen, T. Tanabe, and M. Kagaya, "Sub-Terahertz Imaging of Defects in Building Blocks", NDT&E International, Vol. 42, No. 1, pp. 28-33, January, 2008.

[7] W. Chen, S. Kirihara, and Y. Miyamoto, "Fabrication and Measurement of Micro Three-Dimensional Photonic Crystals of $SiO_2$ Ceramic for Terahertz Wave Applications", J. Am. Ceram. Soc, Vol. 90, No. 7, pp. 2078-2081, July, 2007.

[8] W. Chen, S. Kirihara, and Y. Miyamoto, "Three-dimensional Microphotonic Crystals of $ZrO_2$ Toughened $Al_2O_3$ for Terahertz wave applications", Applied Physics Letter , Vol. 91, No. 15, 153507-1-3, October, 2007.

[9] W. Chen, S. Kirihara, and Y. Miyamoto, "Fabrication of Three-Dimensional Micro Photonic Crystals of Resin-Incorporating $TiO_2$ Particles and their Terahertz Wave Properties", J. Am.

Ceram. Soc., Vol. 90, No. 1, pp. 92-96, January, 2007.

[10] W. Chen, S. Kirihara, Y. Miyamoto, "Static Tuning Band Gaps of Three-dimensional Photonic Crystals in Subterahertz Frequencies", Applied Physics Letters, Vol. 92, pp. 183504-1-3, May, 2008.

[11] H. Kanaoka, S. Kirihara, and Y. Miyamoto "Terahertz Wave Properties of Alumina Microphotonic Crystals with a Diamond Structure", J. Matls. Res., Vol. 23, No. 4, pp. 1036-1041, April, 2008.

[12] Y. Miyamoto, H. Kanaoka, S. Kirihara "Terahertz Wave Localization at a Three-dimensional Ceramic Fractal Cavity in Photonic Crystals", J. Applied Physics, Vol. 103, pp. 103106-1-5, May, 2008.

[13] S. Kirihara and Y. Miyamoto: "Terahertz Wave Control Using Ceramic Photonic Crystals with Diamond Structure Including Plane Defects Fabricated by Micro-stereolithography", Intl. J. Applied Ceram. Tech., Vol. 6, No. 1, pp. 41-44 January, 2009.

[14] S. Kirihara, T. Niki, and M. Kaneko, "Terahertz Wave Behaviors in Ceramic and Metal Structures Fabricated by Spatial Joining of Micro-stereolithography, J. Physics, in print.

[15] S. Kirihara, T. Niki, M. Kaneko, "Three-dimensional Material Tectonics for Electromagnetic Wave Control by Using Micoro-stereolithography", Ferroelectrics, in print.

[16] S. Kirihara, K. Tsutsumi, Y. Miyamoto: "Localization Behavior of Microwaves in Three-dimensional Menger Sponge Fractals Fabricated from Metallodielectric Cu/polyester Media", Science of Advanced Materials, in print.

THERMAL STABILITY OF $Zr_{55}Cu_{30}Ni_5Al_{10}$ METALLIC GLASS IN CONTACT WITH ALUMINUM

Masakatsu Maeda[1], Yasuo Takahashi[1], and Akihisa Inoue[2]

[1]Osaka University , Suita, Osaka, Japan
[2]Tohoku University ,Sendai, Miyagi, Japan

ABSTRACT

The interfacial microstructure and thermal stability of interfaces between $Zr_{55}Cu_{30}Ni_5Al_{10}$ metallic glass and Al formed by ultrasonic bonding were investigated. Metallic glass can be bonded to Al under low ultrasonic power, in a short time, and with low bonding force compared to bonding $Zr_{55}Cu_{30}Ni_5Al_{10}$ metallic glass to itself. The interfacial temperature during ultrasonic bonding does not reach the glass transition temperature. Owing to this, the interface formed by ultrasonic bonding is free of devitrification. However, thermal analysis by differential scanning calorimetry at a heating rate of 0.05 $Ks^{-1}$ indicates that the devitrification of the joint starts at 688 K, which is 42 K lower than the devitrification temperature of the metallic glass. The deterioration of the thermal stability is induced by the formation of the interface between the metallic glass and the Al. After annealing the joint at 700 K for 900 s in vacuum, a layered reaction product is formed between the metallic glass and the Al, while the interior of the metallic glass remains amorphous.

INTRODUCTION

Metallic glasses provide unique mechanical[1,2] and functional[3-5] properties which are important for industrial products in the next generation. Welding technology of metallic glasses is demanded for manufacturing complicated or large-sized components. Thus, establishment of welding technology feasible for metallic glasses is now recognized as one of the key to broaden application fields of the material. Among the welding techniques which have been examined to date, ultrasonic bonding can be one of the most suitable welding technique for metallic glasses, since it can weld materials at low temperature in a very short time. Recently, Maeda et al.[6] have proven that $Zr_{55}Cu_{30}Ni_5Al_{10}$ metallic glass foils can be successfully welded to each other by ultrasonic bonding. Furthermore, the thermal stability of amorphous state is not deteriorated by the welding process. This achievement suggests that ultrasonic bonding creates interfaces free of heat-affected zone (HAZ). Such HAZ-free interface is not only preferable for practical application, but also necessary for fundamental research of metallic glasses.

Welding metallic glasses with dissimilar materials will be necessary for designing the products to exhibit effectively the merits of metallic glasses compensating the demerits. However, the thermal stability of metallic glasses in contact with dissimilar crystalline materials is not clearly understood. Only a few reports have suggested that metallic glasses can crystallize (devitrify) even at temperatures lower than the glass transition point ($T_g$) in the case where they are in contact with dissimilar crystalline materials[7]. The metallic glass in the vicinity of the interface of a crystalline phase will be unstable compared to the bulk interior due to the excessive interfacial energy[8,9] and the high mobility of the constituent atoms at the interface. Both of these factors dominate the devitrification behavior of metallic glasses, since they are related to the driving force and the frequency of atomic diffusion[10], respectively. Therefore, it is important to investigate the thermal stability of metallic glasses in the vicinity of the interface with crystalline phases.

In the present study, the thermal stability of the $Zr_{55}Cu_{30}Ni_5Al_{10}$ metallic glass in contact with aluminum has been investigated. Aluminum was selected out of the four constituents of the metallic glass to simplify the consideration of the phenomenon. It has the lowest melting temperature, i.e. the highest atomic mobility in the condensed state among the four. Ultrasonic bonding has been utilized to form the interface.

## EXPERIMENTAL PROCEDURE

$Zr_{55}Cu_{30}Ni_5Al_{10}$ metallic glass specimens were cut to a 10-mm-long, 3-mm-wide rectangular shape out of 50-μm-thick foils produced by rapid roll-quench method. The surfaces to be bonded were polished using 0.25 μm diamond paste spread on a polishing cloth. Then the specimens were degreased with an acetone-bath ultrasonic cleaner just before bonding. Al specimens were cut to a 3-mm-long, 2-mm-wide rectangular shape from 200-μm-thick fully-annealed ribbons of which nominal purity was 99.99 %.

Ultrasonic bonding was implemented under the ultrasonic frequency of 75.0 kHz without external heating. The ultrasonic power, the vibration applying time, and the bonding force were set at 25.00 W, 500 ms, and 15.00 N, respectively. The metallic glass specimens were fixed on the workbench, while the Al specimens were vibrated by a flat-faced wedge-tool. The ultrasonic vibration was applied to a direction parallel to the interface. Temperature measurement during bonding was implemented at the center of the bonding interface using fine K-type thermocouple under the sampling rate of $2\times10^4$ $s^{-1}$.

The interfaces of the joints were analyzed by scanning electron microscopy (SEM), energy-dispersive X-ray spectroscopy (EDS) and micro-area X-ray diffraction (MAXRD). The SEM observation was implemented on the cross-section of the interface parallel to the applied direction of the ultrasonic vibration. Compositional contrast images of back-scattered electrons were taken. The collimation diameter for MAXRD was 50 μm.

The thermal stability was estimated using differential scanning calorimetry (DSC). The samples were heated up to 800 K at a heating rate of 0.01 $Ks^{-1}$ in 99.999% Ar atmosphere.

## RESULTS AND DISCUSSION

$Zr_{55}Cu_{30}Ni_5Al_{10}$ metallic glass foils can be bonded to Al ribbons successfully under a low ultrasonic power, short vibration applying time, and low bonding force condition compared to that for bonding $Zr_{55}Cu_{30}Ni_5Al_{10}$ metallic glass foils to each other[6]. Figure 1 depicts the SEM micrograph of an interface of $Zr_{55}Cu_{30}Ni_5Al_{10}$ metallic glass and Al formed by ultrasonic bonding under the condition of 25.00 W, 500 ms, and 15.00 N. In Figure 1(a), a wide bonded area of 620 μm in length is observed. The metallic glass in the vicinity of the interface shows homogeneous contrast indicating that phase separation by devitrification has not occurred at the interface even with a high magnification depicted in Figure 1(b). Figure 1(c) shows the MAXRD pattern taken from the interfacial area shown in Figure 1(b). All diffraction peaks in the pattern are identified as those from the metallic glass and Al, *i.e.* no peak of the phases related to the devitrified metallic glass is found. These results indicate that the metallic glass does not devitrify by ultrasonic bonding. Figure 2 shows the thermal history during ultrasonic bonding of a $Zr_{55}Cu_{30}Ni_5Al_{10}$ metallic glass foil and an Al ribbon at 25.00 W, 500 ms, and 15.00 N, measured at the center of the bonding area. The rapid increase and decrease of the temperature appearing at the start of application of the ultrasonic vibration is considered that they correspond to the friction heating between the thermocouple and the metallic glass foil instead of that between the

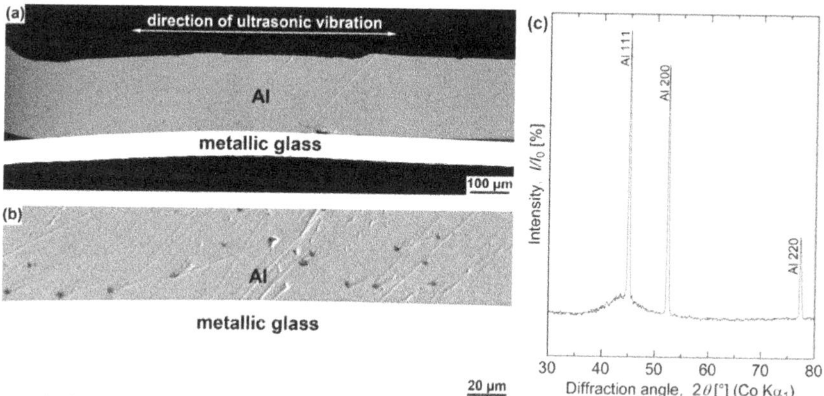


Figure 1. Cross-section of the interface between $Zr_{55}Cu_{30}Ni_5Al_{10}$ metallic glass and Al bonded at 25.00 W, 500 ms, 15.00 N. (a) Overview of the interface, (b) close-up of the bonded area, (c) micro-area X-ray diffraction pattern taken from the interface shown in Figure 1(b).

metallic glass specimen and the Al ribbon[6]. The temperature increases gradually and saturates at 388 K, which is far lower than $T_g$. Thus, the thermal stability of the metallic glass is hardly considered to have been affected by the thermal history during ultrasonic bonding.

Figure 3 shows the DSC curve of a joint of $Zr_{55}Cu_{30}Ni_5Al_{10}$ metallic glass foil and Al ribbon bonded at 25.00 W, 500 ms, and 15.00 N, in comparison with a base material of $Zr_{55}Cu_{30}Ni_5Al_{10}$ metallic glass. The DSC curve of the joint has two distinct differences from that of as-prepared metallic glass. One is the glass transition being unclear. This might be due to the insufficient mass of the specimen and the slow heating. The other difference is the appearance of two exothermic reactions starting at 688 K and 733 K as marked on the figure with $T_{x1}$ and $T_{x2}$, respectively. The peak of the reaction starting at $T_{x1}$ appears broad. The appearance of the peak starting at $T_{x2}$ is similar to that at $T_x$ of the metallic glass, although the peak height is low. This

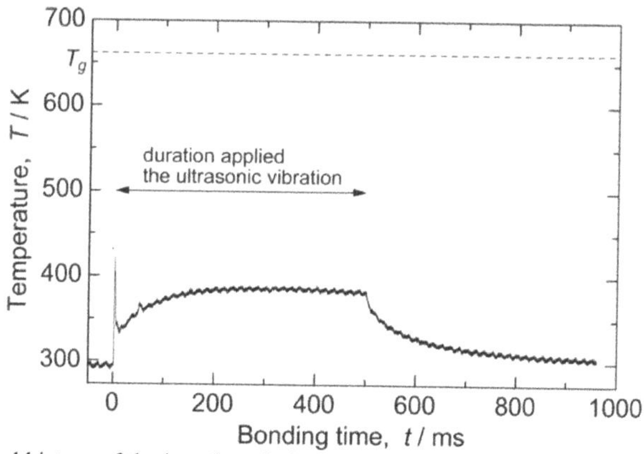


Figure 2. Thermal history of the interface during ultrasonic bonding at 25.00W, 500 ms, 15.00 N.

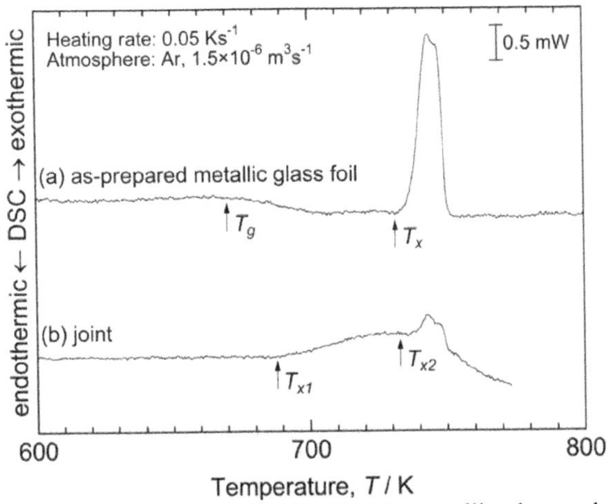


Figure 3. DSC curves of an as-prepared $Zr_{55}Cu_{30}Ni_5Al_{10}$ metallic glass and a joint formed at 25.00 W, 500 ms, 15.00 N.

result indicates that the reaction corresponds to the devitrification of the metallic glass interior.

The DSC curve of the joint of metallic glass and Al taken in the present study is very similar to that of marginal glass-formers or nanocomposites presented by D.V. Louzguine-Luzgin and Inoue[11]. They have suggested that the broad peak starting at $T_{x1}$ ($T_{bg}$ in their paper) corresponds to the growth of the pre-existing nuclei or nanoparticles in the amorphous matrix. In the case of the present study, the peak starting at $T_{x1}$ will be related to the devitrification proceeding at the glass / crystalline interface. The devitrification of the metallic glass before reaching $T_x$ reduces the height of the peak starting at $T_{x2}$ by consuming the glass matrix. Thus the peak height lowering shown in Figure 3(b) supports this consideration.

To obtain a further evidence of devitrification at temperatures lower than $T_{x2}$, a joint has been annealed at 700 K for 900 s in vacuum of $5.4\times10^{-3}$ Pa. This heat treatment condition is sufficient for advancing the reaction starting at $T_{x1}$ and does not devitrify the $Zr_{55}Cu_{30}Ni_5Al_{10}$ metallic glass[6]. Figure 4 shows the interface after the annealing. In the micrograph shown in Figure 4(a), a 4.1-μm-thick layered reaction product is formed between the metallic glass and Al. The compositional contrasts at both ends of the layer appear dark in comparison with that at the center, indicating that there is a compositional gradient in the layer. The composition profile along the line X–Y drawn on Figure 4(a) is presented in Figure 4(b). The compositions of Zr and Ni decrease monotonously from the metallic glass to Al. On the other hand, the composition of Cu decreases sharply at the interface between the metallic glass and the reaction layer and then increases gradually toward the interface between the reaction layer and Al. Considerable amount of Cu diffuses also into Al matrix. The major constituent in the reaction layer is Al. Therefore, it is likely that the devitrification of the metallic glass has been induced primarily by dissolution of the constituents of the metallic glass into Al and subsequent nucleation-and-growth of stable crystalline phases. In addition, the composition of the metallic glass in the vicinity of the interface with the reaction layer fluctuates. This phenomenon is reported as the preceding behavior of metallic glasses in their devitrification[10]. It should be noted that the fluctuation is

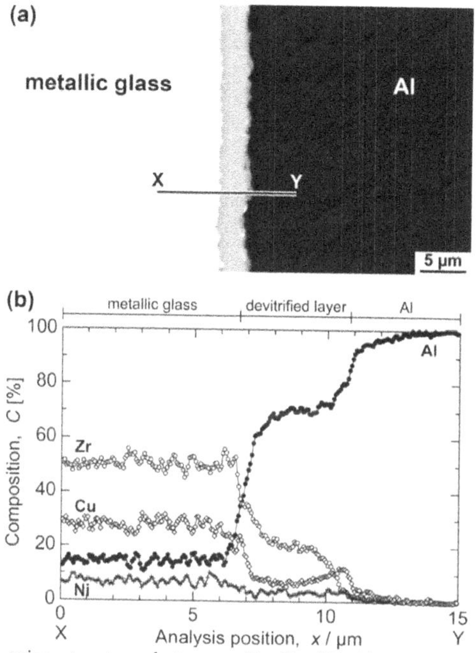


Figure 4. Interfacial microstructure between $Zr_{55}Cu_{30}Ni_5Al_{10}$ metallic glass and Al after annealing the joint at 700 K for 900 s. (a) SEM micrograph, (b) compositional distribution profile along the line from X to Y marked in the micrograph.

occurring only in the vicinity of the interface: the composition in the metallic glass interior is retained homogeneous. Thus the fluctuation is considered as one of the preceding phenomena in devitrification induced by the existence of the interface with crystalline phases.

## CONCLUSION

The interfacial microstructure and thermal stability of interfaces between $Zr_{55}Cu_{30}Ni_5Al_{10}$ metallic glass and Al formed by ultrasonic bonding were investigated by SEM, EDS, and DSC. The following points were determined.

1) A wide bonded area is formed under the ultrasonic power of 25.00 W, vibration applying time of 500 ms, and bonding force of 15.00 N. The temperature at the interface is 388 K at the highest during ultrasonic bonding. The metallic glass does not devitrify during the bonding process.
2) The devitrification of the joint starts at 688 K, which is 42 K lower than the devitrification temperature of the metallic glass. The devitrification proceeds from the interface forming a layered reaction product.

## ACKNOWLEDGMENTS

The authors express their gratitude to Prof. M. Fukuhara, and Prof. X. Wang, Institute for Materials Research, Tohoku University, for their fruitful discussions and supply of high-quality metallic glass samples. This study was financially supported by the Japan society for promotion of science through grant-in-aid for scientific research (project No. 18206076 and 21560746) and by the ministry of education, culture, sports, science and technology of Japan through grant-in-aid for cooperative research project of nationwide joint-use research institutes on development base of joining technology for new metallic glasses and inorganic materials.

REFERENCES

[1] A. Inoue, B.L. Shen, H. Koshiba, H. Kato, and A.R. Yavari, "Ultra-high strength above 5000 MPa and soft magnetic properties of Co-Fe-Ta-B bulk glassy alloys", *Acta Mater.*, **52**, 1631-37 (2004).

[2] A. Inoue, B. Shen, and A. Takeuchi, "Fabrication, properties and applications of bulk glassy alloys in late transition metal-based systems", *Mater. Sci. Eng. A*, **441**, 18-25 (2006).

[3] C. Qin, K. Asami, T. Zhang, W. Zhang, and A. Inoue, "Corrosion behavior of Cu-Zr-Ti-Nb bulk glassy alloys", *Mater. Trans.*, **44**, 749-53 (2003).

[4] S. Yamaura, S. Nakata, H. Kimura, Y. Shimpo, M. Nishida, and A. Inoue, "Hydrogen Permeation of the Melt-Spun Ni–X–Zr Amorphous Membranes", *Mater. Trans.*, **46**, 1768-70 (2005).

[5] B. Shen, M. Akiba, and A. Inoue, "Excellent soft-ferromagnetic bulk glassy alloys with high saturation magnetization", *Appl. Phys. Lett.*, **88**, 131907 (2006).

[6] M. Maeda, T. Yamasaki, Y. Takahashi, and A. Inoue, "Interfacial microstructure and thermal stability of $Zr_{55}Cu_{30}Ni_5Al_{10}$ metallic glass joints formed by ultrasonic bonding", *Mater. Trans.*, **50**, 1263-68 (2009).

[7] D.G. Papageorgiou and G.A. Evangelakis, "Adlayer deposition induced surface crystallization of $Cu_{46}Zr_{54}$ bulk metallic glass", *Surface Sci.*, **602**, 1486-91 (2008).

[8] Y. Nishi, K. Mikagi, and F. Kanazaki, "Solid-liquid interfacial energy on crystallization in Fe-Si-B alloy glass", *J. Mater. Sci.*, **21**, 3497-500 (1986).

[9] F. Sommer, R.N. Singh, and E.J. Mittemeijer, "Interface thermodynamics of nano-sized crystalline, amorphous and liquid metallic systems", *J. Alloy. Compound.*, **467**, 142-53 (2009).

[10] A.K. Gangopadhyay, T.K. Croat, and K.F. Kelton, "The effect of phase separation on subsequent crystallization in $Al_{88}Gd_6La_2Ni_4$", *Acta Mater.*, **48**, 4035-43 (2000).

[11] D.V. Louzguine-Luzgin and A. Inoue, "Nano-Devitrification of Glassy Alloys", *J. Nanosci. Nanotechnol.*, **5**, 999-1014 (2005).

# LASER WELDING OF $Zr_{55}Cu_{30}Al_{10}Ni_5$ BULK METALLIC GLASS

M. Ashour[1], S. Katayama[1], T. Kuroda[1], and H. Kimura[2]
1- Joining and Welding Research Institute (JWRI), Osaka University, Ibaraki, Osaka, Japan
2- Institute for Materials Research (IMR), Tohoku University, Sendai, Miyagi, Japan

## ABSTRACT

Disk and diode laser sources were used for bead-on-plate welding of $Zr_{55}Cu_{30}Al_{10}Ni_5$ bulk metallic glass. It was found that the disk laser could produce an amorphous metal weld without crystallization in the heat affected zone using a laser power of 3 kW and a scanning speed of 72 m/min. Meanwhile, using the same laser power, an amorphous metal weld could be produced with the diode laser using a scanning speed of 20 m/min. However, crystallization in the heat affected zone could not be fully suppressed, even with a higher scanning speed.

## INTRODUCTION

Since being developed, bulk metallic glasses (BMGs) have attracted the interest of materials scientists and engineering researchers. This is attributed to their unique properties, such as high specific strength, high corrosion resistance and high strain-rate super plasticity in the super cooled liquid region[1-3]. However, due to size limitations, the application of BMGs has not reached the expected level yet[2, 4]. Moreover, it is necessary to integrate BMGs components into an engineering structure. Therefore, some researches are devoted to developing a reliable welding process for BMGs.

The main concerns in welding BMGs are to maintain the glassy structure in the metal weld, and to suppress crystallization in the heat affected zone (HAZ). These can be realized by using welding process with a high energy density heat source to achieve rapid heating and cooling rates. Another possibility is solid-state welding, which utilizes the super plasticity of BMGs in the super cooled liquid state. Reports on successful welding of BMGs using laser welding, electron beam welding, spark welding, friction welding, and explosive welding have recently been published[4-9]. Among these qualified processes, laser welding is very attractive because of its reproducible high quality, fast and precise performance, and the capability for robotic or beam scanner delivery. In particular, high power fiber and disk lasers with higher beam quality, higher efficiency and fiber delivery have been developed to realize the advantages of easier operation, higher productivity, and cost reduction[10].

In this study, newly-developed high power disk and diode lasers were utilized to weld $Zr_{55}Cu_{30}Al_{10}Ni_5$ BMG. The effects of different conditions, including laser power and welding speed on the reformation of the glassy structure was investigated using the bead-on-plate welding technique.

## EXPERIMENTAL PROCEDURE

$Zr_{55}Cu_{30}Al_{10}Ni_5$ BMG was prepared by arc melting a mixture of pure elements in Ar. The alloy was cast in the form of a 3 mm thick plate. Specimens 2 mm in thickness, 25 mm in width, and 25 mm in length were prepared by machining the cast plate. Two different types of laser sources were used to weld $Zr_{55}Cu_{30}Al_{10}Ni_5$ BMG samples in this study. The first laser source is a 10 kW disk laser (TruDisk 10003). The second one is 3 kW diode laser (LDL 160-3000). Figure 1 shows the power density profile of both the disk and diode laser beams. To avoid the deterioration in the glass forming ability of BMG upon oxidation, welding was done in a chamber filled with Ar gas. A schematic of experimental set-up is shown in Figure 2. The laser power and welding speed for welding with the disk and diode laser sources are given in Table 1.

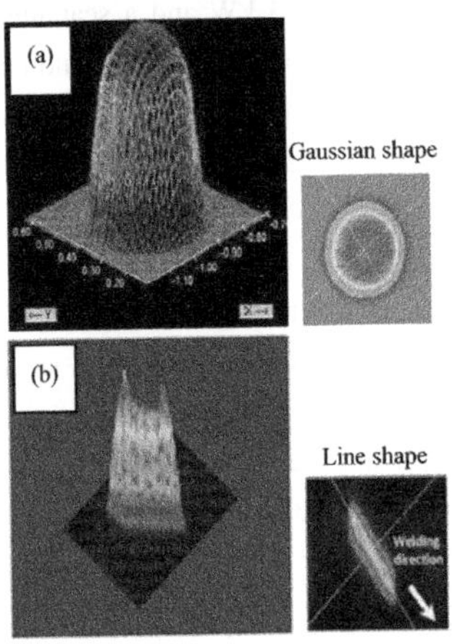


Fig. 1 Power density profile of laser beam. (a) Disk laser. (b) Diode laser.

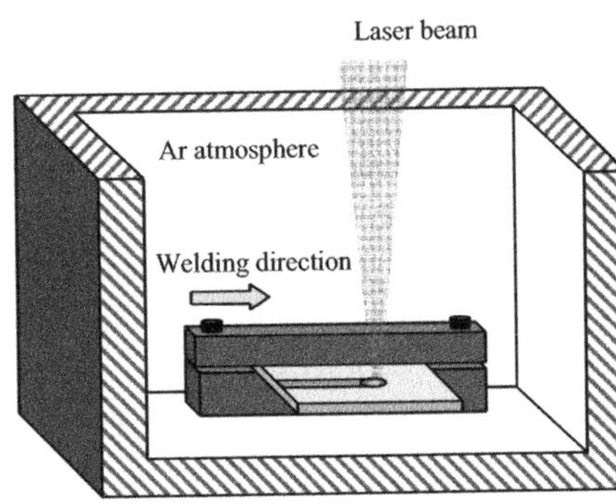


Fig. 2 Schematic drawing of experimental setup.

Table 1 Welding parameters

| Laser source | Power, kW | Welding speed, m/min. |
|---|---|---|
| Disk laser | 3 | 20, 50, 72 |
| Diode laser | 3 | 10, 15, 20, 25 |

The microstructure of welded samples was examined by optical microscopy and scanning electron microscopy (SEM). Micro-focused X-ray diffractometry (XRD) was used to investigate the formation of glassy structure upon welding.

## RESULTS AND DISCUSSION

### Disk laser welding results

Figure 3 shows cross-sections of $Zr_{55}Cu_{30}Al_{10}Ni_5$ BMG bead-on-plate welds made using the disk laser. Samples were irradiated with 3 kW laser power at different welding speeds. A full penetration weld could be obtained at the welding speed of 20 m/min as shown in Figure 3(a). As the welding speed increased to 50m/min, as seen in Figure 3(b), less heat was absorbed in the work piece, resulting in shallower penetration. At even higher welding speed, 72 m/min, in Figure 3(c), weld penetration was further decreased, and undercuts were formed on

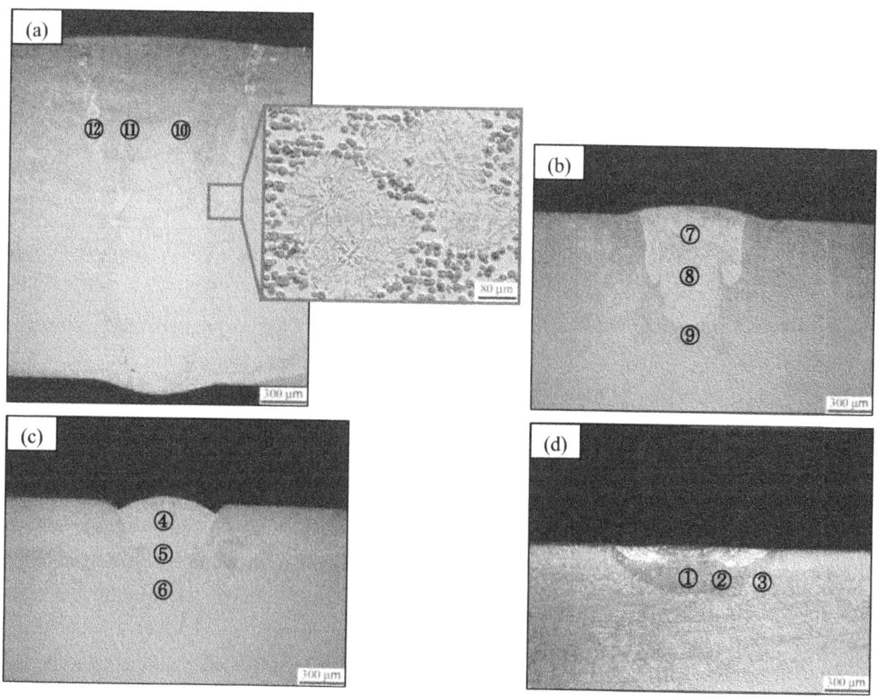


Fig. 3 Transverse sections of welds produced with 3 kW disk laser at a focal distance and scanning speeds of 20m/min (a), 50 m/min (b) and 72 m/min (c), and at +5 mm defocused distance and 72 m/min (d).

both sides of the weld bead. The undercut formation may be attributed to the insufficient time for strong melt flow behind the keyhole towards the center to redistribute over the sides, and is hence frozen as undercuts. Applying a defocused distance of +5 mm, it was possible to avoid the formation of undercuts, as shown in Figure 3(d).

Looking into Figure 3(a) again, a uniform width HAZ can be easily distinguished on both sides of the weld metal. At higher magnification, this HAZ was recognized to have some crystalline structure. This crystallization indicates that the thermal cycle was not rapid enough to maintain the glassy structure upon welding. The absence of such a distinguished HAZ microstructure in other welded samples might be a sign of crystallization suppression.

Therefore, micro-focused XRD was used to confirm the glassy structure formation. Figure 4 shows the XRD patterns of welded samples. As can be seen, crystalline phases were detected in samples welded at 20 and 50 m/min. in both weld metal and HAZ. On the other hand, XRD patterns from specimens welded at 72 m/min indicate the existence of only glassy phase.

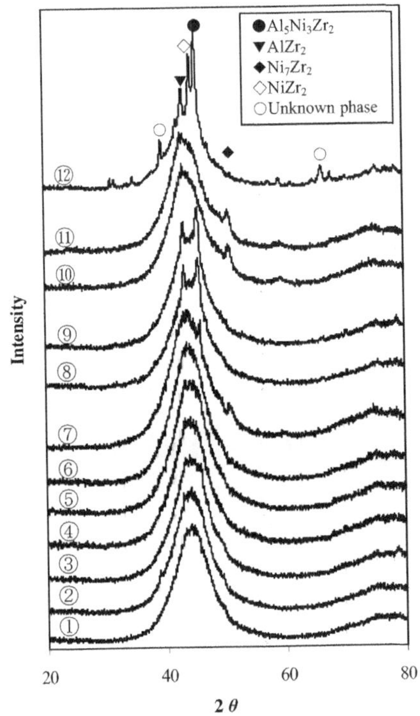


Fig. 4 Micro-focused XRD patterns from specimens in Fig. 3.

Diode laser welding results

Diode laser source, used in this study, is characterized by its rectangular beam shape with dimensions of 0.6 x 11 mm which gives moderate power density resulting in conduction welding. On contrary to disk laser which has circular beam shape with 0.3 mm diameter resulting in high power density. This high power density can be utilized to achieve a keyhole or a deep penetration welding mode. In conduction mode, weld pool size is limited by the conduction of the heat away from laser beam irradiation point on the workpiece surface. It, in turn, results in wider and shallower welds, as shown in Figure 5.

The XRD patterns of welded samples are shown in Figure 6. It is understood that an amorphous weld metal could be obtained only at the welding speed of 20m/min. However, it was

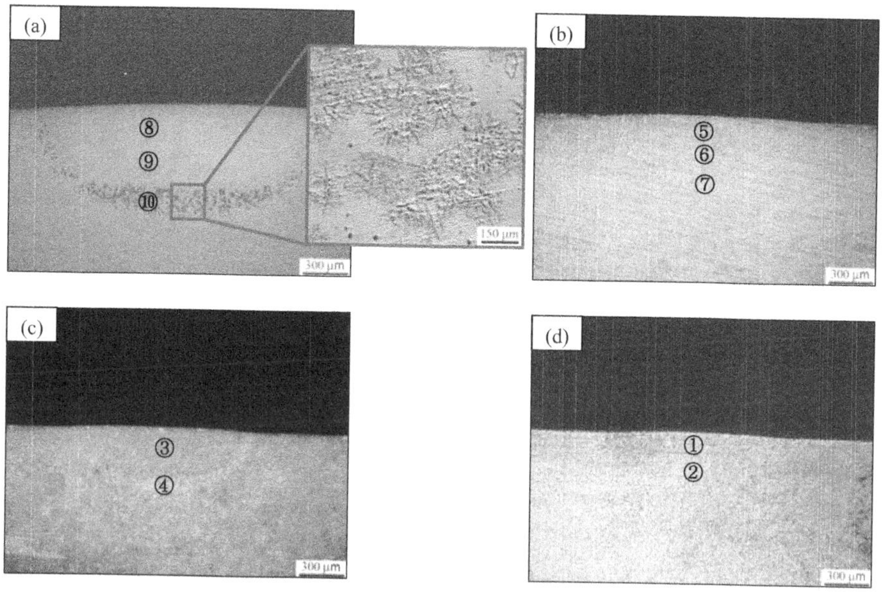


Fig. 5 Transverse sections of diode laser welds with 3 kW at scanning speed of 10m/min (a), 15 m/min (b), 20 m/min (c) and 25 m/min (d).

not possible to maintain the glassy structure in the HAZ. It means that the HAZ is crystallized more easily than the weld metal. In the HAZ, therefore, extremely fast thermal cycles are required to form a glassy phase because the bulk in the zone has once experienced a cooling during production of the specimen. In other words, the cooling history for the production of the specimen should play an important role in whether or not crystals are absent or present in the HAZ. Faster welding speeds should be applied to avoid crystallization. On the other hand, in increasing the welding speed above 25 m/min, the heat is not concentrated enough to produce a weld or a re-melted zone.

Based on the above results, we tried to apply a 2-stage welding and treatment technique using a disk laser. The first stage is to produce a full penetration weld by applying 3 kW laser power using a 20 m/min welding speed. The second stage is to produce an amorphous layer on the surface of the crystallized weld metal by two passes of 3 kW laser power using 72 m/min welding speed. A transverse section of this weldment is shown in Figure 7.

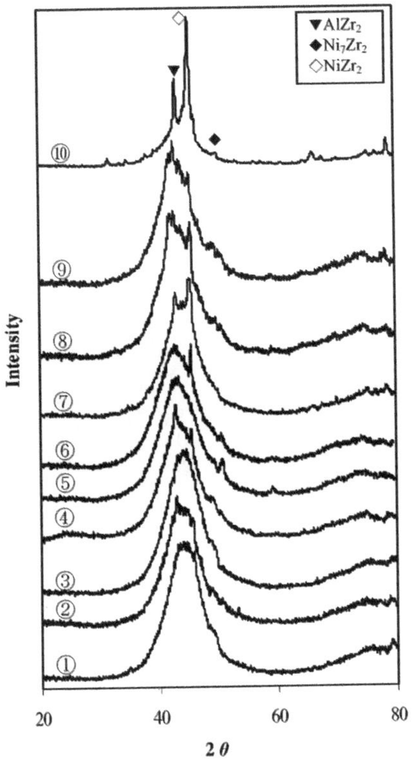

Fig. 6 Micro-focused XRD patterns from specimens in Fig. 3.

CONCLUSIONS

The weldability of $Zr_{55}Cu_{30}Al_{10}Ni_5$ BMG with disk and diode lasers was investigated. The results are summarized as follows:

1- A fully amorphous weld metal without visible defects was produced using a disk laser with a power of 3 kW, and a scanning speed 72 m/min, at +5 mm defocused distance.

2- Due to its moderate power density, the diode laser produced an amorphous weld metal with a smooth surface using a power of 3 kW and a scanning speed of 20 m/min., but with a partially crystallized HAZ, even at higher scanning speed.

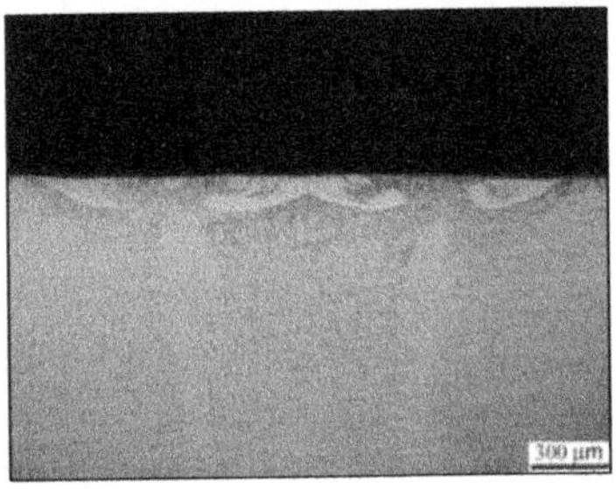

Fig. 7 Transverse section of amorphous layer made by disk laser.

REFERENCES

[1]A. Inoue, "Stabilization of Metallic Supercooled Liquid and Bulk Amorphous Alloys", *Acta Mater.*, 48, 279-306 (2000).

[2]W.H. Wang, C. Dong, C.H. Shek, "Bulk Metallic Glasses", *Mater. Sci. Eng.*, R 44, 45-89 (2004).

[3]A. Inoue, "Bulk Amorphous Alloys: Preparation and Fundamental Characteristics", *Trans Tech Publications*, Switzerland, 1998.

[4]J. Kim, D. Lee, S. Shin, C. Lee, "Phase Evolution in $Cu_{54}Ni_6Zr_{22}Ti_{18}$ Bulk Metallic Glass Nd:YAG Laser Weld", *Mater. Sci. Eng.*, A 434, 194-201 (2006).
[5]B. Li, Z.Y. Li, J.G. Xiong, L. Xing, D. Wang, Y. Li, "Laser Welding of $Zr_{45}Cu_{48}Al_7$ Bulk Glassy Alloy+, *J. Alloys and Compounds*, 413, 118-121 (2006).
[6]Y. Yokoyama, N. Abe, K. Fukura, T. Shinohara, A. Inoue, "Electron-Beam Welding of $Zr_{50}Cu_{30}Ni_{10}Al_{10}$ Bulk Glassy Alloys", *Mater. Sci. Eng.*, A 375-377, 422-426 (2004).
[7]Y. Kawamura, Y. Ohno, "Spark Welding of $Zr_{55}Al_{10}Ni_5Cu_{30}$ Bulk Metallic Glasses", *Sci. Mater.*, 45, 127-132 (2001).
[8]T. Shoji, Y. Kawamura, Y. Ohno, "Friction Welding of Bulk Metallic Glasses to Different Ones", *Mater. Sci. Eng.*, A 375-377, 394-398 (2004).
[9]Y. Kawamura, "Explosive Welding of Ni-Based Amorphous Foils for Micro-Tooling Applications", *Mater. Sci. Eng.*, A 375-377, 112-119 (2004).
[10]S. Katayama, in *New Developments in Advanced Welding*, ed. N. Ahmed., Woodhead Publications, Cambridge, 2005, Chapter 6.

# Nano-Particles and Powders

NANOSTRUCTURED $(Y_{1-x}Gd_x)_2O_3:Eu^{3+}$ POWDERS OBTAINED THROUGH AEROSOL SYNTHESIS

Katarina R. Marinkovic[1], Lidija T. Mancic[1], Maria Eugenia Rabanal[2], Luz Gomez[2], Miroslav Dramicanin[3], and Olivera B. Milosevic[1]

[1]Institute of Technical Sciences of Serbian Academy of Sciences and Arts, Knez Mihajlova 35/IV, 11000 Belgrade, Serbia

[2]Universidad Carlos III de Madrid, Avda de la Universidad 30, 28911 Leganés, Madrid, Spain

[3]Institute "Vinca", P.O.Box 522, 11001 Belgrade, Serbia

ABSTRACT

In this study, nanophase, spherical, polycrystalline $(Y_{1-x}Gd_x)_2O_3:Eu^{3+}$ phosphor particles were synthesized from aerosols of corresponding nitrate solutions, ultrasonically generated at 1.3 MHz frequency, and thermally decomposed at 900 and 1100°C. Structural and morphological analyses done by X-ray powder diffraction (XRPD) and transmission electron microscopy analysis (TEM) implied the formation of non-aggregated spherical, submicronic particles with smooth particle surfaces and filled morphology. The phase development and structural changes, determined using the Topas program, implied a nanocrystalline inner structure (crystallites < 20 nm), that was also confirmed by SAED analysis. A bcc *Ia-3* cubic phase was identified in all as-prepared samples, apart from the mixed oxide with the gadolinium content > 75%, where the existence of a secondary, fcc *Fm-3m* cubic phase is determined. Only *Ia-3* cubic phase was identified in thermally treated samples (1100 °C), while the powder morphological features were maintained, followed by increase in powder crystallinity and phase homogeneity. Functional properties were analyzed by means of photoluminescent analysis. Emission spectra showed typical $Eu^{3+}$ $^5D_0 \rightarrow {}^7F_i$ (i = 0, 1, 2, 3, 4) transitions with the dominant red emission peak at 611 nm.

INTRODUCTION

$Y_2O_3$ and $Gd_2O_3$ represent good host materials for $Eu^{3+}$ activated phosphor materials, and have been widely studied for application as red light emitting sources in modern display devices such as plasma and field emission display[1-3]. These oxides can be used in their composite structure (i.e., solid solution) since it improves their luminescent properties[4]. In addition, $(Y,Gd)_2O_3:Eu^{3+}$ is a very efficient X-ray phosphor, and as such, it is one of the representatives of ceramic scintilators specially designed for medical and industrial X-ray detectors in CT scanners[5]. For utilization in modern display devices, phosphor particles having advanced properties such as fine size, spherical morphology and non-agglomeration characteristics are desirable since such properties ensure higher brightness and resolution[6]. Production of advanced phosphor is hard to achieve with conventional synthesis methods like solid state reaction, since this yields large particle sizes with irregular morphologies and hard agglomerates[7]. In general, conventional processes have disadvantages in controlling the morphology and maintaining the uniformity in composition of phosphor particles. Therefore, synthesis methods that have straightforward control over the particle properties are of great interest. Spray pyrolysis is a feasible method for achieving these objectives, since the particle formation mechanism inherent to spray pyrolysis assures the formation of powders consisting of particles with the properties mentioned above. Namely, atomized precursor solutions are fed into a furnace, leading to the successive solvent evaporation, drying, solute precipitation and chemical decomposition on a droplet level, ensuring the formation of particles with required compositional and structural characteristics[8-10]. Also, one more advantage of this process is that it represents a continuous process that can easily be commercialized.

In this work we aim to synthesize $(Y,Gd)_2O_3:Eu^{3+}$ powders with different gadolinium content through aerosol synthesis, and to elucidate the real particle structure. Proper evaluation of particle structural and morphological features, and correlation with their optical properties is the first step towards processing advanced phosphors.

## EXPERIMENTAL

Nanostructured powders of $(Y_{1-x}Gd_x)_2O_3:Eu^{3+}$ systems, with different gadolinium content x=0.25, 0.50, 0.75, were obtained through spray pyrolysis method. The synthesis was carried out of 0.1 M water precursor solution with stoichiometric amounts of $Y(NO_3)_3 \cdot 6H_2O$, $Gd(NO_3)_3 \cdot 6H_2O$ and $Eu(NO_3)_3 \cdot 5H_2O$. The doping concentration of $Eu^{3+}$ ions was set at 5 at %. Final precursor solution having the following characteristics: pH=4.4, density $\rho$=1.02561 g/cm$^3$, surface tension $\gamma$=68.2 mN/m and viscosity $\mu$=1.055 mPas were atomized by an ultrasonic atomizer operating at frequency of 1.3MHz (RBI, France). Based on these values the average droplet size[11] and the mean particle size[12] were calculated to be 3.39 μm and 567 nm, respectively. Obtained aerosol was fed by an air flow (1 l/min) into a triple-zone tubular quartz reactor with preset temperatures at 200-700-900°C. The as-prepared powders were collected and additionally thermally treated at 1100°C for 12h.

X-ray diffraction analysis (XRD) using a Philips X'Pert MPD diffractometer, operating with Cu K$\alpha$ radiation at 40mA and 40kV, was used to identify the structural composition. The 2$\theta$ range analyzed was 10° to 100° with a step scan of 0.02 and a counting time 15s per each step. Microstructural parameters were obtained through Rietveld refinement preformed by Topas Academic software[13]. The Rietveld method is a powerful method for the extraction of structural information from powder diffraction patterns. In the Rietveld analysis the least squares refinements method are carried out until the best fit is obtained (achieved when goodness of fitting – gof ~ 1) between the calculated patterns, which is based on the refined structure models, and the observed diffraction pattern. The refinement was carried out through FWHM based calculation, where the contribution of Gaussian and Lorentzian type broadening was used for the calculation of crystallite size (cs) while purely Lorentzian profile was assumed for the calculation of microstrain (ms). Morphological features of the synthesized samples were investigated on a JEOL-JEM 2000EX transmission electron microscope operating at 200kV and it was combined with SAED analysis. Optical properties were measured by a spectrofluorometer system Fluorolog-3 Model FL3-221 (HORIBA JOBIN-Yvon). A xenon lamp (450W) was used for excitation of the examined phosphor particles in the UV region.

Precursor solutions were characterized at 20°C utilizing the following equipment: pH value – Orion research pH/milivoltmeter 611, density – AP-PAAR density meter DMA 55, surface tension – digital tensionmeter K10T Kruss.

## RESULTS AND DISCUSSION

XRD patterns of the as-prepared $(Y_{1-x}Gd_x)_2O_3:Eu^{3+}$ (x=0.25,0.50,0.75) powders, presented in Figure 1a revealed that the samples posses cubic, bixbyte type of structure with space group *Ia-3*, corresponding to the structure of $Y_2O_3$ (JCPDS 41-1105, a=10.60Å). These patterns depict that as-prepared sample with highest concentration of gadolinium, i.e., $(Y_{0.25}Gd_{0.75})_2O_3:Eu^{3+}$, apart of the main body-centered *Ia-3* phase, contains a second one, ascribed to face-centered *Fm-3m* phase. The inset in the picture shows the position of peaks and their relevance according to gadolinium content in $(Y_{1-x}Gd_x)_2O_3:Eu^{3+}$. Namely, the increase in gadolinium content leads to the increase of lattice parameters and consequent pick shift towards lower 2θ angles.

Phase development was followed in detail through Rietveled refinement in the Topas Academic program, and is presented in Figure 1b for the case of $(Y_{0.75}Gd_{0.25})_2O_3:Eu^{3+}$ as-prepared sample. Derived lattice parameters together with values of crystallite size and microstrain are summarized for all the investigated samples in Table I. The refinement confirmed that the samples with x=0.25 and 0.50 posses cubic *Ia-3* structure and that the sample with 75 % of gadolinium is composed out of a primary *Ia-3* phase and of one intermediate phases, cubic *Fm-3m* (23.34 wt%), structurally similar to the $Gd_2Te_6O_{15}$ (file card 37-1400, $a$ = 5.611 Å). The presence of cubic *Fm-3m* phase for the sample with the highest content of gadolinium is in accordance to our previous results related to the pure $Gd_2O_3:Eu^{3+}$ system.

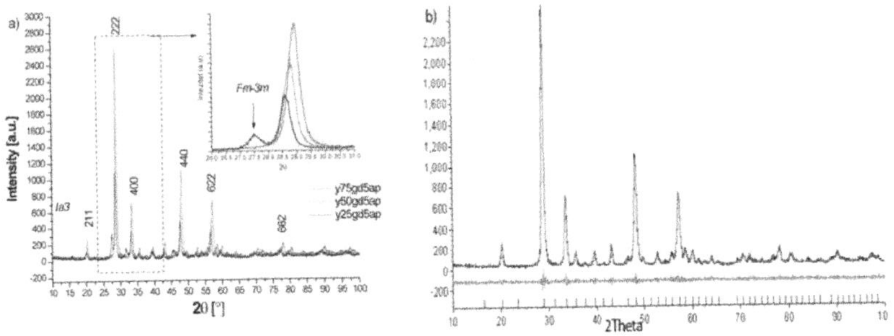

Figure 1. XRD patterns of $(Y_{1-x}Gd_x)_2O_3:Eu^{3+}$ as-prepared samples showing peak shift depending on gadolinium content and b) Rietveld refinement for as-prepared $(Y_{0.75}Gd_{0.25})_2O_3:Eu^{3+}$ sample

We have shown that when it synthesized through spray pyrolysis method under similar synthesis conditions, as-prepared powders of $Gd_2O_3:Eu^{3+}$ systems contain *Fm-3m* phase that disappear afterwards during thermal treating[14,15]. Structural refinement of *Fm-3m* phase was done based on the structural parameters that correspond to $CaF_2$ structure (ICSD 76045). Based on the diffractograms and on the refinement results it is evident that the thermal treatment at 1100°C/12h led to the formation of solely *Ia-3* phase. When analyzing the data presented in Table I a linear increase in the value of lattice parameters with the increase of gadolinium content can be observed, due to relation between different ionic radius of the constituent ions and crystal lattice parameters of pure oxides ($Y^{3+}$=0.90 Å, $Gd^{3+}$=0.938 Å and $Eu^{3+}$=0.947 Å; $a_{Y_2O_3}$=10.60 Å, $a_{Gd_2O_3}$=10.81 Å)[14,16]. The crystallite size in the as-prepared samples is around 20nm for all mixed oxides. In the case of thermaly treated samples at 1100°C/12h, higher energy supply improved crystallite growth evident from the higher crystallite values. Observing the microstrain values the lower ones that correspond to thermally treated samples are in agreement with the fact that additional thermal treating led to relaxation of crystal lattice.

From the transmission electron microscopy images presented in Figure 2, it is evident that uniform precursor precipitation on the droplet level during the synthesis process led to the formation of solid spherical particles. The particles are nanoporous and consist of primary particles that represent the inner structure of secondary, spherical particles. Selected area electron diffraction patterns confirmed the XRD analysis and showed the presence of *Ia-3* cubic phase in as prepared samples where x=0.25 and 0.50, while for x=0.75 apart of *Ia-3* phase the second *Fm-3m* one was confirmed. Weak SAED rings in Figure 2d correspond to body centered (bcc) phase with *Ia-3* space group and lattice cell parameter

Table I. Results of structural refinement for $(Y_{1-x}Gd_x)_2O_3:Eu^{3+}$ system

| sample | gof | $r_{brag}$ | a, Å | cs, nm | *ms*,% |
|---|---|---|---|---|---|
| **$y_{75}gd_5$ ap** | 1.059 | 1.042 | 10.672 ± 0.0005 | 21.49 ± 0.36 | 0.479 ± 0.017 |
| **$y_{50}gd_5$ ap** | 1.043 | 1.574 | 10.729 ± 0.0006 | 23.49 ± 0.55 | 0.596 ± 0.022 |
| **$y_{50}gd_{10}$ ap** | 1.058 | 1.517 | 10.732 ± 0.0007 | 23.89 ± 0.56 | 0.708 ± 0.024 |
| **$y_{25}gd_5$ ap** | 1.441 | 3.766 | 10.773 ± 0.001 | 28.47 ± 1.56 | 0.589 ± 0.039 |
| **$y_{75}gd_5$ 1100** | 1.354 | 3.365 | 10.667 ± 0.0001 | 155.94 ± 5.86 | 0.0329 ± 0.0032 |
| **$y_{50}gd_5$ 1100** | 1.164 | 1.282 | 10.724 ±0.0002 | 202.63 ±20.24 | 0.0563 ±0.0070 |
| **$y_{50}gd_{10}$ 1100** | 1.196 | 2.629 | 10.730 ± 0.0002 | 191.36 ±17.02 | 0.0195 ±0.0096 |
| **$y_{25}gd_5$ 1100** | 1.194 | 1.708 | 10.771 ± 0.0004 | 55.25 ± 2.72 | 0.0333 ± 0.014 |

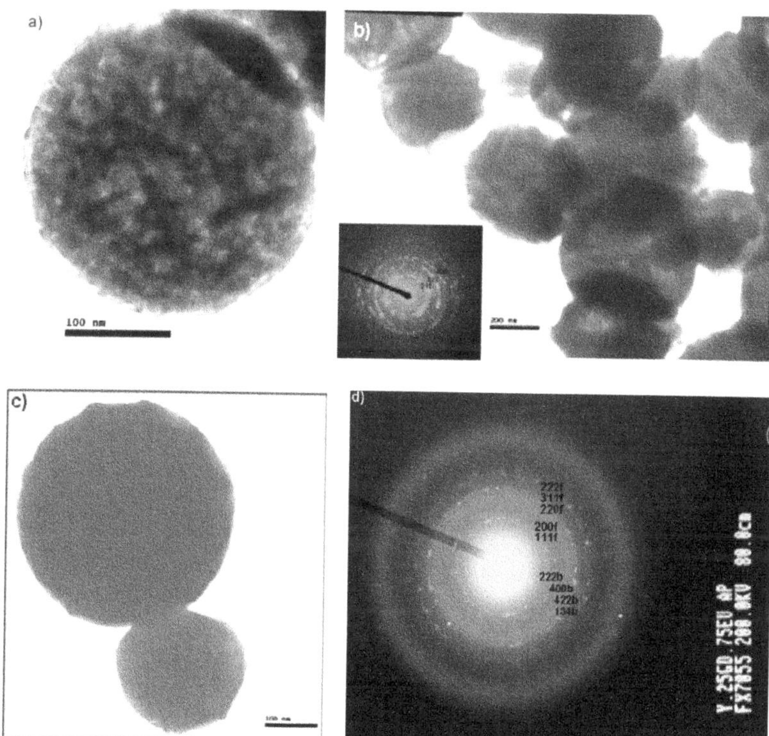


Figure 2. TEM micrographs of: a) as-prepared $(Y_{0.75}Gd_{0.25})_2O_3:Eu^{3+}$; b) as-prepared $(Y_{0.50}Gd_{0.50})_2O_3:Eu^{3+}$ with SAED indentifying *Ia-3* phase; c) as-prepared $(Y_{0.25}Gd_{0.75})_2O_3:Eu^{3+}$; d) the corresponding SAED image $(Y_{0.25}Gd_{0.75})_2O_3:Eu^{3+}$ identifying *Ia-3* and *Fm-3m* phases

a=10.77 Å, while the stronger ring correspond to, face centered secondary *Fm-3m* phase with cell parameter a=5.60 Å. These findings are in good agreement with XRD analysis (Table I). The ring patterns of SAED images presented within Fig. 2 imply polycrystalline nature of these materials while their span suggests high defect concentration.

Analysis of photoluminescent properties was done for the samples thermally treated at 1100°C. Excitation and emission spectra are presented in Figure 3 for the case of $(Y_{25}Gd_{75})_2O_3:Eu^{3+}$ sample. Excitation spectra corresponds to 611 nm emission wavelength and is obtained for wavelength range from 310 to 560 nm. Two intense peaks are denoted, one for $^7F_0 \rightarrow ^5L_6$ excitation at 393 nm and the other for $^7F_0 \rightarrow ^5D_2$ transition at 464.5 nm (Fig. 3a). Emission spectra was obtained in the wavelength range from 500 to 725 nm under the excitation of $Eu^{3+}$ ion into $^5L_6$ level with 393 nm wavelength (Fig. 3b). It represents a typical emission spectrum of the all analyzed samples and it corresponds to characteristic emissions of $Eu^{3+}$ ion and its *f-f* spin forbidden transitions from $^5D_1$ and $^5D_0$ to $^7Fj$ (j=0, 1, 2, 3, 4) levels. Sharp emission lines indicate high material crystallinity, while the main line emission, corresponding to clear red emission, is located at 611nm and belongs to $^5D_0 \rightarrow ^7F_2$ transition. These spectra confirm that $Eu^{3+}$ ion enters the lattice of $(Y_{1-x}Gd_x)_2O_3$ as an isomorphic substitution for yttrium i.e. gadolinium without charge compensation. There are two different types of substitutional sites in the unit cell (three of point group symmetry $C_2$ and one of the point group symmetry $S_6$)[17,18]. Because the $S_6$ sites have inversion symmetry, electric dipole transitions are forbidden and in the emission spectra all the lines belong to $Eu^{3+}$ incorporated into $C_2$ site while only one week intensity line located at 581.6nm belongs to $Eu^{3+}$ ion incorporated into $S_6$ site (Fig 3b).

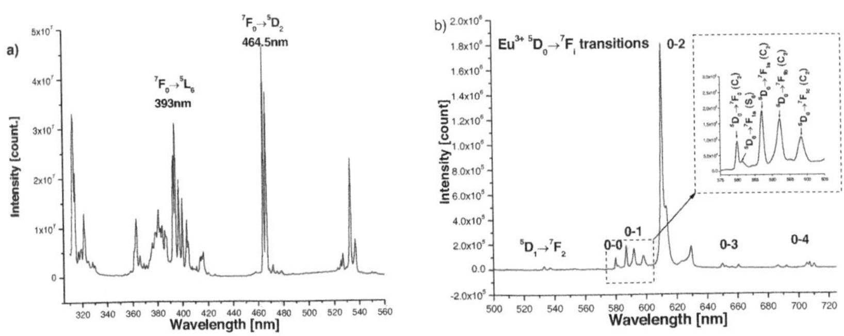


Figure 3. Excitation and emission spectra of mixed $(Y_{0.25}Gd_{0.75})_2O_3:Eu^{3+}$ oxide thermaly treated at 1100°C/12h

## CONCLUSION

Mixed $(Y,Gd)_2O_3:Eu^{3+}$ phosphors with advanced properties were synthesized through spray pyrolysis. Nanostructural powders consisting of fine spherical, non-agglomerated particles having filled morphology and polycrystalline nature (crystallite size of 20 nm) were obtained. The change of gadolinium content had influence on the structural phase characteristics. As-prepared samples with composition x=0.25, 0.50 only had the *Ia-3* phase, while the sample compositionally closest to pure $Gd_2O_3$, with x=0.75, showed the same characteristics as pure gadolinium oxide synthesized under similar conditions. Apart from the primary cubic bcc *Ia-3* phase, a second cubic fcc phase with *Fm-3m* space group was confirmed. In all of the thermally treated samples, only the *Ia-3* phase prevailed. Incorporation of the $Eu^{3+}$ ion within the $(Y,Gd)_2O_3$ host lattice was also proved through emission spectra having dominant red emission peak at 611nm.

## ACKNOWLEDGEMENT

The research is financially supported through the Project #142010 of the Ministry of Science and Technological Development of Serbia and COST 539 Action.

## REFERENCES


1. S. Jung, Y.C. Kang, J.H. Kim, "Generation of phosphor particles for photoluminescence applications by spray pyrolysis", *J.Mat.Sci,* **42** 9783–9794(2007).
2. W.N. Wang, W. Widiyastuti, T. Ogi, I.W. Lenggoro, K. Okuyama, "Correlations between Crystallite/Particle Size and Photoluminescence Properties of Submicrometer Phosphors", *Chem. Mater.* **19**, 1723-1730 (2007).
3. V. Bedekar, D.P. Dutta, M. Mohapatra, S. V. Godbole, R. Ghildiyal and A.K. Tyagi, "Rare-earth doped gadolinia based phosphors for potential multicolor and white light emitting deep", UV LEDs, *Nanotechnology* **20** 125707-125716 (2009).
4. H.S. Roh, Y.C. Kang, S.B. Park, "Morphology and Luminescence of $(GdY)_2O_3$:Eu Particles Prepared by Colloidal Seed-Assisted Spray Pyrolysis", *J. Colloid Interf. Sci.*, **228,** 195–199 (2000).
5. Y.K. Kim, H.K. Kim, G. Cho, D.K. Kim, "Effect of yttria substitution on the light output of $(Gd,Y)_2O_3$:Eu ceramic scintillator", *Nucl. Inst. Meth. B,* **225,** 392-396 (2004).
6. Y.C. Kang, H.S. Roh S.B. Park, "Preparation of $Y_2O_3$:Eu phosphor particles with filled morphology at high precursor concentrations by spray pyrolysis", *Adv.Mater.,* **12,** 451-453 (2000).
7. D.J. Seo, Y.C. Kang, S.B. Park, „The synthesis of *(Y1−xGdx)2O3*:Eu phosphor particles by flame spray pyrolysis with LiCl flux", *Appl. Phys. A,* **77** 659–663 (2003).

8. T.T. Kodas, M.J. Hampden-Smith, Aerosol Processing of Materials, Wiley-VCH, 1999.
9. G.L. Messing, S.C. Zhang, G.V. Jayanthi, "Ceramic Powder Synthesis by Spray Pyrolysis", *J.Amer.Ceram.Soc.*, **76**, 2707-2726 (1993).
10. O. Milosevic, L. Mancic, M.E. Rabanal, J.M. Torralba, B. Yang, P. Towsend, „Structural and Luminescent properties of $Gd_2O_3$:Eu and $Y_3Al_5O_{12}$:Ce Phosphor Particles Synthesized *via* Aerosol", *J.Electrochem.Soc.*, **152 (9)** G707-G713(2005).
11. R.J. Lang, "Ultrasonic atomization of liquid", *J.Acous.Sci*, **43,** 6-8 (1962).
12. T.Q. Liu, O. Sakurai, N. Mizutani, M. Kato, "Preparation of spherical fine ZnO particles by the spray pyrolysis method using ultrasonic atomization techniques", *J.Mater.Sci.*, **21,** 3698-3702 (1986).
13. A.A. Coelho, Topas-Academic, 2004.
14. Y. Wang, O. Milosevic, L. Gomez, M.E. Rabanal, J.M. Torralba, B. Yang, P.D. Towsend, "Thermoluminescence responses from europium doped gadolinium oxide", *J. Phys.: Condens. Matter*, **18,** 9257-9272 (2006).
15. M.E. Rabanal, L.S. Gomez, A. Khalifa, J.M. Torralba, L. Mancic, O. Miloasevic, "Structural properties of europia-doped-gadolinia synthesized through aerosol", *J.Eur.Cer. Soc.*, **27,** 4325-4328 (2007).
16. M.V. Nazarov, J.H. Kang, D.Y. Jeon, E.J. Popovici, L. Muresan, B.S. Tsukerblat, "Lattice parameter and luminescence properties of europium activated yttrium oxide", *Solid State Comm.*, **133,** 183-186 (2005).
17. E. Antic-Fidancev, J. Hölsä, M. Lastusaari, "Crystal field energy levels of $Eu^{3+}$ and $Yb^{3+}$ in the $C_2$ and $S_6$ sites of the cubic C-type $R_2O_3$", *J. Phys.: Condens. Matter*, **15** 863-876 (2003).
18. M. Buijs, A. Meyerink, G. Blasse, „Energy transfer between $Eu^{3+}$ ions in a lattice with two different crystallographic sites: $Y_2O_3:Eu^{3+}$, $Gd_2O_3:Eu^{3+}$ and $Eu_2O_3$", *J.Luminesc.*, **37,** 9-20 (1987).

# A SIMPLE APPROACH TO FORM HYDROTHERMALLY STABLE TEMPLATED HOLLOW SILICA NANOPARTICLES

Raymond V. Rivera Virtudazo, Hideo Watanabe, Masayoshi Fuji* and Minoru Takahashi
Ceramics Research Laboratory, Nagoya Institute of Technology,
3-101-1 Honmachi, Tajimi, Gifu, 507-0033, Japan. *E-mail : fuji@nitech.ac.jp

ABSTRACT

Using an ammonia-hydrothermal templated sol-gel process, amorphous, stable, nano-size (< 100nm), hollow, silica particles were successfully prepared via the hydrolysis and condensation of tetraethylorthosilicate (TEOS) and ammonia water ($NH_4OH$), with inorganic calcium carbonate ($CaCO_3$) as the template. The synthesized $CaCO_3$ core-$SiO_2$ shell nanoparticles were stirred at ambient temperature for 2 h, and ammonia-hydrothermally treated at 90°C for 3 h, 9 h, 1 day and 10 days. This is a simple approach to stabilize silica networks within an amorphous silica shell and preserve the morphologically stable features, and it is a good way to form templated hollow composite glass ceramic materials with functionalized interiors.

## INTRODUCTION

For the past years, nano-sized hollow silica particles have received much attention because they have low density with high specific surface area. Hollow nanoparticles paved the way for applications in catalysis, as the stationary phase for selective separation, controlled (drug) delivery, artificial cells, light fillers, low dielectric-constant prosthetic materials, or photonic crystals [1-5]. Still, there are many challenges in the synthesis of hollow silica materials, such as controlling the porosity of the shell, and functionalizing the interiors with a stable core-shell that enhances the feasibility of several good quality applications. In this research work, hydrothermally stable amorphous nano-size hollow silica has been synthesized via low temperature (90°C) ammonia-hydrothermal sol-gel template ($CaCO_3$) method. Ammonia-hydrothermal treatment accelerates the polymerization and reconstruction of silica frameworks that could help increase the stability of the silica amorphous shell. This is supported by previous studies on sol-gel process using TEOS, ethanol (EtOH) and ammonia- hydrothermal reaction [6-9]. This method does not need special requirements such as manipulation of a complex composition with the experimental parameters and toxic reagents that are commonly used to produce micro / nano-sized hollow particles.

In this study, the effect of aging time (up to 10 days) during ammonia-hydrothermal treatment, and the physicochemical properties of the nano-materials, such as the thermal properties, crystallinity, and morphology of the synthesized hydrothermally stable hollow silica were investigated.

## EXPERIMENTAL

### Materials

A cubic calcium carbonate with an average size of around 60nm, ($CaCO_3$, Nittetsu Mining) was used as core particles. Tetraethoxysilane (TEOS, Wako Pure Chemical) was used as the precursor for the silica shell. Ammonia water ($NH_4OH$ 28%, Wako Pure Chemical) was used as a catalyst. Ethanol (EtOH, Wako Pure Chemical) was used as a solvent for the hydrothermal sol-gel template reaction.

### Synthesis of hydrothermally stable hollow silica nanoparticles

Nano-sized hollow silica particles were synthesized based on the previous method of Fuji *et al* using calcium carbonate as template [1]. Then during the synthesis, the solution was subjected to a hydrothermal process at a low temperature (90°C), and aged for a period time. Calcium carbonate ($CaCO_3$), ethanol (EtOH), tetraethoxysilane (TEOS) and ammonium aqueous solution ($NH_4OH$) were mixed and stirred for two hours at ambient temperature with the molar ratio of 7.3:58.9:1:7.8:3.4 ($CaCO_3$: EtOH: TEOS: $H_2O$:$NH_3$). Core-shell particles were formed by hydrolysis of TEOS and then condensation of the silica coated over the core calcium carbonate particles. The reaction product of a white gel was divided into equal parts: one part was for no hydrothermal (HS0dAH0) and the other part was equally divided and autoclaved for 90 °C, aged for 3 h (HS3hrs90H1), 9 h (HS9hrs90H1), 24 h (HS24hrs90H1) and 10 days (HS10d90H1). For autoclaving, the white gel solution was divided into four parts and placed in a 50 mL capacity autoclaveble Teflon-lined stainless steel vessel at an estimated maximum pressure of 50 kgf/cm$^2$, and heated to 90°C for 3 h, 9 h, 24 h and 10 days. The synthesized product was filtered and washed several times (until it becomes neutral) with ethanol, then dried in a vacuum oven at 90°C for 3 h. To remove the calcium carbonate core, an acid treatment (3 mol/L, HCl) was done for 8 h while stirring. The sample was then filtered and washed several times

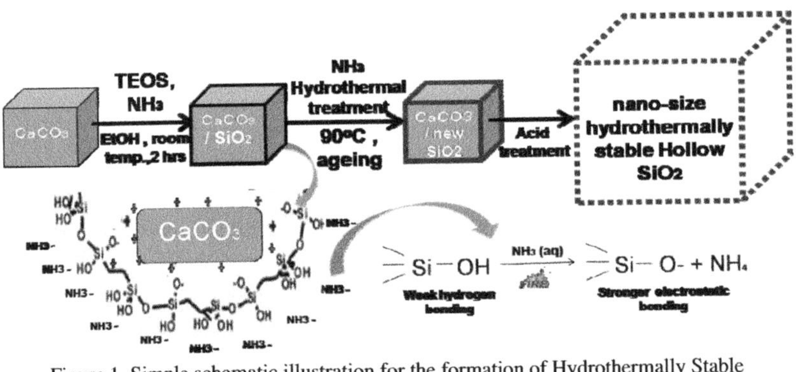


Figure 1. Simple schematic illustration for the formation of Hydrothermally Stable Hollow Silica nanoparticles [15, 16 & 17].

(until it becomes neutral) with distilled water. Finally, the sample was dried at 90°C for 3 h or 1 day, and hydrothermally stable amorphous hollow silica nano-size particles were obtained. A schematic of the formation of hydrothermally stable hollow silica nanoparticles is shown in Figure 1.

Characterization

The end product was characterized by powder X-ray diffraction (XRD, Model RINT 1100, Rigaku) with Cu $K\alpha_1$ radiation ($\lambda$ = 1.54056 Å), with data collected between $5^\circ \leq 2\theta \leq 60^\circ$ in steps of $0.02^\circ$ at room temperature in an operating voltage of 40 kV and an emission current of 30 mA. The thermal properties of the sample were investigated using the thermogravimetry analysis (TG, TG-8120, Rigaku, Japan) in an oxygen atmosphere. A heating rate of 10°C/min was used to heat the sample from 22 to 1000°C. The morphology and microstructure of the hollow particles were examined using scanning electron microscopy (SEM; JSM-7000F, JEOL).

## RESULTS AND DISCUSSION

### Crystal structure analysis by X-ray Diffraction

The XRD pattern of the non-hydrothermal sample (HS0dAH0, Figure 2b) and ammonia-hydrothermal treated sample aged for 10 days (Figure 2c) showed a diffraction angle identical or identified as calcite (core nanoparticles $CaCO_3$) as shown in Figure 2a [10]. Sharp peaks indicate the crystalline structure of cubic $CaCO_3$, in this case (hypothetically) the cubic core nanoparticles of the $CaCO_3$ template that was coated by amorphous silica shell. To confirm that the

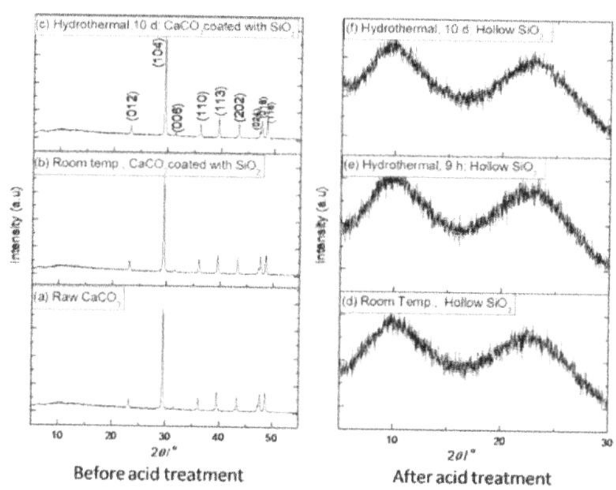


Figure 2. XRD pattern before acid treatment showed a cubic $CaCO_3$ (RMHA, (a)) which has the same results obtained by the samples synthesized at room temperature (HS0dAH0, (b)) and samples hydrothermally treated for 10 d (HS10d90H1, (c)). Then after acid treatment, samples synthesized at room temperature (HS0dAH0ac, (d)); hydrothermally treated for 9 h (HS9hrs90H1ac, (e)) and 10 d (HS10d90H1ac, (f)) showed an amorphous hollow $SiO_2$.

nano-sized calcium carbonate core particles were indeed coated by amorphous silica networks; an acid treatment was done on all the experimental samples. As expected, after treating with acid for 8 h and washed four times with distilled water and dried, no visual peaks (no other crystalline phases, (Figure 2d, 2e and 2f)) were observed for the non-hydrothermal and all ammonia-hydrothermal treated samples. (see Figure 3 for morphological observation.).

Morphological analysis by scanning electron microscopy

SEM (scanning electron microscopy) images revealed that the samples synthesized at room temperature (HS0dAH0) and ammonia-hydrothermal treated at 90 °C for 10 days (HS10d90H1) have the same morphological appearance from the original raw materials cubic $CaCO_3$ (RMHA) as shown in Figure (3a, 3b and 3c). This indicates that the ammonia-hydrothermal treatment samples have no alteration in terms of morphological shape and structural profile of the particles. This was evidently preserved and observed after 10 days aging as shown clearly in Figure 3c.

After acid treatment, 30 to 100nm hollow nanoparticles were clearly observed for the samples synthesized in room temperature and ammonia-hydrothermal treatment. *{Note that in the high –energy field emission SEM observation, hollow silica particles shows a transparency shell, hence any metal or carbon elements can be easily detected in the observe hollow nano-size particles (Figure 3d and 3e)}.* It has a thin wall whose thickness appears to be uniform / smooth outer core-shell surface as shown in Figure 3d and 3e. This indicates the formation of hydrothermally stable amorphous silicate sheet in the surface of cubic $CaCO_3$ after autoclaving at 90 °C for certain aging time [8, 9, 15 & 17].

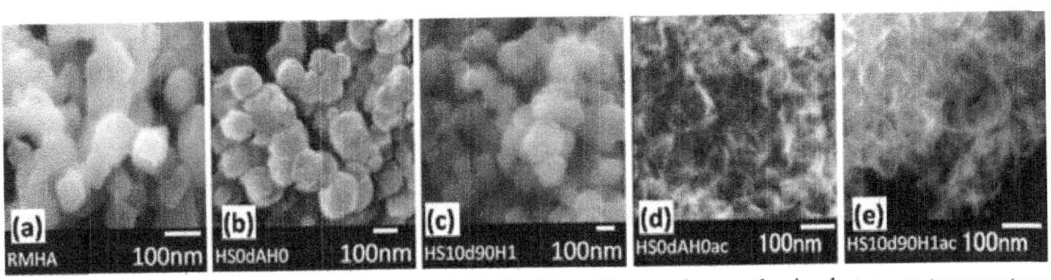


Figure 3. SEM images of (a) raw $CaCO_3$ (RMHA); (b) cubic samples synthesized at room temperature (HS0dAH0) and (c) hydrothermally treated for 10 days (HS10d90H1) showed cubic $CaCO_3$ before the acid treatment. After the acid treatment, the SEM images of (d) samples synthesized at room temperature (HS0dAH0ac); nanoparticles samples (e) hydrothermally treated for 10 days (HS10d90H1ac) at 90°C showed amorphous hollow $SiO_2$ nanoparticles.

Thermal analysis

In order to verify the results, thermal properties were investigated using TG-DTA analysis as shown in Figure 4. There were three significant temperature regions observed in TG-DTA curves. These were Region I (25°C to 150°C): removal of bound water [11] due to the unattached molecular water which causes the loss of weight (TG curve) and an endothermic peak (DTA peak; Region II (200°C to 500°C): slight weight loss, mainly attributed to some loss of $OH^-$ groups [11]; and Region III (above 520°C): characteristic of structural change and decomposition of $OH^-$ groups. The effect of ammonia-hydrothermal treatment is clearly shown in Figure (4A and 4B); the large gap of weight loss due to removal $OH^-$ groups was very evident comparing to that of non-hydrothermally treated samples. This proves that ammonia-hydrothermal treatment accelerates the removal of $H^-$ and $OH^-$ ions which could slightly re-arranged or structurally re-orientated the amorphous silica shell (Figure 4B(c2 and b2)). This enhances the hydrothermal stability of the hollow particles due to the pore walls strengthening as well as due to the presence of hydrophobic groups in the core-wall which improve the integrity of the microporous structure on the core-shell surface [9, 12, 13, 16 & 17].

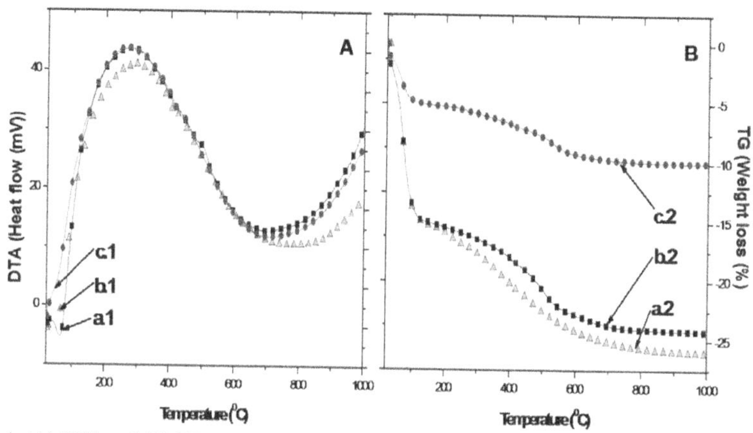


Figure 4. (A) DTA and (B) TG thermogram of nano-size hollow silica after acid treatment (a) ambient temperature [HS0dAH0ac: (a1~ DTA and a2~ TG) data], (b) hydrothermally treated for 9 h [HS9d90H1ac: (b1~ DTA and b2~ TG) data] and (c) hydrothermally treated for 10 days [HS10d90H1ac: (c1 DTA and c2 TG) data].

Mechanism for the formation of hydrothermally-stable hollow nanoparticles

By treating the as-synthesized hollow silica particles using the ammonia-hydrothermal method, the interaction between the silica matrices could transform from weaker $Si^+OH^-$ electrostatic interaction to a stronger $Si^+O^-$ interaction (Figure 1) [14, 16 &17]. This process not only restructures the nanostructure of the silica networks into a semi-ordered nanostructure, but also increases the thermal and hydrothermal stability of the nanostructure [17]. For better understanding of the mechanism, an extensive study of the characterization of the synthesized hollow silica nanoparticles such as with Small Angle X-ray diffraction (SAXD), BET / BJH surface analysis and FT-IR would be recommended.

CONCLUSIONS

In this study, a simple strategy to enhance the stability of hollow nano-silica particles was demonstrated via an ammonia-hydrothermal template process. When monodispersed $CaCO_3$ core-$SiO_2$ shell nano-particles were ammonia-hydrothermally treated at 90°C for a period of time (10 days) and then acid treated (in 3 mol/L, HCl), hydrothermally stable nano-hollow silica particles were directly obtained. This eventually increases the wall strength and extensive growth of silicate units inside the amorphous shell. Stabilized hollow silica nano-materials are expected to be suitable for various applications such as ion exchangers and catalyst supports, and are good future prospects for functionalizing the core shell with bioactive material.

REFERENCES

[1]M. Fuji, C.Takai, Y. Tarutani, T. Takei, and M. Takahashi, "Surface Properties of Nanosize Hollow Silica Particles on the Molecular Level", *Advanced Powder Technol.*, **18**, 81-91(2007).

[2]S. Hyukim, U. Jeong, Y. Xia, "Polymer Hollow Particles with Controllable Holes in Their Surfaces", *Nat. Mater.*, **4,** 671(2005).

[3]X-H. Li, D-H. Zhang, and J-S. Chen, "Synthesis of Amphiphilic Superparamagnetic Ferrite/Block Copolymer Hollow Submicrospheres", *J. Am. Chem. Soc.*, **128**, 8382-8383(2006).

[4]H. Zou, S. Wu, and J. Shen, "Polymer/Silica Nanocomposites: Preparation, Characterization, Properties, and Applications", *Chem. Rev.*, **108 (9)**, 3893-3957(2008).

[5]A.P.R. Johnston, B.J. Battersby, G.A. Lawrie, and M. Trau, "Porous Functionalized Silica Particles: A Potential Platform for Biomolecular Screening", *Chem. Commun.*, 848-850(2005).

[6]K. Byrappa, *Handbook of Hydrothermal Handbook: A Technology for Crystal Growth and Material Processing*, **387**, 870(2001).

[7]R. Mokaya, "Improving the Stability of Mesoporous MCM-41 Silica via Thicker More Highly Condensed Pore Walls", *J. Phys. Chem. B*, **103 (46)**, 10204-10208(1999).

[8]C. Sanchez, G. J. D. A. A. Soler-Illia, F. Ribot and D. Grosso, "Design of Functional Nano-Structured Materials Through the Use of Controlled Hybrid Organic–Inorganic Interfaces", *C. R. Chimie*, **6**, 1131–115(2003).

[9]Q. Wei, F. Wang, Z-R. Nie, C-L. Song, Y-L. Wang, and Q-Y. Li, "Highly Hydrothermally Stable Microporous Silica Membranes for Hydrogen Separation", *J. Phys. Chem. B*, 112 (31), 9354-9359(2008).

[10]JCPDS-International Center for Diffraction Data. PCPDFWIN **v. 2.00** (1998).

[11]J. A. Kim, J-K. Suh, S-Y. Jeong, J-M. Lee, and S-K. Ryu, "Hydration Reaction in Synthesis of Crystalline-Layered Sodium Silicate", *J. Ind. Eng. Chem.*, **6**, 4, 219-225(2000).

[12]D-H. Park, N. Nishiyama, Y. Egashira, and K. Ueyama, "Enhancement of Hydrothermal Stability and Hydrophobicity of a Silica MCM-48 Membrane by Silylation", *Ind. Eng. Chem. Res.* **40**, 6105-6110(2001).

[13]H. P. Lin, C. Y. Mou, S. B. Liu. "Ammonia Hydrothermal Treatment on Mesoporous Silica Structure Synthesized from Acidic Route", *Chem. Lett.* **28**, 1341-1342 (1999).

[14]C-Y. Mou and H-P. Lin, Control of Morphology in Synthesizing Mesoporous Silica", *Pure Appl. Chem.*,**72**, (1–2), 137–146(2000).
[15]H-P. Lin and C-Y. Mou, "Salt Effect in Post-Synthesis Hydrothermal Treatment of MCM-41", *Microporous & Mesoporous Materials*, **55**, 69(2002).
[16]V .A. Parfenov and S. D. Kirik, "Effect of Reaction Medium on Hydrothermal Stability of Mesostructured Silicate Material MCM-41", *Chemistry for Sustainable Development*,**11**, 735–740(2003).
[17]H. P. Lin, C. Y. Mou, S. B. Liu, "Formation of Mesoporous Silica Nanotubes", *Adv. Mater*. **12**, 1, 103-106 (2000).

# PHOTOCATALYTIC PROPERTIES OF TITANIA/SILICA COMPOSITE NANOPARTICLES AND GELS

Masanori Hirano, Keisuke Ota, and Shunsuke Oya
Department of Applied Chemistry, Faculty of Engineering, Aichi Institute of Technology
Toyota, Aichi, 470-0392 Japan

## ABSTRACT

Titania/silica composite gels consisting of silica and titania nanoprticles with various structures were synthesized through a sol-gel process using sodium metasilicate ($Na_2SiO_3$) and titania nanotubes or titania nanoparticles with anatase, rutile, and brookite-type structures. Titania nanotubes (TNT) and brookite-type nanoparticles were formed under basic hydrothermal conditions in the presence of NaOH. Rutile- and anatase-type titania nanoparticles were hydrothermally synthesized by the hydrolysis of acidic $TiCl_4$ and $TiOSO_4$ solutions, respectively. Their photocatalytic activity and adsorptivity were evaluated by measuring the methylene blue (MB) concentration remaining in the solution after being maintained in the dark or under UV-light irradiation. The titania/silica composite gels containing 20 mol% anatase- and brookite-type titania showed relatively good photocatalytic activity.

## INTRODUCTION

Titanium(IV) oxide (titania, $TiO_2$) is attractive as an effective catalyst for photooxidizing a variety of hazardous organic chemicals in air and in water at room temperature.[1] Among the main three polymorphs: anatase (metastable), rutile (stable), and brookite (metastable), the anatase-type $TiO_2$ has been known to exhibit good photocatalytic activities. The properties of titania photocatalysts can be modified by doping with various components.[2] The addition of $SiO_2$ (amorphous silica) to titania has been reported to be effective for enhancing the phase stability [3] and photocatalytic activity [4] of anatase-type titania. We have applied thermal hydrolysis of an acidic aqueous $TiOSO_4$ solution in order to directly load nanometer-sized anatase-type titania on the silica gel particles [5] and natural porous silica, pumice [6] under mild hydrothermal conditions. Aqueous solution routes are useful for preparing functional ceramic materials with distinctive features, e.g. titania nanotubes that were synthesized under basic hydrothermal conditions in the presence of NaOH or KOH.[7] Much work on the synthesis of titania nanoparticles with various structures, e.g., anatase, rutile, and brookite via aqueous solution routes have been reported.[8, 9]

The present study was concerned with the investigation of the synthesis of various kinds of titania/silica composite gels consisting of amorphous silica and titania nano-sized crystals of single phase and various polymorphs. Single phase titania nanoparticles with various structures were selectively synthesized as sols under hydrothermal conditions. Titania nanotubes (TNT) and brookite-type nanoparticles were formed under basic hydrothermal conditions in the presence of NaOH. Additionally, rutile- and anatase-type titania nanoparticles were hydrothermally synthesized under acidic conditions by the hydrolysis of acidic $TiCl_4$ and $TiOSO_4$ solutions, respectively. Using sodium metasilicate ($Na_2SiO_3$) and titania sols with nanotube structures, anatase, rutile, and brookite-type, titania/silica composite gels were synthesized through sol-gel processing. The structure, morphology, and photocatalytic properties of the products were investigated.

## EXPERIMENTAL

Aqueous solutions of reagent-grade $TiOSO_4$ or $TiCl_4$, and solution mixtures of $TiOSO_4$ and NaOH prepared in Teflon containers were placed in stainless-steel vessels, respectively. After the vessels were tightly sealed, they were heated at 120 °C or 180-240 °C for 24 h. After hydrothermal treatment, thus obtained titania nanoparticles and nanotubes were mixed with sodium metasilicate ($Na_2SiO_3$) in the presence of $H_2SO_4$ under controlled pH conditions with stirring. The composite gels were washed with distilled water, separated, and dried. The phases of the as-prepared samples were examined by X-ray diffractometry (XRD). The morphology of the as-prepared samples was observed under transmission electron microscopy (TEM). The crystallite size of titania was estimated from the line broadening of diffraction peaks, according to the Scherrer equation. The specific surface area of the prepared samples was calculated based on the BET method.

The photocatalytic activity and adsorptivity of these prepared samples were estimated from the change in the concentration of methylene blue ($C_{16}H_{18}N_3S$, MB) both under ultraviolet ray (UV) irradiation from black light (20 W) and in the dark, respectively. To 250 $cm^3$ of aqueous MB solution ($5.0\times10^{-5}$ mol/$dm^3$), 0.05 g of sample powders were dispersed via ultrasonic stirring for 5 minutes, and maintained in the dark and under irradiation of ultraviolet ray with an intensity of 1mW/$cm^2$ while stirring for 0-5 h. The UV-light irradiation time dependence of MB decomposition was estimated from the absorbance change with the spectrophotometer.

## RESULTS AND DISCUSSION

### Formation Titania/Silica Composite Gels

A hydrothermal treatment was completed to prepare single phase of titania nano-sized crystals with various polymorphs and titania nanotubes (TNT). XRD patterns of as-prepared titania samples that were precipitated under various controlled hydrothermal conditions are shown in Fig. 1. Three types of titania nanoparticles with anatase-, rutile-, and brookite-type structures and titania nanotubes were directly synthesized as single phase, respectively. Since rutile-type titania coexisting with anatase-type titania was synthesized under hydrothermal conditions at 180 °C by the hydrolysis of acidic $TiCl_4$ solution, a hydrothermal treatment was carried out at high temperature of 240 °C to prepare single phase rutile-type titania (R 240). Anatase-type titania (A 180) was hydrothermally synthesized by the hydrolysis of acidic $TiOSO_4$ solution at 180°C. Brookite-type titania (B 180) was formed under basic hydrothermal conditions at 180°C in the presence of NaOH. Titania nanotubes (TNT 120) were obtained from $TiOSO_4$ solution under basic hydrothermal condition at 120 °C using NaOH. The crystallite sizes of anatase-, rutile-, and brookite-type titania were 16, 29, and 19 nm, respectively. Figure 2 shows TEM images of the as-prepared titania samples that were formed under various hydrothermal conditions. The anatase-type titania particles were of relatively uniform size. Since the rutile-type titania was prepared at higher temperature of 240 °C than the other titania polymorphs, the morphology of rutile particles was rod-like with good crystallinity. The brookite-type titania and TNT were spindle-like and fibrous, respectively. The crystallite sizes of anatase, rutile, and brookite estimated from the XRD line broadening in the samples seemed to be smaller than the particle sizes of the samples observed in the TEM images.

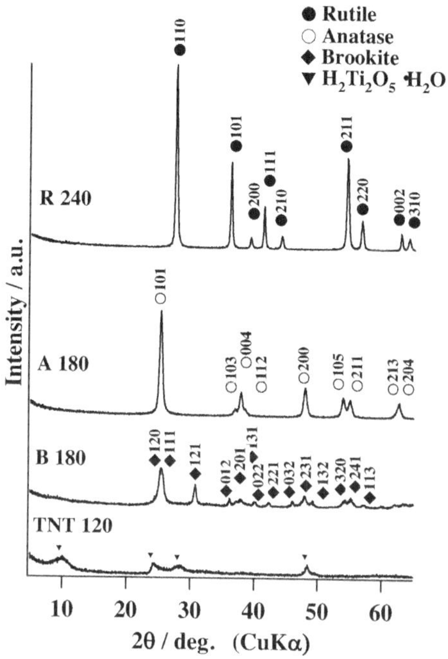


Fig. 1. X-ray diffraction patterns of precipitates obtained at various hydrothermal conditions.

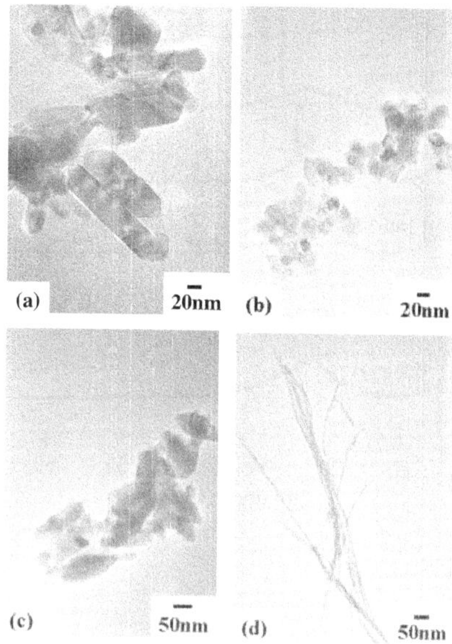


Fig. 2. TEM images of precipitates obtained under various hydrothermal conditions.
(a) R 240, (b) A 180, (c) B 180, (d) TNT 120.

Titania/silica composite gels were synthesized through sol-gel processing using sodium metasilicate ($Na_2SiO_3$) and hydrothermally prepared titania. XRD patterns of titania/silica composite gel samples containing 20 mol% rutile-, anatase-, and brookite-type titania and titania nanotube in comparison with silica gel are shown in Fig. 3 and corresponds to sample names R 240/20, A 180/20, B 180/20, and TNT 120/20, respectively. The XRD patterns show that the composite gels are composed of amorphous silica having a broad peak and titania crystals with various polymorphs. TEM images of the titania/silica composite gels are shown in Fig. 4. Titania particles and tubes with various kinds of morphology dispersed in the amorphous silica matrix are observed. The BET surface area of the titania/silica composite gels containing 20 mol% anatase-, rutile-, and brookite-type titania were 602, 680, 411 $m^2/g$, respectively. In this study, titania/silica composite gels were successfully synthesized through sol-gel processing using sodium metasilicate ($Na_2SiO_3$) and hydrothermally prepared titania nanotubes or titania nanoparticles with anatase, rutile, and brookite-type structures.

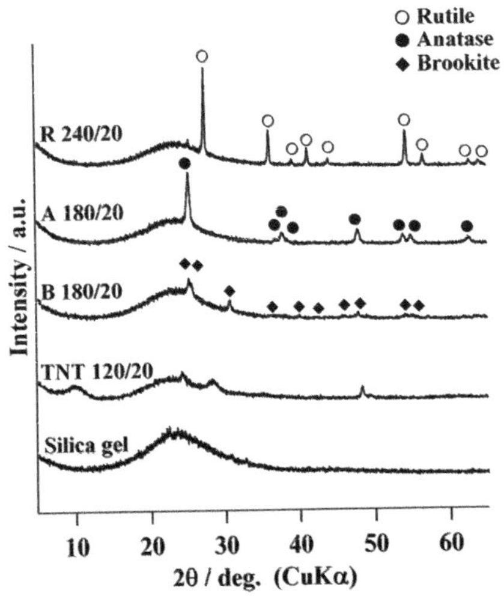


Fig. 3. X-ray diffraction patterns of titania/silica composite gels and silica gel.

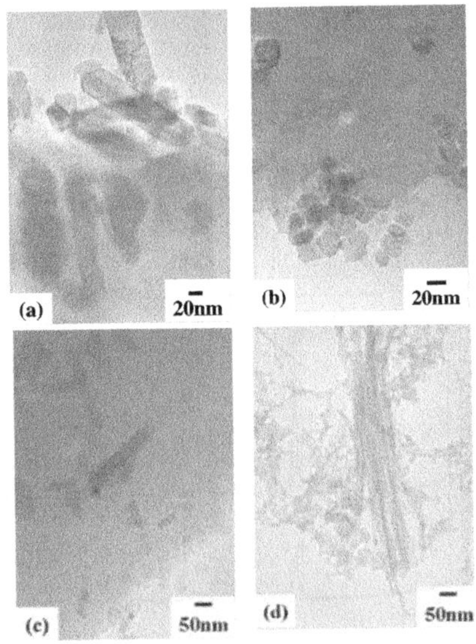


Fig. 4. TEM images of titania/silica composite gels. (a) R 240/20, (b) A 180/20, (c) B 180/20, (d) TNT 120/20.

Photocatalytic Activity of Titania/Silica Composite Gels

The decrease in the MB concentration of the solution containing anatase-type titania (A 180) or titania/silica composite gel containing 20 mol% anatase-type titania (A 180/20) after 0~5 h in the dark and under UV irradiation is shown in Fig. 5 as examples. The decrease in the concentration of MB in the dark corresponds to the adsorptivity of MB by the sample. As indicated in Fig. 5, the composite gel samples containing amorphous silica with high specific surface area showed higher adsorption of MB in the dark than those for titania nanoparticles. The difference between the decrease on MB concentration in the dark and under UV irradiation corresponds to the photocatalytic decomposition of MB by the sample, that is to say photocatalytic activity. The photocatalytic decomposition of MB was observed to gradually increase with increase of maintenance time.

The photocatalytic activity of titania with different structures prepared under various hydrothermal conditions and titania/silica composite gels that were prepared by the sol-gel method using titania hydrothermally synthesized, which is plotted as changes in the photocatalytic decomposition of MB against time in the dark and under UV irradiation, is shown in Fig. 6. The anatase-type titania obtained at 180 and 240 °C and brookite-type titania synthesized at 240 °C showed better photocatalytic activity than others. The photocatalytic decomposition of MB in the samples R 240, B 180, and TNT 120 did not increase so much with increasing exposure time.

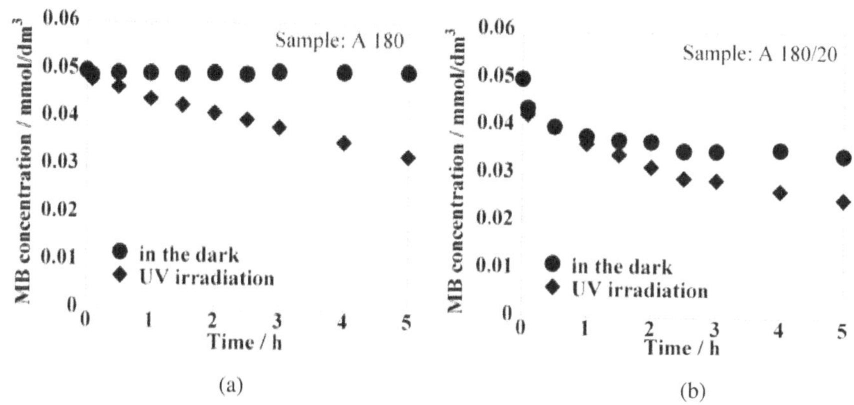


Figure 5. Change in MB concentration of the solution in the presence of (a) anatase (A180) and (b) titania/silica composite gel containing 20 mol% anatase-type titania (A 180/20) with stirring time in the dark and ultraviolet irradiation time.

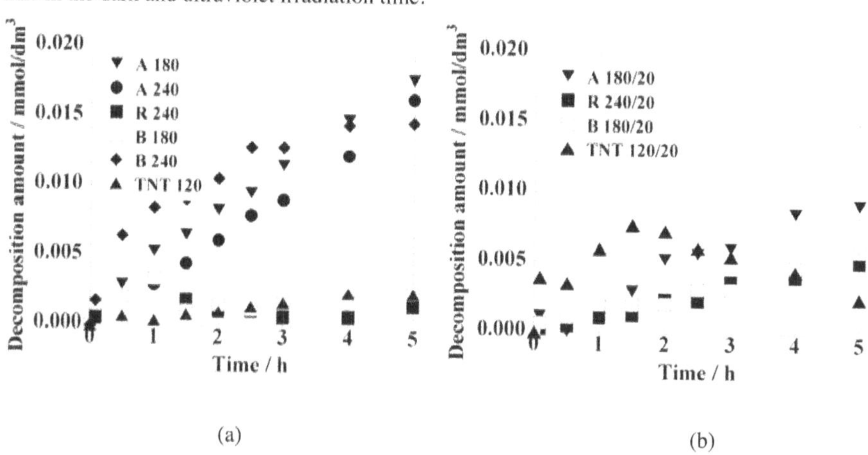


Figure 6. Change in photocatalytic decomposition amount of MB for the solution in the presence of (a) titania nanoparticles and (b) titania/silica composite gels containing 20 mol% titania having various structures with maintenance time.

The MB concentration of the solution containing titania/silica composite gel decreased markedly at the beginning of the stirring time in the dark, compared to the case of pure titania (Fig. 5). This decrease in the concentration of MB in the dark, which corresponds to the adsorptivity of MB by the sample may be ascribed to the formation of composite gel consisting of 20 mol% titania and 80 mol% silica. Although the content of photoactive titania is 20 mol% in the composite gel samples compared to the case of pure titania, it can be seen from the figure that the composite gel containing anatase-type titania (sample: A 180/20) has relatively high photocatalytic activity. The photocatalytic activity of titania/silica composite gels containing 20 mol% titania synthesized under hydrothermal conditions at 240 °C was also estimated. The titania/silica composite gels containing anatase-, brookite-, and rutile-type titania showed photocatalytic activity in that order. As a result, it was shown that the titania/silica composite gels containing 20 mol% anatase-type titania had good photocatalytic activity.

## SUMMARY

Titania/silica composite gels were formed by the sol-gel method from sodium metasilicate and titania prepared under hydrothermal conditions. To investigate the effect of the structure of titania on the photocatalytic activity of titania/silica composite gels, titania with various structures including anatase, rutile, brookite, and nanotubes were synthesized under hydrothermal conditions. Rutile- and anatase-type titania nanoparticles were hydrothermally synthesized by the hydrolysis of acidic $TiCl_4$ and $TiOSO_4$ solutions, respectively. Brookite-type titania nanoparticles and titania nanotubes were formed under basic hydrothermal conditions in the presence of NaOH. The adsorptivity, which was evaluated by the measurement of the concentration of methylene blue remaining in the solution in the dark, markedly increased by the formation of composite gels consisting of amorphous silica and photoactive titania, compared to the case of pure titania. The composite gels showed good photocatalytic activity under UV-light irradiation notwithstanding the low content (20 mol%) of photoactive titania in the samples compared to the case of pure titania.

## REFERENCES

1. M. A. Fox, M. T. Dulay, "Heterogeneous Photocatalysis", *Chem. Rev.*, **93**, 341-357 (1993).
2. M. Hirano and K Matsushima, "Effect of Niobium on the Structure and Photoactivity of Anatase ($TiO_2$) Nanoparticles", *J. Nanosci. Nanotechnol.*, **6**, 762-770 (2006).
3. M. Hirano, K. Ota, and H. Iwata, "Direct Formation of Anatase ($TiO_2$) / Silica ($SiO_2$) Composite Nanoparticles with High Phase Stability of 1300°C from Acidic Solution by Hydrolysis under Hydrothermal Condition", *Chem. Mater.*, **16**, 3725-3732 (2004).
4. M. Hirano, K. Ota, M. Inagaki, and H. Iwata, "Hydrothermal Synthesis of $TiO_2/SiO_2$ Composite Nanoparticles and Their Photocatalytic Activities", *J. Ceram. Soc. Jpn.*, **112**, 143-148 (2004).
5. M. Hirano and K. Ota, "Preparation of Photoactive Anatase-Type $TiO_2$/Silica Gel by Direct Loading Anatase-Type $TiO_2$ Nanoparticles in Acidic Aqueous Solutions by Thermal Hydrolysis", *J. Mater. Sci.*, **39**, 1841-1844 (2004).
6. X.-Y. Chuan, M. Hirano, and M. Inagaki, "Preparation and Photocatalytic Performance of Anatase-Mounted Natural Porous Silica, Pumice, by Hydrolysis under Hydrothermal Conditions", *Appl. Catal. B: Environ.*, **51**, 255-260 (2004).
7. T. Kasuga, M. Hiramatsu, A. Hoson, T. Sekino, and K. Niihara, "Formation of Titanium Oxide Nanotube", *Langmuir*, **14**, 3160-3163 (1998).
8. J. Livage, M. Henry, J. P. Jolivet, and C. Sanchez, "Chemical Synthesis of Fine Powders", *MRS Bull.*, 18-25 (1990).
9. Y. Zheng, E. Shi, S. Cui, W. Li, and X. Hu, "Hydrothermal preparation of Nanosized Brookite Powders", *J. Am. Ceram. Soc.*, **83**, 2634-36 (2000).

# PRESSURE FILTRATION OF PHENYLALANINE-ADSORBED SUBMICROMETER SILICON CARBIDE PARTICLES

Shinichi Baba, Yoshihiro Hirata, Naoki Matsunaga and Soichiro Sameshima
Department of Chemistry, Biotechnology, and Chemical Engineering, Kagoshima University, 1-21-40 Korimoto, Kagoshima 890-0065, Japan

## ABSTRACT

The adsorption of phenylalanine ($C_6H_5CH_2CHNH_3^+COO^-$ at pH 7) on 800 nm diameter negatively charged SiC particles was investigated with a UV-VIS spectrophotometer at a wavelength of 258 nm using an aqueous 30 vol% SiC suspension at pH 7. When 0.37–10 mg phenylalanine was added to 1 $m^2$ of SiC surface, about 1/10 of phenylalanine added was adsorbed on the SiC surface. The applied pressure and suspension height during the consolidation of the SiC suspensions with and without phenylalanine were continuously recorded using a pressure filtration apparatus with a constant crosshead speed (0.2 mm/minute). The final packing density achieved at 19 MPa of applied pressure increased in the following order of the amount of phenylalanine added: 0 mg/$m^2$ (61.0%) $\approx$ 3.74 mg/$m^2$ (60.6%) < 0.37 mg/$m^2$ (69.1%) < 0.037 mg/$m^2$ (71.3%). After calcination at 1000°C in Ar, the density changed to 61.0, 64.2 and 65.4% with additions of phenylalanine of 0, 0.37 and 0.037 mg/$m^2$, respectively. The addition of a small amount of phenylalanine was effective in increasing the packing density before and after calcination.

## INTRODUCTION

Forming technology for ceramic powders is divided into dry and wet processing methods. Uniaxial pressing and isostatic pressing are typical convenient dry forming methods. More uniform microstructure green compacts are produced by wet forming methods such as filtration, pressure filtration, electrophoretic deposition or doctor blading. Pressure filtration can reduce the consolidation time of a colloidal suspension as compared with the filtration method using a gypsum mold because of a higher compressive pressure. The capillary tube suction pressure of a gypsum mold is 50–100 kPa, but the applied pressure during pressure filtration can be as high as 80 MPa when stainless steel equipment is used[1]. In previous papers[2-8], we analyzed the consolidation behavior of aqueous suspensions of nanoparticles in a size range from 20 to 800 nm with and without polyelectrolyte dispersants using a newly-developed pressure filtration apparatus at a constant crosshead speed or at a constant compressive pressure of a piston. The addition of a small amount of polyacrylic ammonium to 20–800 nm particles changed the consolidation behavior and the resultant packing density because of the steric or electrosteric stabilization effect under an applied pressure[4, 7-9]. However, little information has been reported for the adsorption and the influence of low molecular weight dispersants on the consolidation of nanoparticles. In our previous papers[10-12], we studied the interaction between ceramic particles and amino acid with both carboxyl and amino groups in aqueous solutions. In this paper, the influence of phenylalanine adsorbed on 800 nm SiC was studied at a constant crosshead speed of the piston. The measured results were analyzed based on the established filtration theory for dispersed particles and the newly-developed filtration theory for flocculated particles.

## EXPERIMENTAL

A following high purity α-SiC powder supplied by Yakushima Electric Industry Co., Ltd.,

Kagoshima, Japan, was used in this experiment : SiC 98.90 mass%, $SiO_2$ 0.66 mass%, Al 0.004 mass%, Fe 0.013 mass%, free C 0.37 mass%, median size 800 nm, specific surface area 13.4 $m^2/g$. As-received α-SiC powder was dispersed at 0.1 or 30 vol% in an aqueous solution containing phenylalanine at pH 7 ($C_6H_5CH_2CHNH_3^+COO^-$, molecular weight 165.2) of 0.05–5 mass% against SiC. A 1.0 M-HCl solution and a 1.0 M-$NH_4OH$ solution were used for pH adjustment. The above SiC suspensions were stirred for 24 h at room temperature. For the measurement of amount of phenylalanine adsorbed on the SiC surface, a 0.1 vol% SiC suspension stirred for 24 h was filtered through 5 sheets of membrane filter of 100 nm pores. The concentration of free phenylalanine in the filtrate was determined using the UV spectra at 258 nm of wavelength. The dilute 0.1 vol% SiC suspensions were also used to measure the zeta potential at a constant ionic strength of 0.01 M-$NH_4NO_3$ (Rank Mark II, Rank Brothers Ltd., Cambridge, UK). The rheological behavior of the SiC suspensions was measured by a cone and plate type viscometer (DV II+Pro, Brookfield Eng. Lab., Massachusetts, USA).

Figure 1 shows the schematic illustration of the developed pressure filtration apparatus. The suspension in a closed cylinder was filtered through a glass filter of 20 μm pore diameter and three sheets of a membrane filter of 0.1 μm pore diameter. These filters were attached to the bottom of the piston (polymeric resin) moving at a crosshead speed of 0.2 mm/min. When the suspension in a closed cylinder was compressed by the piston, the filtrate flowed into and through the pore channels formed in the upper piston. The applied load and the height of the piston were continuously recorded (Tensilon RTC, A & D Co. Ltd., Tokyo, Japan). The consolidated SiC compact was taken out of the cylinder and dried at 100°C in air for 24 h. The dried compact was heated at 1000°C in Ar atmosphere for 1 h to give enough strength for the measurement of bulk density by the Archimedes method using kerosene.

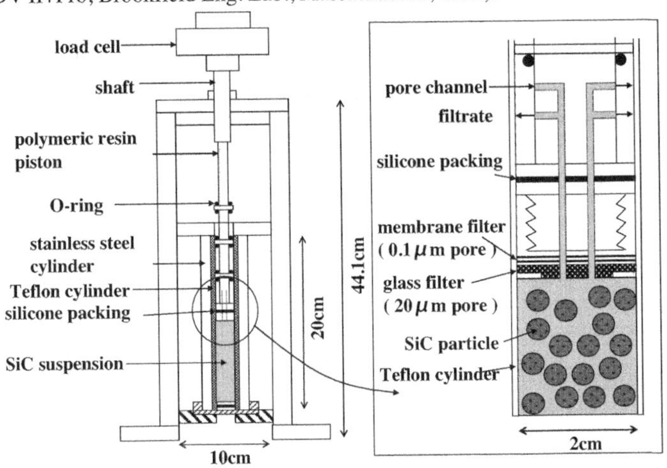


Figure 1. Schematic illustration of pressure filtration apparatus recording the applied load and suspension height. The filtrate overflows in the open spaces between the piston and cylinder through the glass and membrane filters attached to the bottom of the piston.

RESUTLS AND DISCUSSION

Zeta potential and rheology of SiC suspensions

Figure 2 shows the zeta potential of SiC powder in the suspensions at pH 7. The SiC powder without phenylalanine was charged −34.2 mV. Addition of 0.05 or 5 mass%

phenylalanine to the SiC increased slightly the zeta potential toward the negative values. However, no change of zeta potential was measured for 0.5 mass% addition of phenylalanine. The surface of SiC is coated with thin $SiO_2$ film[13]. The $SiO_2$ film reacts with the solution to form $SiOH_2^+$ sites ($SiO_2 + 2H_2O \rightarrow Si(OH)_4$, $Si(OH)_4 + H^+ \rightarrow Si(OH)_3OH_2^+$) or $SiO^-$ sites ($Si(OH)_4 + OH^- \rightarrow Si(OH)_3O^- + H_2O$). The number of $SiO^-$ sites is greater than that of $SiOH_2^+$ sites, explaining the negatively charged surface at pH 7. Phenylalanine ($C_6H_5CH_2CHNH_2COOH$) of 99.3% dissociates to form $NH_3^+$ and $COO^-$ groups in the molecule. The adsorption of electrically neutral phenylalanine with equal numbers of $COO^-$ and $NH_3^+$ groups on SiC surface results in no change in the zeta potential. However, the result in Fig. 2 may be explained by the shield effect of phenyl group with no charge. When 0.05 mass% phenylalanine is added, the $SiOH_2^+$ sites of SiC surface are covered with neutral phenyl group through the adsorption of $COO^-$ group of phenylalanine, resulting in the increased total potential to negative value. With the further addition of phenylalanine (0.05–0.5 mass%), the negatively charged $SiO^-$ sites are covered with neutral phenyl group through the electrostatic interaction of $NH_3^+$ group. As a result, the covered SiC surface is changed from negative one ($SiO^-$ site) to neutral one (phenyl group), increasing the total surface potential to the positive value with addition of phenylalanine. This tendency is observed in the range from 0.05 to 0.5 mass% addition of phenylalanine. At 0.5 mass% phenylalanine, the competitive reactions ($SiOH_2^+$ site-$COO^-$ group, $SiO^-$ site-$NH_4^+$ group) reach a steady-state. When the zeta potential of SiC surface is shifted to a positive value, the adsorption of $COO^-$ group on $SiOH_2^+$ site becomes easily. As a result the SiC surface is again charged negatively (> 0.5 mass% phenylalanine) because of the shield effect of phenylalanine adsorbed. Therefore the charged sites of SiC surface interacting with the ionized groups of phenylalanine may change at about 0.5 mass% addition of phenylalanine.

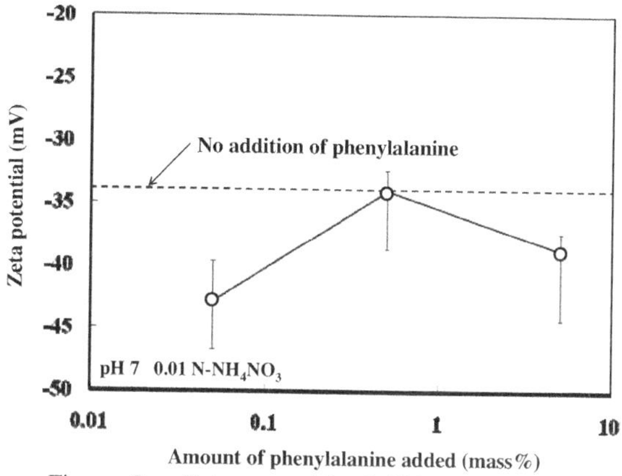


Figure 2. Zeta potential of SiC powders with phenylalanine in the suspensions at pH 7.

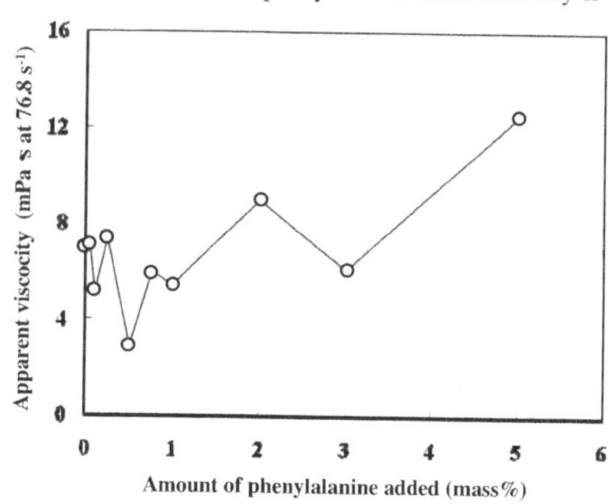


Figure 3. Apparent viscosity of 30 vol% SiC suspension at a shear rate of 76.8 $s^{-1}$ as a function of the amount of phenylalanine added.

Figure 3 shows the influence of dispersant on the rheological properties of 30 vol% SiC suspension at pH 7. The SiC suspension without dispersant at pH 7 showed a pseudoplastic flow but the apparent viscosity was relatively low because of the enough electrostatic repulsion. The addition of phenylalanine gave a small influence on the viscosity as seen in Fig. 3.

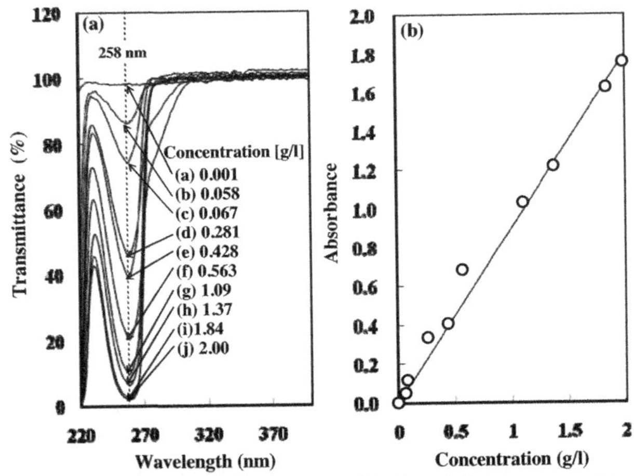


Figure 4. (a) UV spectra of phenylalanine aqueous solutions and (b) the relation between absorbance and concentration of phenylalanine at 258 nm of wavelength.

Adsorption of phenylalanine on SiC surface

The zeta potential of SiC particles measured in Figure 2 indicates the adsorption of phenylalanine on negatively charged SiC surface at pH 7. The amount of free phenylalanine in the SiC suspension was quantified by UV spectra of phenylalanine in the filtrate. Figure 4(a) shows the UV spectra of phenylalanine solutions. Phenylalanine absorbs the photon at 258 nm of wavelength. Figure 4(b) shows the Lambert-Beer rule where the absorbance A is related to the transmittance (T), concentration (C) of phenylalanine, thickness (L) of cell and constant k by Eq. (1). As seen in Figure 4(b), the concentration of phenylalanine is well determined from the absorbance of phenylalanine.

$$A = -\log T = kCL \qquad (1)$$

Figure 5 shows the relationship between added and adsorbed amounts of phenylalanine in 30 vol% SiC suspensions at pH 7. The amount of phenylalanine adsorbed increases when the concentration of phenylalanine is increased. However, about 1/10 of phenylalanine added was adsorbed in the range from 0.37 to 10 $mg/m^2$-SiC. The adsorption behavior of phenylalanine is different from that of PAA (polyacrylic ammonium). A small amount of PAA added is completely adsorbed on $Al_2O_3$ or SiC surface, and free PAA coexists in a concentrated PAA solution[14-16].

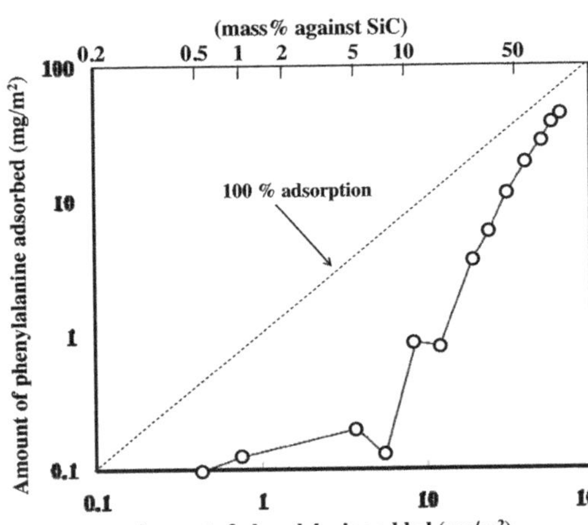


Figure 5. Relation of amounts between added and adsorbed phenylalanine in SiC suspensions at pH 7.

Pressure filtration of SiC suspension with phenylalanine

Figure 6 shows the applied pressure for 30 vol% SiC suspension at a crosshead speed of 0.2 mm/min as a function of normalized volume of filtrate. The SiC suspension without dispersant was filtered at a relatively low applied pressure. According to the filtration theory of well dispersed particles[6,7], the applied pressure is proportional to the volume of filtrate under a constant crosshead speed. However, an almost plateau region of applied pressure was measured for SiC suspension without dispersant. This consolidation behavior was analyzed in our previous papers[5-7] and related to the phase transition from dispersed to flocculated suspension at a critical pressure. The filtration theory for the flocculated suspension is also proposed in our previous papers[5-7]. This consolidation of flocculated SiC particles resulted in 61% of packing at 19 MPa of applied pressure. After heating at 1000°C, no change was measured for the packing density. On the other hand, the addition PAA and phenylalanine to the SiC suspensions suppressed the phase transition to a flocculated suspension. The applied pressure increased linearly with an increase in the volume of filtrate. This result indicates the adsorbed and free dispersant in the suspension contribute to enhance the dispersion of SiC particles. In addition, the packing density of SiC particles at 19 MPa was greatly increased to 68–71%. These values reflect only the packing densities of SiC particles without dispersant. After heating at 1000°C, the packing density generally decreased because of the disappearance of dispersant. As seen in Figure 6, the excess addition of phenylalanine (5 mass%) decreased the packing density at 19 MPa and shape of the compact was broken when the compact was taken out of the filtration apparatus. Addition of a very small amount of phenylalanine is recommended to form a high density green compact.

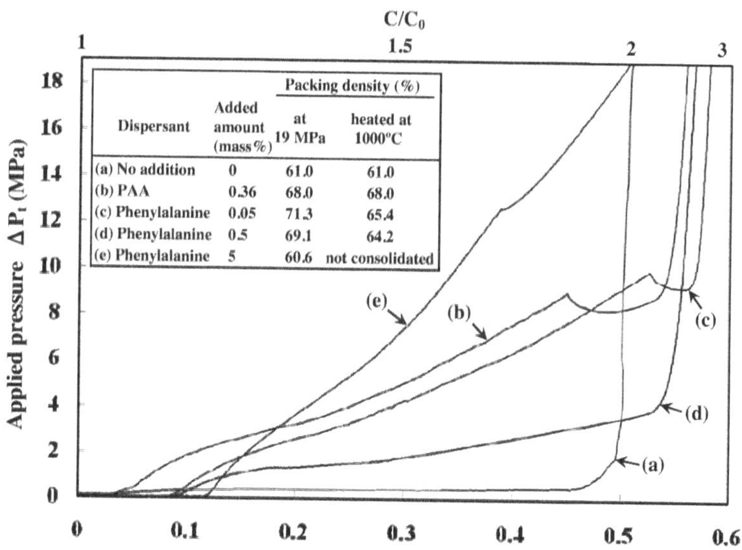


Figure 6. Relation between normalized volume of filtrate ($V_0$: volume of initial SiC suspension) and pressure applied at a crosshead speed of 0.2 mm/min for 30 vol% SiC suspensions.

## CONCLUSIONS

The addition of phenylalanine (0.05–5 mass%) to negatively charged SiC particles in aqueous solutions at pH 7 caused: (1) an increase of zeta potential to a negative value; (2) little change in the apparent viscosity; and (3) adsorption of phenylalanine of less than 1/10 of added amount. The adsorbed and free phenylalanine in the SiC suspension suppressed the phase transition to a flocculated state during the pressure filtration. A high packing density (> 70%) was achieved by the addition of a very small amount of phenylalanine (0.05 mass% relative to SiC ).

## REFERENCES

[1]F. F. Lange and K. T. Miller, "Pressure Filtration: Consolidation Kinetics and Mechanics", *Am. Ceram. Soc. Bull.*, **66 (10),** 1498-1504 (1987).

[2]K. Kishigawa and Y. Hirata, "Packing Density and Consolidation Energy of Flocculated Aqueous SiC Suspension", *J. Eur. Ceram. Soc.*, **26 (1-2)**, 217-221 (2006).

[3] K. Kishigawa and Y. Hirata, "Forming of Aqueous SiC Suspension by Pressure Filtration", *J. Ceram. Soc. Japan*, **114 (3)**, 259-264 (2006).

[4]Y. Hirata, M. Miyamoto, Y. Tanaka and X. H. Wang, "Colloidal Consolidation of Ceramic Nanoparticles by Pressure Filtration", *J. Am. Ceram. Soc.*, **89 (6),** 1883-1889 (2006).

[5]Y. Tanaka, Y. Hirata, N. Matsunaga, M. Nakamura, S. Sameshima and T. Yoshidome, "Pressure Filtration of Nanometer-Sized SiC Powder", *J. Ceram. Soc. Japan*, **115 (11)**, 786-791 (2007).

[6]Y. Hirata and Y. Tanaka, "Pressure Filtration Model of Ceramic Nanoparticles", *J. Am. Ceram. Soc.*, **91 (3)**,819-824 (2008).

[7]Y. Hirata, H. Uchima, Y. Tanaka and N. Matsunaga, "The Effect of Electric Field on Pressure Filtration of Ceramic Suspension", *J. Am. Ceram. Soc.*, **92 (51)**, 557-562 (2009).

[8]Y. Hirata, Y. Tanaka, S. Nakagawa and N. Matsunaga, "Pressure Filtration of Colloidal SiC Suspension", *J. Ceram. Process. Res.*, **10 (3)**, 311-318 (2009).

[9]X. H. Wang and Y. Hirata, "Influence of Polyacrylic Acid on Rheology of SiC Suspension and Mechanical Properties of Densified SiC", *Ceram. Inter.*, **31 (5)**, 677-681 (2005).

[10]M. Chaen and Y. Hirata, "Interaction between Nanometer-Sized Hydroxyapatite Particles and Amino Acid", *Trans. Mater. Res. Soc. Japan*, **29 (5)**, 2379-2382 (2004).

[11]H. Takahashi, H. Shimazu, Y. Hirata and S. Sameshima, "Interaction between Ultrafine Ceria Particles and Glycine", *J. Ceram. Process. Res.*, **5 (1)**, 25-29 (2004).

[12]Y. Hirata, H. Takahashi, H. Shimazu and S. Sameshima, "Dispersibility of Nanometer-Sized Ceria Particles", *Ceramic Transactions*, **152**, Colloidal Ceramic Processing of Nano-, Micro-, and Macro-Particulate System, The Am. Ceram. Soc., Westerville, Ohio, 2004, Edited by W. H. Shih, Y. Hirata and W. Carty, pp. 1-10.

[13]Y. Hirata, K. Miyano, S. Sameshima and Y. Kamino, "Reaction between SiC Surface and Aqueous Solutions Containing Al Ions", *Colloids Surf. A: Phys. Eng. Asp.*, **133**, 183-189 (1998).

[14]Y. Hirata, J. Kamikakimoto, A. Nishimoto and Y. Ishihara, "Interaction between α-Alumina Surface and Polyacrylic Acid", *J. Ceram. Soc. Japan*, **100 (1),** 7-12 (1992).

[15]Y. Hirata, A. Nishimoto and Y. Ishihara, "Effects of Addition of Polyacrylic Ammonium on Colloidal Processing of α-Alumina", *J. Ceram. Soc. Japan*, **100 (8),** 989-990 (1992).

[16]Y. Hirata, S. Tabata and J. Ideue, "Interaction of the Silicon Carbide-Polyacrylic Acid-Yttrium Ions System", *J. Am. Ceram. Soc.*, **86 (1),** 5-11 (2003).

# CHARACTERIZATION OF THE TRIBOLOGICAL INTERFACE IN LOW WEAR MAGNELI TYPE CERAMICS

Rolf Waesche, Mathias Woydt, and Birgit Rehmer
BAM Federal Institute for Materials Research and Testing
Berlin, Germany

## ABSTRACT

The tribological behaviour under dry sliding conditions of Magneli phase $Ti_4O_7$ and e-phase $Ti_2Cr_2O_7$ with α-alumina as a sliding partner has been characterised at room temperature by oscillating sliding. The mechanical properties as well as Vickers hardness measurements do not reveal big differences between the materials, and 4-point bending strengths are around 200 MPa for both materials. Although the friction coefficients are in the medium range of 0.6 and do not differ much between the materials, the wear behaviour is very different. At room temperature, the wear coefficient of $Ti_2O_7$ is strongly dependent on relative humidity, and wear is highest under dry conditions when relative humidity is low as it is typically found for many ceramic materials. $Ti_2Cr_2O_7$ shows remarkably different wear behaviour with the lowest wear in dry conditions. The measured coefficients of wear at low relative humidity of 4% are more than 2 orders of magnitude lower than those for $Ti_2O_7$.

## INTRODUCTION

Titanium dioxide forms a group of substoichiometric oxides (Magneli phases) of the general formula $Ti_nO_{2n-1}$ in decreasing oxygen partial pressures that are stable at room temperature but reoxidize above ~360°C. The reoxidation process can be suppressed by the introduction of transition metals, like Cr, into the $TiO_2$ crystal lattice and the formation of ternary so called Anderson phases[1]. Oxide ceramic materials like alumina or zirconia are among the most wear resistant materials, often due to their high hardness and/or fracture toughness. In these cases the wear resistance is mainly attributed to their good mechanical properties and largely dependent on a well developed and flawless microstructure. However, since mechanical properties are dependent on temperature, the wear resistance decreases with increasing temperatures. Titania based ceramics exhibit good wear properties too, but not based primarily on excellent mechanical properties but rather on the influence on the modification of the sliding interface. This is the case for in-situ generated oxide layers due to sliding of non-oxide ceramics in air as for instance with (Ti,Mo)(C,N)28Co, a cermet[2] or SiC-TiC-$TiB_2$ a ternary particulate ceramic[3]. The so called lubricious oxides[2-6] provide a good wear resistance not only at room temperature but at higher temperatures as well. Among these the Magneli phases $Ti_nO_{2n-1}$ and Anderson phases $Ti_{n-2}Cr_2O_{2n-1}$ are interesting materials with acceptable mechanical properties so that they can be applied directly to build components for high temperature sliding applications. However, they can be applied as plasma sprayed coatings with good success as well[7-9]. They show extremely low wear coefficients under dry sliding conditions of $10^{-8}$ to $10^{-10}$ $mm^3 \cdot N^{-1}m^{-1}$, corresponding to practically zero wear properties[10].

In this paper we present some tribological results obtained under dry oscillating sliding conditions and some corresponding pictures of the sliding interface of the Anderson phase $Ti_2Cr_2O_7$ made with scanning electron microscopy.

## EXPERIMENTAL

### Materials

The materials are $Ti_4O_7$ and $Ti_2Cr_2O_7$ which were prepared by conventional hot pressing at 1600°C in a graphite die. The density of $Ti_4O_7$ is about 99% of theoretical density, that one of $Ti_2Cr_2O_7$ is 97%. The mechanical properties are listed in Table 1. The 4-point bending strength was determined according to European standard DIN EN 843-1:1995, Young's modulus and Poisson number were determined according to DIN-EN 843-2. Hardness was measured by Vickers indentation with 200 g load. Prior to tribotesting the surface was polished with a surface roughness of $R_a$ = 53 nm for $Ti_4O_7$ and $R_a$ = 19 nm for $Ti_2Cr_2O_7$, resp.

The polished surfaces have been cleaned before the tribotesting with ethanol. After the tribotests the surfaces have not been treated by any cleansing procedure and wear particles were not removed prior to SEM investigations.

Table 1: Mechanical properties

| | Density [%] | Bending Strength [MPa] | Hardness $H_{V\,0.2}$ | Youngs Modulus [GPa] | Poisson number |
|---|---|---|---|---|---|
| $Ti_4O_7$ | 99 | 212(22) | 1200(30) | 223.3 | 0.29 |
| $Ti_2Cr_2O_7$ | 97 | 290(34) | 1200(50) | 240 | 0.30 |

### Tribotesting

The tribometer used in this investigation is described in more detail elsewhere[11]. A ball (10mm diameter) is fixed at the top of a lever with an integrated load cell for the measurement of friction force. The ball (not rotating) is positioned on the disk that is fixed on a vibrating table providing a stroke of 0.2 mm and is loaded by a dead weight acting as the normal force perpendicular to the sliding direction. By means of a blade spring with strain gauges the linear wear can be measured continuously during the test. The linear wear $W_l$ is the change of distance between ball and disk perpendicular to the disk surface. The Tribotest parameters are compiled in Table 2.

The friction behaviour is described by the coefficient of friction (COF), defined as the ratio of friction force $F_f$ and the normal force $F_n$. For reason of comparison the stationary average value of the second half of the test is selected. The wear behaviour is described by the coefficient of wear, k, that is defined as the ratio of volumetric wear to the product of load $F_n$ and the sliding distance s. The volumetric wear is calculated from the size d of the wear scars at the ball and the planimetric wear $W_q$.

The wear scar at the ball is nearly circular, while at the disk the scar is slightly elongated in the direction of sliding. The wear scar profile is recorded by tactile measurement with diamond stylus and used for the calculation of the wear volume.

All calculations of volumetric wear at both bodies in contact are based on equations compiled eslewhere[12].

Table 2: Tribotest parameters

| | |
|---|---|
| Disk | $Ti_4O_7$ and $Ti_2Cr_2O_7$ |
| Ball (counterbody) | $Al_2O_3$-Kugel, ∅ = 10 mm |
| Tribological movement | Oscillating sliding |
| Stroke | 0.2 mm |
| Frequency | 20 Hz |
| Normal force | 10 N |
| Number of cycles | 100 000 |
| Temperature | 23 °C |
| Hertzian contact pressure | $Ti_4O_7$: 757 MPa; $Ti_2Cr_2O_7$: 778 MPa |

## RESULTS AND DISCUSSION

Figure 1 shows the coefficient of friction and the coefficient of wear for both ceramic materials as function of relative humidity over the range from dry conditions of about 4 % relative humidity to moist conditions of about 100 % relative humidity. Regarding the friction both materials show practically identical behaviour with a coefficient of friction of about 0.6 over the whole humidity range. However, the wear behaviour shows distinctive differences. For medium and moist environmental conditions the wear coefficient of the two materials is similar, however, for dry conditions of 4% relative humidity the wear coefficient of $Ti_2Cr_2O_7$ is about two orders of magnitude lower than that one of $Ti_4O_7$. The wear scar profiles, as presented in Figure 2, reveal that practically all wear takes place at the disk, therefore the system wear coefficients here presented are identical with the wear coefficients of the ceramic disk materials.

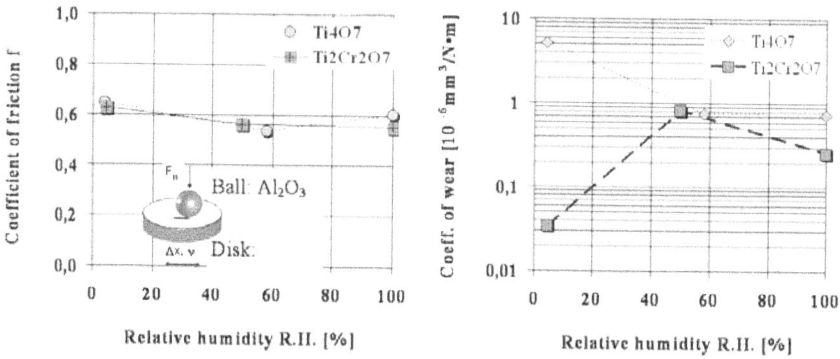


Figure 1. Coefficient of friction (left) coefficient of wear (right) in dependence of relative humidity for 2 materials $Ti_4O_7$ and $Ti_2Cr_2O_7$

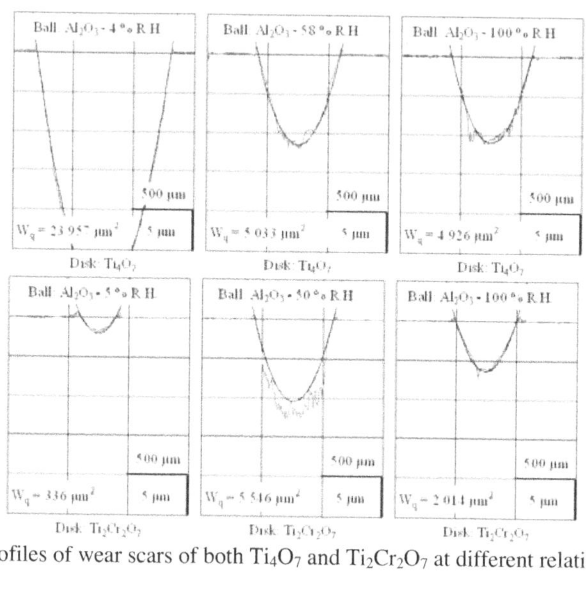


Figure 2. Profiles of wear scars of both $Ti_4O_7$ and $Ti_2Cr_2O_7$ at different relative humidity.

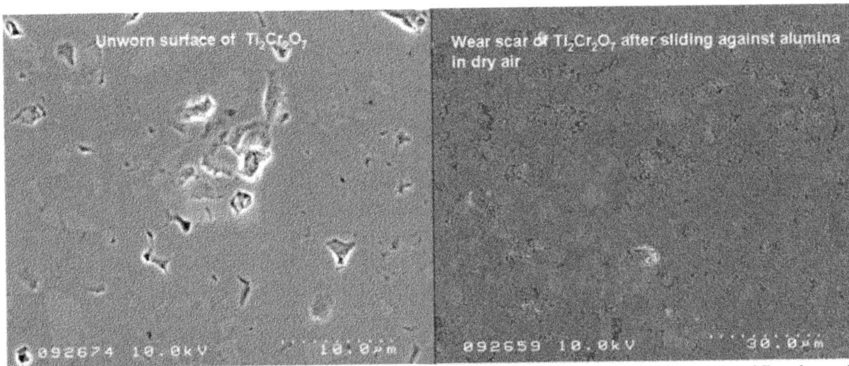


Figure 3. Scanning electron microscopic images of $Ti_2Cr_2O_7$ wear scar at a magnification of 3000 (left) and 1000 times, resp (right).

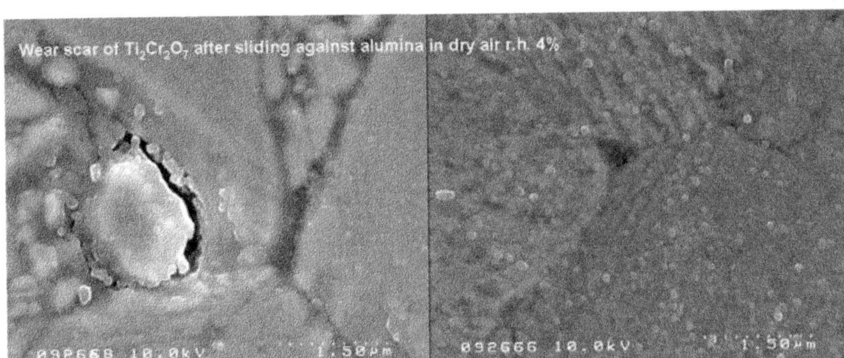


Figure 4. Scanning electron microscopic images of $Ti_2Cr_2O_7$ wear scar at a magnification of 20000 times.

Figure 5. Scanning electron microscopic images of $Ti_2Cr_2O_7$ wear scar at a magnification of 1000 (left) and 20000 times, resp (right).

Figure 3 shows the wear scar at a magnification of 1000 times and the unworn surface at 3000 times. The microstructure shows some remaining porosity and a brighter second phase which was analysed by EDX as a chromium rich phase. Characteristically the worn surface is almost free from wear debris. Although rather flat, the worn surface shows some cracks and exhibits a stressed state. At a higher magnification of 20000 times in Figure 4, the surface cracks become more evident but a thin glassy film can also be seen. This film is obviously transparent since it allows observing the cracks beneath surface. It also fills the grooves and cracks in the surface. In the micrograph on the right side small sphere like particles with the dimension of about 100 nm are covering the surface.

Figure 5 shows the worn surface of $Ti_2Cr_2O_7$ after sliding in normal air with relative humidity of about 50%. The surface is not as smooth and is covered by wear debris. However, in the micrograph at the right side at higher magnification the transparent film may be seen also. It looks like compacted wear debris and appears to be much thicker.

On the basis of the results obtained so far the reason for the high wear resistance can only be hypothesized. At first, the reason might be seen in the binary phase composition of the material. This may lead to a soft $TiO_2$-$Cr_2O_3$ oxide mixture in the interface with a much less abrasiveness. This would be similar to the interface modification in SiC-TiC-$TiB_2$ ceramics[3,13]. At second explanation could be seen in the difference in oxidation stability of $Ti_4O_7$ and $Ti_2Cr_2O_7$. $Ti_4O_7$ as a Magneli phase is not oxidation resistant and tends to reoxidize to form $TiO_2$ in the interface during sliding. In this way it would loose the ability to provide shear planes otherwise present and known as effective sliding systems in the oxide sub-lattice[4,5]. $Ti_2Cr_2O_7$ however is highly oxidation resistant and does not reoxidize even at temperatures as high as 1000°C. In this way the shear planes also present in the oxygen sub-lattice would be also effective in dry air and provide the observed higher wear resistance. The contact pressure as calculated according to the Hertzian theory and shown in Table 2 is too similar for both materials and does not give any hints with regard to why the wear behaviour is so different. Therefore, to prove the validity of the two hypotheses more in depth micro analytical investigations are necessary.

## SUMMARY AND CONCLUSIONS

The wear resistance of $Ti_4O_7$ and $Ti_2Cr_2O_7$ as two Magneli type ceramics has been evaluated. The investigation of the wear scar as obtained after the tribo test revealed a smooth surface for $Ti_2Cr_2O_7$ with a characteristic thin glassy film with an estimated thickness of less than 100 nm. The nature of this film is not clear yet and its investigation needs more analytical effort like TEM. It is possible that the higher wear resistance of $Ti_2Cr_2O_7$ results from the binary composition and the soft mixture of the two oxides in the interface. A second possibility could be seen in the different oxidation stability of the two ceramics and the provision of shear planes in the oxygen sub-lattice in the case of $Ti_2Cr_2O_7$.

## Acknowledgement

The authors would like to express their gratitude to Mrs. B. Strauss for SEM micrographs, to S. Glaubitz for measurement of mechanical properties, to A. Krause for tribological measurements and to C. Neumann for profilometry.

## REFERENCES

[1]H.D. Werner, "Phase relations in the system $TiO_2$-$CrO_{1.5}$", *N.Jahrbuch f. Mineralogie Monatshefte* 219-234 (1974)

[2]M. Woydt, A. Skopp, I. Dörfel, K. Witte, "Wear engineering oxides/Anti-wear oxides", *Wear*, **218**, 84-95 (1998)

[3]R. Waesche, D. Klaffke, "In-situ formation of tribologically effective oxide interfaces in SiC-based ceramics during dry oscillating sliding", *Trib. Lett*, **5**[2,3], 173-190 (1998)
[4]M.N. Gardos, "The effect of ion vacancies on the tribological properties of rutile ($TiO_{2-x}$) ". *Tribology Transactions*, **32**, 30-31 (1989)
[5]M.N. Gardos, "The effect of Magneli phases on the tribological properties of polycrystalline rutile", Proceedings 6th Int. Congress on Tribology, Yokohama, Japan, 1993, Vol.3, 201-206
[6]"Review on Lubricious Oxides and their Practical Importance". In: Handbook of Surface Modifications and processing: Physical & Chemical Tribological Methodologies. Edited by: G.E. Totten, Marcel Dekker, New York, 2004. ISBN: 0-9247-4872-7
[7]M. Woydt, "Triboactive materials for dry high speed sliding applications up to 800°C", Proc. 15th Int. Colloquium on Tribology, J. Bartz(Ed).TAE Esslingen, 17-19- Jan 2006, Esslingen, Germany 2006
[8]M. Berger, S. Saaro, C.C. Stahr, S. Thiele, M. Woydt, "Development of ceramic coatings in the $Cr_2O_3$-$TiO_2$ system", *Thermal Spray Bulletin,* **1**, 64-77 (2009).
[9]J. Landa, I. Illarramendi, N. Kelling, M. Woydt, A. Skopp, M. Hartelt, "Potential of thermal sprayed $Ti_nO_{2n-1}$ coatings for substituting Molybdenum based ring coatings", *Industrial Lubrication and Tribology*, **59**, 217-229 (2007)
[10]D. Spaltmann, M. Hartelt, M. Woydt, "Triboactive materials for dry reciprocating sliding motion at ultra-high frequency", *Wear,* **266**, 167-174 (2009)
[11]Klaffke, D. "On the Influence of Test Parameters on Friction and Wear of Ceramics in Oscillating Sliding Contacts", *Tribotest journal,* **1-4**, 311-320 (1995)
[12]Klaffke, D. "Fretting wear of ceramics", *Tribology international*, **22**, 89-101 (1989)
[13]R. Waesche, D. Klaffke. M. Woydt, I. Doerfel, "Structure of the interface in SiC based ceramic composites formed by tribo-oxidation under dry sliding conditions", *Ceramic Transactions*, **146**, 401-407 (2004)

# DISPERSION BEHAVIOR AND SURFACE INTERACTION CONTROL OF SiC NANOPARTICLES IN AQUEOUS MEDIA BY USING POLYMERIC DISPERSANTS

Yosuke Nomura, Motoyuki Iijima, and Hidehiro Kamiya
Graduate School of Bio-Applications and systems Engineering, Tokyo University of Agriculture and Technology, Koganei, Tokyo, Japan

## ABSTRACT

To control the dispersibility of silicon carbide (SiC) nanoparticles in an aqueous solution, branched polyethylenimine (PEI) with different molecular weights (MWs), i.e., 1,800 and 10,000, was used as a dispersant. The effect of the pH of a suspension and the molecular weight of the dispersant on the amount of steric repulsive force generated by the adsorption of PEI was examined. First, the effect of the suspension pH on the viscosity of the SiC aqueous suspension was examined at pH 4, 7, and 11. For SiC aqueous suspensions without dispersant and with PEI of MW 1,800, no viscosity changes were observed at each of the aforementioned pH values. On the other hand, the viscosity of the SiC aqueous suspension decreased remarkably when PEI with MW 10,000 was added. It was also found that the suspension viscosity decreased with an increase in suspension pH, even though the absolute value of the zeta potential of the SiC nanoparticles was low. In order to discuss these viscosity changes, the hydrodynamic size of the PEI molecules and the surface interactions between the SiC nanoparticles were measured for each of the aforementioned pH values using dynamic light scattering (DLS) and colloid probe AFM, respectively. An increase in the hydrodynamic size of the PEI and the steric repulsive force between SiC nanoparticles was measured at high pH values; this increase resulted in an effective decrease in the suspension viscosity.

## INTRODUCTION

SiC nanoparticles synthesized by the plasma CVD method are one of the promising materials that can be used as sintering additives[1], in power devices[2], and as optical emission materials[3-4] because of their high purity and sinterability. In the process of manufacturing ceramics, it is important to improve the dispersibility of raw powder materials in aqueous solvents in order to inhibit the formation of defects in products, and to improve functionality. However, it is quite difficult to improve the dispersibility of SiC nanoparticles in aqueous media because of their hydrophobicity. Polymer dispersants such as styrene/maleic acid copolymer[5-6] and PEI have been used to generate a steric repulsive force and improve the dispersibility of SiC particles of sizes greater than submicron. However, the active mechanism of these dispersants has not yet been well established. Furthermore, it is expected that the optimum molecular weight or/and the additive conditions required to disperse nanoparticles differ from those required to disperse large size particles, because the surface distance between nanoparticles is considerably shorter than that in the case of large size particles[7].

In this study, the effect of the molecular weight of a dispersant, and the value of the suspension pH on the dispersibility of SiC nanoparticles was examined to control the dispersibility of SiC nanoparticles in an aqueous solution. The dispersibility and the interaction between SiC nanoparticles in an aqueous solution with and without PEI were estimated by determining the apparent viscosity of the suspension and by the colloid probe AFM method, respectively.

## EXPERIMENTAL PROCEDURE

### Preparation and characterization of SiC aqueous suspensions

β-SiC nanoparticles synthesized by the thermal plasma CVD method were purchased from Sumitomo Osaka Cement Co., Ltd. (diameter of approximately 30 nm determined by the FE-SEM image analysis, and BET surface area of 48.8 $m^2/g$). Branched PEI with molecular weights (MWs) of 10,000 and 1,800 (Wako Pure Chemical Ind., Ltd.) was used as a dispersant. First, PEIs were dissolved in ultrapure water such that their concentration was 1.0 $mg/m^2$, for suspension. Nitric acid and ammonium nitrate were added to the solution to control the pH and counter-ion concentration of the suspension. The counter-ion concentration was controlled to be 0.133 M in order to reduce the electrostatic repulsive force and to obtain changes in the steric repulsive force by PEI. Then, SiC nanoparticles with 5.0 vol% were mixed with this solution by using an ultrasonic homogenizer. After stirring the resulting solution for 24 h using a magnetic stirrer, the suspension was deformed by a using planetary mixer (MX-201, THINKY Co., Ltd.) for 5 min. In order to macroscopically evaluate the aggregation and dispersion behavior of the SiC nanoparticles in aqueous media, the apparent viscosity of the suspension was determined by using a coaxial cylindrical viscometer at a shear rate ranging from 1 to 300 $s^{-1}$. Zeta potentials of the SiC nanoparticles were measured by a microscopic electrophoresis method.

### 2.2 Measurement of particle interaction forces between SiC nanoparticles by colloid probe AFM

In order to analyze the changes in the interaction between SiC nanoparticles adsorbed with PEI, the steric repulsive force between SiC nanoparticles in the aqueous media was examined by the AFM (Nanoscope E, Digital Instruments, Inc.) colloid probe method[8]. The interaction between the SiC colloid probe and the sintered SiC substrate was investigated. The colloid probe and substrate were prepared using a method similar to that used in previous studies[7,9]; however, in the method used in this study, the neck growth condition occurred at 1343 K under Ar. The SiC colloid probe and substrate were soaked in the dispersant solution for 24 h so that the polymer dispersant could be adsorbed on them. Then, the particle interaction forces were measured in the solution that was used for soaking in the previous procedure.

Measurement of hydrodynamic size of PEIs

The hydrodynamic size of the PEIs that have MWs of 1,800 and 10,000 was determined by the dynamic light scattering method in water. These polymer solutions were prepared under the same condition as that under which the SiC suspension was prepared.

## RESULTS AND DISCUSSION

### Characterization of SiC aqueous suspensions

Figure 1 shows the zeta potential of SiC nanoparticles with and without PEI at different pH values. When PEI was not added, the isoelectric point was approximately at pH 3, and the zeta potential decreased with an increase in the pH values. On the other hand, when PEI with each molecular weight was added to the suspension, the zeta potential shifted to the positive values for all pH values. It is considered that cationic PEI effectively adsorbed on the negatively charged surface of SiC nanoparticles. As shown in Figure 1, it can also be observed that the absolute value of the zeta potential of the SiC nanoparticles with adsorbed PEI decreased at pH 11. This phenomenon was more typical when PEI with MW 1,800 was added.

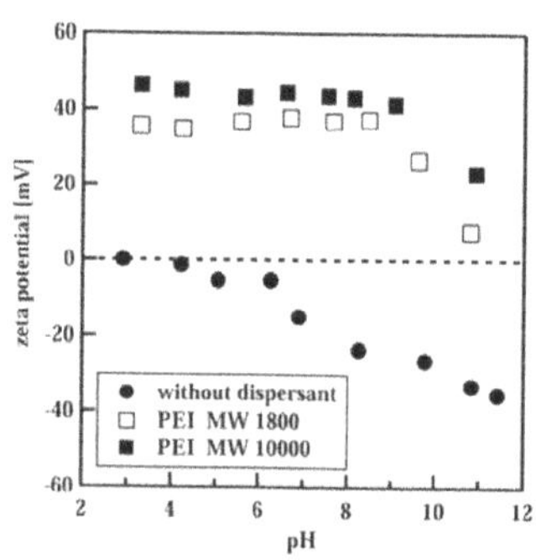


Figure 1. Zeta potential of SiC nanoparticles with and without PEI.

### Effect of pH on the viscosity of SiC aqueous suspensions with PEIs

The apparent viscosity of SiC suspensions (5.0 vol%) measured at pH 4, 7, and 11 is shown in Figure 2. Without the dispersant, there were no significant differences in apparent viscosity for each of the aforementioned pH values, even though the values of the zeta potential changed (Fig. 1). When PEI with MW 1,800 was added, the apparent viscosity of SiC suspensions also did not change with pH. In addition, the apparent viscosity of SiC suspensions with PEI of MW 1,800 was higher than that of the SiC suspensions without dispersants. In contrast, when PEI with MW 10,000 was added, the apparent viscosity of the SiC suspension decreased remarkably. The decrease in the SiC suspension viscosity was significant when the pH value was high.

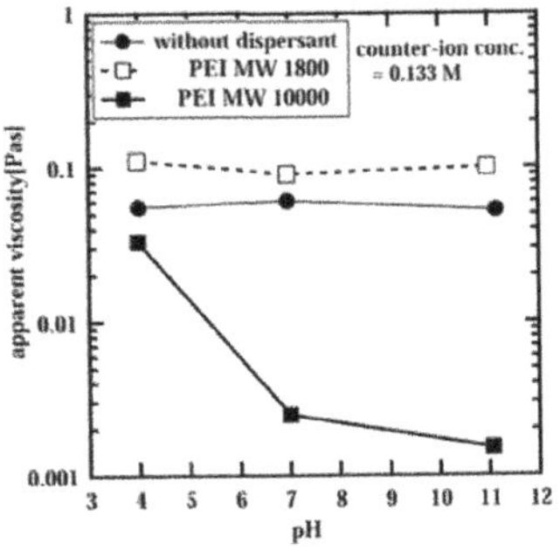


Figure 2. Effect of pH on the apparent viscosity at 300 $s^{-1}$ of SiC aqueous suspensions.

Particle interaction forces between the SiC nanoparticles by colloid probe AFM

In order to show the adsorption behavior of PEI on the SiC surface, the force curves between the SiC colloid probe and the SiC substrate in aqueous solution at each pH values during the approach process were measured by using the colloid probe AFM method (Fig. 3). In the absence of the dispersant, it was observed that the attractive forces at all the pH values were observed at a distance of approximately 5 nm. The fact that electrostatic repulsive forces were not generated is attributed to a very high counter-ion concentration (0.133 M). When PEI with MW 1,800 was added, an attractive force was generated between particles during the approaching process, as observed in the case when dispersants were not added. On the other hand, when PEI with MW 10,000 was added, a repulsive force was observed between particles at a distance of approximately 5 nm. This repulsive force increased with the pH values. This steric repulsive force is generated by adsorption of PEI on the SiC surface, because the generation of an electrostatic force is inhibited at a very high counter-ion concentration. The apparent viscosity of the SiC suspension decreased due to this steric repulsive force.

3.4 Measurement of hydrodynamic size of PEIs

Figure 4 shows the hydrodynamic size distribution of PEI in aqueous solution at each pH values. In the case of PEIs with MWs 1,800 and 10,000, it was found that the size of PEI increases with the pH values. Zhu et al.[10] have shown that the degree of protonation of PEI increased with a decrease in the pH values. It is probable that the molecular size of PEI decreased with the dispersion of PEI molecules at low pH values because of the generation of the electrostatic repulsive force among PEI molecules. On the other hand, at high pH values, it was found that the size of PEI increased with molecular assembly because the electrostatic repulsive force was not generated between PEI molecules. Therefore, it is estimated that when PEI with MW 1,800 was added, the effect of the viscosity decrease of the SiC suspension with the steric repulsive force was not observed (Fig. 3), because the adsorbed layer of PEI on the SiC surface was not sufficiently thick for particle dispersion. Though it is difficult to explain the reason why the apparent viscosity of the SiC suspension with PEI of MW 1,800 was higher than that of the SiC suspension without the dispersant from our presented results, a plausible reason could be that the amount of adsorbed PEI with MW of 1,800 was low and free PEI were present in the suspension. These free PEI have the capability to cause depletion aggregation and increase the suspension viscosity. However, when PEI with MW 10,000 was added, the generation of the steric repulsive force between the SiC nanoparticles (Fig. 3) and the effective decrease in the suspension viscosity (Fig. 2) were observed because the adsorbed layer of PEI on the SiC surface was sufficiently thick. Further, at high pH values, an effective steric repulsive force was generated between the SiC nanoparticles because of the molecular assembly of PEI, which caused a decrease in the suspension viscosity.

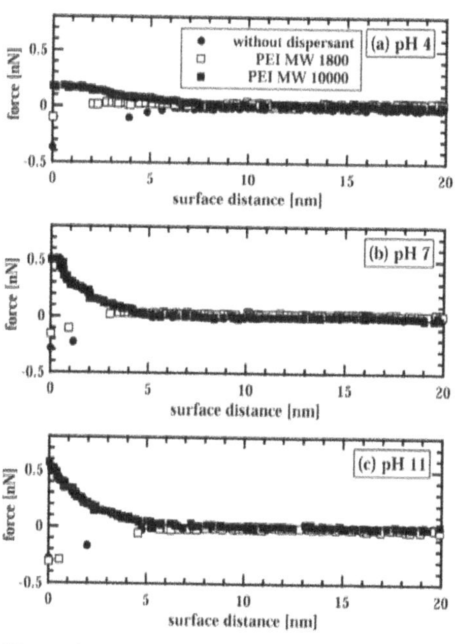


Figure 3. Effect of pH on the force curves between the SiC colloid probe and the substrate in water with and without PEI.

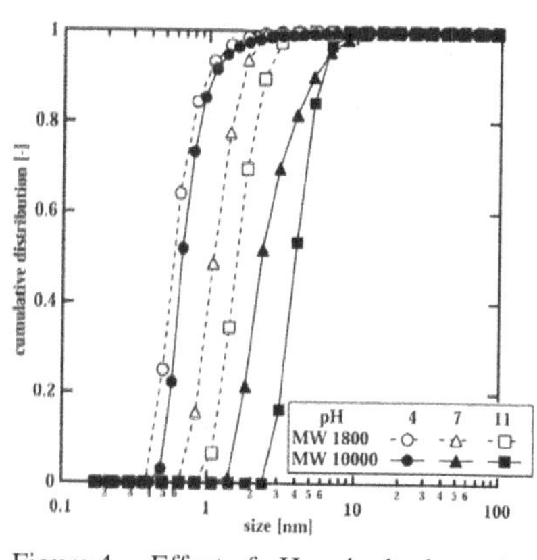


Figure 4. Effect of pH on hydrodynamic size distribution of PEIs measured by dynamic light scattering.

## CONCLUSIONS

The effect of the molecular weight of PEI and the suspension pH values on the dispersibility of the SiC nanoparticles in aqueous media was examined. When PEI with MW 10,000 was added, the viscosity of SiC suspension decreased remarkably with an increase in the suspension pH; however, significant changes in the viscosity of the SiC suspension with changes in the pH for both the cases, i.e., when PEI with MW 1,800 was added and when no dispersant was added, were not observed. Results of the colloid probe AFM measurements showed that when PEI with MW 10,000 was added, a large steric repulsive force was generated at high pH. This force effectively decreased the viscosity of the SiC suspension. In the case of PEIs with MWs 1,800 and 10,000, the hydrodynamic size of PEI increased at high pH because of the molecular assembly of PEI. It is estimated that this increase in the hydrodynamic size of PEI promoted the generation of the steric repulsive force between particles.

## REFERENCES

[1]N. Hidaka and Y. Hirata, "Sintering and mechanical properties of SiC using nanometer-size powder", *Ceramic Transactions*, **190**, 115-128 (2006)

[2]V. Derycke, P. G. Soukiassian, F. Amy, Y. J. Chabal, M. D. D'angelo, H. B. Enriquez, and M. G. Silly, "Nanochemistry at the atomic scale revealed in hydrogen-induced semiconductor surface metallization", *NatureMmaterials*, **2**, 253-258 (2003)

[3]J. Y. Fan, X. L. Wu, F. Kong, T. Qiu, and G. S. Huang, "Luminescent silicon carbide nanocrystallites in 3C-SiC/polystyrene films", *Applied Physics Letters*, **86**, 171903 (2005)

[4]J. Boucle, N. Herlin-Boime, and A. Kassiba, "Influence of silicon and carbon excesses on the aqueous dispersion of SiC nanocrystals for optical application", *J. Nanoparticle Res.*, **7**, 275-283 (2005)

[5]K. Sato, M. Hasegawa, T. Kakui, M. Tsukada, S. Omi, and H. Kamiya, "Influence of the Molecular Structure of a Polymer Dispersant on Concentrated SiC Aqueous Suspensions", *Ceramic Transactions*, **146**, 51-57 (2005)

[6]S. Obata, K. Oda, H. Kinugasa, H. Asano, O. Sakurada, M. Hashiba, K. Hiramatsu, and Y. Nurishi, "Slip casting of basic aqueous SiC slurry containing styrene/maleic copolymer", *Key Engineering Materials*, **159-160**, 157-162 (1999)

[7]K. Sato, S. Kondo, M. Tsukada, T. Ishigaki, and H. Kamiya, "Influence of Solid Fraction on the Optimum Molecular Weight of Polymer Dispersant in Aqueous $TiO_2$ Nanoparticle Suspensions", *J. Am. Ceram. Soc.*, **90**, 3401-3406 (2007)

[8]W. A. Ducker, T. J. Senden, and R. M. Pashley, "Direct measurement of colloidal forces using an atomic force microscope", *Nature*, **353**, 239-241 (1991)

[9]M. Iijima, M. Tsukada, and H. Kamiya, "Effect of surface interaction of silica nanoparticles modified by silane coupling agents on viscosity of methylethylketone suspension", *Colloid and Interface Science.*, **305**, 315-323 (2007)

[10]X. Zhu, F. Tang, T. S. Suzuki, & Y. Sakka, "Role of the Initial Degree of Ionization of Polyethylenimine in the Dispersion of Silicon Carbide Nanoparticles", *J. Am. Ceram. Soc.*, **86**, 189-191 (2003)

# FIBROUS COMPOSITE POWDER COMPACTS FOR THERMAL INSULATION AT HIGH TEMPERATURE

T. Ohmura[1], Y. Ito[1], I. Abe[1], H. Abe[2], and M. Naito[2]
[1] Hamamatsu Research Center, NICHIAS Corporation, 1-8-1 Shin-miyakoda, Kita-ku, Hamamatsu, Shizuoka, 431-2103 Japan
[2] Joining and Welding Research Institute, Osaka University, Osaka University, Osaka University 567-0047, Japan

## ABSTRACT

A nanoparticle bonding process was used to make composite fibers coated with a porous fumed silica layer in the dry phase to developed highly efficient thermal insulation materials. SiC or $TiO_2$ powder was also added as an opacifier by using the bonding process. The composites were compacted into a board by dry pressing. Their good mechanical strength made it possible to machine the compacts for various applications. The resultant board, containing SiC powder, has a very low thermal conductivity, ranging from 0.026 to 0.030 W/(m·K) in the temperature from 100 to 600°C. Additional, an equation proposed by the authors to predict thermal conductivity considering solid, radiation, and gas heat conduction was applied to discuss the effect of each opacifier. The equation determined that the SiC powder contributed by reducing the radiation heat conduction.

## INTRODUCTION

Recently, because escalating energy costs, interest in thermal insulation materials has intensified globally. In this study, a nanoparticle bonding process was used to make composite fibers coated with porous fumed silica layer in the dry phase. These composites were compacted into a board by dry pressing. This technique has the advantage of preventing contact between the fibers in the compacts due to the presence of the coating layer. In addition, since fumed silica was fixed on the fibers, segregation rarely occurred during forming. Therefore, a highly uniform dispersion of fibers in the compact could be easily achieved[1-4]. Furthermore, the thermal conductivity of the fumed silica compacts at high temperature could also be kept at a low value by adding SiC powder as an opacifier[5]. However, the effect of the properties of the opacifiers on the thermal conductivity has not been fully determined.

In this study, SiC and $TiO_2$ particles were evaluated to determine their effect on thermal conductivity, along with their effect on the mechanical strength of the compact, which is an important parameter for machining for various applications. A predictive equation for thermal conductivity proposed by the authors[5] was applied to discuss the effect of each opacifier on thermal conductivity in more detail.

## EXPERIMENT AND ANALYSIS

### Test specimen

Table I shows the recipe and the materials used. 15 mass % SiC and $TiO_2$ powders were added, aiming at reducing the radiation heat conduction. Ceramic fibers 11 μm in average diameter were used to reinforce the specimen. The fiber composition is 65 mass %$SiO_2$ and 25%

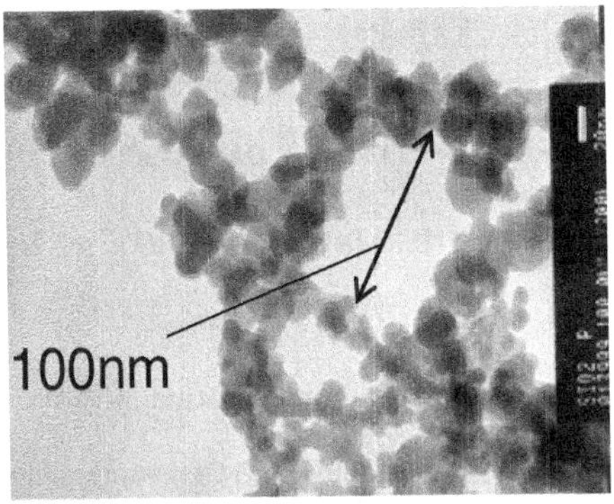


Figure 1. TEM image of nanoparticles.

Figure 2. SEM image of SiC powder.

$Al_2O_3$. Fumed silica nanoparticles were bonded on the surface of the ceramic fibers, and SiC or $TiO_2$ powders using a nanoparticle bonding process[3-4]. Figure 1 shows a TEM image of nanoparticles. They have a three-dimensional nano-scale chain-like aggregated structure with voids in sizes of several tens of nanometers. Figures 2 and 3 show SEM images of the SiC and $TiO_2$ powders, respectively. The primary particle size of $TiO_2$ powder was 250 nm. However, it had an aggregated particle structure as shown in Figure 3.

The composites were compacted into a board by dry pressing. The dimensions of the board were 100×150×15 to 30 mm thick.

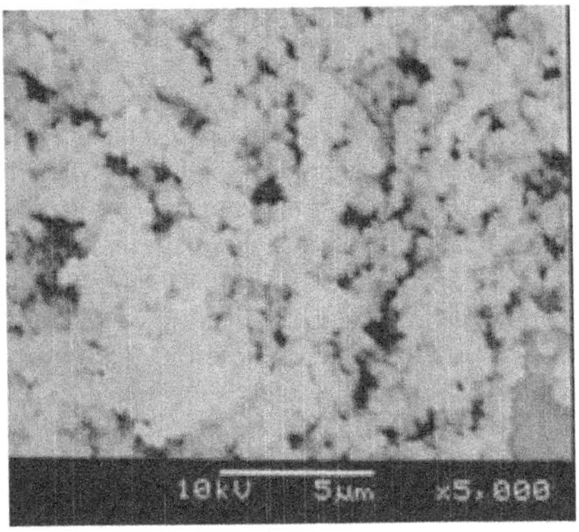


Figure 3. SEM image of $TiO_2$.

Measurement method of bending strength

A three-point bending strength test[10] was applied for the specimens with dimension 30×100×15 mm. The loading velocity was set at 1 mm/min. The distance between loading points in the bending test apparatus was 80 mm, and the radius of curvature of the loading rod was 5 mm.

Table I. Chemical composition

| Material | Compounding ratio [mass %] | Average diameter |
|---|---|---|
| Fumed silica particle | 60 | 7 nm |
| SiC particle | 15 | 0.15 to 25 μm |
| $TiO_2$ particle | | 0.25 μm |
| Ceramic fiber | 25 | 11 μm |

The three-point bending strength was obtained by the following equation:

$$\sigma = \frac{3 \cdot P \cdot s}{2 \cdot w \cdot d^2} \quad (1)$$

Here, (Pa) is bending strength, $d$ (m) is the thickness of specimen, $P$ (N) is compressive fracture load, $s$ (m) is the distance between loading points in bending test apparatus, and $w$ (m) is the width of the specimen.

Measurement and analysis of thermal conductivity

Thermal conductivity of the compacted board was measured using the cyclic heat method[6,7]. It is well known that the cyclic heat method provides two ways of measuring thermal diffusivity; one is to measure the phase lag of the temperature wave generated that propagates through a specimen, the other way is to measure the amplitude decay of the temperature wave. The physical value obtained from these two ways is the thermal diffusivity of a specimen, so the thermal conductivity of the specimen is obtained by multiplying the specific heat and the bulk density by the thermal diffusivity.

The following equation to estimate the thermal conductivity of nanoporous materials was proposed by the authors[5]:

$$\lambda = A\rho + \frac{B}{\rho} T^3 + \lambda_g \quad (2)$$

Here, (W/(m·K)) is the thermal conductivity of the specimen, and $_g$ (W/(m·K)) is the thermal conductivity of gas. (kg/m$^3$) is bulk density, and $A$ (W·m$^2$/(kg·K)) is the coefficient based on the solid conduction. $B$ (W·kg/(m$^4$·K$^4$)) is the coefficient based on the radiation heat conduction, and $T$ [K] is absolute temperature. In the case of the nanoporous material, $_g$ has already been known as the following value[5].

$$\lambda_g = 9.3 \times 10^{-3}\ \mathrm{W/(m \cdot K)} \quad (3)$$

The following equation is derived from Eq. (2).

$$\lambda - \lambda_g = A\rho + \frac{B}{\rho} T^3 \quad (4)$$

From Eq. (4), it is obvious that $\lambda - \lambda_g$ is in proportion to the third power of the absolute temperature $T$. Therefore, assuming that the $y$-axis and $x$-axis are $\lambda - \lambda_g$ and $T^3$ in $x$-$y$ system respectively, a straight line is obtained. According to Eq. (4), the intercept of the line with the $y$-axis gives $A\rho$, and the slope of the line gives $B/\rho$. Therefore, the coefficients $A$ and $B$ are easily obtained from the bulk density $\rho$ of the specimens.

RESULTS AND DISCUSSION

Bending strength and porosity

Table II shows the porosities and bending strengths of two kinds of specimens; ones having 15 mass % of the SiC powder, and the other having 15 mass % of the $TiO_2$ powder. For testing both properties, the number of specimens containing SiC and $TiO_2$ powders were 13 and 11, respectively.

It was obvious that the bending strength of the specimens having the SiC powder was higher than in those specimens having $TiO_2$ powder. The strength is quite acceptable for machining various shapes[2]. The $TiO_2$ powder had an aggregated structure, as shown in Figure 3. It is supposed that some hard aggregates still remained in the board, although the powder

Table II. Porosity and Bending strength

| Specimen | Average Porosity [-] | Standard deviation of porosity [-] | Average bending strength [Mpa] | Standard deviation of bending strength [Mpa] |
|---|---|---|---|---|
| SiC 15 mass% | 0.85 | 0.01 | 0.49 | 0.08 |
| $TiO_2$ 15 mass% | 0.84 | 0.01 | 0.36 | 0.05 |

composites were processed using the particle bonding process. These aggregates, having a larger size than that of SiC powder, may have led to the lower bending strength of the specimen. Further analysis of the fracture surface of the specimen should be conducted in the future.

Thermal conductivity

Figure 4 shows the results of measured thermal conductivity of two kinds of specimens: one having 15 mass % SiC powder and a bulk density of 455 $kg/m^3$, and the other having 15 mass % $TiO_2$ powder and the bulk density of 453 $kg/m^3$. The standard deviation from the average values measured was about 5%. The thermal conductivity of the specimen having SiC powder increases very slightly with the temperature. However, it achieved very low thermal conductivity ranging from 0.026 to 0.030 W/(m·K) in the temperature from 100 to 600°C. This suggests that SiC powder prevents radiation heat conduction at high temperature.

By comparison, the thermal conductivity of the specimen having $TiO_2$ powder was almost in proportion to the temperature, although its refractive index is almost the same as that of SiC. To better understand the differences, an equation to predict thermal conductivity, as shown in Eqs.(2)-(4), was applied.

Figure 5 shows the thermal conductivity obtained by subtracting the thermal conductivity of gas $\lambda_g$ from those $\lambda$ measured using the cyclic heat method. Symbols ○ and △

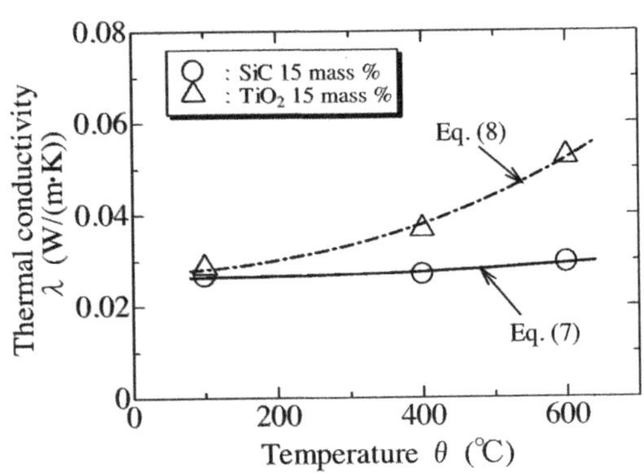


Figure 4. Thermal conductivity for both specimens with temperature.

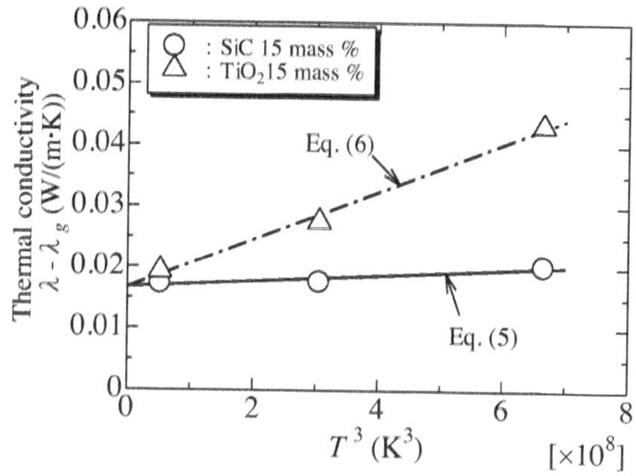


Figure 5. Relation between the third power of absolute temperature and thermal conductivity of the fibrous fumed silica compacts.

indicate the measured values of thermal conductivity for the specimens having SiC powder or $TiO_2$ powder, respectively. Solid and dashed-and-dotted lines indicate the calculated values of thermal conductivity derived from Eq. (4).

$$\text{SiC (15 mass \%)}: \lambda - \lambda_g = 4.90\times 10^{-12} T^3 + 0.0167 \quad (5)$$

$$\text{TiO}_2\text{ (15 mass \%)}: \lambda - \lambda_g = 3.93\times 10^{-11} T^3 + 0.0167 \quad (6)$$

Therefore, the following equation on the relation between bulk density and thermal conductivity was derived by using Eqs. (2) – (6):

$$\text{SiC (15 mass \%)}: \lambda = 3.7\times 10^{-5}\rho + \frac{2.2\times 10^{-9}}{\rho}T^3 + 9.3\times 10^{-3} \quad (7)$$

$$\text{TiO}_2\text{ (15 mass \%)}: \lambda = 3.7\times 10^{-5}\rho + \frac{1.8\times 10^{-8}}{\rho}T^3 + 9.3\times 10^{-3} \quad (8)$$

Eqs. (7) and (8) mean that both coefficients *A* agreed well in case of considering the accuracy of the experiment. However, there was a large difference between both coefficients *B*. It means that solid heat conductions were almost the same for both boards, but that the radiation heat conductions were noticeably different. In case of the board having SiC powder, SiC worked well to prevent the radiation heat conduction. In case of the board having $TiO_2$ powder, they form large aggregate particles as starting material. However, during the particle bonding operation, it is supposed that some of aggregates were dispersed into small size of $TiO_2$ particles. The primary particle size was 250nm which is too small to effectively prevent heat radiation conduction. Therefore, some small size dispersed particles may not have worked well as an opacifier, which led to increasing the thermal conductivity with temperature. Further analysis of the microstructure of the board should be completed to understand this issue better.

CONCLUSIONS

1) The bending strength of the specimen having SiC powder was larger than that having $TiO_2$ powder, although the primary particle size of the $TiO_2$ was smaller, and the porosity of both specimens was the same.

2) The specimen having SiC powder achieved very low thermal conductivity ranging from 0.026 to 0.030 W/(m·K) in the temperature from 100 to 600°C; however, the thermal conductivity of the specimen having $TiO_2$ powder increased with temperature, even though its refractive index is almost the same as that of SiC.

3) In the composite having the $TiO_2$ powder, the morphology of $TiO_2$ changed during the particle bonding operation. It is suggested that aggregates were present, some of which were broken down to their primary particle size, while some still remained as large aggregates in the compact, leading to the lower bending strength and the higher thermal conductivity with temperature.

ACKNOWLEDGEMENTS

This study was partially supported by the New Energy and Industrial Technology Development Organization (NEDO) of Japan.

REFERENCES

[1] M. Naito, A. Kondo and T. Yokoyama, "Applications of Comminution Techniques for the Surface Modification of Powder Materials", ISIJ International, Vol.33, 915-924 (1993)

[2] D. Tahara, Y. Itoh, T. Ohmura, H. Abe and M. Naito, "Formation of Nanostructure Composites Using Advanced Mechanical Processing", The First International Conference on the Characterization and Control of Interfaces for High Quality Advanced Materials,

[3] H. Abe, I. Abe, K. Sato and M. Naito, "Dry Powder Processing of Fibrous Fumed Silica Compacts for Thermal Insulation", J. Am. Ceram. Soc., 88 [5], pp. 1359-1361 (2005)

[4] I. Abe, K. Sato, H. Abe and M. Naito, "Formation of Porous Fumed Silica Coating on the Surface of Glass Fibers by a Dry Mechanical Processing Technique", Advanced Powder Technology, 19, pp. 311-320 (2008)

[5] T. Ohmura, I. Abe, Y. Ito, K. Sato, H. Abe and M. Naito, "Development and Evaluation of Characteristics of Nanoporous Materials", *J. Soc. Powder Technol., Japan*, 46-6, 57-62 (2009)

[6] T. Ohmura, M. Tsuboi, M. Onodera and T. Tomimura, "Estimation of Plane Directional Thermal Conductivity of Fibrous Insulation", Proc. of the Sixth ASME-JSME Thermal Engineering Joint Conference., TED-AJ03-1046 (2003)

[7] H. S. Carslow and J. C. Jaeger, Conduction of Heat in Solid, Oxford Science Publications, Second Edition, pp. 105-109 (1959)

[8] T. Ohmura, M. Tsuboi, M. Onodera and T. Tomimura, "Specific Heat Measurement of High Temperature Thermal Insulations by Drop Calorimeter Method", Int. J. Thermophysics., **24**-2, pp. 559-575 (2002)

[9] T. Ohmura, M. Tsuboi, M. Onodera and T. Tomimura, "Study on Effective Thermal Conductivity of Fibrous Insulation", Reports of Institute of Advanced Material Study Kyushu University, 16, pp. 13-17 (2002)

[10] ASTM D 5934 – 96, "Standard Test Method for Determination of Modulus for Rigid and Semi-Rigid Plastic Specimens by Controlled rate of Loading Using Three-Point Bending"

# FABRICATION OF NON-FIRING CERAMICS BY SURFACE PARTICLE ACTIVATION USING A PLANETARY BALL MILL

Apiluck Eiad-ua, Takashi Shirai, Hideo Watanabe, Masayoshi Fuji*, Koji Orito and Minoru Takahashi
Ceramics Research Laboratory, Nagoya Institute of Technology
3-101-1, Hommachi, Tajimi, Gifu 507-0033, Japan; E-mail address: fuji@nitech.ac.jp (M. Fuji).

## ABSTRACT

We have developed a novel non-firing ceramic process using planetary ball milling with an alkaline hydroxide binding reaction. This can reduce the manufacturing cost (e.g., energy cost), especially during post-forming processing of the desired product. In this study, the surfaces of a raw powder material were mechano-chemically activated using a planetary ball mill, with the addition of a certain amount of alkaline solution that solidified and produced chemical bonding between particles. This was induced to elute cations off of the surface of the particle by milling-assisted activation. In this research paper, the main raw material for the non-firing ceramic product was fabricated using waste sludge from the paper manufacturing industry. The effect of the operational conditions in the planetary ball mill, such as rotational speed and milling time, on the mechanical properties of the solidified body were varied and investigated.

## INTRODUCTION

In recent years, sludge from the paper industry has been a serious environmental problem and concern. Paper sludge is usually disposed of in landfills or burned to produce ash (paper sludge ash, PSA) that has been increasing annually in industrial countries. Lately, research has been completed using PSA as an additive raw material for the cement industry. However, only a small amount of PSA can be used as an additive in the cement manufacturing processes [1-2], and as a possible material for artificial aggregates [3]. The rest is mostly disposed of by dumping, which could give rise to serious environmental pollution (e.g., water pollution). Hence, a research study to investigate developing a technique to recycle and utilized PSA waste materials could be significant. In previous studies, it has been reported that PSA was composed of several oxides, namely alumina, silica and calcium [3-4]. These inorganic oxide materials are the basic components for manufacturing ceramic materials, and could be utilized as an initial material for ceramic products if properly investigated.

Recently, various methods such as pyrolysis [5], microwave [6] and calcination [7] have been used as non-firing ceramic processes to prepare\utilize PSA. However, for industrial applications, there are some issues that need to be addressed, such as the wide size distribution, low yields, and high production costs [5-7]. For this reason, we proposed a novel method of preparing non-firing ceramics from PSA by mechano-chemical (MC) processing. In this process, the MC treatment leads to activation of the particle surfaces, which are responsible for forming strong bonds between the particles after chemical solidification using an alkaline solution. The possibility to convert waste materials into valuable ceramics by chemical solidification assisted by surface activation of the powder using the MC treatment at ambient temperature was investigated. Also in this process, one can easily control the manufacturing process, and it has a lower production cost compared to other existing methods [8].

In this study, the effects of the operational conditions in the MC treatment process such as mill rotational speed and milling time on the mechanical properties of the solidified body were investigated in fabricating non-firing ceramic products.

## EXPERIMENTAL PROCEDURE

The sludge ash powder from the paper industry was used as a raw material. A planetary ball mill was used for MC treatment under air at room temperature. Each milling was carried out with a 100 g sample in a 500 ml capacity ceramic zirconia pot with zirconia balls of 10 mm in diameter. The mill was rotated at different speeds up to 300 rpm for 60 min and for different times up to 60 min at 300 rpm. The treated powder was mixed with 3 mol $dm^{-3}$ of KOH at 2,000 rpm for 5 min with an electric mixer. After the mixing, the slurry was poured into the Teflon mold and kept at room temperature until it becomes solid, ready for demolding. Then the specimens were dried at 25°C (RH 90%) for 7 days in a steam oven.

X-ray diffraction, XRD (Model RINT 1000, Rigaku, Cu Kα, and 30kV) was used for the qualitative determination of the composition and crystalline phases present in the paper sludge ash. Surface microstructure and grain size observation were carried out using a field emission scanning electron microscopy, FE-SEM (JSM-7000F). This technique was used for PSA to reveal the microstructures in the secondary electron imaging (SEI) mode. Specimens were coated with Osmium using osmium plasma coater (OPC 60A; Filgen). Particle size distribution, PSD of powder was determined using a particle size analyzer (MicroTrac MT3300II Series). The specific surface area of the paper sludge ash was determined by $N_2$ adsorption using the BET method at 77K using an automatic gas adsorption-desorption apparatus (Belsorp-max, Japan). The mechanical property was evaluated by diametral compression [9] using a universal testing machine (Shimadzu AGS-G). All tests were performed at room temperature. The test results were recorded using an average value of three measurements.

## RESULTS AND DISCUSSION

This study is a part of our research work that deals with the PSA characterization namely the X-ray diffraction (XRD), field emission scanning electron microscopy (FE-SEM), particle size distribution (PSD), specific surface area (BET), and mechanical strength were discussed in this research study.

Figure 1 shows that the XRD patterns of the raw material and MC treated materials at different rotational speeds and milling times. As shown in the graph, the raw material and MC treated materials are mainly composed of gehlenite ($Ca_2Al_2SiO_7$), calcium disilicate ($CaSi_2O_5$), dicalcium silicide ($Ca_2Si$) and calcium aluminum silicide (CaAlSi). Intensity peaks of crystalline PSA materials were broad, smooth with decreasing intensity as the rotation speeds and milling time increased. This indicates that the particles were in the state of miniaturization and amorphization, which is due to the influence of the high-energy action of the planetary ball mill. In this case, the effectiveness of surface activation of PSA by mechano-chemical process can be assessed. On the other hand, the peak intensity of the solid specimen in figure 2 shows a significant increase of calcium disilicate minerals near (2θ) 30°, if compared with Figure 1 (treated powder) in both rotation speed and milling time. It is clearly shown that the calcium disilicate peaks are directly proportional to alkaline reaction ($K^+$) as an effect of the chemical reaction, attributable to the forming of Si-O bonds from $-SiO^-M^+$ (M=K) group, or Si-O-Al bonds within a highly Al and Si-substituted PSA structure (consistent with the assumption that the Al-rich and Si-rich of PSA leaching from mechano-chemical activation should efficiently

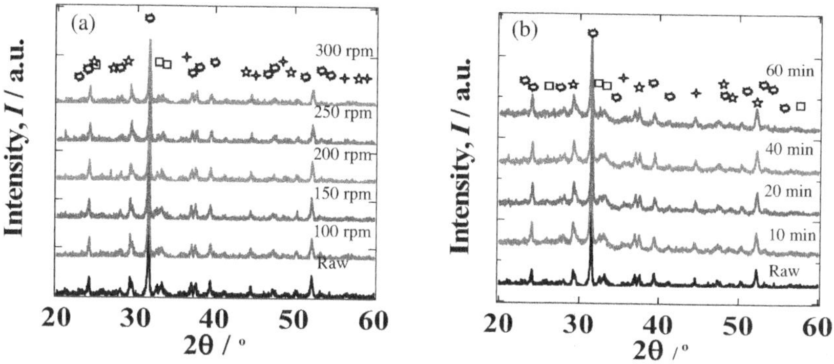


Fig. 1 XRD patterns of mechano-chemical treated material by planetary ball mill

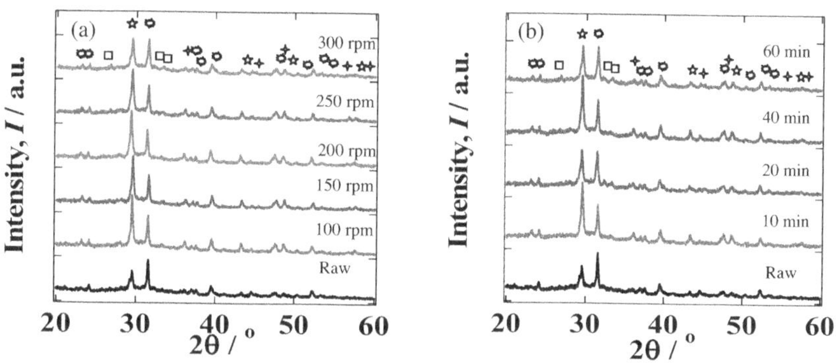


Fig. 2 XRD patterns of solidified material;

(a) different rotation speed, (b) different milling time
( ✪ ($Ca_2Al_2SiO_7$), ✫ ($CaSi_2O_5$), ✦ ($Ca_2Si$), □ (CaAlSi) )

react with alkaline solution and form new chemical bonds [10]. At the same time, the intensity of the peaks of the crystalline phases in Figure 2 decrease with increasing rotational speed and milling time as previous mentioned.

The microstructures of the samples were examined to investigate their development and the influence of the MC treatment on the structure. The characteristic surface microstructures of raw material without and with MC treatment at different rotation speeds and milling time using planetary ball mill are shown in Figure 3 and 4, respectively. Figure 3 shows a scanning electron micrograph of raw PSA and a PSA treated with different rotational speeds. Figure 4 shows a scanning electron micrograph of treated PSA after different milling times. Based on the observed micrograph, changes of micro-surface structure were achieved using the planetary ball mill with

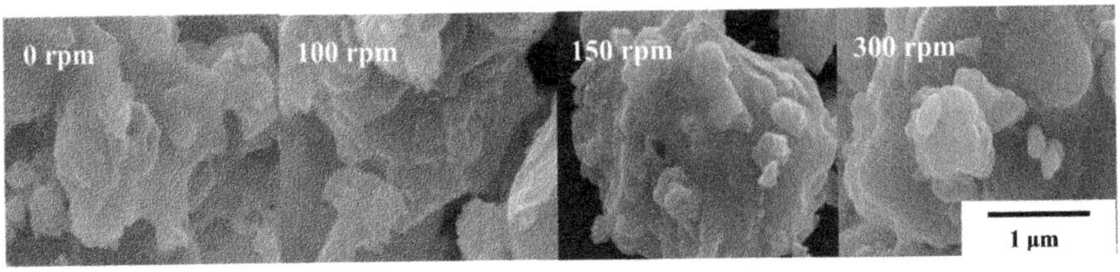


Fig.3 SEM images of initial material treated at different rotation speed.

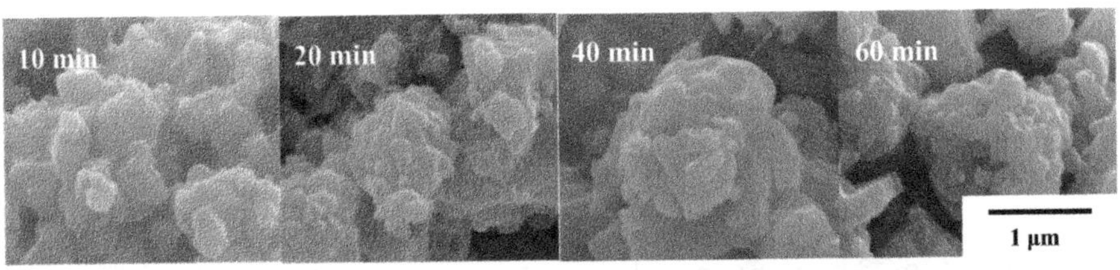


Fig.4 SEM images of initial material treated at different milling time.

increasing rotational speeds and milling time. The morphologies show that the particle size of MC treated sample also decreases at a certain change of rotational speed and milling time by planetary ball milling. But one should be cautious on increasing the rotational speed and the milling time because the small particles (MC treated) tend to agglomerate to form larger particles at high energy during higher rotational speeds and longer milling times.

Figure 5 shows the relationship between the mean particle sizes of samples after MC treatment under various conditions. As shown, the rotation speed and milling time of MC treatment had a great effect on particle size of the raw material. As observed, the particle size decreases at a certain point when the rotational speed and milling time ware varied, which coincides with the SEM analysis in Figure 3 and 4. The results show that average mean particle size ($d_{50}$) is approximately 4 to 28 μm after the samples are subjected to planetary ball milling as compared to the raw powder. But with a higher rotational speed and increasing the milling time, (longer than 150 rpm and 10 min respectively), the mean particle size is almost the same (Figure 5 a and b). Then after 250 rpm (Figure 5a) and after 40 min of milling time (Figure 5b), the particle size tends to increased as shown in the Figure 5 (a and b). In this manner, the reduction of particle size can decrease the distance between vacancy sites and enhance the vacancy diffusion to the external surface and help form new phase and contribute to grain growth. The particle size of PSA only changes a little when the rotation speed increase from 150 to 300 rpm and the milling time increases from 10 to 60 min. The very fine particle size that may have a small negative effect on milling because of agglomeration of fine particles.

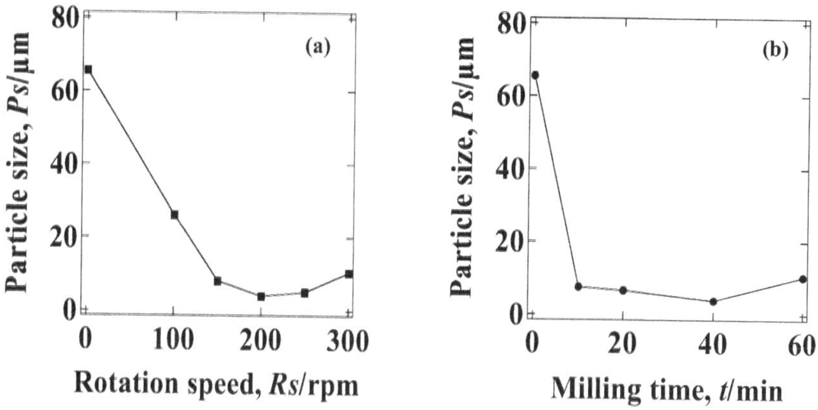


Fig.5 Mean particle size ($d_{50}$) distribution;

(a) different rotation speed, (b) different milling time

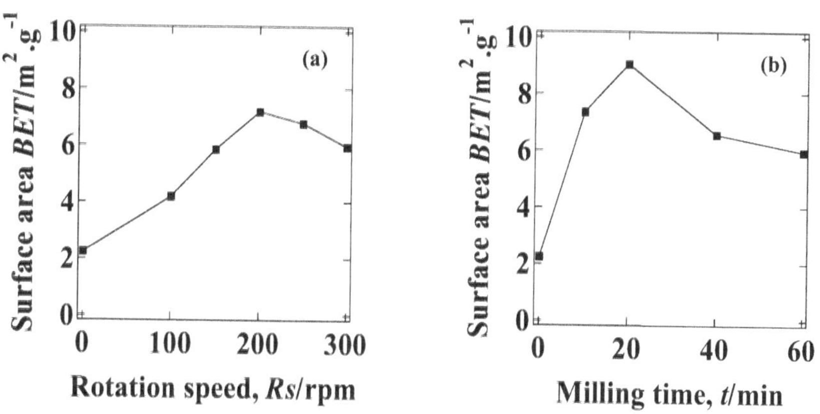


Fig.6 BET surface area;

(a) different rotation speed, (b) different milling time

Figure 6 shows the specific surface area (BET) of samples after MC treatment at different conditions such as the rotational speed and milling time. Based on the obtained results, the surface area increases with increasing the rotational speed and milling time. The surface area increases to a maximum value is 7.18 $m^2g^{-1}$ after milling at 200 rpm, and 9.04 $m^2g^{-1}$ after 20 min of milling. As observed, a larger surface area can be obtained during the milling of PSA using the planetary ball mill. But with further milling at 300 rpm and 60 minutes, the surface area of the sample decrease to about 5.95 $m^2g^{-1}$. This is a typical specific surface area change during mechano-chemical treatment. The increase in the BET during the first period of milling simply is a result of particle fracturing induced by ball impacts, and the later reduction in BET surface area is linked to the agglomeration effects at high energy. It should be noted also that adsorption on particle surfaces was found after the mechano-chemical treatment. These surface area results coincide again with the results of the particles size, SEM analysis and XRD analysis that there is indeed a change in treated PSA particle surfaces.

To observe the mechanical strength of the experimental sample, compression strength was characterized. Figure 7 shows the compression strength of specimens prepared with different MC treatment conditions. It can be observed that the compression strength of MC-treated specimen is clearly stronger than that of non-MC-treated specimen. These specimens show that the compression strength is significantly increasing with MC treatment. In general as observed in the obtained results, the strength of the specimens depends on the total number bonds between the interfaced surfaces of materials after the MC treatment. In this case, it is tentatively considered that the MC treatment enhances surface activation and chemical reaction between the interfaced surfaces, resulting in a product layer bonding between adjacent particles. But, in the cases of high rotation speed above 200 rpm, and long milling time, the compression strength decreases. It was considered that high energy milling leads to powder agglomeration or the density distribution of green body that contributed to reduction of mechanical properties. From the results, one can conclude that the MC treatment could be a powerful technique to obtained activated surface of powder. The advantage of the proposed fabrication method is that the mechano-chemical treatment strongly affects solidification and the mechanical properties of solidified body.

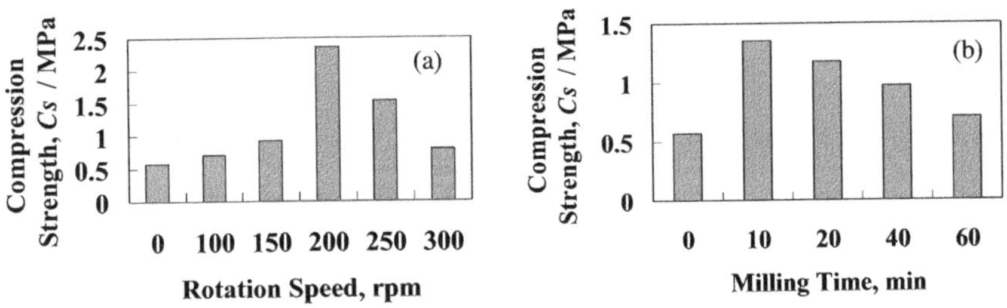


Fig.7 Compression strength of specimens;

(a) different rotation speed, (b) different milling time

CONCLUSION

Non-firing ceramics from paper sludge ash were successfully fabricated by mechano-chemically treating the raw powder with an alkaline-chemical reaction process. The mechano-chemical treatment in a planetary ball mill affects the fabrication of non-firing ceramics when the rotational speed and treatment time are varied. The alkaline solution was used as a binder between the ceramic particles, forming a strong green body at room temperature. The chemical reaction depends on several factors, such as the particle size distribution and the mineral composition of paper sludge ash, the kinds and concentration of the activator, reaction time, reaction temperature, etc. This process is an attractive route for shaping waste materials into valuable ceramics without a firing process.

ACKNOWLEDGEMENT

This research was partially supported by the Ministry of Education, Culture, Sports, Science and Technology (MEXT), Project "Cooperation for Innovative Technology and Advanced Research in Evolution Area".

REFERENCES

[1]H. Eroglu, O. Ucuncu, H. H. Acar, "The effect of sludge addition supplied from pulp mill on the compressive strength of cement", J. Uni. Chem. Tech. and Met., 42, 169-174 (2007).
[2]S. P. Mun, B. J. Ahn, "Chemical conversion of paper sludge incineration ash into synthetic zeolite", J. Ind. Eng. Chem., 7, 292-298 (2001).
[3]Y. Kim, J.H. Kim, K.G. Lee, S.G. Kang, "Recycling of dust wastes as lightweight aggregates", J. Ceram. Proc. Res., 6, 91-94 (2005).
[4]K. Okada, Y. Onoa, Y. Kameshima, A. Nakajima, K. J.D. MacKenzie, "Simultaneous uptake of ammonium and phosphate ions by compounds prepared from paper sludge ash", J. Hazard. Mat., 141, 622-629 (2007).
[5]K. Vyshnyakova, G. Yushin, L. Pereselentseva and Yury Gegotsi, "Formation of porous SiC ceramics by pyrolysis of wood impregnated with silica", J. Appl. Ceram. Tech., 3, 485-490 (2006).
[6]P. Parhi, V. Manivannan, "Microwave metathetic approach for synthesis and characterization of $ZnCr_2O_4$" J. Eur. Ceram. Soc., 28, 1665-1670 (2008).
[7]Y.X. Pang and X. Bao, "Influence of temperature, ripening time and calcinations on the morphology and crystallinity of hydroxyapatite nanoparticles", J. Eur. Ceram. Soc., 23, 1697-1704 (2003).
[8]Z. Zhihui, L. Nan, "Influence of mechanical activation of $Al_2O_3$ on synthesis of magnesium aluminate spinel", Sci. Sinter., 36, 73-79 (2004).
[9]P. Jonsen, H-A. Haggblad, K. Sommer, "Tensile strength and fracture energy of pressed metal powder by diameter compression test", Powd. Tech., 176, 148-155 (2007).
[10] W.K.W. Lee, J.S.J. van Deventer, "Chemical interactions between siliceous aggregates and low-Ca alkali-activated cements", Cem. Con Res., 37, 844-855 (2007).

# DISPERSION BEHAVIOR OF CARBON BLACK NANOPARTICLES WITH A SILANE COUPLING AGENT IN AN ORGANIC SOLVENT

Kenji Takebayashi, Yoshinori Tsuji and Shuji Sasabe
Hosokawa Micron Corporation
1-9 Shoudai, Tajika, Hirakata, Osaka, Japan 573-1132

Motoyuki Iijima and Hidehiro Kamiya
Tokyo University of Agriculture and Technology
2-24-16 Koganei, Tokyo, Japan 184-8588

## ABSTRACT

The dispersion behavior of carbon black (CB) nanoparticles in N-Methylpyrrolidone (NMP) was controlled using a silane coupling agent. A silane coupling agent (Phenyl-triethoxysilane: PTES) with a phenyl group was used to modify the surface of CB powder, and the surface treated CB powder was dispersed by bead milling in NMP. The dispersion behavior was characterized by measuring the aggregate size distribution, rheological behavior, and adsorption ratio of PTES. Through mechanical dispersion using agitation milling with small beads (30μm diameter), the aggregate size distribution of surface treated CB nanoparticles in NMP was reduced to several tens of nanometers. The optimum amount of PTES to add was determined from the measured average diameter of aggregates, and the rheological behavior of the suspension. A Newtonian CB/NMP suspension with a small aggregation size was successfully obtained with a PTES content of 1.7 to 4.8 $\mu mol/m^2$.

## INTRODUCTION

Carbon black (CB) is widely used as a conductive polymer filler, a reinforcing agent in rubber, a pigment, and in printing inks[1,2]. For these many applications, it is necessary to prepare a stable colloidal suspension. Since CB tends to aggregate due to its high specific surface area, various kinds of surface modifications have been tried on CB. For example, surface oxidation[3], oxide coatings, and adsorbed polymer dispersants[4] have been used for dispersion control of CB in aqueous media. However, such methods result in damage below the surface of the CB, and it is difficult to control the thickness of a surface oxidized layer. Alternatively, silane coupling agents are well known surface modifiers for dispersing oxide particles. The purpose of a silane coupling treatment is to improve the dispersion stability of hydrophilic particles in organic solvents and polymers[5,6]. There are also studies of using silane coupling agents on hydrophobic particles such as CB[7,8]. However, only a few studies have examined the dispersion and adsorption behavior of silane coupling agent for the surface treatments on CB particles.

In order to disperse the hard aggregates of nanoparticles in liquid, some mechanical processing methods have been developed, including agitation milling with small beads, ultrasonication[9], and wet type jet mill. In the bead milling process, the mean size of the beads is down to 30 μm in diameter. Recently, agitation milling with such small beads was reported to disperse $TiO_2$ and other oxide nanoparticles in aqueous media[10]. Surface modification of CB with a silane coupling agent using agitation milling could be a promising process to redisperse aggregated.

In this study, the adsorption and reaction behavior of a silane coupling agent with a phenyl group on CB, and the effect of the surface modification on the behavior of CB in an organic solvent were investigated. The effects of bead size and the amount of silane coupling agent additive on dispersion behavior of CB in a NMP suspension were investigated. The dispersion behavior of CB in NMP was characterized using aggregate size distribution, rheological behavior, and adsorption ratio of the silane coupling agent.

## EXPERIMENTAL PROCEDURE

### Preparation of surface treated CB/NMP suspension

CB nanoparticles (Tokai Carbon Co., Ltd. #2150, mean diameter: 25 nm, specific surface area (BET method): 220 $m^2/g$) and Phenyl-triethoxysilane (PTES, KBE-103, Shin-Etsu Chemical Co., Ltd.) were used as the raw material and silane coupling agent, respectively. After putting the PTES coupling agent in an acid aqueous solution with 2 wt% of CB, the mixed suspension was stirred at room temperature for 24 H, and dried at 423 K for 24 H. The amount of PTES added per unit surface area of CB was varied between 0.8 and 12.9 $\mu mol/m^2$. Surface treated CB powder was dispersed in N-Methylpyrrolidone (NMP, Kanto Chemical Co., Ltd.) using a bead mill (MSC-50, Nippon Coke & Engineering Co., Ltd.). The solid mass fraction of modified CB in the NMP suspension was fixed at 5.0 wt%. The circumferential velocity of the agitating rotor, and the circulated suspension volume flow rate were fixed at 10 m/s and 200 ml/min, respectively. $ZrO_2$ beads of 30, 100, and 300 μm average diameter were used for milling. The apparent packing density of beads was maintained at 55 % of the vessel volume.

### Characterization of surface treated CB/NMP suspension

The CB particle aggregate size distribution in the NMP suspension was measured by dynamic light scattering (DLS, FPAIR-1000, Otsuka Electronics Co., Ltd.). A macroscopic evaluation of the aggregation was determined from the dispersion behavior of the CB particles in the NMP suspension measured using a cone-plate type rheometer (MCR101, Anton Paar) at a shear rate ranging from 0.01 to 1000 $s^{-1}$ at 298 K. Particle morphology was observed by transmission electron microscopy (TEM, JEM-2010, JEOL Ltd.). The amount of surface adsorbed and chemisorbed silane coupling agent on CB was determined using the following process. Modified CB before milling and CB/NMP suspension after milling were washed by rinsing with toluene, and were separated by centrifugation in order to remove weakly adsorbed coupling agent. This process was conducted two times. The washed CB nanoparticles with the remaining strongly adsorbed coupling agent were dried under vacuum at 393 K for 24 H. To analyze the amount of adsorbed PTES on CB, the Si (silica) content on CB surface was measured using a fixed quantity method of X-ray Fluorescence Analysis (XRF, ZSX Primus, Rigaku Co.).

## RESULTS AND DISCUSSION

### The effect of bead size on the stability of CB in NMP

The effect of the bead size in the bead mill on the aggregate size distribution of the treated CB in NMP suspension was investigated. The PTES content and the processing time were fixed at 4.8 $\mu mol/m^2$ and 1 H, respectively. Fig.1 shows the aggregate size distribution of CB in NMP after milling. The aggregate size distribution of the treated CB in NMP suspension shifted to smaller size with decreasing $ZrO_2$ bead size. The aggregate size at 10, 50 and 90 % cumulative frequency is shown Table I. It is found that the aggregate size of CB at 10, 50 and 90 % of cumulative frequency reduced about 30 % when 30 μm beads were used compared to 300 μm beads. It is expected that smaller microscopic beads have larger chances to contact the CB aggregates with high collision frequencies. Thus, a highly dispersed CB/NMP suspension was obtained by using microscopic beads.

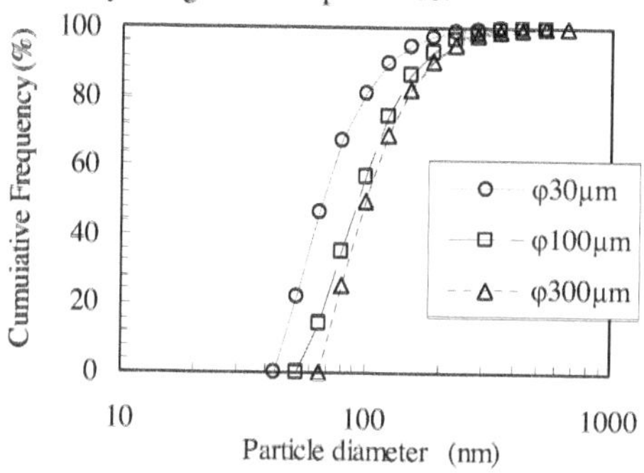


Table I. The particle size at 10, 50 and 90 % cumulative frequency

| Bead size | $d_{10}$ (nm) | $d_{50}$ (nm) | $d_{90}$ (nm) |
|---|---|---|---|
| φ30μm | 48 | 68 | 126 |
| φ100μm | 62 | 94 | 171 |
| φ300μm | 71 | 100 | 186 |

Figure 1. The aggregate particle diameter distribution of CB in NMP after milling with different beads sizes.

The effect of coupling agent additive content on the stability of CB in NMP

The effect of the PTES additive content on the stability of CB/NMP suspension after milling process was examined using DLS. The PTES content ranged from 0.8 to 12.9 μmol/m$^2$. CB was dispersed into NMP by bead milling with 30 μm $ZrO_2$ beads. The average aggregate size as a function of PTES content is shown in Fig. 2. The average aggregate size of CB was 180 nm when the PTES content was 0.8 μmol/m$^2$, and decreased drastically to below 85 nm as the PTES content was increased to 1.7-4.8 μmol/m$^2$. The minimum diameter was obtained when the PTES content was 4.8 μmol/m$^2$. A slight increase of the average diameter was observed when the additive content was increased to an excess amount.

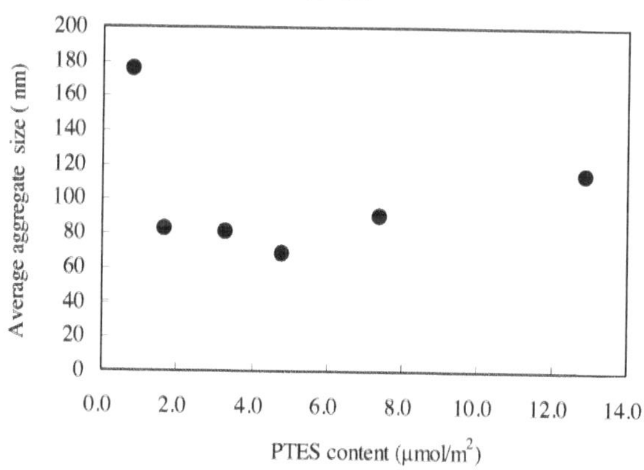


Figure 2. Average aggregate size of CB/NMP suspension prepared with different PTES content.

Fig.3 shows TEM images of the non-treated and the treated CB with an addition of 0.8 and 3.3 μmol/m$^2$ of PTES. Large aggregates were observed in the non-treated CB, and the

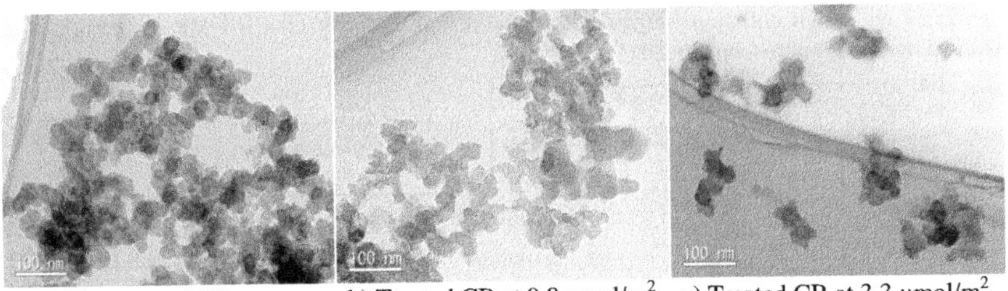


a) Non-treated CB b) Treated CB at 0.8 μmol/m$^2$ c) Treated CB at 3.3 μmol/m$^2$

Figure 3. TEM images of the non-treated and the treated CB in NMP.

aggregate size was reduced after bead milling with PTES. Similarly to the DLS results, it is found that the aggregate size was slightly reduced when the PTES additive content was 0.8 μmol/m$^2$ and was more effectively reduced when the PTES content was 3.3 μmol/m$^2$.

The flow properties and the apparent viscosity at shear rate of 0.1 and 1000 s$^{-1}$ for the CB/NMP suspension treated with different amounts of PTES are shown in Fig. 4 and Table II, respectively. When the PTES content was 0.8, 7.4, and 12.9 μmol/m$^2$, the viscosity changed only slightly compared to the NMP suspension with raw CB. The apparent viscosity was high at low shear rate and low at high shear rate. A shear thinning behavior was observed in these samples. However, when the PTES content was between 1.7 and 4.8 μmol/m$^2$, the viscosity of the CB/NMP suspension significantly decreased, and the value of apparent viscosity was same at low and high shear rate. The CB/NMP suspension prepared by the optimum PTES additive condition possessed Newtonian behavior.

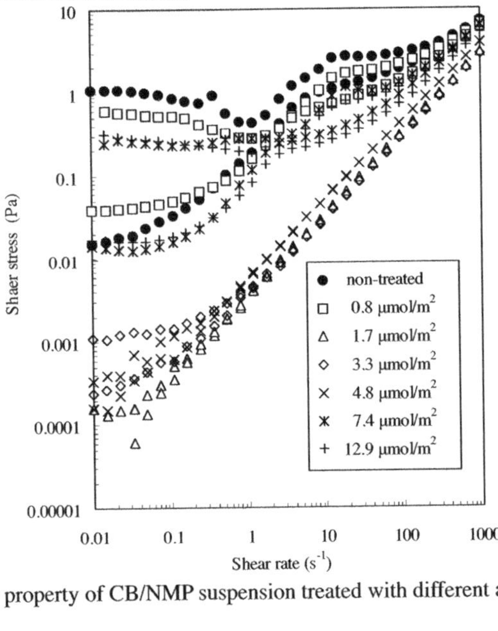


Figure 4. Flow property of CB/NMP suspension treated with different amounts of PTES

Table II. Apparent viscosity of CB/NMP suspensions

| PTES content ($\mu mol/m^2$) | Shear rate at $0.1s^{-1}$ Viscosity (mPas) | Shear rate at $1000s^{-1}$ Viscosity (mPas) |
|---|---|---|
| non-treated | $7.9\times10^3$ | 7.6 |
| 0.8 | $4.8\times10^3$ | 6.7 |
| 1.7 | 4.7 | 3.0 |
| 3.3 | 5.3 | 2.8 |
| 4.8 | 5.7 | 3.9 |
| 7.4 | $2.1\times10^3$ | 6.4 |
| 12.9 | $2.3\times10^3$ | 5.3 |

Evaluation of the adsorption behavior and the role of PTES

The amount of adsorbed PTES per unit surface area of CB for different amounts of PTES before and after milling is shown in Fig.5. For PTES modified CB without milling, the adsorbed PTES on CB increased with PTES content. Thus, it is expected that PTES formed a multilayer on CB before milling. In contrast, for the CB bead milled in NMP and further washed with toluene, the adsorption of PTES on the CB surface showed a saturation value near 0.8 $\mu mol/m^2$. It is expected that the weakly adsorbed PTES was detached during the milling procedure and that only the PTES strongly adsorbed on the CB by π-π interactions[11] of phenyl ring and CB was measured.

From the aforementioned DLS, viscosity, and adsorption measurements, the effect of PTES on the stability of CB in NMP with milling is described as follows. When CB was treated with a low (0.8 $\mu mol/m^2$) PTES additive content, the aggregation occurred between the remaining non treated CB particles. With an excess (7.4, 12.9 $\mu mol/m^2$) of PTES, a bridging effect between particles occurred due to the increase non-adsorbed PTES, which also decreased the stability of CB in NMP. By optimizing the PTES content, which was between 1.7 and 4.8 $\mu mol/m^2$, the suitable amount of PTES adsorbed on the CB contributed to the steric stabilization of CB particles, and improved their stability in NMP.

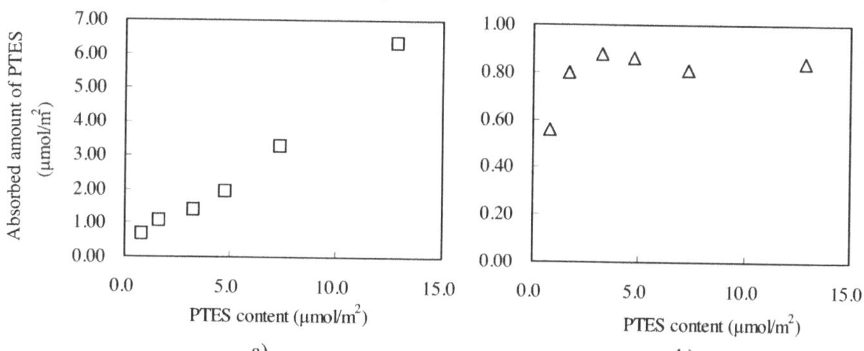


Figure 5. Adsorption behavior of PTES on CB surface a) before and b) after milling.

## CONCLUSION

A CB/NMP suspension with a small aggregate size distribution was prepared by the surface modification of CB with PTES and by further bead milling. The effect of milling bead size and the PTES additive content were investigated. It was found that a smaller aggregate size distribution of CB can be obtained by using smaller beads (30μm) instead of the larger beads (100, 300μm). Furthermore, the PTES content affected the dispersion stability of CB in NMP after milling. A small CB aggregation size and CB/NMP suspension with Newtonian behavior is successfully obtained when the PTES content is 1.7 to 4.8 $\mu mol/m^2$. In contrast, the suspension shows shear thinning behavior with large CB aggregate size distribution when the CB is not modified or the CB is treated with a small (0.8 $\mu mol/m^2$) or excess (7.4, 12.9 $\mu mol/m^2$) amount of PTES. The optimum amount of PTES adsorbed on CB contributes to the steric stabilization of CB particles, and improves their stability in NMP.

## ACKNOWLEDGEMENT

This study was partially supported by Nanotech challenge program from New Energy Development Organization (NEDO) of Japan.

## REFERENCES

[1]K. Miyasaka, K. Watanabe, E. Jojima, H. Aida, M. Sumita and K. Ishikawa, "Electrical conductivity of carbon-polymer composites as a function of carbon black", *J. Mater. Sci.*, **17**, 1610-1616 (1982)

[2]D. C. Edwards, "Polymer-filler interactions in rubber reinforcement", *J. Mater. Sci.*, **25**, 4175-4185 (1982)

[3]M. V. Lopez-Ramon, F. Stoeckli, C. Moreno-Castilla and F. Carrasco-Marin, "On the characterization of acidic and basic surface sites on carbons by various techniques", *Carbon*, **37**, 1215-1221 (1991)

[4]Y. Lin, T. W. Smith and P. Alexandridis, "Adsorption of a siloxane surfactant on carbon black particles dispersed in mixtures of water with polar organic solvents", *J. Colloid Interface Sci.*, **255**, 1-9 (2002)

[5]M. Iijima, M. Tsukada and H. Kamiya, "Effect of surface interaction of silica nanoparticles modified by silane couoling agents on viscosity of methylethylketone suspension", *J. Colloid Interface Sci.*, **305**, 315-323 (2007)

[6]X. Q. Xie, S. V. Ranada and A. T. Dibenedetto, "A solid state NMR study of polycarbonate oligomer grafted onto the surface of amorphous silica", *polymer*, **40**, 6297-6307 (1999)

[7]M. Inaba, Y. Kimura, K. Nagata, K. Inomata and K. Takahashi, "Electrical resistivity of PP/PE composite filled with amino-silane treated carbon black and its thermal behavior", *J. Soc. Reol. Jpn.*, **32**, 129-132 (2004)

[8]T. Amari, K. Koseki and K. Sasaki, "Nonlinear rheological properties of suspension of surface treated carbon black", *J. Soc. Reol. Jpn.*, **76**, 55-62 (2003)

[9]K. Sato, J. -G. Li, H. Kamiya and T. Ishigaki, "Ultrasonic dispersion of $TiO_2$ nanoparticles in aqueous suspension", *J. Am. Ceram. Soc.*, **91**, 2481-2487 (2008)

[10]M. Inkyo, T. Tahara, T. Iwaki, F. Iskandar, C. -J. Hogan Jr. and K. Okuyama, "Experimental investigation of nanoparticle dispersion by beads milling with centrifugal bead separation", *J. Colloid Interface Sci.*, **304**, 535-540 (2006)

[11]R. Coughlin and F. S. Ezra, "Role of surface acidity in the adsorption of organic pollutants on the surface carbon", *Environ Sci. Technol.*, **2**, 291-297 (1968)

MORPHOLOGICAL CHANGE DURING DRYING IN THE SYNTHESIS OF ZnO MICROTUBES FROM AN AQUEOUS SOLUTION

Seiji Yamashita, H. Watanabe, T. Shirai, M. Fuji, and M. Takahashi.
Ceramics Research Laboratory, Nagoya Institute of Technology
Honmachi 3-101-1, Tajimi, Gifu 507-0033, Japan

ABSTRACT

We have successfully synthesized hexagonal-faceted zinc oxide (ZnO) microparticles using a novel simple method that includes the introduction of ammonia bubbles into a $ZnCl_2$ aqueous solution to form white precipitates, followed by filtration and drying of the precipitates to produce ZnO particles. The amount of distilled water or filtrate added to the precipitates before drying determines the morphology of the ZnO particle produced. ZnO microsheets 10 μm in diameter and 0.5 μm thick, and micro- rods or tubes 1~3 μm in length and 0.5 μm in diameter can be synthesized by controlling the amount of water added before drying. In this work, in order to study the transformation of the ZnO during the drying process, the crystalline phase and morphology of the precipitate precursors to the ZnO particles have been investigated using XRD analysis and SEM observation.

INTRODUCTION

Zinc oxide (ZnO) is a semiconductor with a wide energy band gap (3.37 eV) and a large excitation binding energy (about 60 meV) that has extensive applications, including as a varistor, light emitting diode, solar cell, catalyst, sensor and cantilever[1]. Recently, the synthesis of nano- or micro- ZnO particles with various shapes have been reported, such as rod-like[2], flower-like[3,4], tower-like[4], tubular[2] and microbox[5]. These particles have peculiar physical properties derived from their characteristic shapes. For example, a well-controlling array of ZnO rod or tower of tube-like particle on a substrate such as silicon, sapphire ($\alpha$-$Al_2O_3$) or ITO (indium tin-oxide) is expected to have application as a photoelectric device and gas sensor element because of its peculiar light-scattering property and high- specific area. The synthesis of these ZnO particles have been demonstrated by vapor phase methods such as thermal evaporation[6], metal organic chemical vapor deposition (MOCVD)[7] and molecular beam epitaxy (MBE)[8]. Recently, aqueous solution methods have been widely investigated for the synthesis of ZnO particles with various shapes. Aqueous solution methods have advantages such as being simple operations, and having high cost-efficiency; hence, it would be suitable for large-scale synthesis as compared to the vapor phase methods. The previous work includes use of preseeded substrates[9], surfactant agents such as polyethylene glycol[10] and cetyltrimethylammonium bromide (CTAB)[11], and complexing agents such as citric acid[12].

We have recently developed a new and simple method to synthesize single-crystalline ZnO microtubes, in which ammonia bubbles are introduced into a $ZnCl_2$ aqueous solution to form a white precipitate, followed by filtration and drying to produce ZnO particles[13-15]. The advantages of this new method are that the synthesis reaction occurs at a relative low temperature and ambient pressure, and there no need for any template or surfactant as compared to other previous methods. In our previous work, the hexagonal layer-like $Zn_x(OH)_yO_z$-$H_2O$ precursor was formed by reaction of the $ZnCl_2$ aqueous solution and ammonia in the first step. Subsequently, the precursor was dried at 90°C to form ZnO microtubes. According to the previous study, the dissolution of the precursor, and re-precipitation to form ZnO particles might occur during the drying process, and be responsible for the change in morphology of the ZnO particle[16]. However, the important factor and the mechanism for the formation of the tubular structure has not been understood yet; hence, control of the ZnO shape and size has not been well established. In this study, we have investigated the effect of the existence of water and/or other chemicals in the precipitate before drying, which seem to have significant influence on the morphological change of the ZnO particle. Therefore, the amount of water or filtrate added to the precipitate to be dried was changed, and the subsequent morphological change in the ZnO particles was investigated using scanning electron microscopy. We also investigated the change in the crystalline phase of the precipitate to ZnO during drying using XRD analysis.

## EXPERIMENTAL PROCEDURE

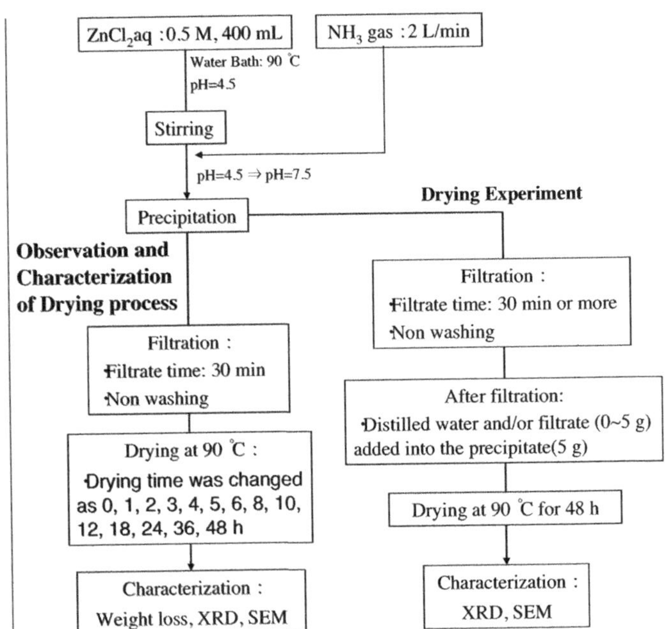


Figure 1: Flowchart of the experimental procedure.

Figure 1 shows a flowchart of the experimental procedure. An Aqueous solution of $ZnCl_2$ (0.5 mol/L) was prepared in a conical glass beaker by adding 27.26 g of $ZnCl_2$ (Wako Pure Chemical Industries Ltd.) into 400 mL distilled water and then was heated up to 90°C while stirring with a magnetic bar. The pH of $ZnCl_2$ solution was 4.5. Ammonia gas (2 L/min) was introduced into the $ZnCl_2$ solution until the pH reached 7.5 to form white precipitate. The precipitate was separated from the solution by suction filtration with a membrane filter for 30 minutes, and then was dried at 90°C. The crystal structure, morphology, and weight loss of the sample at each drying time were characterized by a powder X-ray diffraction (XRD, RINT 1000, Rigaku) and a field emission scanning electron microscope (FE-SEM, JSM-7000FO, JEOL). In order to investigate the effects of the drying conditions on the morphology of the ZnO particles, filtrate or distilled water (0, 1, 2, 3, 4, 5 g) was added to the precipitate (5 g). The precipitate was synthesized at a final pH=7.5, and the suction filtration time was adjusted 30 minutes or more to minimize the solution remaining within the precipitate. The morphology, size, and crystal structure of the obtained dry particles was characterized by SEM and XRD

RESULTS AND DISCUSSION

Figures 2 shows SEM images of samples after different drying times. Sample (a) in Figure 2 (drying time at 90°C: 0h) indicates hexagonal layered particles with a diameter of 50 μm or more. When the drying time increase to 2 and 10 h (Sample (b), (c) in Figure 1), the hexagonal layer particles transformed to tubular or rod-like particle. In Sample (d) (drying time: 24 h), the morphology of the particles is mostly tubes or rods, and is not very different from sample (c).

Figures 2: SEM images of samples prepared after different drying times (a) 0 h, (b) 2 h, (c) 10 h, (d) 24 h at 90°C.

Figure 3 (a) shows XRD patterns of the samples after different drying times, and Figure 3 (b) shows the intensity ratio calculated from the XRD peaks of Figure 3 (a), and the weight loss during the drying process. From Figure 3 (a), the precursor $Zn_x(OH)_yO_z$-$H_2O$ peak appears in the sample after 0 h in drying at 90°C, and the ZnO peak appears in the sample after1 h of drying time. From Figure 3 (b), the intensity ratio of the precursor decreases to 0.2 after drying for 2 h. But it increases again after drying for 2 to 10 h. In contrast, the intensity ratio of the ZnO decreases again after 2 to 10 h. After drying for 10 h, significant changes are not seen. The weight of the precipitate during the drying process in Figure 3 (b) decreases -46 % after 10 h, but does not change much after that. The behavior of the crystalline phase and weight of the sample after 2 to 10 hours indicates that dissolution and re-precipitation from the precursor that is a hexagonal layered $Zn_x(OH)_yO_z$-$H_2O$ particle occurs during the drying process. And these make a change in the morphology of the particle from a hexagonal layered particle to a tubular or rod-like particle.

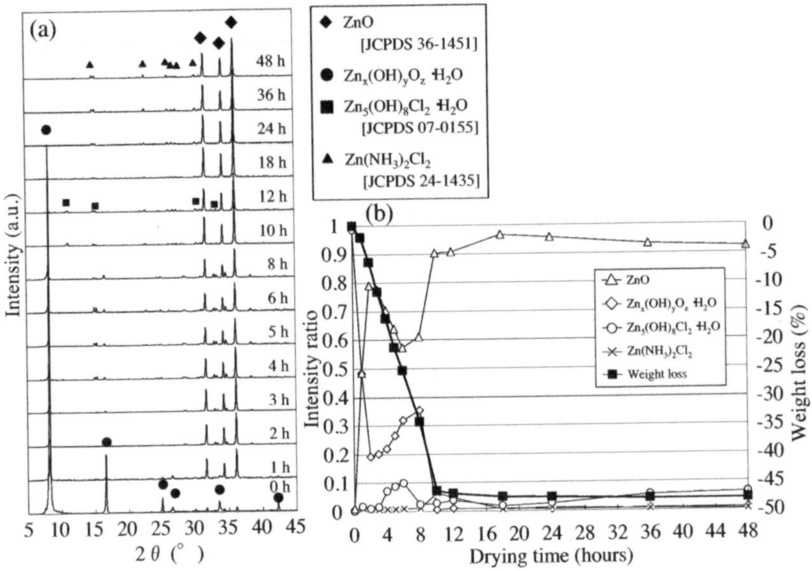


Figures 3 (a) XRD patterns and (b) Intensity ratio calculated from XRD peaks of samples prepared after different drying times.

Figures 4 shows the SEM images and XRD patterns of particles obtained after the drying experiment in which different amounts of filtrate (0, 1, 2, 3, 4 or 5 g) were added to the precipitate. Sample (a) and (b) (0 or 1 g of filtrate added) are layer-like particles with a diameter of around 10 μm or more. When the amount of filtrate added increases to 2, 3, 4 g ((c), (d) and (e) in Figure 4), the particle shape becomes tubular or rod-like with a diameter of around 0.5 μm and a length of around 1 μm. In contrast, sample (f) in Figure 4 (5 g of filtrate added) includes many irregular shaped particles. From the XRD patterns in Figure 4, the diffraction peaks of ZnO appear in all of the samples. But other phase peaks like $Zn_5(OH)_8Cl_2$-$H_2O$ and $Zn(NH_3)_2Cl_2$ increase with increasing amount of filtrate added. The morphology of particles changes to tubular or rod-like from a layer-like particle with increasing amount of additional filtrate.

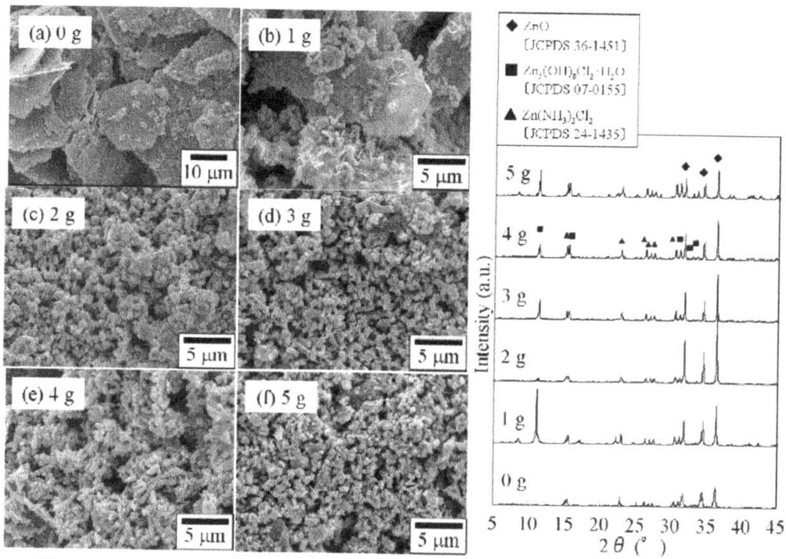


Figures 4: SEM images and XRD patterns of particles obtained after the drying experiments with different amounts of filtrate. (a) 0, (b) 1, (c) 2, (d) 3, (e) 4, (f) 5 g of the filtrate added to the 5 g of the precipitate prior to drying, respectively.

In order to see the effect of the ions in the filtrate, drying experiments were carried out with the addition of distilled water instead of the filtrate. In Figure 5, the sample (a) (no distilled water added) is a layer-like particle with a diameter of around 10 μm. Sample (b) in Figure 5 (1 g of distilled water added) is a tubular or rod like particle with a diameter of around 500 nm and a length of around 1 μm which is similar to the samples (c) and (d) in Figure 4. As the amount of distilled water added increase, the length of the tubes or rods can be seen to increase in the samples (c) and (d) in Figure 5. In sample (e) and (f) in Figure 5, no more growth in the length direction can be seen. In sample (d) in Figure 5, some tubular ZnO particles with a hollow structure can be observed. In the XRD patterns as shown in Figure 5, ZnO peaks appear in all of the samples, and the amount of distilled water added resulted in no significant changes to the XRD patterns.

Based on the drying experiments with additional filtrate (Figure 4) or distilled water (Figure 5), tubular ZnO particle can be prepared by adding a smaller amount of distilled water (1 g) as shown in Figure.5 (b) rather than filtrate (2 or 3 g) as shown in Figure 4 (c) and (d). The difference between the filtrate and the distilled water should be related to the presence of ions. The filtrate contains some kinds of ions such as $NH_4^+$, $Cl^-$, $[Zn(NH_3)_4]^{2+}$, $[Zn(OH)_4]^{2-}$ since the reaction yield based on the initial $ZnCl_2$ is around 60%. This would suggest that dissolution of the precipitate as a precursor and re-deposition of ZnO may occur during the drying process due to the presence of the additional water, and the concentration of the ions might affect the final morphology of the ZnO particle in the dry state.

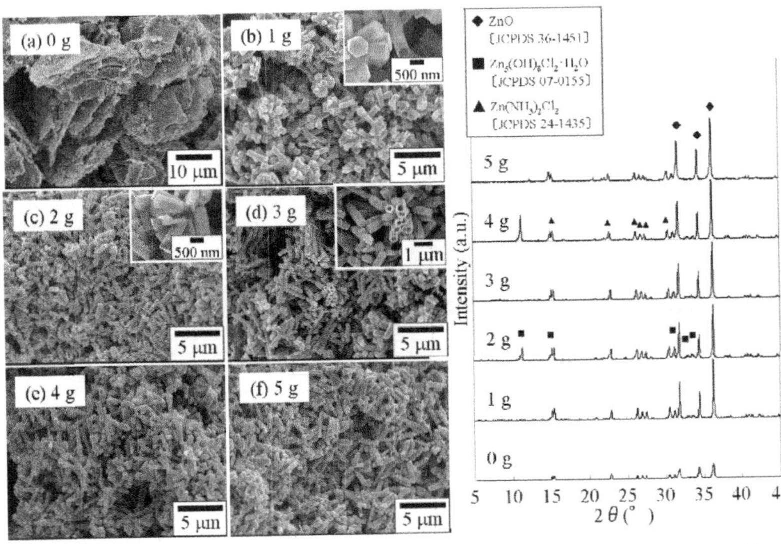


Figures 5. SEM images and XRD patterns of particles obtained after drying experiments with different amounts of distilled water. (a) 0, (b) 1, (c) 2, (d) 3, (e) 4, (f) 5 g of the distilled water added to the 5 g of the precipitate prior to drying, respectively.

## CONCLUSIONS

In summary, dissolution and re-precipitation occurs during the drying process and then the morphology of the particles is transformed from a hexagonal layered precursor to a tubular or rod-like ZnO particle. This is experimentally supported by change in the weight loss and the crystal phase during the drying process. And the shape of ZnO particle can be controlled from layer-like to a tube or rod by increasing the amount of additional filtrate or distilled water to the precipitate prior to drying. ZnO microtubes with a diameter of around 500 nm and length of around 1~3 µm were synthesized by adding filtrate or distilled water in the amount of 40 or 20 mass % of the precipitate to be dried. The length of the tubes or rods increased with increasing the amount of distilled water added. It has been found that water and/or ions might play an important role to change the shape of ZnO particle during the drying process, and the addition of distilled water or filtrate prior to drying is useful to get microtubes selectively.

REFERENCES

[1]L. Schmidt-Mende and J. L. MacManus-Driscoll, "ZnO – nanostructures, defects, and devices," *Science*, **10** [5] 40-48 (2007)

[2]Q. Li, V. Kumar, Y. Li, H. Zhang, T. J. Marks, and R. P. H. Chang, "Fabrication of ZnO Nanorods and Nanotubes in Aqueous Solutions" *Chem. Mater.*, **17** 1001-1006 (2005)

[3]J. Zhang, L. Sun, J. Yin, H. Su, C. Liao and C. Yan, "Control of ZnO Morphology via a Simple Solution Route" *Chem. Mater.*, **14** 4172-4177 (2002)

[4]Z. Wang, X-f Qian, J. Yin and Z-k Zhu, "Large-Scale Fabrication of Tower-like, Flower-like, and Tube-like ZnO Arrays by a Simple Chemical Solution Route" *Langmuir*, **20** 3441-3448 (2004)

[5]R-C Wang, C-P Liua and J-L Huang, "ZnO hexagonal microboxes enclosed only by {0001} facets with epitaxial nanowalls" *Appl. Phys. Lett.*, **89** 173121-3 (2006)

[6]S. L. Mensah and V. K. Kayastha, "Formation of single crystalline ZnO nanotubes without catalysts and templates" *Appl. Phys. Lett.*, **90** 113108-3 (2007)

[7]B. P. Zhang, N. T. Binh, K. Wakatsuki, and Y. Segawa, "Formation of highly aligned ZnO tubes on sapphire {0001} substrates" *Appl. Phys. Lett.*, **84** [20] 4098-4100 (2004)

[8]Liang, H. W, Lu, Y. M, Shen D. Z, Li B. H, Zhang Z. Z, Shan C. X, Zhang J. Y, Fan X. W, and Du G. T, "Growth of vertically aligned single crystal ZnO nanotubes by plasma-molecular beam epitaxy" *Solid State Comm.*, 137(4), 182-186(2006)

[9]S. P. Garcia, and S. Semancik, "Controlling the Morphology of Zinc Oxide Nanorods Crystallized from Aqueous Solutions: The Effect of Crystal Growth Modifiers on Aspect Ratio" *Chem. Mater.* 19. 16. **2007**

[10]J. Liu and X. Huang, "A Low-temperature synthesis of ultraviolet-light-emitting ZnO nanotubes and tubular whiskers" *Journal of Solid State Chemistry* 179 843–848 (2006)

[11]L. Shen, N. Bao, K. Yanagisawa, K. Domen, C. A. Grimes and A. Gupta, "Organic Molecule-Assisted Hydrothermal Self-Assembly of Size-Controled Tubular ZnO Nanostructure" *J. Phys. Chem. C.*, **111** 7280-7287 (2007)

[12]S. P. Garcia and S. Semancik, "Controlling the Morphology of Zinc Oxide Nanorods Crystallized from Aqueous Solutions: The Effect of Crystal Growth Modifiers on Aspect Ratio" *Chem. Mater.*, Vol. 19, No. 16, (2007)

[13]Y. S. Han, L. W. Lin, M. Fuji, and M. Takahashi, "A Novel One-step Solution Approach to Synthesize Tubular ZnO Nanostructures" *Chem. Lett.*, Vol.**36** No.8 (2007)

[14]L. Lin, Y. Han, M. Fuji, T. Endo, H. Watanabe and M. Takahashi, "A Facile Method to Synthesize ZnO Tubes by Involving Ammonia Bubbles" *Ceramic Transactions*, **198** 269-274 (2007)

[15]L. Lin, Y. Han, M. Fuji, T. Endo, X. Wang and M. Takahashi, "A Novel One-step Solution Approach to Synthesize Tubular ZnO Nanostructures" *J. Ceram. Soc. Jpn.*, **116** [2] 168-200 (2008)

[16]L. Lin, H. Watanabe, M. Fuji and M. Takahashi, "Morphological Control of ZnO particles synthesized via a New and Facile Aqueous Solution Route, *Advance Powder Technology*, **20** [2], (2009).

# PREPARATION OF $ZnO/Al_2O_3$ NANOCOMPOSITE PARTICLES BY MECHANICAL TREATMENT

Satoko Tasaki[1], Junichi Tatami[1], Hiromi Nakan0, Toru Wakihara[1], Katsutoshi Komeya[1] and Takeshi Meguro[1]

1, Graduate School of Environment and Information Sciences, Yokohama National University, Yokohama 240-8501

2, Toyohashi University of Technology, Toyohashi, Aichi 441-8580

## ABSTRACT

ZnO ceramics have excellent electrical properties that make it useful as a varistor, sensor, piezoelectric device, and so on. In this study, nanocoposite particles of ZnO/nano $Al_2O_3$ prepared by mechanical treatment were used to fabricate ZnO ceramics. TEM observations show that the ZnO and nano $Al_2O_3$ particles are directly joined in the powder mixture prepared by mechanical treatment. The relative density of the ZnO ceramics fabricated from the nano-composite particles prepared by mechanical treatment was about 98.9%, which is as high as that of ZnO ceramics fabricated from the powder mixture prepared ball milling.

## INTRODUCTION

ZnO is a typical electrical device. Taking advantage of the particular properties of the ZnO ceramics, it has application as a varistors[1], sensors[2], thermoelectric transducer[3], transparent conductive film[4], light emitting diode[5], and so on. The electrical properties of ZnO ceramics are controlled by small amounts of additives. Since $Al_2O_3$ dissolves in ZnO, the carrier concentration can be controlled (controlled valency); therefore, Al doped ZnO improves the electrical conductivity [6,7].

Many researchers have fabricated the ZnO ceramics by solid state reaction using mixing powders prepared by ball milling. In order to make better ZnO ceramics, a homogeneous dispersion of the additives is one of the most important factors in the powder processing. Homogeneous mixing of additives can be achieved by using nanoparticles as the raw material in comparison to larger particles. However, the use of a wet milling process with a liquid dispersion media, such as water and ethanol, might lead to reagglomeration of nanoparticles during drying. Consequently, the amount of mixed material obtained might not be enough to take advantage of the characteristics of the nanoparticles. Therefore, we have focused on the fabrication of ceramics by mechanical treatment, in which nanoparticles are mixed only by the dry mixing process. By using this mechanical treatment process, nanocomposite particles can be obtained; in this process, nanoparticles are bonded to submicron particles by applying external mechanical force, especially shear force.

We have reported that $TiO_2$ particles can be directly bonded to $Si_3N_4$ particle by mechanical treatment, making it is possible to fabricate nanoparticle dispersion $Si_3N_4$ ceramics [8]. In other materials, it has been shown that the mechanical treatment process improves the properties of the ceramics[9,10]. In this study, we developed Al doped ZnO ceramics using nanocomposite particles composed of $\gamma$-$Al_2O_3$ nanoparticles and ZnO submicron particles prepared by mechanical treatment.

## EXPERIMENTAL AND PROCEDURE

ZnO powder (grain size: 0.2–0.6 μm, JIS 1, Hakusui tech Ltd.) and $\gamma$-$Al_2O_3$ powder (grain size: 7 nm, TM-300, TAIMEI CHEMICALS Co., Ltd.) were used in this study. The amount of $\gamma$-$Al_2O_3$ used was 0.1, 1.0, and 3.7mol%. These powders were premixed for 30 min by dry ball milling. Milling was carried out using $Al_2O_3$ balls of $\phi$10 mm. The premixed powder was mechanically treated for 10 min using a powder composer (Nobilta NOB-130, Hosokawa Micron Co., Japan) working at 2000 rpm for 10 minutes to prepare composite powders. For reference, raw material powders of the same compositions were prepared by wet ball milling in ethanol using an $Al_2O_3$ ball.

In order to make granules, paraffin (4mass%, melting point: 46–48°C, Junsei Chemical Co., Japan), dioctyl phthalate (DOP, 2mass%, Wako Pure Chem. Ind., Ltd., Japan), and cyclohexane (S grade, Wako Pure Chem. Ind., Ltd., Japan) were added to these mixed powders followed by sieving the powder mixture using a 250μm-mesh nylon sieve. They were then molded into cylindrical pellets by uniaxial pressing at 50 MPa followed by cold isostatic pressing at 200 MPa. After the binder was burned out in air at 500°C for 3 h, the green bodies were fired at 1400°C in air for 2 h.

The mechanically treated powders were observed using a transmission electron microscope (TEM, JEM-3000F, JEOL, Japan), and their specific surface area was measured by the BET method using a surface area analyzer (Quantasorb, Quantachrome Instruments, USA).

## RESULTS AND DISCUSSIONS

Figure 1 shows TEM images of the $ZnO$-$Al_2O_3$ powder mixture after mechanical treatment and wet ball milling. The angular pieces of grains in Figure 1 (A) and (C) are the ZnO grains. Smaller particles existed on the surface of the ZnO particles (Figure.1 (B)). Figures 1 (B) and (D) are enlarged TEM photographs. The size of the smaller particles on the ZnO particles is less than 10 nm, which is almost the same as that of the $\gamma$-$Al_2O_3$ used in this study (Fig.1 (B)). Therefore, it seems that $\gamma$-$Al_2O_3$ nanoparticles were directly bonded to a ZnO particle. On the other hand, there is also TEM observation of agglomerated $\gamma$-$Al_2O_3$ nanoparticles (Fig.1 (D)) where the $\gamma$-$Al_2O_3$ nanoparticles don't exist on the ZnO surface.

Figure 2 shows the XRD profiles of the $ZnO$-$Al_2O_3$ powder before and after mechanical treatment, and the raw materials. The characterization of the composite powder was carried out by X-ray diffraction using Cokα radiation. It was observed that the phases present in the composite powder consisted of only ZnO and $\gamma$-$Al_2O_3$, which is the same as the raw materials. Thus, we inferred that $\gamma$-$Al_2O_3$ had bonded to the ZnO grain surface due to mechanical treatment. Abe et al. reported on the formation of $SiO_2$ inter-nanoparticle bonds and the mechanical process used to fabricate the $SiO_2$ nanoparticles and $SiO_2$ fiber composites[9]. Tatami et al. reported that $TiO_2$

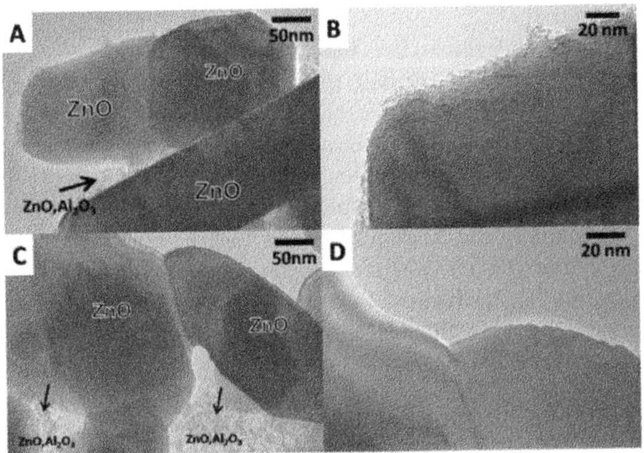

Figure 1. TEM images of the composite particle prepared by mechanical treatment. (0.37mol%$Al_2O_3$-ZnO): (A)Powder mixture after mechanical treatment; (B) Powder mixture after wet ball milling; (C) and (D) enlarged views

nanoparticles directly bond to submicron $Si_3N_4$ particles by mechanical treatment[8]. In this experiment, we were also able to establish the phenomenon reported above. Nogi et al. estimated the temperature at the particle interface in shear fields[11]. To apply these results to our study, the temperature of the particle interface must be maintained below 700°C. It is known that sintering shrinkage of ZnO starts at about 600°C[12]. It should be noted that $\gamma$-$Al_2O_3$ is phase transformed to $\alpha$-$Al_2O_3$ with an increase in grain growth with increasing temperature. Therefore, the temperature of the $\gamma$-$Al_2O_3$ and ZnO particles themselves should be very low, and only the temperature of the interface between ZnO and $\gamma$-$Al_2O_3$ should be slightly higher temperature due to the sheer force.

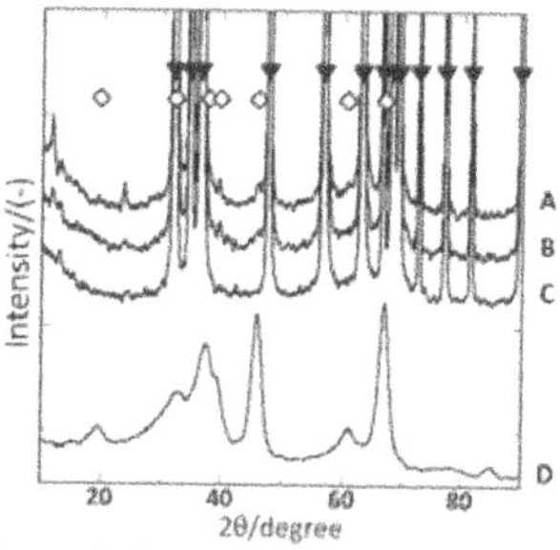

Figure 2. XRD pattern of powders of: A, Mixed powder before mechanical treatment; B, After treatment; C, ZnO powder; D, $\gamma$-$Al_2O_3$ powder. ▼ : ZnO, ◇ : $\gamma$-$Al_2O_3$,

Table 1 lists the specific surface area of the $ZnO$-$Al_2O_3$ powder measured by the BET method after mechanical treatment. For reference, the specific surface area of raw materials ZnO and $Al_2O_3$ is also listed in the table. The specific surface area of the ZnO and $\gamma$-$Al_2O_3$ was 4.2 $m^2/g$ and 196 $m^2/g$, respectively. When the $ZnO{:}Al_2O_3$ ratio was 1:0.037, the specific surface area was 12.1 $m^2/g$ before mechanical treatment, while it decreased to 10.3 $m^2/g$ after mechanical treatment. When 1.0mol% and 0.1mol% of $Al_2O_3$ was added, the specific surface area also changed from 6.7 $m^2/g$ to 5.5 $m^2/g$ and from 4.6 $m^2/g$ to 4.2 $m^2/g$, respectively. The above decrease in the specific surface area is assumed to be due to the formation of bonding area between ZnO particles and $\gamma$-$Al_2O_3$ during mechanical treatment.

Table.1, Change in the specific surface area for each sample.

| sample | specific surface($m^2/g$) | | Estimated value($m^2/g$) |
|---|---|---|---|
| ,-$Al_2O_3$ | 196 | | |
| ZnO | 4.2 | | |
| | mechanical treatment | | |
| | **before** | **after** | |
| 3.7mol%$Al_2O_3$ | 12.1 | 10.3 | 9.8 |
| 1.0mol%$Al_2O_3$ | 6.7 | 5.5 | 5.8 |
| 0.1mol%$Al_2O_3$ | 4.6 | 4.2 | 4.4 |

In order to analyze the change in the specific surface area, we used a geometric model, as shown in Figure 3. In this model, spherical submicron particles (ZnO) with a diameter of *D* and cubic nanoparticles ($\gamma$-$Al_2O_3$) with an edge of *L*. The specific surface area $S_0$ before mechanical treatment is estimated as follows:

$$S_0 = S_{submicron}(1-f)+S_{nano}f \qquad (1)$$

where $f$ is the weight fraction of the nanoparticle, $S_{submicron}$ and $S_{nano}$ are specific surface areas of the submicron-sized particle and the nanosized particle, respectively. If a cubic-shaped nanoparticle of a phase bonds to the spherical submicron particle, the total surface area should reduce by $2L^2$. When all of the nanoparticles bond on the submicron particles, the decrease in surface area $\Delta S$ per unit of weight is as follows:

$$\Delta S = \frac{2L^2}{6L^2}S_{nano}f = \frac{1}{3}S_{nano}f \qquad (2)$$

Hence, the specific surface area after bonding, S, is estimated as follows

$$S = S_{submicron}(1-f)+\frac{2}{3}S_{nano}f = S_{submicron} + \left(\frac{2}{3}S_{nano} - S_{submicron}\right)f \qquad (3)$$

The values of the specific surface area after mechanical treatment calculated from Eq. (3) are also listed in Table 1. The calculated values of the specific surface area are a good match to the experimental values. Thus, it is believed that the specific surface area decreased due to bonding between the ZnO and $\gamma$-$Al_2O_3$ particles during mechanical treatment, and almost totally all of the $\gamma$-$Al_2O_3$ added was bonded on the ZnO particles. These results agree well with that obtained by TEM analysis.

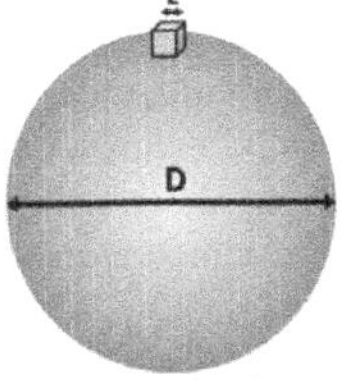

Figure.3. Schematic of composite powder geometry model

CONCLUSIONS

We fabricated ZnO/nano-$Al_2O_3$ composite particles by mechanical treatment. After mechanical treatment, $\gamma$-$Al_2O_3$ nanoparticles were directly bonded on ZnO particles. The ZnO ceramics fabricated by using ZnO/nano-$Al_2O_3$ composite particles had uniform and fine microstructure.

REFERENCES

[1] T. Ogasawara, R. Tanaka, K. Sugawara and T. Sugawara, "Effect of the Additives on the Nonlinear Coefficient of ZnO Varistor", *J. Ceram. Soc. Japan,* **108** [6] 565-570 (2000).

[2] J. Xu, Q. Pan, Y. Shun and Z. Tian., "Grain size control and gas sensing properties of ZnO gas sensor." *Sens. Actuators B*, **66** 277-279 (2000).

[3] M. Ohtaki, T. Tsubota, K. Eguchi and H. Arai, "High-temperature thermoelectric properties of $(Zn_{1-x}Al_x)O$", *J. Appl. Phys.*, **79** [3] 1816-1818 (1996).

[4] M. Nakamura, K. Ikuta, T. Aoki, A. Suzuki, T. Matsushita and M. Okuda, "Ultra Thin Al-doped Transparent Conducting Zinc Oxide Films Fabricated by Pulsed Laser Deposition Method", *J.Vac. Soc. Jpn.*, **51** [5] 323-325 (2008).

[5] A. Tsukazaki, M. Kubota, A. Ohtomo, T. Onuma, K. Ohtani, H. Ohno, S. F. Chichibu and M. Kawasaki, "Blue Light-Emitting Diode Based on ZnO", *Jpn. J. Appl. Phys.*, **44** [21] L643-L645 (2005).

[6] M. Takata, D. Tsubone and H. Yanagida, "Electrical Conductivity of Sintered Bodies of ZnO Doped with $Al_2O_3$", *yogyo-kyokai-shi*, **82** [5] 271-277 (1974).

[7] H. Kaga, Y. Kinemuchi, H. Yilmaz, K. Watari, H. Nakano, H. Nakano, S. Tanaka, A. Makiya, Z. Kato, K. Uematsu, "Orientation dependence of transport property and microstructural characterization of Al-doped ZnO ceramics", *Acta Materialia*, **55,** 4753–4757, (2007).

[8] J. Tatami, E. Kodama, H. Watanabe, H. Nakano, T. Wakihara, K. Komeya, T. Meguro and A. Azushima., "Fabrication and wear properties of TiN nanoparticle dispersed $Si_3N_4$ ceramics", *J. Ceram. Soc. Japan,* **116** [6] 749-754 (2008).

[9] H. Abe, I. Abe, K. Sato, and M. Naito, "Dry Powder Processing of Fibrous Fumed Silica Compacts for Thermal Insulation", *J. Am. Ceram. Soc.*, **88** [5] 1359–1361 (2005).

[10] M. Naito, A. Kondo and T. Yokoyama, "Applications of Comminution Techniques for the Surface Modification of Powder Materials", *ISIJ International.*, 33 [9] 915-924 (1993).

[11] K. Nogi, M. Naito, A. Kondo, A. Nakahira, K. Niihara and T. Yokoyama, "New Method for Elucidation of Temperature at the Interface Between Particles under Mechanical Stirring", *J. Jpn. Soc. Powder Metallurgy.*, **43**[3], 396-401 (1996).

[12] K.G. Ewsuk, D.T. Ellerby and C.B. DiAntonio, "Analysis of nanocrystalline and microcrystalline ZnO sintering using master sintering curves", *J. Am. Ceram. Soc.*, **89** [6] 2003–2009 (2006).

# Interface Characterization and Control

# THERMAL PROPERTIES OF MATERIALS AT INTERFACES FOR ELECTRONIC APPLICATION

Seongjin Hwang, Dongsun Kim, Hyungsun Kim
School of Materials Engineering, Inha University, 253 Younghyun-dong, Incheon metropolitan, Incheon, South Korea

## ABSTRACT

The effect of interfaces on material characteristics is of critical importance in electronic device applications because of reliability and performance requirements. The mismatch in the coefficient of thermal expansion (CTE) affects the fracture of modes at various interfaces between a glass substrate and in certain devices such as dielectrics, barrier ribs, and sealing materials in plasma display panels (PDPs) and solar cells. This paper presents results related to various interfaces in devices, and their effect on the performance of devices for electronic applications. In addition, we introduce a design for the material interface with suitable thermal properties for electronics.

## CURRENT ELECTRONIC DEVICES USING GLASS MATERIALS

Electronic devices such as information displays (liquid crystal displays, organic light emission displays, cathode ray tubes, plasma display panels (PDPs) and vacuum fluorescent displays), and solar cells (dye-sensitized solar cell, DSSC) need low melting glasses either for system materials or packaging materials. Material requirements are related to physical, mechanical, thermal, electrical, chemical and optical properties. The interface control between the glass substrate (or Si wafer) and glass is one of the main research areas in electronic applications because the physical and chemical properties of glasses are related to the reliability of devices. For instance, in Si-solar cells, the thermal properties of the frit are closely related to the electrical properties of the cells, the Ag crystalline growth, and the distribution of the crystalline Ag[1-2]. In PDPs, such properties include high transmittance for transparent dielectric, stable dielectric constant, chemical stability of the Ag electrode during firing, compatibility of the coefficient of thermal expansion (CTE) with the glass substrate, high resistance to humidity infiltration, durability of the ,white dielectric material to etching and a stable etching rate of rib material[3-4]. In addition to the intrinsic material properties, environmental releases and transfers of toxic chemicals from industrial facilities also need to be accounted for in order to develop good system performance.

Glass containing lead shows a high refractive index, low melting temperature and has been used as electronic materials and optical glass materials. Two decades ago, the most commonly used glass system in electronic components was the $PbO-ZnO-SiO_2-B_2O_3$ glasses for low firing temperature applications[5]. Now, with the impending introduction of the EU RoHS (European Restriction of Hazardous Substances) regulations in March 2010 for electronic applications[6], lead-free glass frits such as $Bi_2O_3$, $B_2O_3$ and $P_2O_5$ systems have been investigated as substitutes for lead glass[3-4]. However, lead-free compositions cause many dielectric problems, which are associated with electrodes in display units. The problems include the formation of bubbles at the interfaces between the glass substrate and dielectric, and between the dielectric and Ag powder, resulting in edge curling and a yellowing phenomenon[7]. Thus, interface control has become an important topic when a glass substrate is used for electronic devices.

During firing from room temperature to the maximum firing temperature, the phase of glass powders changes from solid to liquid through the processes of binder burn out and viscous flow. The glass sintering phenomenon is strongly dependent on the physical and thermal properties of glass powders, as shown in Figure 1. In this paper, the thermal properties of low melting glasses (bismate, borate, phosphate, and other glasses) such as surface tension, viscosity, CTE, glass transition temperature (Tg, $\eta=10^{13.3}$ dPa), dilatometric softening point (Tdsp, $\eta=10^{12}$ dPa), and Littleton softening point (Ts, $\eta=10^{7.6}$ dPa) are discussed for several glasses which have been worked in our laboratory over the last decade. Specifically, the relationship between CTE and Tg (Tdsp, Ts) for oxide glasses is introduced with some empirical equations for the prediction of the CTE and Ts of glasses.

## USAGE OF FRITS IN ELECTRONIC DEVICES AND PROBLEMS AT THE INTERFACES

### Screen-Printed Crystalline Silicon Solar Cells

A recent survey suggested that the efficiency of screen-printed silicon solar cells should be mainly decided by good quality front contacts[1-2]. Moreover, several studies have presented the model of current transport with the ohmic contacts as shown in Figure 2 (a) [1-2]. The possibility of

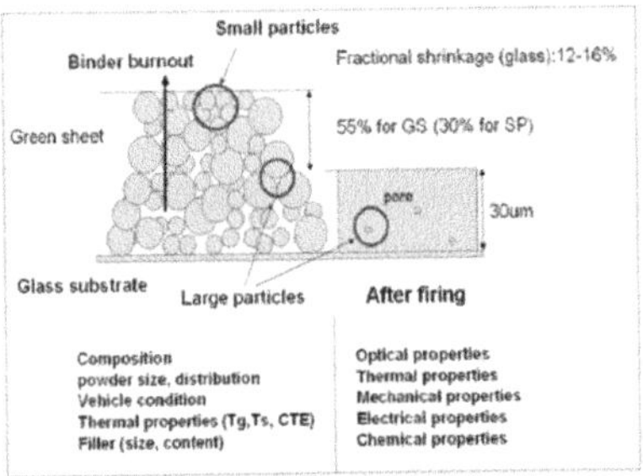


Figure 1. Diagram showing the factors affecting the glass sintering from glass powders (frits) to a sintered body on a glass substrate for the application of either the screen printing (SP) or green sheet (GS) process. With increasing firing time (temperature), the following occur: moisture removal, binder burnout and fusion (viscous flow).

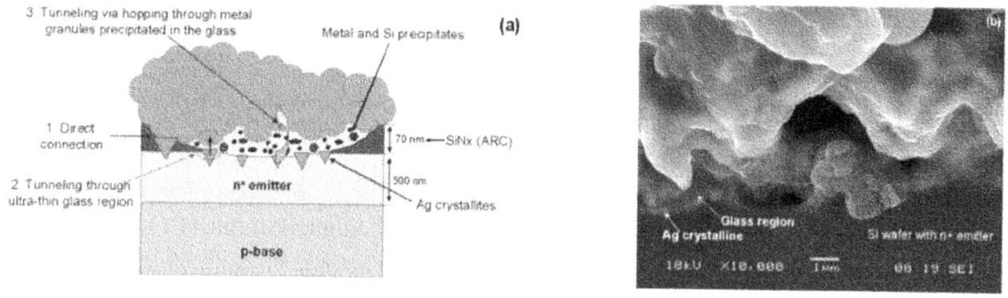


Figure 2. Diagram of the model of the current transport from the n+ emitter silicon to the Ag bulk. The three transport mechanisms: direct interconnection, tunneling through ultra-thin glass regions, and tunneling via discontinuous metal crystals in the glass layer (a) [1-2], and a SEM image if the interface between the Ag electrode and silicon wafer in a screen-printed silicon solar cell (b).

the current transport has generated wide interest in three methods: direct interconnection at isolated spots between the Ag crystallites and the Ag bulk, tunneling through ultra-thin glass regions, and conduction through the glass layer by tunneling via discontinuous metal crystals in the glass layer. Therefore, the ohmic contacts are achieved by the Ag paste, and the frit in the Ag paste plays a crucial role in the contact formation[1-2]. Figure 2 (b) shows the microstructure of the screen-printed, crystalline silicon solar cell.

Lead borosilicate glasses have mostly been used to fabricate screen-printed silicon solar cells by the rapid thermal process (RTP) furnace due to the low melting point (Mp) and a reaction with the silicon wafer. Recently, however, the use of lead oxide has been regulated by worldwide trade negotiations (RoHS, WEEE)[6]. Furthermore, few studies have investigated the frits in Ag pastes for solar cells and low-temperature applications (>600°C). The model of current transport with ohmic contact is based on the lead borosilicate glass in Ag paste for silicon solar cells[1-2]. Therefore, the thermophysical property of the frit is one of the important factors for optimizing the ohmic contact between Ag and the Si wafer.

Dye-Sensitized Solar Cells (DSSC)

The issue of DSSC has always been important for high efficiency, scale-up processes, and high reliability. Recently, some researchers have achieved efficiencies over 10% for small DSSC[9]. However, it is difficult to achieve an efficiency over 10% with large DSSC due to the resistance limitation of the transparent conductive oxide (TCO) glass[9], and white soda-lime-silica (SLS) glasses of low iron content are used, but also white borosilicate glasses are assembled[10]. Therefore, as shown in Figure 3, a silver grid was used as a carrier collector on the transparent electrode with a passivation layer or sealing materials for the DSSC module[11]. Several sealing materials have been used, such as epoxy glue, water glass (sodium silicate), the ionomer resin Surlyn®, aluminum foil laminated with polymer foil, the vacuum sealant Torr Seal®, and glass soldering[11]. Sealing DSSCs has long been a serious problem because of the corrosive and volatile liquid iodide electrolyte used in the cells. These are correlated to the long-term stability of the cells. The requirements of sealing materials are to be leak-proof to the electrolyte components, impermeable to both ambient oxygen and water vapor, and well adhered to the TCO glass.

Several recent studies showed a large area, dye solar module using a glass frit sealing with high stability[9,11]. However, they did not report the glass composition or glass system and only mentioned the glass frit's advantages (cost-effective application via SP, very stable thermal, chemical and mechanical properties) and disadvantages (high processing temperature of around 500°C, related to the deformation of the glass substrate). Moreover, they showed that the well-known low melting glasses, i.e., lead glasses, contaminated the platinum catalyst at the counter electrode of DSSC[9]. A patent has been issued for a sealing glass frit, phosphate glass system[12] and paper of a passivation layer, borosilicate glass[13]. Therefore, lead-free low melting glasses are essential as sealing materials to ensure the long-term stability of the cells.

Displays

Glass frits are used in many parts of PDPs, such as transparent dielectrics, white dielectrics, barrier ribs, sealants and electrodes. PbO-based glasses were widely used in commercial glass frit owing to their low firing temperatures (below 580°C), which are necessary to prevent the deformation of the glass substrate. However, these materials contain PbO (50~85%), which is deleterious to both health and the environment. Thus, ZnO, $Bi_2O_3$, $P_2O_5$, $V_2O_5$ and $B_2O_3$ have been widely used to develop low melting temperature, Pb-free glasses[3-4].

Although phosphate-based ($P_2O_5$) glass is a well-known substitute for lead glass, its practical use for frit glass has lagged far behind due to its poor water resistance and chemical stability. Other substances (bismuth oxide) have also been studied as substitutes for PbO. BaO-ZnO-$B_2O_3$ and BaO-ZnO-$B_2O_3$-$SiO_2$ systems are used because of their good hosting capability of refractory ceramic fillers, low softening temperature, comparable CTE to aluminum borosilicate glass panels such as PD200 ($83x10^{-7}$/K; Asahi Glass, Tokyo, Japan), low dielectric constant, and high optical reflectance[3-4].

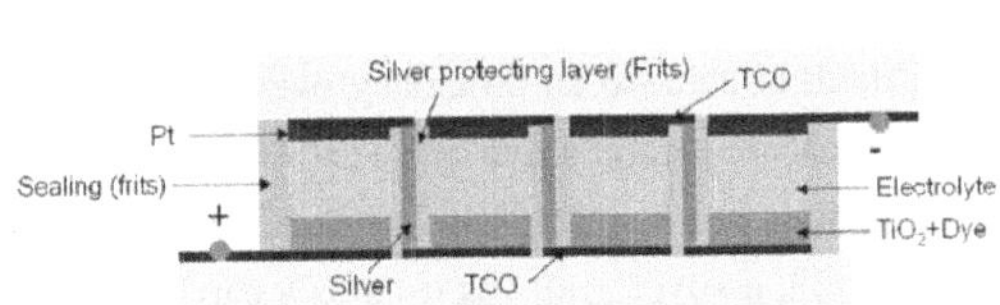


Figure 3. Schematic of the cross section of DSSC (Z type) with two glass plates.

Table I. Typical properties for glass powders for screen printing (SP) on glass substrates.

| Materials | Properties | Unit |
|---|---|---|
| Glass powder | Glass density | $g/cm^3$ |
| | Moisture content | % |
| | Particle size distribution | $d_{10}$, $d_{50}$, $d_{90}$, $d_{max}$ |
| | Dilatometric softening point ($T_{dsp}$) | °C |
| | Littleton softening point ($T_s$) | °C |
| | Glass transition temperature ($T_g$) | °C |
| | Coefficient of thermal expansion | x $10^{-7}$/K |
| | Fluidity at firing temperature | mm |

THERMAL PROPERTIES OF LOW FIRING TEMPERATURE GLASS

The most important factor affecting the glass sintering on glass substrates is the thermal properties such as CTE, Tg, Tdsp, Ts, thermal conductivity, surface tension, and viscosity. As shown in Figure 1, a thick layer on a glass substrate is controlled by the following factors: composition of glass powders, powder size, powder size distribution, vehicle condition or filler properties (size, types, content), and thermal properties of the glass powder. Regarding the paste composed of glass frits and vehicles, the shrinkage rate of the thick film is different from the GS or SP process methods owing to the vehicle content: a shrinkage rate for SP and GS processing of about 30 or 50%, respectively. Table I summarizes the thermal and physical properties of the glasses for application to electronic parts.

Relationship Between CTE and Tg

Figure 4 presents a number of glass systems in terms of the CTE and Tg of those glasses that are useful and reliable compositions in the range of 70-90 ppm/K for CTE and less than 500°C for Tg in electronic applications. The SLS glass substrates have a low firing temperature of less than 500°C when used for the forming thick films. The relationship between Tg and CTE gives CTE∝1/Tg, a constant value of CTE x Tg for each oxide glass. This is discussed in detail below.

Relationship Between Tg and Tdsp (Ts)

The empirical formulae Tdsp∝Tg and Ts∝Tg from our work in Figure 5a are close to the other formulae obtained based on the literature survey of Figure 5b[14]. The slope (*a, b*) from Table II must be interpreted carefully. Recently, Som and Karmakarw reported an empirical relation, Ts/Tg=1.5, for the $K_2O$-$B_2O_3$-$Sb_2O_3$-ZnO glasses[15]. However, this relation seems to be

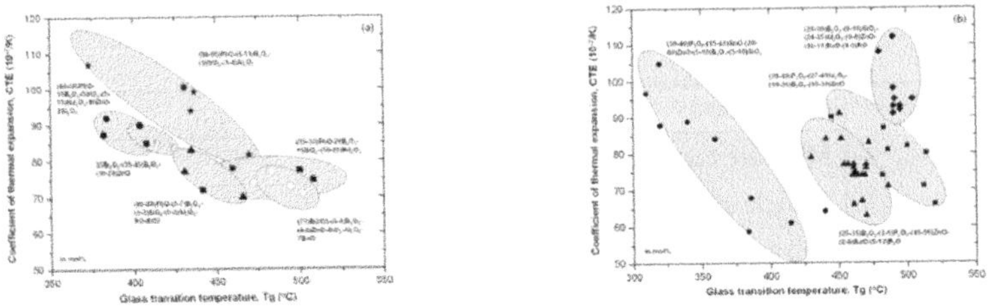


Figure 4. Relationship between CTE and Tg: (a) PbO (closed symbols) and $Bi_2O_3$ oxides (open symbols), and (b) non Pb and Bi oxides, based on our research works.

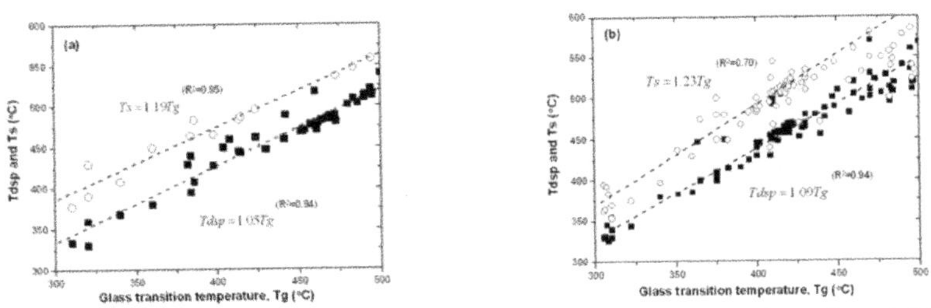


Figure 5. Relationship between Tg and Tdsp (Ts) from (a) our research work, and (b) the Intergrad 6.0's database (Tg ≤ 500°C and CTE < $120x10^{-7}$/K)[14].

only valid in the short range, 252<Tg< 273°C, and 279<Ts<351°C for the high antimony-containing glasses, which would not be applicable to other glasses. We propose that the *c* value (Ts/Tg=*c*) changes slightly from the low value of 1.23 at high Tgs to the high value of 1.24 at low Tgs, according to Fig. 5b and Table II. Thus, Tdsp/Tg and Ts/Tg give uncertain values that are close to 1.05~1.24, rather than a constant value in low firing temperature glasses.

Table II. Liner relationship between Tdsp (Ts) and Tg, as obtained from Figure 5.

| Tg ranges | Figure 5a | Figure 5b | Figure 5a | Figure 5b |
|---|---|---|---|---|
| (300°C-) | Tdsp = *a* Tg | | Ts = *b* Tg | |
| 400°C | *a* = 1.09 | *a* = 1.09 | *b* = 1.23 | *b* = 1.24 |
| $R^2$ | 0.88 | 0.92 | 0.83 | 0.68 |
| 450°C | *a* = 1.08 | *a* = 1.10 | *b* = 1.21 | *b* = 1.24 |
| $R^2$ | 0.91 | 0.95 | 0.89 | 0.72 |
| 500°C | *a* = 1.05 | *a* = 1.09 | *b* = 1.19 | *b* = 1.23 |
| $R^2$ | 0.94 | 0.94 | 0.95 | 0.70 |

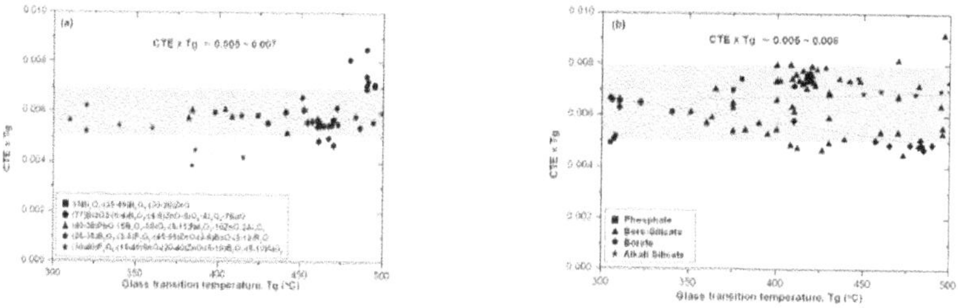


Figure 6. Relationship between (CTE x Tg) and Tg from (a) our database and (b) Intergrad 6.0's database (Tg ≤ 500°C and CTE < $120\times10^{-7}$/K)[14].

Relationship Between (CTE x Tg) and Tg

Low firing temperature glasses have a constant value of CTE x Tg below around 500°C of around 0.006, as shown in Figure 6a. According to Van Uitert *et al.*, the product, CTE x Mp on single oxides, nitrides and carbides, tends to be a constant for a given structure of 0.016 for close-packed structures and 0.027 for rectilinear arrays[16]. In this work, we use the product of CTE x Tg instead of Mp because the Tg is used in order to gain more reliable data compared to Mp, which is difficult to measure in glasses and Mp's data is quite limited in several glass databases. The physical and thermal properties of multi oxides, Tg, Tdsp and Ts of the oxides exhibited certain relationships, as shown in Figure 5. With increasing Tg of the glasses, product of CTE x Tg showed a slightly higher value (Figure 6b). The physical meaning of the high and lower values of the product will be reported in future studies.

## CONCLUSION

Glass sintering on a glass substrate is controlled by the interface reaction, which is itself strongly dependent on the physical and thermal properties of the glass powders used in electronic devices. From the thermal properties of low melting glasses such as CTE, Tg, Tdsp, Ts, the relationships between CTE and Tg, Tg and Ts (Tdsp) for oxide glasses are described with some empirical equations to predict the CTE of the glasses. The information to predict CTE or Ts using Tg will be useful in the design and development of electronic devices.

## ACKNOWLEDGEMENT

This work is the outcome of a Manpower Development Program for Energy & Resources supported by the Ministry of Knowledge and Economy (MKE).

## REFERENCES

[1] D. Kim, S. Shim, S. Hwang, and H. Kim, "Electrical Properties of Screen Printed Silicon Solar

Cell Dependent upon Thermophysical Behavior of Glass Frits in Ag Pastes", *Jpn. J. Appl. Phys.*, **48**, 05EC06-1-3 (2009).

[2] D. Margadonna and F. Ferrazza, "The Status of Crystalline Si Modules", *Renew. Energ.*, **15**, 83-8 (1998).

[3] E. Chong, S. Hwang, W. Sung, and H. Kim, "Dielectric Constant and Optical Reflectance of Ceramic Filler-Added $BaO$-$ZnO$-$B_2O_3$-$P_2O_5$ Glass Composites", *Int. J. Appl. Ceram. Tec.*, **6**, 295-301 (2009).

[4] R. Morena, "Phosphate Glass as Alternatives to Pb-Based Sealing Frits", *J. Non-Cryst. Solids.*, **263&264**, 382-7 (2000).

[5] A.K. Varshneya, S.C. Cherukuri, "Low Firing Temperature Glasses for Electronic Applications", In: K.M. Nair, Editors, *Glasses for electronic applications, ceramic transaction* **20**, The American Ceramic Society, Inc., Westerville, Ohio, 387-95 (1990).

[6] JEM News June 2009, *J. Environ. Monitor.*, **11**, 1127-34 (2009).

[7] S. Lee, M. Park, S. Hwang, and H. Kim, "Shape Change of Ag Electrode with Shrinkage Difference Between Electrode and Dielectric in PDP", *Displays*, **29**, 345-50 (2008).

[8] M.A. Green, K. Emery, Y. Hishikawa, and W. Warta, "Solar Cell Efficiency Tables" (Version 32), *Prog. Photovolt: Res. Appl.*, **16**, 435-40 (2008).

[9] R. Sastrawan, J. Beier, U. Belledin, S. Hemming, A. Hinsch, R. Kern, C. Vetter, F.M. Petrat, A. Prodi-Schwab, P. Lechner, and W. Hoffmann, "New Interdigital Design for Large Area Dye Solar Modules Using a Lead-free Glass Frit Sealing", *Prog. Photovolt: Res. Appl.*, **14**, 697-709 (2006).

[10] R.K. Brow and M.L. Schmitt, "A Survey of Energy and Environmental Applications of Glass", *J. Eur. Ceram. Soc.*, **29**, 1193-2019 (2009).

[11] J. Halme, Dye-Sensitized Nanostructured and Organic Photovoltaic Cells: Technical Review and Preliminary Tests, Helsinki University of Technology Master's thesis, 54-55 (2002).

[12] USA patents No. US 2009/0136766 A1.

[13] D.H. Yeon, E.Y. Lee, K.G. Kim, N.G. Park, and Y.S. Cho, "Zinc Borosilicate Thick Films as a Ag-Protective Layer for Dye-Sensitized Solar Cells", *J. Korean Ceram. Soc.*, **46**, 313-6 (2009).

[14] New Glass Forum, International Glass Database, INTERGLAS ver.6.0, http://www.ngf.or.jp/.

[15] T. Som and B. Karmakar, "Structure and Properties of Low Phonon Antimony Glasses in the $K_2O$-$B_2O_3$-$Sb_2O_3$-$ZnO$ System", *J. Am. Ceram. Soc.*, in press (2009), DOI: 10.1111/j.1551-2916.2009.03227.x.

[16] L.G. Van Uitert, H.M. O'Bryan, M.E. Lines, H.J. Guggenheim, and G. Zydzik, "Thermal Expansion - An Empirical Correlation", *Mater. Res. Bull.*, **12**, 261-8 (1977).

# GENERAL ANALYTICAL CONCEPT AND DESIGN METHODOLOGY TO PRODUCING A CLAY – BASED POLYMER NANOCOMPOSITE

Refat A. El-Sheikhy and Mosleh A. Al-Shamrani
Civil Eng. Dept., King Saud Univ., Saudi Arabia.

## ABSTRACT

Due to the natural properties of low cost, environmentally friendly clay minerals as nano-materials with a high aspect ratio and large surface area, clay platelets are very convenient filler and reinforcing materials in matrix materials such as polymers to produce advanced nano-composites with unique properties. Clay-based polymer nano-composites can have new and different properties that make them suitable for many different industrial and commercial applications, such as in automotive, medicine, electronics, sensors, construction, packaging, coatings, drugs, biomedical engineering, and others. Production of these nano-composite materials involves several steps of material preparation, mixing, and processing. Basically, the design of these nano-composites depends on a simple concept and the mechanism of strengthening a material with weak properties by adding a small amount of another material with strong properties. Reinforcing the main material (polymer) with a certain ratio (5%- 15%) of filler (nano-clay), will transform the polymer into an advanced functional material. The morphology, characterization, and analysis of these materials are based on that concept. There have been problems and difficulties in manufacturing these nano-composites. In the current study, a new point of view is taken involving a new concept, a new mechanism, new techniques for mixing and processing, and new methods of analysis and characterization. This study utilizes a new approach to understand and control the interfacial bond strength between the clay nano-fibers and the polymer, voids and inclusions, the de-bonding process between the fibers and the polymer, fiber orientation, fiber distribution and intensity, fracture, and cracks. This study uses natural Saudi nano-clay and polymers to solve the problems and difficulties with clay/polymer nano-composites that previous studies encountered.

## INTRODUCTION

Over the past 20 years, many researchers have studied nano-composites to develop high quality and high performance materials[1-12]. Basically, the design of these nano-composites depends on the simple concept and mechanism of strengthening of a material with weak properties by adding a small amount of another material with strong properties. Reinforcing the main material (polymer) with a certain ratio (5%- 15%) of filler material (nano-clay) will change the properties of main material (polymer) to an advanced material. The morphology, characterization, and analysis are based on this concept. Several problems and difficulties were faced in the manufacturing of these nano-composites.

Some materials such as nano-clay have layered structure of thickness of (1.0 nm) with aspect ratio of about (200-400) which can improve the fracture and mechanical properties of other materials[1-12]. The challenges in producing these advanced materials relates to the challenges in the field to improve techniques to mix the reinforcing nano-particles with the main material, and the processing of these nano-particles with the polymer to achieve the required nano-composite properties. Previous studies faced several problems in producing these clay/polymer nano-composites which made the manufacturing of these important and promising materials difficult to achieve with the required quality. Previous studies did not recognize all of the difficulties and problems. The current paper discusses the problems with the intent to find suitable solutions. In this paper a new concept, mechanism, method of analyses, and new processing and characterization techniques are introduced.

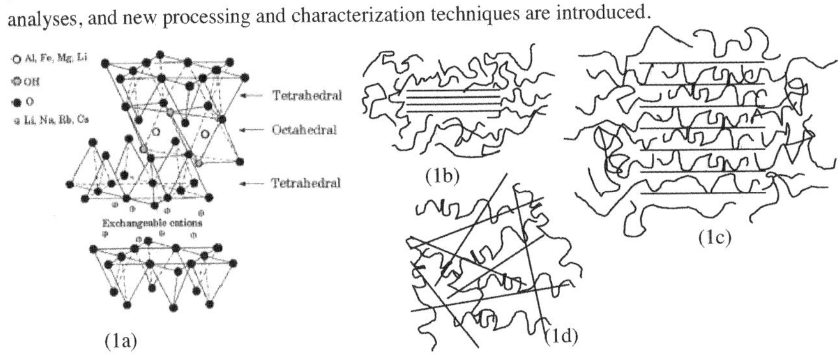


Figure 1: 1a. Structure of nano clay mineral[1-2], 1b. Dispersed agglomerated clay, 1c. Chemically exfoliated clay, and 1d. Mechanically exfoliated clay

PROBLEMS, DIFFICULTIES AND SOLUTIONS

The manufacturing of clay-based polymer nano-composites had several problems and difficulties which include: 1) Non uniform distribution of nano-clay in the polymer matrix; 2) Nano-clay dispersion in polymer; 3) Non homogeneity; 4) Uncontrolled orientation of clay platelets; 5) Agglomeration of the nano-powder into micro and macro particles producing stress concentration zones around their vicinities producing cracks; 6) Difficulty of exfoliation of the nano-clay layers; 7) Difficulty of intercalation and dispersion of the polymer matrix between the exfoliated nano-layers; 8) Weak interfacial bonding between the nano-clay layers and the polymer matrix; 9) Breaking of clay platelets. 10) De-bonding at the bi-interface between clay layers and the polymer matrix; 11) Producing nano, micro, and macro voids during processing; 12) Producing nano-cracks and nano-defects during the processing; 13) Nano-voids, nano-defects and nano-cracks that produce cracks and stress concentration that reduce lifetime and efficiency; 14) Difficulty in producing unidirectional, orthotropic or anisotropic nano-composites; and 15) No control over the distribution, uniformity, orientation, bond, dispersion, and exfoliation of the nano-clay layers.

Figure (1a)[1, 2] shows the structure of the clay platelet while Figure (1b) shows the dispersion of agglomerated clay particles in a polymer. Figure (1c) shows dispersion and chemical exfoliation of clay particles, and Figure (1d) shows mechanical exfoliation by mixing. Figures 2 and 3[1-3] show

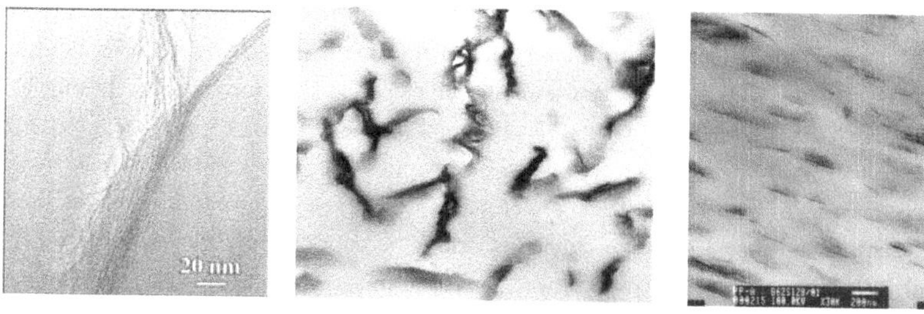


Figure 2. TEM micrographs of a Montimirillonite MMT – Polymer nano-composites[1-3].

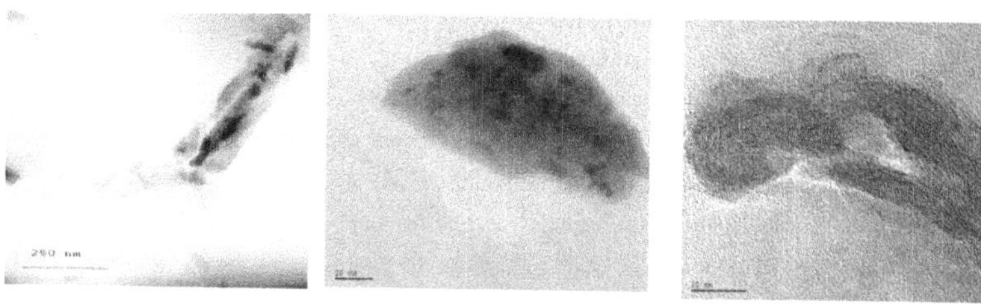

Figure 3. TEM micrographs of Nano-clay – Polymer nanocomposites[2].

TEM micrographs of dispersed nano-clay in polymers showing problems with distribution, orientation, agglomeration, etc.

As solutions to the dispersion and exfoliation problems, the clay must be swollen artificially or naturally. Synthesized clay such as Laponite clays should be treated with a certain swelling agent such as maleic anhydride polypropylene or maleic anhydride polypropylene to exfoliate the clay layers. Natural swelling clay which is difficult to produce, is better than synthesized clay. The current study takes advantage of the natural swelling of nano-clay which is a newly introduced material. Clay particles can be found in nano-size only when dissolved in water with a certain ratio. Otherwise it will be agglomerated into micro and macro sizes. This is one of the most important characteristics of clay.

Since clay minerals are inorganic materials, while polymer is organic, the interfacial bond between them is a big problem because of incompatibility. For treating this main problem, the clays can be converted to organic by creating a very thin layer of organic material on the clay particles surfaces. This can be done by submersing the clay particles in an organic solution such as Octadecylamine. After drying the clay particles, the clay (surface) will become organic and can form an interfacial contact bond with polymers.

To avoid the biggest difficulties with mixing and processing, such as clay particle agglomeration and exfoliation and dispersion, we proposed a new technique for mixing and extruding in addition to using swelling clay. Mechanical-chemical-mechanical exfoliation is possible by mechanically mixing swollen clay particles with polymer, then swelling of clay, and then mechanically mixing again. Exfoliation is achieved through dry mixing, swelling, and mixing again with a high shear mixer, then extruding, pelletizing, extruding again for several times using both single and twin extruder systems. After each mix and each extruding, some samples should be taken for characterization and testing. Strength, fracture toughness, fracture properties, crack locations, crack dimensions, voids and interfacial de-bonding cracks should be characterized rather than fiber orientation and distribution which should be characterized using SEM, TEM and XRD.

## CLAY-BASED POLYMER NANOCOMPOSITE APPROACH

The original and basic approach of producing clay-based polymer nano-composites depends on mixing the nano-material as reinforcement instead of using conventional fillers with the polymer matrix to produce a composite with new properties. Actually due to manufacturing difficulties, the results were not satisfactory because of producing materials with properties similar to the properties of the main material. This original approach was built on the basic theory of dispersion and exfoliation without taking into consideration other important aspects such as fracture, de-bonding, fiber orientation, and voids. A chemical method or agent can be used to exfoliate the nano-clay layers by swelling as in Figure (1c). The second method of exfoliation is mechanically by using a mechanical mixer, such as in Figure (1d). Mechanical high shear can help in the dispersion of the nano-clay in the polymer, but the chemical method is still better because it can be considered as a self dispersion and a self exfoliation process with natural materials.

## PROPOSED APPROACH AND CONCEPT BASED ON FRACTURE MECHANICS

The current approach addresses the difficulties of the previous studies with a new concept, the directional fracture approach DFA[13-17] and dilatational-distortional fracture energy theory[13-17]. Why can nano-particles improve the properties and quality of the matrix material of nano-composites? The answer is by improving fracture toughness, fracture properties, then reducing the failure probability, and extending the lifetime. This understanding of the nano-composite approach represents the foundation of our current study to overcome difficulties in producing a complete material with good mechanical properties, fracture properties, and a long lifetime, in addition to other properties such as gas barrier properties, impact, hardness, etc. This depends on some important procedures.

Material Selection and Preparation

Material selection should be made carefully for both of nano-clay and polymer matrix. In the current study, the materials are the natural clay (Saudi Arabian Montimorillonite MMT) shown in Figures 4-7, and domestic Saudi polymers (Polypropylene PP). The aspect ratio and surface area of the nano-clay platelets are important factors for producing good bonding. The most important factor in the preparation is materials compatibility. Good compatibility between the clay and the polymer is critical. This is because the polymer and clay are completely different. The polymer is an organic and hydrophobic material while the nano-clay is inorganic and hydrophilic. The polymer is a ductile thermoplastic material, while the clay is brittle and thermosetting. The clay surface should be treated to become organic using an organic compatible material such as (Maleic anhydride Polypropylene). Surface modification of MMT will make it compatible with the PP to facilitate interfacial bond. After that, the materials will be ready for the mixing and processing stages.

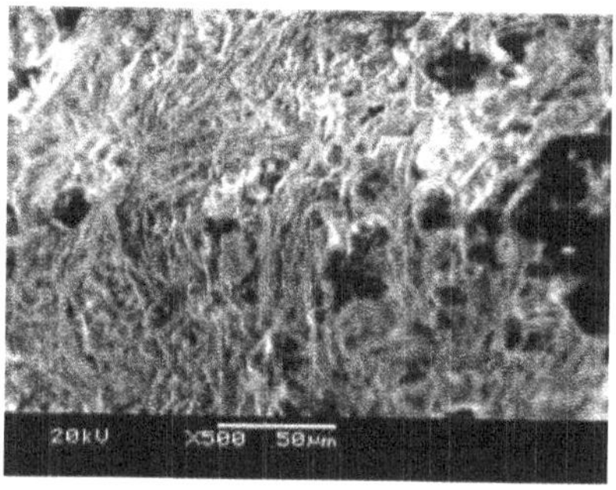


Figure 4. SEM micrograph of natural stone

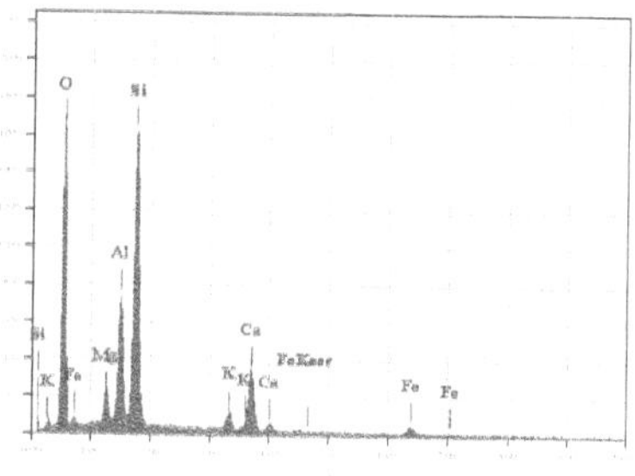


Figure 5. EDS analyses of natural clay stone

Mixing, Extruding, and Processing

After preparation of the nano-clay to have an organic surface, it will be mixed with the polymer using a dry mix method such as roll mill mixer several times until the composite is homogeneous. Then the composite will be mixed again using a high shear mixer. This will facilitate dispersing the clay particles in the polymer matrix through intercalation, exfoliation, and bonding between the clay and the polymer rather to produce a uniform and homogeneous distribution with minimum agglomeration, defects, and voids. This mixing technique should be repeated for several times.

After mixing for several times, the composite will be extruded using both a single extruder and a twin extruder with a length/diameter ratio of 40 and eight different zones of temperature including feed and die zones. These high shear extruders with heating will help in, mixing, exfoliation, distribution, and dispersion of the clay in the polymer. Extruder characteristics such as diameter, length/diameter ratio, angle of the teeth, feed speed and rate, extruder speed, time of extruding, and vacuum will control the processing. These conditions will control the production, the characteristics of the bonds, dispersion, and exfoliation to produce a uniform material. The final nano-composite produced after extruding will be prepared in the form of pellets.

Exfoliation of the nano-clay layers was achieved chemically by swelling as in Figure (1c) using a chemical agent. The second method of exfoliation uses a mechanical mixer as in Figure (1d). High shear can help in dispersing the nano-clay in the polymer, but the chemical method is still better because it enables self dispersion and self exfoliation when using natural materials. The third method (current method) uses the two methods together at the same time.

Figure 6. SEM micrograph of powder of natural Saudi Arabia agglomerated nano-clay fibers

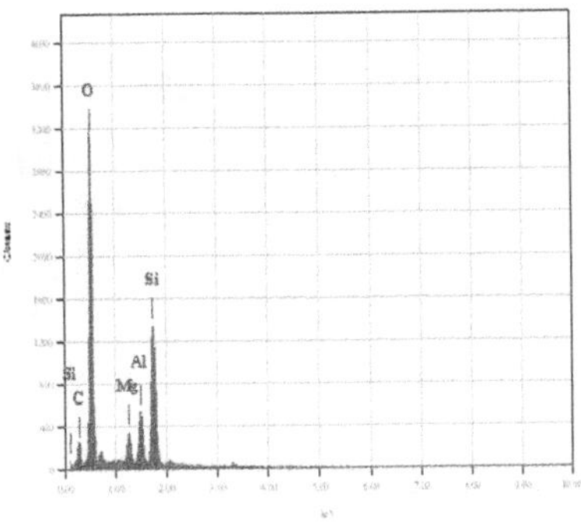


Figure 7. EDS analyses natural nano-powder

Design, Characterization, Analysis Methodology, and Applied Theory

Several mixes of clay/polymers should be made using several ratios of nano-clay powders such as 5%-15% of the polymer. Based on the applied analytical theory [3-17], good characterization, testing, and analysis of the product should be done. Using SEM, TEM & XRD techniques and testing for cracks, interfacial de-bonding, particle agglomeration, and voids, and checking the fracture toughness and fracture properties of the final material should be done as a basic measurement of nano-composite quality. Nano-particle distribution, orientation, exfoliation, dispersion, and agglomeration should all be evaluated.

The applied theory for this analysis is based on the directional fracture approach DFA using dilatational – distortional strain energy theory. While nano-particles are not well suited to being studied by continuum theory, in practice, the nano-composites, behave as a continuum, with consideration of the presence of the voids, cracks, fibers, etc. Therefore, the applied theory should follow the basics of a continuum approach. The theory is dilatational – distortional strain energy theory (MN-theory) based on the directional fracture approach DFA[13-17]. From the mechanical point of view, voids, cracks, defects, and de-bonding at interfaces between fibers and polymers will be produced during processing. The applied theory can deal with this material to investigate the defects and properties. Then, it will assist in improving the production performance, control the material production, and prevent each of the defects, cracks, and voids. The analysis should be done for the actual material, which includes cracks, defects, voids, and inclusions. The polymer itself may contain voids which are full of gases or air due to gases or absence of good vacuum. Voids will produce stress concentration and cracks that propagation to become large cracks that can produce failure or reduce the performance and lifetime. Some nano-particle filler reinforcements may produce inclusions due to the agglomeration of the nano-particles making micro or macro-particles resulting in problems in the nano-composite. If there is a de-bond or weak bond at the bi-interface between the polymer and the clay, it will produce cracks, then failures. Since there is no control of the material structure, it is very difficult to deal with a clay/polymer nano-composite as a regular composite with regular mechanical behavior such as a unidirectional material, orthotropic material, or anisotropic material. A clay/polymer nanocomposite with uniformly distributed nano-clay particles without agglomeration can be considered as an isotropic composite. Therefore, as a general approach, the application of the theory will in an isotropic nano-composite material.

DISCUSSION

Nano-clay (Montimorillonite MMT) as a filler material, and polymer (Polypropylene PP) as matrix are used to produce an advanced nano-composite material (Montimorillonite-Polypropylene MMT-PP). The MMT nano-clay as a filler or reinforcement has the advantages of a large surface area (a large surface area to mass ratio 800 $m^2$/gm compared to a traditional filler material such as conventional clay particles or talc powder), and enhanced mechanical, fracture, thermal, gas barrier, and impact properties. These improved properties can be realized with small loadings. The major difficulties include the difficulty in realizing exfoliation, interfacial bonding, and compatibility of the MMT particles and the PP. Exfoliation is one of the terms used to determine the extent of the dispersion of the clay in the polymer matrix. Montmorillonite MMT is smectite clay with a structure consisting of two tetrahedral layers of silica sandwiching one octahedral layer of aluminum (Figure (1a)). Nano-clay is a hydrophilic, inorganic, thermosetting, brittle material while the polymer is a hydrophobic, organic, and thermoplastic material. Therefore, it was essential to use a third material such as Maleic anhydride polypropylene, to achieve compatibility and bonding between these two different types of materials, and to achieve homogeneity by making the filler clay powder compatible with the polymer matrix. The properties of the polymer and the clay are different, and the clay-polymer nano-composite will produce another material with new properties. A surface treatment of the clay particles to make them organophilic will make the clay compatible with and bond strongly to the polymer. This will result in good mechanical properties since it will reduce the cracks and enhance the fracture properties. If the compatibility agent (surface treatment) could not produce this condition, the properties of pure polymer would be better than the nano-composite properties. If the compatibility agent could not cover all of the clay particles, defects will be created.

Natural Saudi Arabian swelling clay (MMT), as shown in Figures 4-7 was purified. It has some elements which produce swelling of the clay. The natural swelling facilitates the exfoliation and dispersion of clay in the PP, and facilitates intercalation of the PP between MMT platelets to produce a complete nano-composite material. The most important factors in manufacturing clay-polymer nano-composites that affect the properties of the composite are the homogeneous distribution of the nano-clay in the polymer matrix, clay powder size, clay powder shape, aspect ratio, surface area, agglomeration of the clay powder, dispersion of clay in the polymer matrix, exfoliation of the clay particles, dispersion of polymer between the clay particles, the bond between the clay and the polymer, the compatibility agent, the presence of defects, voids, inclusions, and cracks, temperature, the method of mixing, the shear effect during mixing, the method of extruding, the extruder deformed bar, extruding time, and the clay/polymer ratio. The orientation of nano-clay fibers affects the properties greatly. Shear mixing helps with the dispersion of the clay in the polymer, mechanical exfoliation of clay, and intercalation of polymer into the clay layers, but it may break fibers, break bonds and change the fiber orientation. Our new technique utilizes mixing, extruding, and natural swelling of the clay powder. The previous studies had no way to check for exfoliation of the nano-clay, while we use XRD analyses. There was no way to determine the uniformity of the distribution, while we use SEM or TEM. There was no way to characterize the bond between the fibers and polymers. Our approach, based on an understanding of fracture mechanics, provides a good record and assessment of these aspects, as well as the quality of the product by checking the fracture properties and cracks in addition to the other characterization methods.

## CONCLUSION

Traditional techniques for making clay/polymer nano-composites had problems with achieving consistent and quality products and optimal properties. The current research tried to find solutions to these problems by applying a new concept and methodology based on fracture mechanics. Applying the basics of the Directional fracture approach to understand and control the nano-composite structure is a step towards overcoming the problems in nano-composite manufacturing. New techniques for both of mixing and processing were developed. Another important point in this research involves the use of natural nano-clay to produce the clay/polymer nano-composites. Using natural nano-clay will save money, time, and efforts to produce high quality materials. Using domestic Saudi Arabian materials is one of the main factors in finding solutions to the problems encountered by previous researches.

ACKNOWLEDGMENTS

This work is a part of nano technology research project supported by Center of Excellence for Research in Engineering Materials CEREM, Faculty of Eng., King Saud Univ., Saudi Arabia.

REFERENCES

[1]F. Hussain, et. al, "Review article: Polymer-matrix Nano composites, Processing, Manufacturing, and Application: An Overview", *J. of Composite Material*, **Vol. 40, No. 17**, p.1511-1575, (2006)

[2]A. Bhaskar, Polymer silicate and magnetic polymer nanocomposites: processing and characterization, *M. Sc. thesis, University of Florida, USA*, (2003).

[3]Manuel DaSilva , Polymer-clay nanocomposites: the processing challenges of a naturally occurring reinforcement, *Purdue University*, USA, (2005)

[4]D. R. Paul, M. L. Robeson, "Polymer nanotechnology: Nanocomposites", *Polymer,* **Vol. 49**, p.3187-3204, (2008).

[5]Q. H. Zheng, et. al, "Clay–based polymer nanocomposites: Research and commercial Development", *Journal of Nanoscience and Nanotechnology*, **Vol. 5**, p.1574-1592, (2005).

[6]H. Zhang, et. al. "Fracture behavior of in situ Silica nanoparticle-filled epoxy at different temperature", *Polymer* **49**, 3816-3825, (2008).

[7]X. F. Yao, D. Zhou, H. Y. Yeh, "Macro/microscopic fracture characterizations of SiO2/epoxy nanocomposites", *Aerospace science and technology* **12**, 223- 230, (2008).

[8]L. Cui, X. Ma, D. R. Paul, "Morphology and properties of nanocomposites formed from ethylene-vinyl acetate copolymers and organoclays", *polymer* **48**, 6325-6339, (2007).

[9]L. Cui, et. al, "Effect of organoclay purity and degradation on nanocomposite performance, part 2: morphology and properties of nanocomposites", *Polymer* **49** , 3762-3769, (2008).

[10]R. Shah, et.al, "Nanocomposites from poly(ethylene-co-methcrylic acid) ionomers: effect of surfactant structure on morphology and properties", *Polymer*, **46**, 2646-2662, (2005).

[11]Q. Nguyen, D. Baird, "An improved technique for exfoliating and dispersing nanoclay particles into polymer matrices using supercritical carbon dioxide", *Polymer* **48**, 6923-6933, (2007).

[12]J. Park, S. Jana, "The relationship between nano- and micro- structure and mechanical properties in PMMA-epoxy-nanoclay composites", *Polymer* **44**, 2091-2100, (2003).

[13]R. El-Sheikhy, "Fundamental Analysis of the New Version of the New Proposed Fracture Theory", *J. of Material Science Forum*, **Vol. 502**, p. 87-92, (2005).

[14]R. El-Sheikhy. and N. Masaaki, "Analysis of Three Dimensional Failures of Brittle Composites", *J. of Trans. of JWRI*, **Vol. 34, No.1**, p. 67-74, (2005).

[15]R. El-Sheikhy and N. Masaaki, "Determination of the safe history of composites and structures", *J. of Trans. of JWRI*, **Vol. 34, No.1**,p. 99-106, (2005).

[16]R. El-Sheikhy and N. Masaaki, "Fracture under Triaxial Loading", *J. of Trans. of JWRI*, **Vol. 34, No.1**, p. 75-84, (2005).

[17]R. El-Sheikhy and N. Masaaki, "Fracture Analysis of Anisotropic Composites", *J. of Trans. of JWRI*, **Vol. 34, No.1**, p. 85-98, (2005)

# TIME DEPENDENCE OF CONTACT ANGLE BETWEEN SILVER-COPPER-TITANIUM ALLOYS AND BORON NITRIDE

Yoshihisa Sechi[1,2], Taihei Matsumoto[3,4] and Kazuhiro Nakata[4]

[1] Kagoshima Prefectural Institute of Industrial Technology, 1445-1 Oda, Hayato-cho Kirishima, Kagoshima, Japan, 899-5105

[2] Graduate School of Engineering, Osaka University, 2-1, Yamada-oka Suita, Osaka, Japan, 565-0871

[3] Toyo Tanso Co., Ltd., 5-7-12, Takeshima, Nishiyodogawa-ku, Osaka, Osaka, Japan, 555-0011

[4] Joining and Welding Research Institute, Osaka University 11-1, Mihogaoka, Ibaraki, Osaka, Japan, 567-0047

## ABSTRACT

The contact angle between silver-copper-titanium braze alloys with different titanium contents of up to 2.80 mass % and hexagonal boron nitride ceramics was investigated using the sessile drop method at 1123K in Argon. The final contact angle was less than 30° when the Ti content was over 0.41 mass%. Meanwhile, the contact angle curves showed different behavior. When using braze alloy containing 2.80 mass % of titanium, the initial contact angle was acute, just after the melting of the braze. When using brazes containing less than 2.26 mass % titanium, the contact angle was larger than 90° at first, and slowly decreased to an acute angle. A wettable layer of titanium nitride was observed at the interface.

## INTRODUCTION

Successful bonding of ceramics to metals is a common application requirement, and new combinations of joint-material are common requirements for high functionality products in recent years. However, there are some problems such as joint defects due to thermal stress in the joint, and material deterioration due to the dissimilar thermal coefficients in joining two different materials[1]. Among the many joining methods available, brazing has good characteristics for joining dissimilar materials due to the suppression of damage to the base materials.

Boron nitride ceramics have various positive functional characteristics. Of these ceramics, hexagonal boron nitride (h-BN) ceramics have good thermal properties and are solid-lubricants[2], but wettability to the brazing metals is relatively low compared to the other ceramics[2-3] making it difficult to braze h-BN ceramics to other materials.

Because joining characteristics are affected by wettability, and most molten metals show low wettability to h-BN, it is considered that brazing of h-BN ceramics is worthy of research attention.

The addition of titanium as an active element in silver-copper alloy braze makes it possible to braze h-BN ceramics and tungsten carbide directly[4]. Previously, some research[3, 5,-6] was carried out on brazing composition dependence, focusing on the reaction of Ti with h-BN, but there is little work where the Ti content was under 2 mass %. In this study, the time dependence of contact angle between hexagonal boron nitride and silver-copper-titanium alloy braze was characterized using the sessile drop method, and, to investigate the reaction layer of the interface, cross-section observation and elemental analyses of interface layer by electron probe micro-analysis (EPMA) was performed.

## EXPERIMENTAL PROCEDURES

Hexagonal-BN and Ag-Cu-Ti alloy brazes were used in this work. Hexagonal-BN was produced by the hot pressing method using only high purity crystal. Ag-Cu-Ti alloy brazes were made as plate, which include Ti as a major active component for direct ceramic brazing. Nominal compositions and properties are summarized in Table 1 and Table 2. Wetting properties of h-BN with respect to molten Ag-Cu-Ti braze alloy were estimated using the sessile drop method. For this, block shaped Ag-Cu-Ti alloy and an h-BN plate were prepared and degreased as mentioned above. The surface roughness (Ra) of the h-BN plate was approximately 1.2 m because it includes many pores due to low relative density. The Ag-Cu-Ti block was placed on the h-BN plate in a vacuum chamber, and the chamber was evacuated to decrease the pressure to at least $3.7\times10^{-4}$ Pa. The specimen was heated up to 1023K, which is higher than the melting point of the braze alloy and the brazing temperature in vacuum; then the evacuation and Ar (99.999% purity) substitution cycle was repeated at least four times. Finally the specimen was heated to 1123K in Ar atmosphere, and the contact angle of the molten braze alloy and the h-BN was measured under 1atm of Ar atmosphere, which was continuously flowing at a finite flow rate. Samples were subsequently cross-sectioned, ground using SiC paper, and polished with 3 m - 1 m polycrystalline diamond to provide microstructural information. Cross sectional observation and elemental analysis of the interface were

Table 1. Braze used in the experiment.

| Elements (mass %) | | | |
|---|---|---|---|
| No | Ag | Cu | Ti |
| 1 | 72.01 | Bal s . | 0.00 |
| 2 | 71.49 | bal. | 0.28 |
| 3 | 71.53 | bal. | 0.41 |
| 4 | 71.34 | bal. | 0.63 |
| 5 | 71.21 | bal. | 0.85 |
| 6 | 71.07 | bal. | 1.25 |
| 7 | 70.90 | bal. | 1.28 |
| 8 | 70.20 | bal. | 1.70 |
| 9 | 70.15 | bal. | 2.26 |
| 10 | 69.61 | bal. | 2.80 |

Table 2. Boron nitride used in the experiment.

| Nominal composition (mass %) | Bend strength at room temperature (MPa) | Density ($\times10^{-3}$ kg / m$^3$) | Relative Density (%) | Ra ( m) | Size (mm) |
|---|---|---|---|---|---|
| h-BN > 99.993 | 32.5 | 1.93 | 82.5 | 1.19 | 20*20*5 |

performed using electron probe micro-analyzer (JEOL Co. Ltd. JXA-8621MX) and X-ray diffractometer (Bruker AXS Co. Ltd. D8: CoK ). The diameter of the collimator was 50 m to determine the crystallographic phases in the micro region.

## RESULTS

Figure 1 shows the time dependence of the contact angle between h-BN and Ag-Cu-Ti braze at 1123K. In the case of the braze without Ti, the contact angle was about 160°, and time dependence was not observed. With increasing Ti content, the contact angle decreased quickly, and a final contact angle was achieved in a short period. In the case of the 2.80% Ti braze, the contact angle was acute at the melting point. A small addition of Ti increased the rate of change of the contact angle, and the rates were nearly the same for Ti contents from 0.28% to 2.25%. However, when the Ti content increased to 2.80%, the rate of change decreased.

Figure 2 shows the triple line velocity (dR/dt) as a function of instantaneous contact angle for seven Ag-Cu-Ti alloys having different Ti contents. These points are plotted after the dynamic shape change of the specimen from 'cubic' to 'droplet' due to the block shape specimen. The 0.28% Ti, dR/dt results were not plotted because the radius of the molten braze alloy was out of the optical field of view. The spreading rate (dR/dt) versus contact angle shows linear behavior. In addition, 2.80% Ti results in a straight line slope greater than

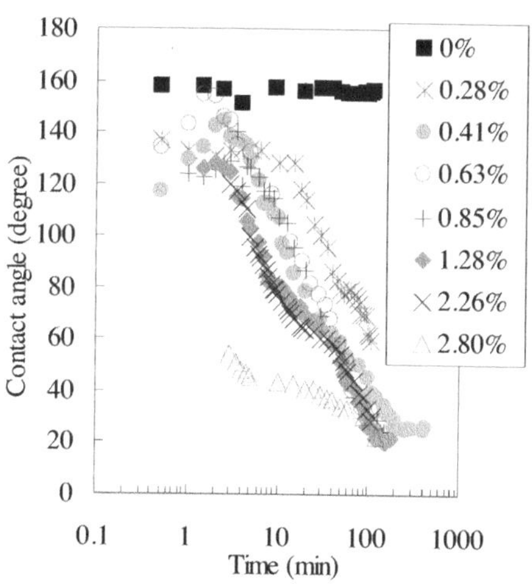


Figure 1. Time dependence of contact angle for Ag-Cu-Ti braze alloys at 1123K.

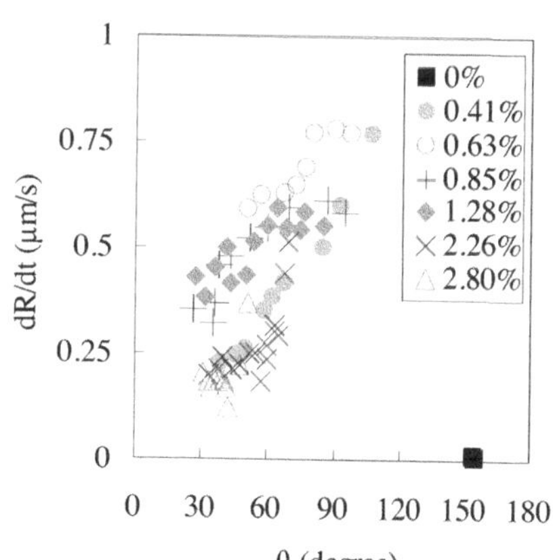


Figure 2. Triple line velocity dR/dt as a function of instantaneous contact angle .

Figures 3 and 4 show an elemental map analyses of the h-BN / Ag - Cu - 1.28%Ti braze interface after contact angle measurement near the edge and central area. From these distributions of elements, the concentration of Ti near the interface, which infiltrated into open pores of h-BN, was observed and thickness of reaction layer was evaluated as 2 - 10 m in Figure 3. The thickness of the reaction layer at the tip of the edge was thinner than that of the inside area as shown in Figure 3. In most cases of Ti content, the concentration of Ti at the edge-interface was lower than that of the inside area as shown in Figure 3. In the central area, a continuous reaction layer was observed on the braze side of the interface. In most cases, the total thickness of the reaction layer continuously increased toward the central area of the interface, so the value at the center was a little larger than that of the inside area of the edge – or nearly equal to it. This was seldom found to vary.

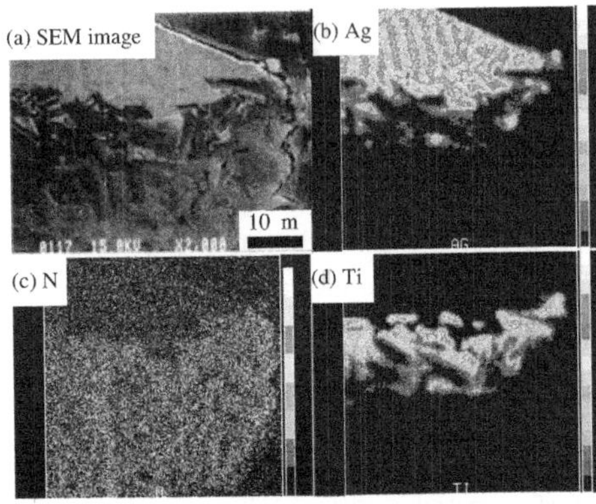


Figure 3. Elemental map analysis of Ag-Cu-1.28%Ti braze alloy / h-BN interface

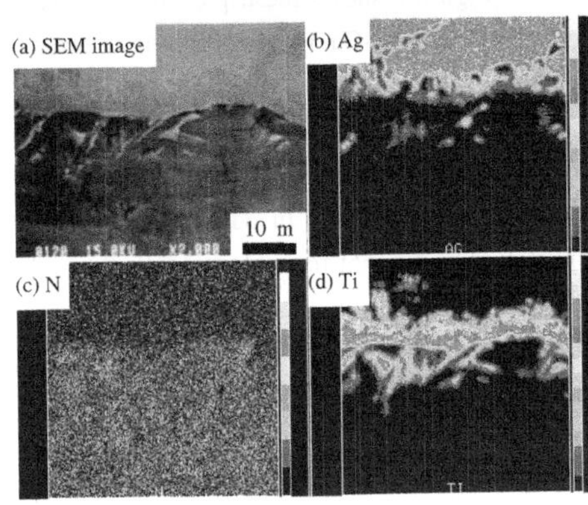


Figure 4. Elemental map analysis of Ag-Cu-1.28%Ti braze alloy / h-BN interface

Figure 5 shows map analysis of Ag - Cu - Ti braze at the braze-center – far from the h-BN/braze interface – after contact angle measurement. In the case of the 0.63% Ti doped material, a eutectic structure of Ag and Cu was observed and a small amount of Ti was distributed in the same area of Cu. While, in the case of 2.80% Ti doped, a relatively large amount of Ti distribution suggests coarse precipitates in the same area of Cu.

Figure 6 shows XRD profiles of the interface and braze after contact angle measurement specimens. Some kinds of titanium nitride existed on both interfaces using 0.63% Ti and 2.80% Ti doped in (a) and (b). While $Cu_3Ti$ existed only in 2.80% Ti doped braze in (b).

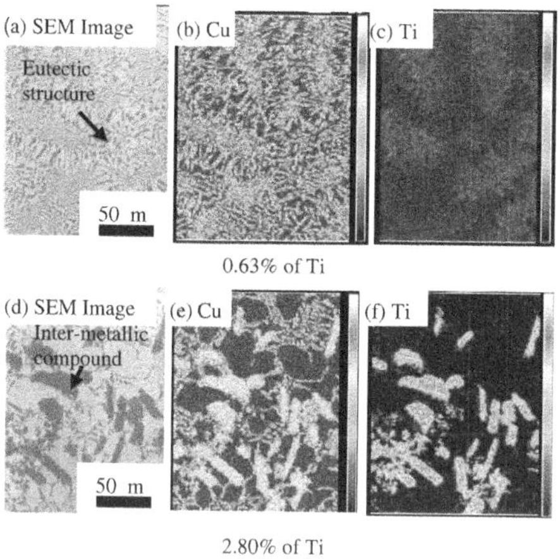


Figure 5. Elemental Map analysis of bulk of Ag-Cu-Ti braze away from h-BN / braze

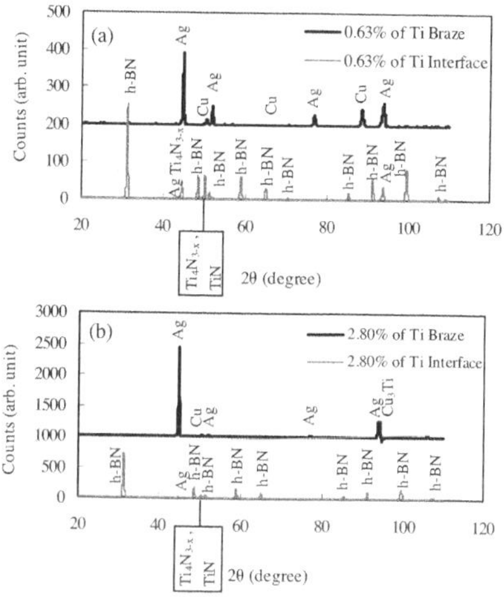


Figure 6. XRD profiles at the interface and bulk braze.

DISCUSSION

In this study, the effect of oxidized titanium on the measurement of contact angle cannot be ignored because Ti is an active ingredient even in low partial pressure of $O_2$ such as on the order of $10^{-23}$ Pa at 1123K[7], which was much lower than that of the experimental atmosphere. Therefore, when Ag-Cu-Ti braze alloy melted, a titanium oxide film was thought to exist on the surface. In addition, it is difficult to remove the oxide film[8] mechanically using the ceramic drop tube method[9-10] because Ti reacts with most ceramics. Therefore, a block-shaped metal specimen on the h-BN plate was used in this study.

From Figure 1, most of the measurements showed dynamic behavior such as an increase of contact angle just after the specimen melted. This was caused by a change of the specimen's shape from cubic to droplet. The difference in the initial contact angle between cases of 0.28%Ti and 2.26%Ti seems to be due to the initial shape of an Ag-Cu-Ti alloy block specimen. In cases including 2.80% Ti, the initial contact angle is acute (46°), and probably caused by high initial content of Ti near the solid-liquid interface sufficient to cause a reaction between Ti and N in h-BN. It seems that content of Ti in Ag-Cu-Ti braze as low as 0.28%Ti is sufficient to reduce contact angle on h-BN in Ar. In addition, with increasing Ti content, contact angle decreased rapidly. It would appear that the difference in the final contact angle between former researches[3,5-6] and this study comes from the difference in experimental atmosphere, initial braze shape, and surface roughness[8-9].

At the interfacial-reaction wetting stage, the spreading kinetics is limited by the growth kinetics of the reaction product at the triple line in a direction parallel to the drop/substrate interface, and this growth can be limited either by the supply of the reactive solute from the drop bulk to the triple line or by the reaction kinetics at the triple line[11-12]. The wetting rate was fast and nearly the same value when the Ti content of metals was 0.28% to 2.26%. This tendency of the wetting rate seems to show that the main reaction mechanism of Ti to h-BN at the interface is the same for the Ti content from 0.28% up to 2.26%, but on the contrary, when the Ti content of metal was 2.80%, its wetting rate was not smaller than that of 0.28% to 2.26%. This difference appears to arise from the transportation of Ti in melting drop of Ag-Cu-Ti alloy. In Figure 2, the spreading rate (dR/dt) versus contact angle shows linear behavior and it seems that spreading kinetics is influenced by the diffusion of titanium to the triple line[11-12].

If the liquid metal contains a low concentration of Ti, the reaction around the triple line depends on the supply of Ti in the liquid by convection and diffusion. On the other hand, with a high content of Ti, the limiting step of the growth kinetics of the reaction layer at the triple line seemed to be limited by the reaction kinetics. The reason is not explained clearly, but another phenomenon may exist because the triple line velocity for the 2.26% Ti seems to converge under 60° and that for the 2.80% Ti under 45°. In Figure 3 (c) and (d), it was found that the higher content of Ti resulted in a thicker interfacial reaction phase, which contained TiN[3,4,13] or other titanium nitrides such as $Ti_3N_{4-x}$[14]. Besides, standard free energies of formations of nitrides[15] describes those of TiN and BN at 1123K are about -220 and -160 [kJ / g mol $N_2$ ], respectively. Thus judging from standard free energies of formations, TiN is more stable than BN around 1123K and reacts with N in h-BN. Figure 4 (c) and (d) showed that a continuous reaction layer was formed in central areas of the interface, which may be the result of a long heating time. On the other hand, the reaction layer in Figure 4(d) was not so increased. These tendencies were observed for all Ti contents. Figure 5 (d) indicated that not all of the Ti reacted with the N in h-BN, and some remained as a coarse inter-metallic compound and as precipitates even after the final contact angle state after a long heating time. It seems that a thick interfacial reaction phase may interfere with the diffusion of nitrogen and further formation of the reaction phase[9]. These precipitates turned out to be $Cu_3Ti$ from XRD profile as shown in Figure 6 (b). On the contrary, most of the Ti was consumed in case of 0.63% Ti braze as shown in Figure 5 (b).

## CONCLUSIONS

The contact angle between h-BN and Ag-Cu-Ti braze, and the effects of the Ti content

in the braze alloy were investigated. Results of the study can be summarized as follow:
(1) The contact angle between h-BN and the braze decreased to less than 30° in the case of the braze with over 0.41mass % Ti, as measured by the sessile drop method at 1123K in Ar.
(2) An excess amount of Ti, which did not react with h-BN, was found to exist in the form of $Cu_3Ti$ in the bulk Ag-Cu-Ti braze region of the contact angle measurement specimens.
(3) In the case where the liquid metal contains a low concentration of Ti under 1.3 mass%, the reaction around the triple line seemed to be depend on the supply of Ti in the liquid by convection and diffusion.

ACKNOWLEDGEMENTS

This work was supported by Joint Research system at the Joining and Welding Research Institute, Osaka University, and Grant-in-Aid for Cooperative Research Project of Nationwide Joint-Use Research Institutes on Development Base of Joining Technology for New Metallic Glasses and Inorganic Materials from The Ministry of Education, Culture, Sports, Science and Technology, Japan.

REFERENCES

[1] W. Wlosinski, "Interfaces in dissimilar materials joints", Fueg. Keram. Gla. Met., 22-36 (1985).

[2] R. H. Biddulph, "Boride and carbide ceramics- The systems", Proc. 1st Euro. Sym. on Eng. Ceram., 45-61 (1985).

[3] M.G. Nicholas, D.A. Mortimer, L.M. Jones and R.M. Crispin, "Some observations on the wetting and bonding of nitride ceramics", J. Mat. Sci., **25,** 2679-89 (1990).

[4] Y. Sechi, A. Takezaki, T. Tsumura and K. Nakata, "Dissimilar laser brazing of boron nitride and tungsten carbide", Smart Process. Tech. **2,** 27-30 (2008).

[5] E. Benko, "Wettability studies of cubic boron nitride by silver-titanium", Ceram. Int., **21**, 303-7 (1995).

[6] E. Benko, E. Bielanska, V.M. Perevertelio and O. B. Loginova, "Formation peculiarites of the interfacial structure during cBN wetting with Ag-Ti, Ag-Zr and Ag-Hf alloys"., Diam. Rel. Mat., **6**, 931-4 (1997).

[7] L.S. Darken, R. W. Gurry, Physical Chemistry of Metals, McGrow Hill, New York, 1953

[8] H. Fujii, H. Nakae and K. Okada, "Interfacial reaction wetting in the boron nitride/molten aluminum system", Acta Metall. mater., **41**, 2963-71 (1993).

[9] H. Fujii and H. Nakae, "Precise measurement of the wetting of ceramics by molten Aluminum", Materia Japan, **34,** 1269-75 (1995).

[10] N. Eustathopoulos, M.G. Ncholas, B. Drevet, Wettability at High Temperatures, Oxford:

Elsevier, (1999)

[11] A. Mortensen, B. Drevet, N. Eustathopoulos, Kinetics of Diffusion-limited spreading of sessile drops in reactive wetting, Scripta Mater., **36**, 645-51 (1997)

[12] R. Voitovich, A. Mortensen, F. Hodaj, N. Eustathopoulos, Diffusion-limited reactive wetting: Study of spreading kinetics of Cu-Cr alloys on carbon substrates, Acta Mater., **47**, 1117-28 (1999)

[13] S.D. Peteves, Joining nitride ceramics. Ceram. Int., **22**, 527-33 (1996).

[14] W. Lengauer and P. Ettmayer, The crystal structure of a new phase in the titanium-nitrogen system, J. Less-Common Met., **120,** 153-9 (1986)

[15] J.F. Elliott, M. Gleiser, Thermochemistry for Steelmaking. Vol. I. Massachusetts: Addison-Wesley; (1960)

HARD X-RAY PHOTOELECTRON SPECTROSCOPY ANALYSIS FOR ORGANIC-INORGANIC HYBRID MATERIALS FORMATION

Ken Cho[1], Kosuke Takenaka[1], Yuichi Setsuhara[1], Masaharu Shiratani[2], Makoto Sekine[3], Masaru Hori[3], Eiji Ikenaga[4], Hiroki Kondo[5], Osamu Nakatsuka[5], and Shigeaki Zaima[5]

[1]Joining and Welding Research Institute, Osaka University, and Japan Science and Technology Agency, CREST, Ibaraki, Osaka, Japan

[2]Department of Electronics, Kyushu University and Japan Science and Technology Agency, CREST Nishi-ku, Fukuoka, Japan

[3]Plasma Nanotechnology Research Center, Graduate School of Engineering, Nagoya University, and Japan Science and Technology Agency, CREST, Chikusa-ku, Nagoya, Japan

[4]Japan Synchrotron Radiation Research Institute, Sayo-cho, Sayo-gun, Hyogo, Japan

[5]Graduate School of Engineering, Nagoya University, Chikusa-ku, Nagoya, Japan

ABSTRACT

Interactions of nitrogen plasmas with polymer surfaces were investigated using hard x-ray photoelectron spectroscopy (HXPES) to complete depth analyses of the chemical bonding states in the nano-surface layer of polymethylmethacrylate (PMMA) films via. The PMMA films were exposed to the nitrogen plasmas sustained via inductive coupling of radio-frequency (RF) power with multiple low-inductance antenna (LIA) modules. The etching rate of the PMMA films was 38 nm/min. The surface roughness of PMMA increased from 0.3 nm to 0.7 nm with increased exposure time. The HXPES was carried out for non-destructive depth analysis of chemical bonding states in the nano-surface layer of PMMA films. The HXPES results indicated that nitrogen functionalities were formed in the shallower regions up to about 27 nm from the surface without showing significant degradation of the molecular structure of PMMA due to nitrogen plasma exposure.

INTRODUCTION

Organic-inorganic hybrid materials, which are composed of inorganic functional films/polymer substrates and vice versa, are expected as key materials for flexible electronics including flexible displays and thin-film photovoltaic cells and advanced ultra-large-scale integration circuits (ULSIs) [low-k interlayer and extreme ultraviolet (EUV) lithography][1-3]. For successful fabrication of these devices, it is required to control chemical bonding structure of polymers surface without suffering degradation of polymer layer. However, compared to the amount of research on the plasma modification of the polymers, few studies indeed have been done to analyze the degradation of the molecular structure of the polymer due to plasma exposure in deeper region up to tens of nm from the surface.[4]

The bond-dissociation energies of polymers are comparatively smaller than that of inorganic materials. For example, the bond-dissociation energies for thermal decomposition in the molecular structure of polymethylmethacrylate (PMMA) ranges from 3.4 eV for the O-C(=O) bond, 3.6 eV for the C-$CH_3$ bond to 8.4 eV for C=O bond.[5] Therefore, in the plasma process, it is considered that ions, radicals, photons and electrons from plasma give significant degradation in chemical bonding structures of polymers beneath the surface and/or the organic-inorganic interface. Thus it is required to develop plasma processing technologies.

As one of the effective plasma sources for polymer processing, inductivity-coupled plasma (ICP) sources have been developed with multiple low-inductance antenna (LIA) modules, which have allowed low-voltage operation of ICPs. Ion energy distribution measured with a mass-separated ion energy analyzer showed significantly reduced ion energy at the sheath edge as low as 5 eV.[6,7]

In this study, interactions of nitrogen plasmas with polymer surfaces have been investigated on the basis of non-destructive depth analyses of chemical bonding states in the nano-surface layer of PMMA films via hard x-ray photoelectron spectroscopy (HXPES). The HXPES analyses have been performed at the SPring-8, the national synchrotron radiation facility in Japan.

## EXPERIMENTAL

Nitrogen plasmas were generated via inductive coupling of radio frequency (RF) power at 13.56 MHz with a set of eight LIA units[6]. The LIA unit consisted of U-shaped antenna conductor, which was fully covered with dielectric tubing for complete isolation from the plasma. Eight LIA units with a 70 mm width and a 160 mm height were mounted on the top flange of the discharge chamber. The chamber had a 500 mm inner diameter and a 200 mm height, which was connected to diffusion chamber made of stainless-steel vessel with a 500 nm inner diameter and a 400 nm height. The chamber was pumped to $9.6\times10^{-3}$ Pa using a high vacuum system with a turbo-molecular pump.[8]

Plasma parameters were measured with a Langmuir probe, which was inserted radially at an axial position of 280 mm below the top flange. Kinetic energy distributions of impinging onto the ground potential (ion energy at the sheath edge) from the plasmas were measured using a mass separated ion energy analyzer (Hiden, EQP500), which was mounted beside the substrate holder. For investigations of the plasma-polymer interaction, PMMA films were exposed to nitrogen plasmas sustained at the RF power of 2000 W at a gas pressure of 1.3 Pa. Here it is noted that the PMMA films were set on water-cooled substrate holder when they were exposed to the plasmas.

Surface morphological changes of the polymer films via plasma exposure were observed with atomic force microscopy (AFM). The measurement error or the height resolution of the AFM employed in the present investigation was 0.3 nm.

Chemical bonding states of nano-surface layer of polymer films exposed to plasmas were analyzed using HXPES. The HXPES analysis was performed at a photon energy of 7940 eV, which was carried out at the national synchrotron radiation facility, SPring-8 of the Japan Synchrotron Radiation Research Institute (JASRI). The inelastic mean free path (IMFP) of 8000 eV electrons in PMMA is approximately 18 nm, which results in a HXPES probing depth of about 54 nm for 8000 eV hard x-ray. This feature with HXPES makes it possible to evaluate the chemical bonding states in the deeper regions via variation of take-off angle (TOA) for photoelectrons detection from TOA = 88° for analysis of deeper regions (about 54 nm) down to TOA = 30° for analysis of shallower regions (about 27 nm) without degradation of the bonding states due to ion milling, which is necessary in the conventional x-ray photoelectron spectroscopy analysis.[9] Here it is noted that surface of the polymer samples for HXPES measurements were coated with Au-Pd layer with thickness of about 4 nm in order to avoid anomalous charging of the sample surface due to photoelectron emission during the measurements.

RESULTS AND DISCUSSION

Ion energy distribution of the $N_2^+$ ions impinging onto the ground potential through the sheath edge of the nitrogen plasmas were measured using the mass-separated ion-energy analyzer. Figure 1 shows the variation of kinetic energy distribution of $N_2^+$ ions impinging to the ground potential for the nitrogen plasma sustained at a gas pressure of 1.3 Pa as a parameter of RF power (500-2000 W). The peak values of ion energy distributions decreased from 17.4 eV to 14.9 eV as the RF power increased from 500 W to 1000W and then were almost constant at 14.9 eV. Theses date demonstrate that ICPs driven by LIA units are considered to be one of the plasma sources suitable for low-damage and high-quality processes of polymers.

For investigations of the surface morphological change of the PMMA films due to nitrogen plasma exposures, the plasma exposed surface was imaged with AFM. Figure 2 shows typical AFM images of the surface of the PMMA films (a) before and (b) after exposure to nitrogen plasmas sustained at the RF power of 2000 W at gas pressure of 1.3 Pa for 5min.

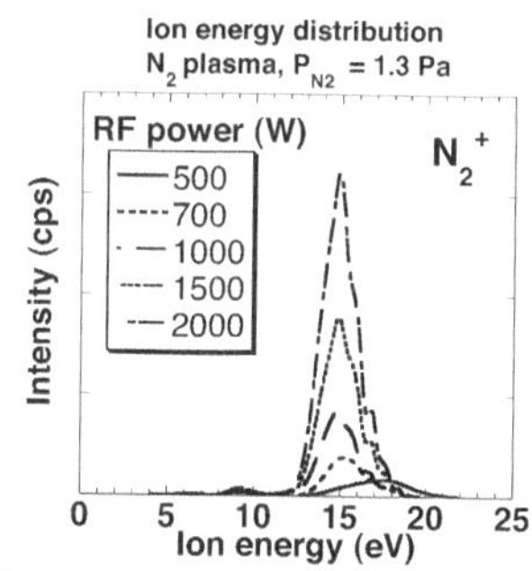


Figure 1. The variation of kinetic energy distribution of $N_2^+$ ions impinging 5 w to the ground potential for the nitrogen plasma as a parameter of RF power.

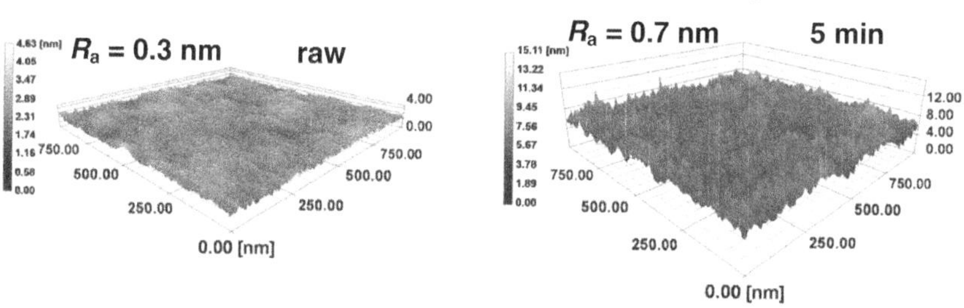


(a) Before the plasma exposure (raw) (b) After exposure to nitrogen plasmas (5 min)

Figure 2. AFM images of the surface of the PMMA films

After exposure to the plasmas, the surface of the PMMA films showed slight increase in surface roughness ($R_a$) evaluated from the AFM images. The surface roughness of PMMA slightly increased from 0.3 nm to 0.7 nm due to the plasma exposed for 5 min. Furthermore, with increasing exposure time, the etching depth increased linearly and the etching rate was 38 nm/min. These results exhibit that the surface modification using nitrogen plasma is attractive as techniques to roughen surface for improving adhesion.

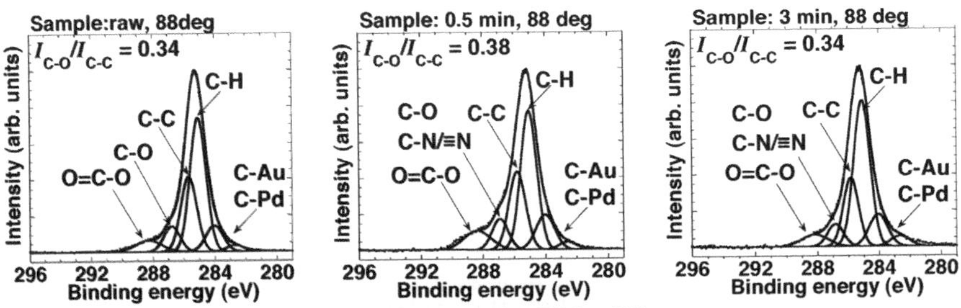


(a) HXPES measured at TOA = 88°

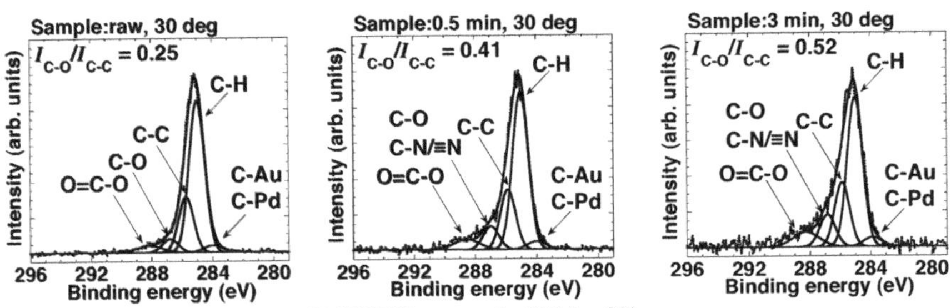


(b) HXPES measured at TOA = 30°

Figure 3. HXPES C1s spectra for PMMA films without and with exposure to nitrogen plasmas sustained at the RF 2000 W at gas pressure of 1.3 Pa for 0.5-5min.

Chemical bonding states of nano-surface layer of polymers exposed to nitrogen plasmas were analyzed using HXPES. Figure 3 shows HXPES C1s spectra for PMMA films without and with exposure to nitrogen plasmas sustained at the RF 2000 W at gas pressure of 1.3 Pa for 0.5 min and 3 min. Here the spectra shown in Figure 3(a) were measured at TOA = 88° for analysis of deeper region up to about 54 nm from the surface and those shown in Figure 3(b) were measured at TOA = 30° for analysis of shallower region up to about 27 nm from the surface. The C1s spectra of the raw PMMA samples can be decomposed into four main components: a component at binding energy of 285.0 eV due to the C-H bond, a component at 285.9 eV due to C-C bond, a component at 286.8 eV due to C-O bond and a component at 289.0 eV due to O=C-O bond[10]. Before discuss about the chemical bonding states of nano-surface layer of PMMA, it is necessary to note the following two points. First, the peak intensities of O=C-O bond are observed to be considerably lower than that those observed in conventional XPS measurement (not show here) because of formation of Au-C=O π back bonding due to Au-Pd coating on the surface of the PMMA sample[11]. Second, in the HXPES C1s spectrum, the peak of C-N bond and C≡N bond are sure to appear at binding energy of 286.3 eV and 287.2 eV, respectively[12-13]. Therefore, the peak of C-O bond (286.8 eV) is formed including the photoelectrons from C-N bond and C≡N bond. The peak-area ratio of C-O bond (include C-N bond and C≡N bond) to C-C bond ($I_{C\text{-}O(C\text{-}N,C\equiv N)}$ / $I_{C\text{-}C}$) evaluated from Figure 3 were summarized in Figure 4. The peak area ratio $I_{C\text{-}O(C\text{-}N,C\equiv N)}$ / $I_{C\text{-}C}$ measured at TOA = 88° are found to be almost constant at 0.34-0.37. Whereas, the peak area ratio $I_{C\text{-}O(C\text{-}N,C\equiv N)}$ / $I_{C\text{-}C}$ measured at TOA = 30° are found to increase from 0.25 to 0.49 with increasing exposure time from at 0 min to 1 min and saturated at almost constant values of 0.49-0.52 after 1 min. These results indicate that the nitrogen functionalities (C-N bond and C≡N bond ) was formed in the shallower region up to about 27 nm from the surface without significant degradation of the molecular structure of the polymer due to nitrogen plasma exposure.

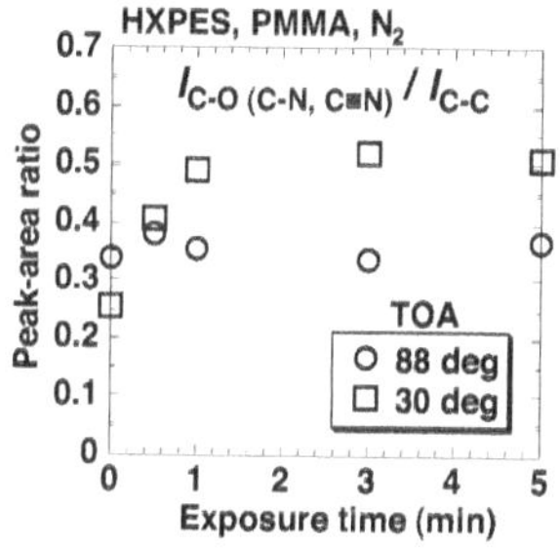


Figure 4. Peak-area ratio of C-O bond to C-C bond.

The results of chemical bonding states analyses with HXPES imply that the enhancement of the surface roughness caused by the nitrogen plasma exposure is due to forming nitrogen functionalities by chemical reaction, rather than due to carbonization induced by dissociation of oxygen that originates in molecular structure of the PMMA.

## CONCLUSIONS

The interactions between polymer and nitrogen plasma were investigated on the basis of HXPES for depth analysis of chemical bonding states in the nano-surface layer of PMMA films, which were exposed to the plasma sustained via RF inductive coupling with multiple LIA modules. The etching rate of the PMMA films was 38 nm/min. The surface roughness of PMMA slightly increased from 0.3 nm to 0.7 nm with increased exposure time from 0 min (raw) to 5 min. The HXPES analysis indicated that the nitrogen functionalities was formed in the shallower region up to about 27 nm from the surface without significant degradation of the molecular structure of the polymer due to nitrogen plasma exposure.

## ACKNOWLEDGEMENT

This work was supported partly by The Global COE Program "Center of Excellence for Advanced Structural and Functional Materials Design" from the Ministry of Education, Culture, Sports, Science and Technology (MEXT), Japan, and Grant-in-Aid for Cooperative Research Project of Nationwide Joint-Use

Research Institutes on Development Base of Joining Technology for New Metallic Glasses and Inorganic Materials from MEXT, Japan. The synchrotron radiation experiments were performed at the SPring-8 with the approval of Japan Synchrotron Radiation Research Institute (JASRI) as Nanotechnology Support Project of the Ministry of Education, Culture, Sports, Science and Technology. (Proposal No. 2008B1847/ BL47XU).

REFERENCES

[1] M.-C. Choi, Y. Kim and C.-S. Ha, "Polymers for flexible displays: From material selection to device applications" *Prog. Polym. Sci.* 33 (2008) 581.

[2] K. L. Chopra, P.D .Paulson and V. Dutta, "Thin-Film Solar Cells: An Overview" *Prog. Photovolt: Res. Appl.* 12 (2004) 69.

[3] International technology roadmap for semiconductors, 2007 edition, Executive summary

[4] J. Meichsner, M. Nitschke, R. Rochotzki and M. Zeuner, "Fundamental investigations in plasma modification of polymers" *Surf. Coat. Tech.* 74-75 (1995) 227-231.

[5] S. I. Stoliarova, P. R. Westmorelamda, M. R. Nydenb and G. P. Forney, "A reactive molecular dynamics model of thermal decomposition in polymers: I. Poly(methyl methacrylate)" *polym.* 44 (2003) 883.

[6] Y. Setsuhara, T. Shoji, A. Ebe, S, Baba, N. Yamamoto, K. Takahashi, K. Ono, and S. Miyake, "Development of Internal-Antenna-Driven Large-Area RF Plasma Sources Using Multiple Low-Inductance Antenna Units" *Surf. Coat. Technol.,* 174 – 175, 33-39 (2003).

[7] K. Takenaka, Y. Setsuhara, K. Nishisaka and A. Ebe, "Characterization of Ion Energy Distribution in Inductively Coupled Argon Plasmas Sustained with Multiple Internal-Antenna Units" Jpn. J. Appl. Phys. 47 (2008) 6900.

[8] Y. Setsuhara, K. Takanaka, A. Ebe, and K. Nishisaka, "Production and Control of Large-Area Plasmas for Meters-Scale Flat-Panel-Display Processing with Multiple Low-Inductance Antenna Modules" *Plasma process. Polym.* 4 (2007).

[9] K. Kobayashi, "Hard X-ray photoemission spectroscopy" *Nucl. Instrum. Meth. Phys. Res. A* 601 (2009) 32.

[10] P. Groning, O. M. Kuttel, M. Collaud-Coen, G. Dietler and L. Schlapback, "Interaction of low-energy ions (< 10 eV) with polymethylmethacrylate during plasma treatment" *Appl. Surf. Sci* 89 (1995) 83.

[11] F. A. Cotton, G. Wilkinson and P.L. Gaus, Basic Inorganic Chemistry, 2nd edn., Wiley, New York 1987.

[12] P. Brant, J. H. Enemark and A. L. Balch, "X-ray photoelectron spectra of palladium and platinum complexes of carbenoid and related ligands" *J. Orga. Chem.* 114, 99 (1976).

[13] M. Barber, J.A. Connor, M.F. Guest, I.H. Hillier, M. Schwartz and M. Staccy, "Bonding in some donor–acceptor complexes involving boron trifluoride. Study by means of ESCA and molecular orbital calculations" *J. Chem. Soc. Faraday Trans. II* 69, 551, (1973).

# NUMERICAL ANALYSIS OF PLASMA KEYHOLE WELDING OF AN ALUMINUM THIN PLATE

Minoru Miyata, Shinichi Tashiro, and Manabu Tanaka
Joining and Welding Research Institute, Osaka University
11-1, Mihogaoka, Ibaraki, Osaka, Japan, 567-0047

## ABSTRACT

In the automotive industry, tailored blanks and aluminum bodies are used to reduce the weight of the car. Laser welding is generally used for the tailored blank and aluminum body. Due to the high cost of laser welding, plasma arc welding (PAW) is being investigated as a more cost efficient alternative. To further investigate the complex phenomena behind PAW, a computational model was developed. In this study, we built the numerical computational model for plasma arc welding and tried to elucidate the keyhole formation mechanism.

## INTRODUCTION

A plasma torch is suitable as a heat source for many applications because it can stabilize high temperature arc plasma easily by employing shielding gas. It has many advantages as a heat source, such as high heating efficiency, highly controllable characteristics, and low equipment investment cost. It is widely utilized for melting, cutting, welding, decomposition, volume reduction, and detoxification of toxic waste [1-2].

A plasma arc is a type of torch. Since it has a water-cooled copper nozzle in front of the cathode to constrict the arc plasma, an arc plasma with high energy density and high arc pressure can be produced. Therefore it is suitable, especially for welding and cutting. The properties of the heat source from the plasma arc can be controlled by the torch geometry, which is governed by the plasma nozzle diameter. In this paper, numerical simulation results on the properties of the heat source for plasma arc welding are reported.

## SIMULATION MODEL

Table 1 shows the simulation conditions. The calculation region for the plasma arc consists of a tungsten cathode with the diameter of 3.2mm and tip angle of 60 degrees, a water-cooled copper nozzle with the inner diameter of 3.2mm and 4.0mm, arc plasma, and a water-cooled copper anode. It is described in a cylindrical coordinate frame of reference with axial symmetry around the arc axis. The electrode gap is set to be 5mm. The arc current is given inside the cathode. The arc current is 100A. The shielding gas is introduced from the inner nozzle on the upper boundary at a flow rate of 0.5 Liters/minute, and from the outer nozzle at the flow rate of 10 Liters/minute. Laminar flow is assumed, and the arc plasma is considered to be in local thermodynamic equilibrium (LTE). The other numerical modeling methods are given in detail in our previous papers[3]. The differential equations (1)-(6) are solved iteratively by the SIMPLER numerical procedure[4].

Table 1. Simulation conditions

| cathode diameter | tip angle | innner copper nozzle diameter | outer copper nozzle diameter |
|---|---|---|---|
| 3.2mm | 60° | 3.2mm | 4.0mm |
| electrode gap | arc current | gas flow rate from innner nozzle | gas flow rate from outer nozzle |
| 5.0mm | 100A | 0.5Liters/minute | 10Liters/minute |

The mass continuity equation is:

$$\frac{1}{r}\frac{\partial}{\partial r}(r\rho v_r)+\frac{\partial}{\partial z}(\rho v_z)=0 \quad (1)$$

The radial momentum conservation equation is:

$$\frac{1}{r}\frac{\partial}{\partial r}(r\rho v_r^2)+\frac{\partial}{\partial z}(\rho v_r v_z)$$
$$=\frac{\partial P}{\partial r}-j_z B_\theta+\frac{1}{r}\frac{\partial}{\partial r}\left(2r\eta\frac{\partial v_r}{\partial r}\right)+\frac{\partial}{\partial z}\left(\eta\frac{\partial v_r}{\partial z}+\eta\frac{\partial v_z}{\partial r}\right)-2\eta\frac{v_r}{r^2} \quad (2)$$

The axial momentum conservation equation is:

$$\frac{1}{r}\frac{\partial}{\partial r}(r\rho v_r v_z)+\frac{\partial}{\partial z}(\rho v_z^2)$$
$$=\frac{\partial P}{\partial z}+j_r B_\theta+\frac{\partial}{\partial z}\left(2\eta\frac{\partial v_z}{\partial z}\right)+\frac{1}{r}\frac{\partial}{\partial r}\left(r\eta\frac{\partial v_r}{\partial z}+r\eta\frac{\partial v_z}{\partial r}\right)+\rho g \quad (3)$$

The energy conservation equation is:

$$\frac{1}{r}\frac{\partial}{\partial r}(r\rho v_r h)+\frac{\partial}{\partial z}(\rho v_z h)$$
$$=\frac{1}{r}\frac{\partial}{\partial r}\left(\frac{r\kappa}{c_p}\frac{\partial h}{\partial r}\right)+\frac{\partial}{\partial z}\left(\frac{\kappa}{c_p}\frac{\partial h}{\partial z}\right)+j_r E_r+j_z E_z-R \quad (4)$$

The current continuity equation is:

$$\frac{1}{r}\frac{\partial}{\partial r}(r j_r)+\frac{\partial}{\partial z}(j_z)=0 \quad (5)$$

Ohm's law is:

$$j_r = -\sigma E_r \, , \, j_z = -\sigma E_z \tag{6}$$

In the above equations, $t$ is time, $h$ is enthalpy, $P$ is pressure, $v_z$ and $v_r$ are the axial and radial velocities, $j_z$ and $j_r$ are the axial and radial component of current density, $g$ is acceleration due to gravity, κ is thermal conductivity, $C_p$ is specific heat, ρ is density, η is viscosity, $\sigma$ is electrical conductivity, $R$ is radiation emission coefficient, $E_r$ and $E_z$ are the radial and axial components of the electric field defined by $E_r = -\partial V/\partial r$ and $E_z = -\partial V/\partial z$, where $V$ is electric potential. The azimuthal magnetic field Bθ induced by arc current is evaluated by Maxwell's equation:

$$\frac{1}{r}\frac{\partial}{\partial r}(rB_\theta) = \mu_0 j_z \tag{7}$$

where, $\mu_0$ is the permeability in a vacuum.

To solve the above equations (1) to (6), the special condition about thermal flux that occurs only at the electrode surface must be considered. An additional energy flux term at the cathode surface is needed. In equation (4) an additional energy flux term at the cathode surface is needed to consider thermionic cooling due to the emission of electrons, ion heating, and radiation cooling. The additional energy flux for the cathode $H_k$ is defined as:

$$H_K = -\varepsilon\alpha T^4 - |j_e|\phi_K + |j_i|V_i \tag{8}$$

where, ε is surface emissivity, α is the Stefan-Boltzmann constant, $\varphi_K$ is work function of the tungsten cathode, $V_i$ is ionization potential of argon, $j_e$ is electron current density, and $j_i$ is ion current density.

As for thermionic emission of electrons at the cathode surface, $j_e$ cannot exceed the Richardson current density $J_R$(5) given by:

$$|j_R| = AT^2 \exp\left(-\frac{e\phi_{Ke}}{k_B T}\right) \tag{9}$$

where, A is the thermionic emission constant for the cathode surface, $\varphi_e$ is effective work function for thermionic emission of the electrode surface at the local surface temperature, and $k_B$ is the Boltzmann's constant. The ion-current density $j_i$ is then assumed to be $|j|-|j_R|$ if $|j|$ is greater thanl $|j_R|$, where $|j|=|j_e|+|j_i|$ is total current density at the cathode surface obtained from equation (5)

Similarly, for the anode surface, equation (4) needs an additional energy flux term for heating by electron condensation (thermionic heating) and radiation cooling. The additional energy flux for the anode, $H_A$ is defined as:

$$H_A = -\varepsilon \alpha T^4 + |j|\phi_A \tag{10}$$

where, $\varphi_A$ is work function of the anode and $|j|$ is current density at the anode surface obtained from equation (5). The term including $\varphi_A$ accounts for heating of the anode by electrons, which delivers energy equal to the work function absorbed at the anode. This term is analogous to the cooling effect that occurs at the cathode when electrons are emitted.

Figure 1 shows the pressure balance of keyhole. The metal vapor pressure of aluminum and magnesium[6] is calculated as:

$$\log P = -163800/T - \log T + 12.32 \quad \text{(mmHg)} \tag{11}$$

$$\log P = -7550/T - 1.41\log T + 12.79 \quad \text{(mmHg)} \tag{12}$$

P represents metal vapor pressure, and T is surface temperature of the keyhole.

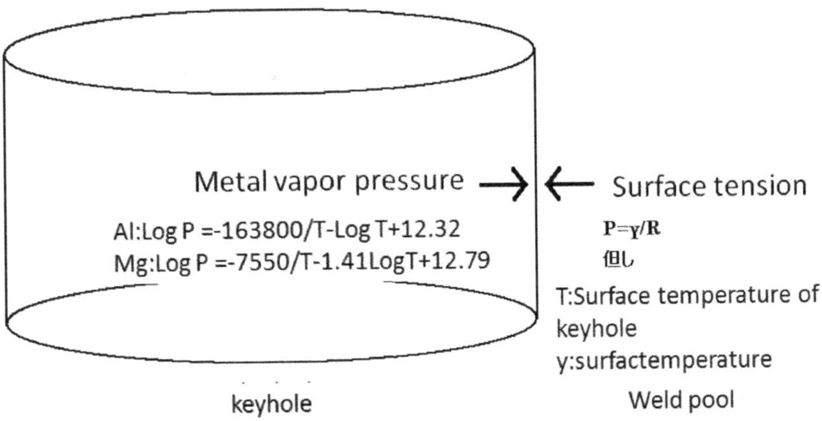


Figure 1. Model of the pressure balance at the keyhole.

## RESULT AND DISCUSSION

Figure 2 shows the temperature distribution of PAW. When nozzle diameter is 4.0mm, maximum temperature in plasma reached 20000k, and maximum axial velocity is 210km/sec. When nozzle diameter is 3.2mm, the maximum temperature in the plasma reachd 21500k, and and maximum axial velocity is 260km/sec. For both the 3.2mm and 4.0mm nozzle, the base metal is covered with plasma, so the weld pool spreads out.

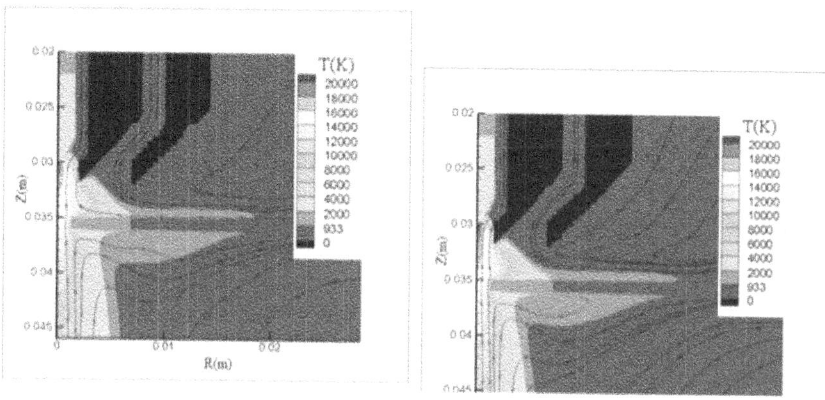


(a) nozzle diameter =3.2 (b) nozzle diameter =4.0

Figure 2. Temperature distribution in the weld pool.

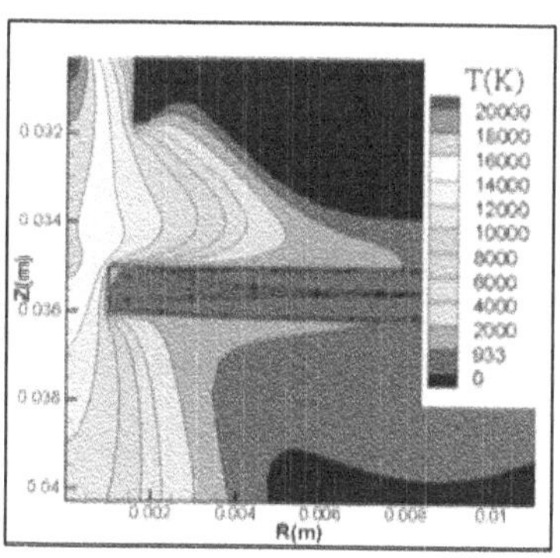


Figure 3. Temperature distribution and convection flow in the weld pool.

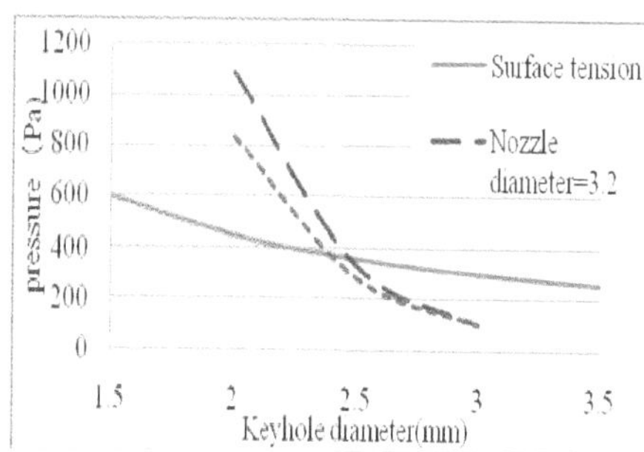


Figure 4. Pressure balance at the keyhole.

Figure 3 shows the temperature distribution and convection flow in the weld pool. The base metal is heated by the arc plasma and there is temperature gradient in the radial direction. In the weld pool, convection is generated by the Maragoni force in the radial direction. The energy is transferred primarily by convection flow and heat conduction. Therefore, the weld pool becomes 14mm.

Figure 4 shows the balance of pressure at the surface of the keyhole. Surface tension shortens the keyhole, and the metal vapor depresses the keyhole. A larger keyhole diameter results in a smaller surface tension. On the other hand, the metal vapor pressure increases as the keyhole diameter becomes larger. If the metal vapor pressure becomes equal to the surface tension, we determine the keyhole diameter. When the nozzle diameter is 3.2mm, the keyhole diameter is 2.2mm.

## CONCLUSION

In this paper, basic characteristics of plasma arc welding have been determined. The main conclusions are summarized as follows.

1) In PAW, the plasma velocity and the plasma temperature increase due to a thermal pinch affect.
2) The heat in the weld pool is taransported outward in a radial direction and downward in an axial direction due to the convection flow in the weld pool generated by the Maragoni force and the shear force by plasmajet.
3) If the keyhole diameter is large, the keyhole is shrunk because the surface tension exceeds the metal vapor pressure and the shear force. On the other hand, if the keyhole diameter is small, the keyhole is expanded by the metal vapor pressure and the shear force.

## REFERENCES

1) M. Ushio, M. Tanaka and J.J. Lowke, "Anode Melting From Free-Burning Argon Arcs", IEEE Trans. Plasma Sci., 32 (2004), 108-117.

2) M. Boulos, P. Fauchais, and E. Pfender: THERMAL PLASMA, Plenum Pub. Corp. (1994), 159.

3) M. Tanaka, et.al., "A Unified Numerical Modeling of Stationary Tungsten-Inert Gas Welding Process", Metal. Trans. A, 33A (2002), 2043-2052.

4) S.V. Patanker, Numerical heat transfer and fluid flow, Hemishpere Publishing Corporation, 1980

5) M. E. Pfender, "Electric Arcs and Arc Gas heaters", Ch. 6, published in M. H. Hirsh and H. J. Oskam, Gaseous Electronics, Academic Press, New York (1978), 291-398.

6) The Japan institute metals, edition no.3 data book of metals, 1993 Tokyo, Maruzen Co., Ltd, (in Japanese).

# A SIMULATION MODEL OF A SHORT ARC LAMP

Y. Tsujimura, K. Yamamoto and M. Tanaka
Joining and Welding Research Institute, Osaka University, Osaka, Japan
11-1 Mihogaoka, Ibaraki, Osaka, Japan

## ABSTRACT

In the lithography used to produce semiconductor products and Liquid Crystal Displays (LCD), a high power short arc lamp is desired as the display enlarges. However, this can result in a high temperature that melts the electrode. It is very difficult to solve this problem because there is still a lot of uncertainty about the detailed operation mechanisms of a short arc lamp. The objective of this study was to determine the mechanisms, like the energy balance and the mass transfer, for a short arc lamp by constructing a numerical analysis model. Because there is limited modeling in this area currently, we extended a numerical analysis model for an atmospheric pressure free burning arc to model a short arc lamp.

## INTRODUCTION

A short arc lamp is part of the lithography equipment used to produce semiconductors and Liquid Crystal Displays[1]. Making the short arc lamp high power is desired as the display enlarges. However, the temperature of the tungsten electrodes can become high and the electrodes can melt if the short arc lamp works with a large current. If the electrodes melt and the form of electrodes change, the period of time that the equipment can be used decreases. Repairs are difficult because the electrodes cannot exchange easily in a short arc lamp.

There are two ways that a short arc lamp can work without melting the electrodes, either by changing the electrode material, or by changing the design of a short arc lamp. The electrodes are made from tungsten because tungsten has a very high melting point. So, changing the electrode material is an impractical idea. That leaves changing the design of the short arc lamp so that the electrodes won't melt when operated under large current. However, solving this problem is very difficult because there is still a lot of uncertainty about the detailed operation mechanisms of a short arc lamp. This study has aimed at determining the mechanisms, like the energy balance and the mass transfer, of the short arc lamp by constructing a numerical analysis model. Because there is little modeling in this area currently, we extended a numerical analysis model for an atmospheric pressure free burning arc[2] to the short arc lamp and tried to construct a numerical analysis model. Then, we studied the intensity and efficiency of the radiation.

## SIMULATION MODEL

The tungsten cathode, the arc plasma, the tungsten anode and the glass wall are described relative to cylindrical coordinates, assuming rotational symmetry around the arc axis. The calculation domain is shown in Figure 1. The diameter of the tungsten cathode is 16 mm with a 60 degrees conical tip, and the diameter of tungsten anode is 34 mm. Tungsten electrodes never melt. The gap between the electrodes is 9.5 mm. The domain for the computation is divided into 95 nodes axially and 90 nodes radially using a non-uniform grid.

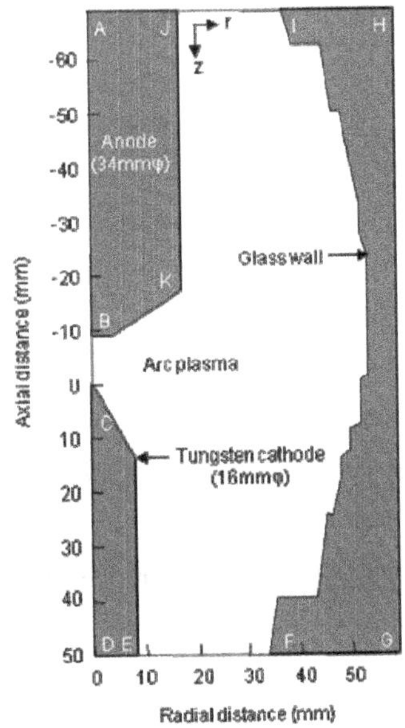

Figure 1. Short arc lamp simulation model.

An atmosphere of argon with 5 percent mercury is closed in a glass wall. Axial velocities are set to be 0 m/s at the upper and lower boundary. The flow is assumed to be a laminar flow, and the arc plasma is assumed to be under the local thermodynamic equilibrium (LTE)[3,4].

A short arc lamp is usually used with an arc current of 100 A to 200 A. The pressure before arc ignition is from 2 to 8 atm, and increases to 20 to 40 atm during use. I n this study, the arc current is set at 186 A, and pressure before ignition of arc is set at 5 atm.

The pressure in a short arc lamp increases with increasing temperature because a short arc lamp is in a closed system. The pressure is determined by the average temperature of the short arc lamp using Charles's law.

$$P = \frac{P_0 T_A}{T_0} \qquad (1)$$

where, $P$ is the pressure in a short arc lamp, $P_0$ is 5 atm as the pressure before ignition of the arc, $T_A$ is the average temperature in the short arc lamp, and $T_0$ is room temperature (300 K) as the temperature before ignition. The properties of the arc plasma depend on pressure[5]. The plasma properties at pressure in the short arc lamp are calculated using a linear approximation based on the properties at 1, 10, 20, 30 and 40 atm.

We conducted time non-dependant two-dimensional calculations. The governing equations used in the model are as follows:

The mass continuity equation is:

$$\frac{1}{r}\frac{\partial}{\partial r}(r\rho v_r)+\frac{\partial}{\partial z}(\rho v_z)=0 \tag{2}$$

The radial momentum conservation equation is:

$$\frac{1}{r}\frac{\partial}{\partial r}(r\rho v_r^2)+\frac{\partial}{\partial z}(\rho v_z v_r)=$$

$$-\frac{\partial p}{\partial r}-j_z B_\theta+\frac{1}{r}\frac{\partial}{\partial r}\left(2r\eta\frac{\partial v_r}{\partial r}\right)+\frac{\partial}{\partial z}\left(\eta\frac{\partial v_r}{\partial z}+\eta\frac{\partial v_z}{\partial r}\right)-2\eta\frac{v_r}{r^2} \tag{3}$$

The axial momentum conservation equation is:

$$\frac{1}{r}\frac{\partial}{\partial r}(r\rho v_r v_z)+\frac{\partial}{\partial z}(\rho v_z^2)=$$

$$-\frac{\partial p}{\partial z}-j_r B_\theta+\frac{\partial}{\partial z}\left(2\eta\frac{\partial v_z}{\partial z}\right)+\frac{1}{r}\frac{\partial}{\partial r}\left(r\eta\frac{\partial v_r}{\partial z}+r\eta\frac{\partial v_z}{\partial r}\right)+\rho g \tag{4}$$

The energy conservation equation is:

$$\frac{1}{r}\frac{\partial}{\partial r}(r\rho v_r h)+\frac{\partial}{\partial z}(\rho v_z h)=$$

$$\frac{1}{r}\frac{\partial}{\partial r}\left(\frac{r\kappa}{c_p}\frac{\partial h}{\partial r}\right)+\frac{\partial}{\partial z}\left(\frac{\kappa}{c_p}\frac{\partial h}{\partial z}\right)+j_r E_r+j_z E_z-U \tag{5}$$

The current continuity equation is:

$$\frac{1}{r}\frac{\partial}{\partial r}(rj_r)+\frac{\partial}{\partial z}(j_z)=0 \tag{6}$$

$$j_r=-\sigma E_r,\ j_z=-\sigma E_z \tag{7}$$

where, $h$ is enthalpy, $p$ is pressure, $v_r$ and $v_z$ are radial and axial velocities, $j_r$ and $j_z$ are the radial and axial components of the current density, $g$ is acceleration due to gravity, $c_p$ is specific heat, $\kappa$ is thermal conductivity, $\rho$ is density, $\eta$ is viscosity, $U$ is radiative emission coefficient, and $\sigma$ is electrical conductivity. $E_r$ and $E_z$ are the radial and axial components of the electric field defined by:

$$E_r=-\frac{\partial V}{\partial r},\quad E_z=-\frac{\partial V}{\partial z} \tag{8}$$

where, $V$ is electric potential.

The azimuthal magnetic field, $B_\theta$ induced by the arc current is evaluated using

Maxwell's equation:

$$\frac{1}{r}\frac{\partial}{\partial r}(rB_\theta) = \mu_0 j_z \qquad (9)$$

where, $\mu_0$ is the permeability of free space.

It is necessary to consider the effects of energy transfer at the electrode surfaces. The additional energy fluxes at the cathode and anode are described as:

$$\text{Cathode}: \; F_K = -\varepsilon\alpha T^4 - |j_e|\phi_K + |j_i|V_i \qquad (10)$$

$$\text{Anode}: \; F_A = -\varepsilon\alpha T^4 - |j|\phi_A \qquad (11)$$

where, $\varepsilon$ is surface emissivity, $\alpha$ is the Stefan-Boltzmann constant, $\phi_K$ is the work function of the tungsten cathode, $V_i$ is the ionization potential of the plasma gas, $j_e$ is the electron current density, $j_i$ is the ion current density, $\phi_A$ is the work function of the anode and $T$ is the temperature. For the cathode surface, $F_K$ needs to be included in Equation (5) to take into account the thermionic cooling by emission of electrons, radiative cooling and ion heating. Similarly, for the anode surface, $F_A$ is required in Equation (5) to take into account radiative cooling and thermionic heating. Furthermore, for the cathode surface, the electron current and the ion current are considered separately and defined based on the Richardson-Dushman equation of thermionic emission as follows:

$$j_R = AT^2 \exp\left(-\frac{e\phi_K}{k_B T}\right) \qquad (12)$$

where, $k_B$ is the Boltzmann's constant and $A$ is the Richardson's constant, which depend on the cathode material. The ion current density, $j_i$ is then assumed to be $j_i = j - j_R$, where the total current density, $j$ is $|j| = |j_e| + |j_i|$.

The major physical properties used in this model are listed in Table 1[5-6].

## CALCULATED RESULT

Figure 2 and Figure 3 show the calculated results of arc voltage, pressure, mass flow and temperature fields for the short arc lamp. The arc voltage becomes 37.9 V, and pressure in the short arc lamp becomes 29.9 atm. These results are agreed with our experiment. In the experiment, arc voltage is 40.2 V and pressure in the short arc lamp is 30 atm. The validity of this model is confirmed by these calculated results.

Table 1 Major physical properties used in this model

| Physical properties of cathode, arc plasma, and anode | | |
|---|---|---|
| Cathode | Work function | 4.5 V |
| | Effective work function for thermionic emission | 2.7 V |
| | Thermionic emission constant for Richardson equation | 16.0 A/cm$^2$K$^2$ |
| | Thermal conductivity | 0.84~0.43 W/cm K |
| | Electrical conductivity | 2.2×10$^5$~5.9×10$^3$ A/Vcm |
| | Specific heat | 0.13~0.22 J/g K |
| | Density | 18.91g/cm$^3$ |
| Arc plasma (Ar+Hg) | Ionization protential of Argon | 15.42 V |
| | Thermal conductivity | 1.7×10$^{-4}$~0.14 W/cm K |
| | Electrical conductivity | 1.0×10$^{-4}$~2.2×10$^2$ A/Vcm |
| | Specific heat | 0.43~11 J/g K |
| | Density | 2.0×10$^{-2}$~2.1×10$^{-4}$ g/cm3 |
| | Viscosity | 2.7× 10$^{-3}$~2.1×10$^{-4}$ g/cm s |
| Anode | Work function | 4.5 V |
| | Thermionic emission constant for Richardson equation | 10.0 A/cm$^2$K$^2$ |
| | Thermal conductivity | 1.67~0.85 W/cm K |
| | Electrical conductivity | 2.2×10$^5$~5.9×10$^3$ A/Vcm |
| | Specific heat | 0.13~0.22 J/g K |
| | Density | 18.91g/cm$^3$ |

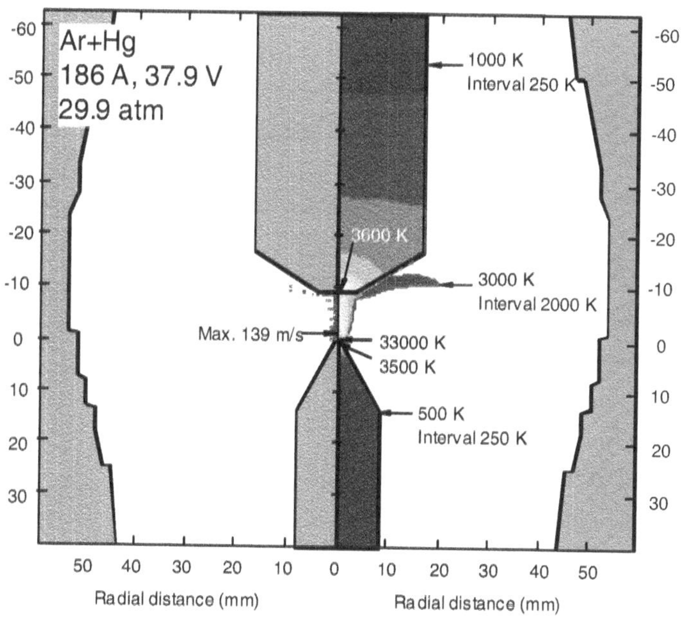


Figure 2. Calculated result of flow speed and temperature field at 186 A.

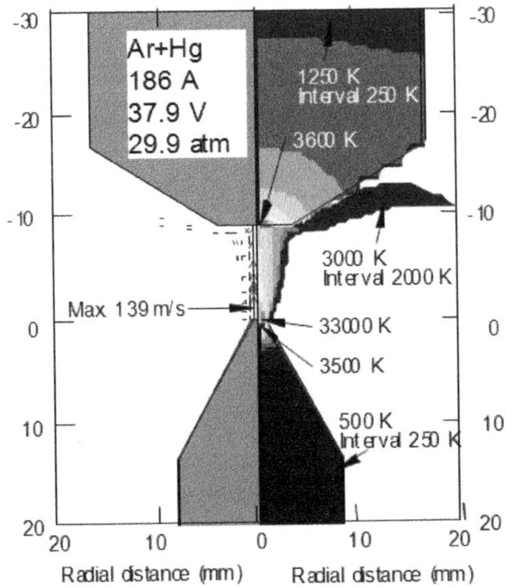


Figure 3. Calculated results magnifying the region near the electrodes.

The maximum temperatures of the electrodes are very high. Especially, the temperature of the tungsten anode is just below the melting point. Because the cathode has enough volume to cool by heat conduction, the temperature of the cathode becomes low, but the temperature becomes high for thermionic emission at the cathode tip[7]. The maximum flow speed becomes 139 m/s near the cathode.

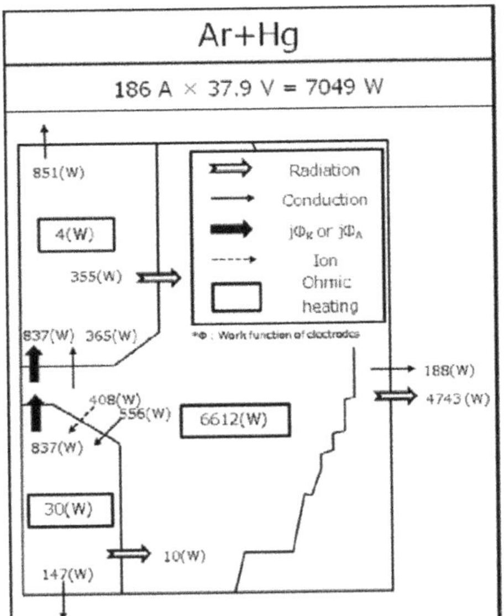


Figure 4. Calculated result of energy balance.

Figure 4 shows the calculated result for the energy balance. Almost all the heat is generated by the Ohmic heating of the arc plasma. This heat is transfer to the tungsten electrodes and outside of the short arc lamp. At the cathode surface, the energy is balanced by ion heating, conduction from arc plasma, thermionic cooling by emission of electrons, and so on. At the anode surface, the energy is balanced by thermionic heating, conduction from the arc plasma, and so on. The energy loss by radiation to the outside become 4743 W, and this energy is very large. The total power lost to radiation is over 70 percent of the input power.

## CONCLUSIONS

(1) A model for a short arc lamp was built by adjusting a numerical analysis model for an atmospheric pressure free burning arc.

(2) The calculated results for arc voltage and pressure in the short arc lamp agree with experiment.

(3) The maximum temperatures of the electrodes are very high. Especially, the temperature of the tungsten anode is just below the melting point.

(4) 70 percent of the input power is used for the radiation.

## REFERENCES

[1]P. Flech, Light and Light Sources, High-Intensity Discharge Lamps, Springer, (2006).

[2]M. Tanaka, J.J. Lowke, "TOPICAL REVIEW, Predictions of weld pool profiles using plasma", J.Phys.D:Appl.Phys.40(2007)R1-R23.

[3]T. Ouji, Introduction of Welding Processes, First edition (1996), SANPO PUBLICATIONS, INC (in Japanese).

[4]Japan Welding Society, Physics of Welding Process (1996), KUROKI PUBLICATION, INC (in Japanese).

[5]A.B. Murphy, et al, "A computational investigation of the effectiveness of different shielding gas mixtures for arc welding", J.Phys.D:Appl.Phys.42(2009),115205(14pp).

[6]"Edition No. 3 Data Book of Metals", (1993), THE JAPAN INSTITUTE METQALS, Tokyo, Maruzen co.,ltd, (in Japanese).

[7]Technical Commission on Welding Processes, Japan Welding Society: Welding Guide Book 6 "Developments in Advanced Welding Processes", (2009), KUROKI PUBLICATION, INC.

# Energy and Environment

# SOME RECENT ADVANCES IN THE UNDERSTANDING OF GAS-SOLID INTERACTIONS DURING THE GASIFICATION OF BROWN COAL AND BIOMASS

Chun-Zhu Li
Curtin Centre for Advanced Energy Science and Engineering, Curtin University of Technology, GPO Box U1987, Perth, WA 6845, Australia

## ABSTRACT

Gasification is the core of many advanced energy technologies that achieve high efficiencies and low emissions of greenhouse gases and other pollutants, particularly with the utilisation of brown coal and biomass. There are strong gas-solid interactions inside a gasifier. Our recent studies have shown that the gas-solid interactions should be an important consideration in the design and operation of a gasifier for brown coal or biomass. The gas-solid interactions could greatly change the char gasification rates by influencing the structure of char as well as the forms of the inherent catalytic species during gasification. In particular, the interactions between volatiles and char influence many aspects of the reactions involving char, and could almost terminate the gasification of char in the worst case. The essence of volatile-char interactions is believed to be the reactions involving radicals, especially H radicals. An overview of the progress in the author's research group in this area will be provided.

## INTRODUCTION

Gasification is the conversion of a solid fuel into a gaseous fuel. Gasification is a purification process. The "dirty" inorganic matter in the solid fuel will be mostly left as ash after gasification. As the gasification product gas can be further cleaned prior to use, gasification offers chances to remove pollutant-forming species (e.g. HCN and $NH_3$) so that the overall process (e.g. IGCC for power generation) using gasification as the "fuel processing step" will have lower emissions of air pollutants compared with the traditional combustion process. Advanced gasification technologies are also particularly suitable to couple with $CO_2$ sequestration technologies to achieve near-zero emissions. Therefore, gasification is the most important part of many advanced energy technologies that achieve high efficiencies and near-zero emissions of greenhouse gases and other pollutants. Gasification continues to find wide applications in power generation (e.g. IGCC and IGFC) and in the production of liquid fuels and chemicals.

Both fossil fuels (e.g. coal and petroleum coke) and renewable biomass (solid wastes) can be gasified. Many low grade fuels that are not suitable for the conventional combustion process may be utilized via gasification[1]. Therefore, gasification will play an important role in our search for additional energy sources to satisfy our ever increasing demand for energy. In fact, gasification is particularly suitable for the utilization of low-rank fuels[1-2] such as brown coal and biomass. Thermodynamically, these fuels can be gasified at a much lower temperature[2] than a high-rank fuel (e.g. anthracite). Therefore, there is a great potential to achieve very high efficiency for the gasification of low-rank fuels provided that high gasification rates are achieved. Kinetically, gasification can be conceptually divided into 2 inter-related steps: pyrolysis (to produce volatiles and char) followed by the subsequent simultaneous reforming of volatiles and gasification of char with gasifying agents. The gasification of char is normally the rate-limiting step and thus achieving rapid gasification of char is the key to realize the potential of low-rank fuel gasification.

A gasifier is a multi-phase reaction system complicated by the interfacial phenomena. Take a fluidised-bed gasifier as an example. As soon as brown coal or biomass particles are fed into the gasifier, they will be heated up rapidly and pyrolyzed to produce volatiles and char. In the presence of gasifying agents (e.g. $O_2/H_2O$), the volatiles will be thermally cracked and reformed, which will produce various radicals (e.g. H radicals) as intermediate reaction products. This is especially true for the volatiles from brown coal and biomass that are rich in oxygen and aliphatic species. The char particles, both within the bed or in the free board, are thus surrounded by gasifying agents, volatiles and the intermediates of volatile reforming, including radicals. Our work has shown that the gas-solid interactions, in addition to the interactions (gasification) between char and gasifying agents, are an important aspect of the reactions within a fluidised-bed gasifier. For example, the volatile-char interactions can alter each of the four factors[3] influencing char reactivity: (1) char structure, (2) catalyst concentration, (3) the physio-chemical form of catalysts, and (4) the distribution/dispersion of catalysts in char. The most important aspect of volatile-char interactions is the interaction between char and radicals. Furthermore, the volatile-char interactions are not unique for fluidised-bed gasifiers because there will always be some degree of volatile-char interactions in any gasifier. Within the space limit, this paper aims to provide a brief overview of some examples of our work on the gas-solid interactions during the gasification of brown coal and biomass.

## EFFECTS OF VOLATILE-CHAR INTERACTIONS ON THE VOLATILISATION OF METALLIC SPECIES

A profound feature of low-rank fuels such as brown coal and biomass is the presence of highly dispersed (even at the atomic scale) alkali and alkaline earth metallic (AAEM) species[4]. Some AAEM species exist as ion-exchangeable cations forming part of carboxylates in the coal/biomass structure. Brown coal and biomass also contain dissolved salts, NaCl and KCl respectively.

Unlike discrete minerals in bituminous coals and anthracite, the finely dispersed ash-forming species in brown coal and biomass can easily volatilize, even at temperatures as low[5] as 300°C. Their volatilisation is a major concern for the use of gasification product gas because, for example, these species could cause erosion/corrosion in downstream turbines. However, the AAEM species retained in char could act as excellent catalysts for the gasification of char. In fact, the catalytic effects of these AAEM species on char gasification make the gasification of char from brown coal and biomass distinctly different from that of high-rank coals. It is essential to understand the mechanisms of the volatilisation of AAEM species during pyrolysis.

We have obtained direct experimental evidence[6] to demonstrate that volatile-char interactions could drastically enhance the volatilisation of AAEM (Na) species. The experiments were carried out in a purposely designed two-stage fluidised-bed/fixed reactor, as is shown in Figure 1a. Three quartz frits were installed in a quartz fluidised-bed reactor to divide one reactor into two distinctively different reaction compartments in tandem. At room temperature, a NaCl-loaded Loy Yang brown coal sample was loaded into the top reactor compartment. The whole reactor was then heated under the slow heating rate conditions during which substantial AAEM species were retained[3,5] in the char. Once the reactor had reached 900°C, an acid-washed Loy Yang brown coal sample (containing virtually no AAEM species, which is termed as H-form coal) was continuously fed into the bottom reactor compartment. The pyrolysis of this H-form coal produced volatiles without releasing any AAEM species into the gas phase. The volatiles would be cracked as they flowed from the bottom to the top reactor compartment. The configuration of this reactor meant that the volatiles were forced to interact with the AAEM-containing char in the top reactor compartment. The extent of volatile-char interactions could be controlled by the concentration of the volatiles (coal feeding rate) and/or the volatile-char interaction time.

Figure 1b shows the retention of AAEM species as a function of volatile-char interaction time. A 100% AAEM retention means that there was no additional volatilisation of AAEM species due to the volatile-char interactions. The volatile-char interactions have definitely led to the release of mono-valent Na from the char while have had little effects on the release of divalent AAEM species (Mg and Ca). These data were the first direct experimental evidence to prove that the volatile-char interactions could lead to drastic volatilisation of Na. The difference between the volatilisation of mono-valent (Na) and that of divalent (Mg and Ca) AAEM species were also seen in other studies[5,7,8] and will be discussed further below.

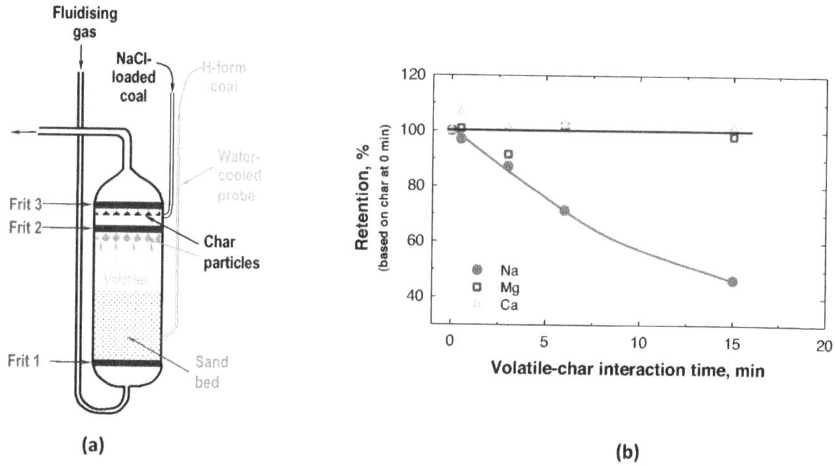


Figure 1. (a), Two-stage fluidised-bed/fixed-bed reactor. (b) Effects of volatile-char interactions on the volatilisation of AAEM species. Modified and reprinted from Ref. 6 with permission from Elsevier. The pyrolysis conditions for the AAEM-rich coal in the upper stage can be very different from and independent of those for the AAEM-free coal in the lower stage in the two-stage fluidised-bed/fixed-bed reactor.

Regardless of the forms of Na as carboxylates or NaCl in coal, Na would be bonded directly to char matrix (CM) most probably via O-containing structures in char at this temperature. The breakage of these bonds with increasing temperature is responsible for the increased volatilisation[8] of AAEM species with increasing temperature. Our study[9] has found that the volatile-char interactions would have profound effects on the volatilisation of Na only at temperatures of 700°C or higher. This is because that the essence of volatile-char interactions is the reactions of (H) radicals with char: substantial extents of thermal cracking of volatiles would only take place[10-11] at temperatures higher than 700°C. During the volatile-char interactions, the char is surrounded by radicals, most likely H radicals. It is believed[6] that the following reaction would take place to result in the release of Na:

$$CM\text{–}Na + H = CM\text{–}H + Na \quad (1)$$

While the release of a mono-valent species (Na or K) requires the breakage of one CM–M (M = AAEM species) bond, the release of divalent AAEM species (Mg and Ca) would require the simultaneous breakage of two CM–M bonds. This explains partly why the mono-valent Na and K are normally more volatile than the divalent Mg and Ca, as is shown in Figure 1.

Reaction 1 for the release of Na is in good agreement with the observation[6,9] that the release of Na due to volatile-char interactions was not necessarily accompanied by discernable weight loss of char.

The release of Na according to Reaction 1 would depend on the strength of bond CM–Na. It is believed[6,8] that the bond CM–M (including CM–Na) would be continuously broken and re-formed during pyrolysis. The newly formed CM-M bond would tend to be stronger than the original CM-M bond. This would in turn mean that the volatile-char interactions would tend to have a stronger effect on the release of Na in a nascent char than that on an "aged" char. Indeed, a lot more Na was released during the rapid pyrolysis of brown coal at a fast heating rate[5,12] than during the interactions of volatiles with an "aged" char[6] produced from the pyrolysis at a slow heating rate.

The effects of volatile-char interactions on the release of AAEM species during the pyrolysis of biomass[13] at similar temperatures are very different from that for brown coal[6,9] shown in Figure 1. In the absence of significant weight loss due to volatile-char interactions, little additional AAEM species would be volatilized during the volatile-char interactions. Two reasons have been proposed[13] to explain the difference between coal and biomass. Firstly, biomass (and thus its char) contains a lot more oxygen than brown coal (and likely its char). The abundant oxygen in the biomass char tends to make the bond CM-M stronger and thus Reaction 1 would be more difficult to take place. Evidence was obtained to show that the volatile-char interactions could also result in significant volatilisation of K (and even Mg/Ca, albeit to a much lesser extent) at high temperatures (e.g. 1000°C), at which the char would have been drastically de-oxygenated due to thermal cracking. Secondly, some biomass can have much higher ash yields (e.g. ~ 7wt% for cane trash and bagasse) than Victorian brown coal (~ 1wt%). Some non-volatile silicates may have formed although no concrete evidence was obtained to show the importance of silicate formation in the volatilisation of AAEM species during volatile-char interactions.

The volatile-char interactions during the pyrolysis of cane trash could lead to significant weight loss[13] because the "gasifying agents" (e.g. $H_2O$ and $CO_2$) produced as a part of volatiles gasified the char. This is the self-gasification. In this case, significant AAEM species could be volatilized during the volatile-char interactions. It has been proposed[13,14] that the cell structure in biomass has important implications to the release of AAEM species. Only when the volatile–char interactions led to carbon removal for the partial destruction of the cell structures in cane trash could the significant volatilisation of Mg and Ca be possible[13].

## EFFECTS OF VOLATILE-CHAR INTERACTIONS ON CHAR STRUCTURE AND REACTIVITY

Figure 2 shows some typical examples of the effects of volatile-char interactions on the reactivity of char[9,15] with air as measured at 400°C in a thermogravimetric analyser (TGA). All volatile-char interaction experiments were carried out using the reactor shown in Figure 1. For a Na-form Loy Yang brown coal (i.e. all carboxylic groups were exchanged to sodium carboxylates), the volatile-char interactions resulted in substantial decreases in reactivity[9], as is shown in Figure 2a. The volatilisation of Na during the volatile-char interactions has definitely contributed to the decreases in reactivity.

However, the plot of reactivity as a function of Na concentration in char for the data in Figure 2a showed[9] that the volatilisation of Na was not the only factor contributing to the decreases in reactivity. It is clear that the char structure must have also changed during the volatile-char interactions. In fact, Na was chemically bonded to the char matrix as an integral part of the char. The volatilisation of significant amounts of Na would need to take place not only with the Na on the char (pore) surface, but also Na from inside the char matrix. Therefore, the volatilisation of Na would imply changes in the char structure.

The changes in char structure due to the volatile-char interactions are best shown[9] in Figure 2b for an acid-washed (termed as H-form) coal in which no AAEM species were present. The decreases in char reactivity due to the volatile-char interactions can only be due to the changes in char structure. The data in Figures 2a and 2b clearly show that the H radicals, largely responsible for the effects of volatile-char interactions, must have penetrated into the inner char matrix to cause char structure to change within the char matrix. The changes in char structure due to volatile-char interactions have been further confirmed[16] by characterising the structural features of char with FT-Raman spectroscopy.

For the Ca-form coal (i.e. all carboxylic groups existing as calcium carboxylates), the volatile-char interactions resulted in decreases in reactivity at lower char conversions but increases at higher char conversions. The increases in char reactivity were due to the changes in Ca dispersion caused by the volatile-char interactions, which reduced the tendency to form[9] "less reactive carbonaceous matter", some of which can only be burnt in air by raising temperatures to 600°C. In fact, the changes in AAEM dispersion/distribution were not a surprise: AAEM species must have been mobilized and therefore their dispersion changed by the volatile-char interactions before AAEM species could be volatilized.

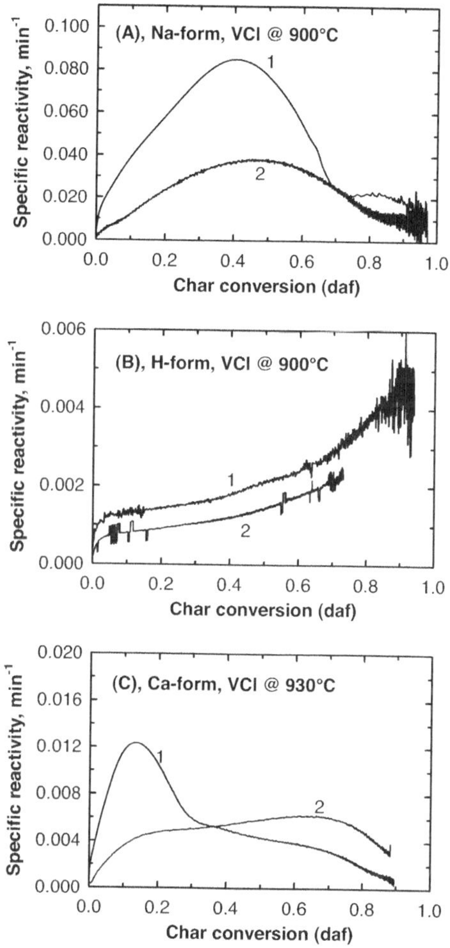


Figure 2. Effects of volatile-char interactions (VCI) on the char reactivity as measured at 400°C in air in a TGA. Curve 1, without VCI; curve 2, with VCI. Modified/reprinted from Ref. 9 with permission from Elsevier.

During the gasification of coal with steam at elevated temperatures, another mechanism of volatile-char interactions to slow down or inhibit the char gasification is the adsorption on char surface of H radicals from the gas phase cracking/reforming of volatiles. Adsorbed H radicals form relatively stable structures on the char surface, preventing the adsorption of steam on the same char surface and thus inhibiting the gasification of char by steam. In a fluidised-bed reactor, the volatilisation of AAEM species and the adsorption of H radicals on the char surface can even combine to result in the almost complete termination[16-17] of char gasification. Once the volatile-char interactions were stopped, the char could still be gasified with steam[16-17].

Largely due to the volatile-char interactions, increasing temperature did not necessarily always lead to increases in gasification rate[18] for the gasification of brown coal.

OTHER ASPECTS OF GAS-SOLID INTERACTIONS

The above examples have illustrated that the gasification of char, e.g. with steam, is much more complicated than the simple $C + H_2O = CO + H_2$ reaction, which is often viewed as a simple reaction at the gas-solid interface. The essence of volatile-char interactions is the interactions of char with radicals, especially H radicals. The H radicals can penetrate deep into the char matrix to initiate/induce further reactions inside the char. Further examples are given below to illustrate some interesting characteristics of the reactions involving H radicals within the char matrix.

In a study[19] on the gasification of cane trash, the biomass was pyrolized by heating up slowly and then steam was introduced to gasify the char in situ. The volatile-char interactions were minimised. Figure 3 shows the changes in Raman band ratios during gasification. Both Raman band area ratios in Figure 3 reflect[20] the relative ratio of small aromatic rings (no more than 6 fused rings) to large ones in the char. It was surprising to note that there were drastic decreases in the ratio of small to large aromatic rings even just after contacting with steam for 20 s. As the char weight loss was very small during this short period of gasification, the losses of small rings were mainly a result of the transformation of small rings into big ones. It is believed[19] that the intermediates of char-steam reactions, especially H, penetrated deep into the char matrix to induce the ring condensation reactions.

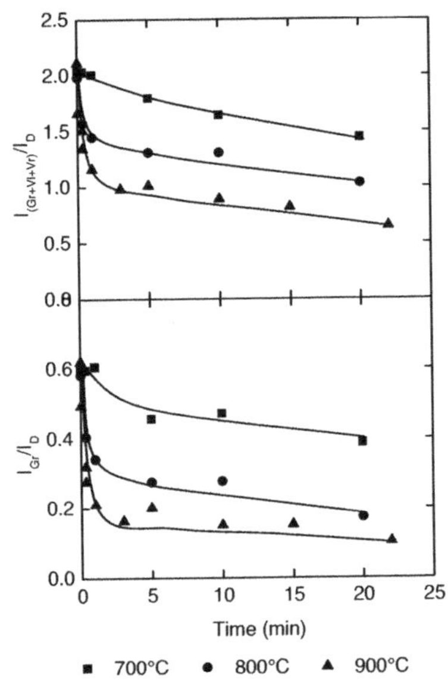


Figure 3. Ratios of Raman band areas as a function of time and temperature of steam gasification. Reprinted from Ref. 19. with permission from Elsevier.

Another example where the gasification is not limited/restricted to the reactions at the gas-solid interface is the formation of $NH_3$ and HCN during pyrolysis and gasification. It has been suggested[21-22] that the volatile-char interactions can supply additional H for the formation of $NH_3$ and HCN from the hydrogenation of coal-N. More recent experiments[23-24] have indicated that the volatile-char interactions could supply H radicals for the hydrogenation of char-N without corresponding decreases in char yield. As nitrogen is uniformly distributed within coal/char, these data clearly indicate that H radicals were able to penetrate deep into the char matrix to react with char-N.

## CONCLUSIONS

The gasification of brown coal and biomass is a very complicated multi-phase reaction system. There are strong gas-solid interactions inside a gasifier. The examples given in this short review have clearly demonstrated that the gas-solid interactions, especially volatile-char interactions, could drastically change the gasification characteristics of brown coal and biomass, for example, char structure, inherent catalyst concentration/distribution in char, gasification rates and even the release of coal-N. The char gasification is not simply limited to the reactions at the gas-solid interface. In particular, H radicals could penetrate into the char matrix to induce reactions within the char. The gas-solid interactions should be an important consideration in the design and operation of a gasifier for brown coal or biomass.


## ACKNOWLEDGEMENTS

The author wishes to thank various government granting bodies and industry partners for their continued support of the research reviewed in this short paper. The author also wishes to thank his co-workers whose contribution has made this review possible.


## REFERENCES


[1] C.-Z. Li, Editorial, Special Issue – "Gasification: A Route to Clean Energy", *Process Safety and Environmental Protection*, **84,** 407-408 (2006).

[2] J.-i. Hayashi, S. Hosokai, and N. Sonoyamah, "Gasification of Low-Rank Solid Fuels with Thermochemical Energy Recuperation for Hydrogen Production and Power Generation", *Process Safety and Environmental Protection*, **84**, 409-419 (2006).

[3] C.-Z. Li, "Some Recent Advances in the Understanding of the Pyrolysis and Gasification Behaviour of Victorian Brown Coal", *Fuel*, **86,** 1664-1683 (2007).

[4] J.-i. Hayashi, and C.-Z. Li, *Chapter 2 Structure and Properties of Victorian Brown Coal*. In: Li, C.-Z (editor). *Advances in the Science of Victorian Brown Coal*; Elsevier: Oxford, 2004; pp 11-84.

[5] D.M. Quyn, H. Wu, S.P. Bhattacharya, and C.-Z. Li, Volatilisation and Catalytic Effects of Alkali and Alkaline Earth Metallic Species During the Pyrolysis and Gasification of Victorian Brown Coal. Part II. Effects of Chemical Form and Valence, *Fuel*, **81,** 151-158 (2002).

[6] H. Wu, D. M. Quyn, and C.-Z. Li, "Volatilization and Catalytic Effects of Alkali and Alkaline Earth Metallic Species During the Pyrolysis and Gasification of Victorian Brown Coal. Part III. The Importance of the Interactions between Volatiles and Char at High Temperature", *Fuel*, **81**, 1033-1039 (2002).

[7] C. Sathe, J.-i. Hayashi, C.-Z. Li, and T. Chiba, "Release of Alkali and Alkaline Earth Metallic Species During Rapid Pyrolysis of a Victorian Brown Coal at Elevated Pressures", *Fuel*, **82**, 1491-1498 (2003).

[8] C.-Z. Li, C. Sathe, J. R. Kershaw, and Y. Pang, "Fates and Roles of Alkali and Alkaline Earth Metals During the Pyrolysis of a Victorian Brown Coal", *Fuel*, **79**, 427-438 (2000).

[9] X. Li, H. Wu, J.-i. Hayashi, and C.-Z. Li, "Volatilization and Catalytic Effects of Alkali and Alkaline Earth Metallic Species during the Pyrolysis and Gasification of Victorian Brown

Coal. Part VI. Further Investigation into the Effects of Volatile-Char Interactions", *Fuel*, **83**, 1273-1279 (2004).
[10]C.-Z. Li, and P. F. Nelson, "Fate of Aromatic Ring Systems During Thermal Cracking of Tars in a Fluidized-Bed Reactor", *Energy & Fuels*, **10**, 1083-1090 (1996).
[11]J.-i. Hayashi, S. Amamoto, K. Kusakabe, and S. Morooka, "Evaluation of Vapor-Phase Reactivity of Primary Tar Produced by Flash Pyrolysis of Coal", *Energy & Fuels*, **9**, 290-294 (1995).
[12]D. M. Quyn, H. Wu, and C.-Z. Li, "Volatilisation and Catalytic Effects of Alkali and Alkaline Earth Metallic Species During the Pyrolysis and Gasification of Victorian Brown Coal. Part I. Volatilization of Na and Cl from a Set of NaCl-Loaded Samples", *Fuel*, **81**, 143-149 (2002).
[13]D. M. Keown, J. I. Hayashi, and C.-Z. Li, "Effects of Volatile-Char Interactions on the Volatilisation of Alkali and Alkaline Earth Metallic Species During the Pyrolysis of Biomass", *Fuel*, **87**, 1187-1194 (2008).
[14]D. M. Keown, G. Favas, J.-i. Hayashi, and C.-Z. Li, "Volatilization of Alkali and Alkaline Earth Metallic Species During the Pyrolysis of Biomass: Differences between Sugar Cane Bagasse and Cane Trash", *Bioresource Technology*, **96**, 1570-1577 (2005).
[15]H. Wu, X. Li, J.-i. Hayashi, T. Chiba, and C.-Z. Li, "Effects of Volatile-Char Interactions on the Reactivity of Chars from Nacl-Loaded Loy Yang Brown Coal", *Fuel*, **84**, 1221-1228 (2005).
[16]S. Zhang, PhD thesis, Monash University, in preparation.
[17]B. Bayarsaikhan, N. Sonoyama, S. Hosokai, T. Shimada, J.-i. Hayashi, C.-Z. Li, and T. Chiba, "Inhibition of Steam Gasification of Char by Volatiles in a Fluidized Bed under Continuous Feeding of a Brown Coal", *Fuel*, **85**, 340-349 (2006).
[18]X. Li, and C.-Z. Li, "Volatilization and Catalytic Effects of Alkali and Alkaline Earth Metallic Species During the Pyrolysis and Gasification of Victorian Brown Coal. Part VIII. Catalysis and Changes in Char Structure During Gasification in Steam", *Fuel*, **85**, 1518-1525 (2006).
[19]D. M. Keown, J.-i. Hayashi, and C.-Z. Li, "Drastic Changes in Biomass Char Structure and Reactivity Upon Contact with Steam", *Fuel*, **87**, 1127-1132 (2008).
[20]X. Li, J.-i. Hayashi, and C.-Z. Li, "FT-Raman Spectroscopic Study of the Evolution of Char Structure During the Pyrolysis of a Victorian Brown Coal", *Fuel*, **85**, 1700-1707 (2006).
[21]L. L. Tan, and C.-Z. Li, "Formation of NOx and SOx Precursors During the Pyrolysis of Coal and Biomass. Part Ii. Effects of Experimental Conditions on the Yields of NOx and SOx Precursors from the Pyrolysis of a Victorian Brown Coal", *Fuel*, **79**, 1891-1897 (2000).
[22]C.-Z. Li, and L. L. Tan, "Formation of NOx and SOx Precursors During the Pyrolysis of Coal and Biomass. Part III. Further Discussion on the Formation of HCN and $NH_3$ during Pyrolysis", *Fuel*, **79**, 1899-1906 (2000).
[23]L. J. McKenzie, F.-J. Tian, and C.-Z. Li, "Effects of Volatile-Char Interaction on the Formation of HCN and $NH_3$ during the Gasification of Victorian Brown Coal in $O_2$ at 500°C", *Fuel*, **85**, 2148-2154 (2006).
[24]F.-J. Tian, S. Zhang, J.-i. Hayashi, and C.-Z. Li, "Formation of NOx precursors during the pyrolysis of coal and biomass. Part X. Effects of volatile-char interactions on the conversion of coal-N during the gasification of a Victorian brown coal in $O_2$ and steam at 800°C", submitted.

# SYNTHESIS AND APPLICATIONS OF POROUS CARBON MATERIALS IN THAILAND: A REVIEW OF RECENT RESEARCH

W. Tanthapanichakoon[1,2], N. Tonanon[1], A. Siyasuk[1], P. Poommarin[1], N. Thongprachan[1], and T. Charinpanitkul[1]

[1] Center of Excellence in Particle Technology, Faculty of Engineering, Chulalongkorn University, Pathumwan, Bangkok 10330 Thailand

[2] SCG Chemicals Co., Bangsue, Bangkok 10800 Thailand

## ABSTRACT

The present review on the synthesis and applications of porous carbon materials covers activated carbons, carbon gels, foams and monoliths with pore sizes ranging from micro- to macro-pores. Activated carbons are widely used as adsorbents in gas- and liquid-phase separation/ treatment processes. The synthesis of meso- and/or macro-porous carbon gels and monoliths enhances the applications of these functional materials. Similarly, macroporous carbon foams made of CNTs should be applicable as chemical sensors and catalyst supports. This review covers publications contributed over the past decade by Thai researchers and their collaborators, in particular by Center of Excellence in Particle Technology at Chulalongkorn University.

## INTRODUCTION

Porous materials have specific surface areas several hundred to thousand folds larger than their non-porous counterparts. The pore sizes are classified as micropore (diameter < 2 nm), mesopore (2-50 nm) and macropore (>50 nm). In addition to pore characteristics, the diverse applications of porous materials also depend on their chemical constituents, morphology, and surface topology. Among porous materials, carbon is one of the commonest substances found in our daily life and countless industries. The present review covers activated carbons (AC), carbon gels (CG), foams (CF) and monoliths (CM) with pore sizes ranging from micro- to macro-pores published over the past decade by Thai researchers and their collaborators, in particular by Center of Excellence in Particle Technology (CEPT) at Chulalongkorn University (CU).

### Research in Thailand over past decade

The main raw materials of interest come from two carbon sources. The first consists of agro-industrial wastes such as Eucalyptus bark, wattle, corncob, nut residues from bio-diesel production, coffee residues and discarded car tires. Though the resulting porous carbon materials (PCM), mostly activated carbons (AC), possess hard-to-control physicochemical properties such as specific surface area, pore size distribution, morphology[1-7], they have found applications in wastewater treatment, adsorption of aqueous phenol and heavy metals etc.[8-13]. The second source consists of chemical substrates such as 1,3-dihydroxylbenzene ($C_6H_4(OH)_2$) with formaldehyde (HCHO), wattle tannin with furfural (TFu), multi-walled carbon nanotubes (MWCNT) with sodium salt solution of carboxymethyl cellulose (CMC) etc. The synthesized porous carbon materials are mostly CG, CF and CM[14-17]. Work done by CEPT, CU, and its collaborators are recapped in the next section.

## RESEARCH IN CEPT, CU, AND ITS COLLABORATORS

### Preparation and regeneration of activated carbons (ACs)

Highly mesoporous ACs made from waste tire, coffee residues, etc. were prepared, characterized and tested[18-21]. Application of batch-wise supercritical water treatment (SWT) to the preparation & regeneration of typical ACs, anthracite and activated anthracite powders was also investigated. In addition, hydrogen peroxide ($H_2O_2$) was employed to enhance the oxidation power of SWT.

Preparation of RF Carbon Gels (CG)

RF solutions were prepared from resorcinol ($C_6H_4(OH)_2$) (*R*), formaldehyde (HCHO) (*F*), sodium carbonate ($Na_2CO_3$) (*C*) as basic catalyst, and distilled water (*W*) as the diluent. After a solution became RF gel, it was washed with t-butanol to replace the solvent at least three times and freeze-dried at 263 K. Then the gel was pyrolyzed at 1023 K under a 200 $cm^3$/min flow of nitrogen to obtain the CG[22-25].

Preparation of Carbon Monoliths (CM)

First, CG was prepared from RF as above and dried with hot air at 348 K until its weight became constant to obtain an interconnected macroporous precursor of CM. Micropores within the interconnected macropores could be developed by two methods: carbonization of the precursor in $N_2$ gas, and $CO_2$ activation. More practically, the hierarchical pore structure of CM was obtained via impregnation of the precursor with a catalyst (calcium nitrate, $Ca(NO_3)_2$), followed by $CO_2$ activation[26-27].

Preparation of Carbon Foams (CF) from MWCNT

MWCNT (0.5, 1.0, 1.5 wt.%) were dispersed in 0.5, 1.0 or 2.0 wt.% surfactant solutions and the resulting suspensions were placed in cylindrical sample holders and frozen via contact freezing with a plate heat exchanger[28]. Initially stabilized at a pre-cooled temperature of 5 °C, the suspensions were cooled at a constant rate (-0.5, or -3.0 K $min^{-1}$) down to 233 K and held constant thereafter. These frozen samples were lyophilized at 253 K for 4 days to obtain the CFs. Cut vertically in halves with respect to the direction of freezing, the porous structures of CF samples and their macropore size distributions were determined by image analysis via SEM micrographs. The electric resistivity of CNT foams was measured at ambient temperature in a vacuum chamber with the four-probe method under controlled pressure of dry air[28]. As the chamber pressure was varied, the change in electric resistance was monitored using a digital multimeter.

Table I summarizes the relevant literature review results.

## RESULTS AND DISCUSSION

Preparation and regeneration of activated carbons

Only the experimental results of SWT and adsorption of phenol are described here. It was found that SWT provided only a slight increase in the surface area and mesopore volume of the various carbon materials tested. In fact, treatment time and pre-treatment with NaOH, KOH, or $H_2O_2$ had little effect on the porous properties of anthracite powders. In addition, as $H_2O_2$ concentration increased, SWT could produce a significant decrease in the micropore volume. In short, SWT turned out to be an ineffective chemical activation method.

Table II shows typical adsorption capacity as well as SWT regeneration yield and efficiency for two ACs (CAL and AA) saturated with phenol. The adsorption capacity of virgin CAL and AA are 174 and 178 mg/g AC, respectively. The corresponding SWT-regenerated adsorption capacities are about 95 and 116 mg/g AC, respectively, whilst the accompanying weight loss is negligible. Nevertheless, ortho-chlorophenol exhausted ACs could effectively be regenerated with SWT in a *continuous* process[29].

It was shown experimentally that, in the absence of $H_2O_2$, phenol decomposition during SWT regeneration rose from 9.96% to 13.70% as the reaction temperature increased from 673 to 723 K. This small decomposition efficiency may be attributed to the oxidizing power of supercritical water and presence of dissolved oxygen in de-ionized water. However, with sufficient $H_2O_2$, phenol decomposition became essentially complete, thereby indicating the oxidation power enhancement of $H_2O_2$.

Table I. Summary of recent literature review results contributed by Thai researchers.

| Type of Porous Carbon Materials | Raw Materials | Synthesis / Activation Method | Characteristics of Carbon Material: Pore size type | BET ($m^2/g$) | Remarks | Application | Ref. |
|---|---|---|---|---|---|---|---|
| **Activated Carbon** ***- Agricultural waste*** | Eucalyptus & Wattle waste | $CO_2$ activation | Micropore | Euca. = 1491; Wattle = 1032 | **Pore vol.** of Euca. = 0.8 $cm^3/g$; Wattle = 0.56 $cm^3/g$ | N/R | 1 |
| | Corncob | $CO_2$ activation | Micropore | 919-986 | Residual volatile matter in char about 17-25% | Monoethylene glycol adsorption | 2 |
| | Corncob, bagasse, sawdust, rice husk, macadamia nut-shell | Activation by $CO_2$ & steam | Micro & Mesopore (Depend on type of raw mat.) | Up to 750 by $CO_2$ and steam activation | Bagasse and sawdust have more mesopore volumes than the other raw materials, including commercial AC | N/R | 3 |
| | Eucalyptus bark | Activation by phosphoric acid + $CO_2$ | Micropore | 1239 | This preparation shortens synthesis time and uses low acid fraction. Bulk density = 0.251 $g/cm^3$, ash content 4.88%, yield 26.2% | Iodine, methylene blue, Cu(II) and Pb(II) adsorption | 4 6 |
| | Longan seed | $CO_2$ activation | Micro& Macropore | 538-1204 | Pore vol. = 0.27-0.76 ($cm^3/g$) at 19 - 60 % burn-off | N/R | 7 |
| | Nut residues | Pyrolysis & activation by KOH, $H_3PO_4$ and $CO_2$ | Mesopore | 532 | Potential use of residues from biodiesel industry | Water purification | 5 |
| | Waste tire | Steam activation | Micro & Mesopore | 737/ 1119 (before/ after HCl treatment) | Pore vol.= 1.09-1.62 $cm^3/g$ , $V_{micro}$ = 0.37 $cm^3/g$ and $V_{meso}$ = 0.79 $cm^3/g$. HCl improves porous property of AC. | Phenol, Black 5 and Red 31 dye adsorption. Regenerated by ethanol | 18 20 |
| | Coffee residue | Activation by $ZnCl_2$ and $CO_2$ or steam | Mesopore | 914 | Amount of formaldehyde adsorption depends on surface chemistry of type of activation | Formaldehyde adsorption | 21 |
| ***- Commercial*** | Commercial AC | Fluid-bed activation by DI water | N/R | N/R | N/R | Cu(II) adsorption | 8 |
| | Coal-based granular | Fluid-bed activation by TCE and air | N/R | 1100-1200 | Bulk density = 0.5 $g/cm^3$ | Regenerated with SDS and DI water | 9 |
| **Carbon Gel** ***- From chemical substrate*** | TFu | Sol gel | Mesopore | Up to 700 | HCl aging can improve porosity of carbon gel. Mole ratio of tannin to either catalyst or water also affects porosity of carbon gel | N/R | 14 15 |
| | RF | Sol gel | Mesopore | 518-550 | Optimal pore size affects immobilized enzyme loading | Enzyme fixation | 16 |
| | RF | Sol gel | Mesopore | Up to 720 | $V_{meso}$ up to 0.9 $cm^3/g$. High conc. of catalyst affects formation of nano-whiskers on gel | N/R | 17 |
| | RF | Sol gel + Ultrasonic irradiation + Microwave drying | Mesopore | Up to 750 | Ultrasonic (US) irradiation improves mesoporosity. Microwave drying maintains porosity at high catalyst conc. and yields 3D-interconnected carbon monolith. | N/R | 23 24 |
| | RF | Sol gel + Ultrasonic emulsification | Mesopore | 510-610 | $V_{meso}$ = 0.86 $cm^3/g$. US emulsification yields submicron mesoporous carbon spheres. | N/R | 25 |
| **Carbon Monolith** ***- From Chemical substrate*** | RF | Sol gel + Ultrasonic irradiation | Macro, meso & micropore | 624 | US irradiation affects carbon monolith. Average $D_{micro}$ = 1.2 μm., $V_{micro}$ = 0.22 and $V_{meso}$ = 0.38 $cm^3/g$. Suitable activation temp. ~ 850 °C. | Potential air and water purification, separation and catalyst support. | 26 27 |
| **Carbon Foam** ***- Carbon nanotubes*** | MWCNTs + CMC | Freeze-drying technique | Macropore | N/R | Porosity up to 97%, bulk density 0.05-0.06 $g/cm^3$ and average pore dia. up to 75 μm. | Gas sensing material | 28 |

** N/R: not reported; RF: 1, 3-dihydroxylbenzene ($C_6H_4(OH)_2$) with formaldehyde (HCHO); MWCNTs + CMC: multi-walled carbon nanotubes with sodium salt solution of carboxymethyl cellulose (CMC); TCE: trichloroethylene; SDS: Sodium dodecyl sulfate; TFu: Wattle tannin–furfural.

Table II. Adsorption capacity, regeneration efficiency and regeneration yield of two representative activated carbons saturated with phenol.

| Type of AC | Trial | Adsorption capacity (mg/g AC) | Regeneration efficiency (%) | Regeneration yield (%) |
|---|---|---|---|---|
| CAL carbon (commercial) | 1 | 173.76 | - | - |
| | 2 | 95.82 | 55.15 | 98.50 |
| | 3 | 94.74 | 98.87 | 98.48 |
| Activated anthracite (AA) | 1 | 178.17 | - | - |
| | 2 | 116.27 | 65.25 | 99.60 |
| | 3 | 115.64 | 99.45 | 97.34 |

Synthesis of RF carbon gels

*Effect of surfactants*

The effect of 3 different surfactants on the carbon gels was investigated. A 10 wt% addition of SPAN 80 surfactant was found to give narrower mesopore distribution than the addition of less than 10 wt%. Therefore, the pore size distribution of carbon cryogel microspheres could be controlled to a certain extent by varying the addition of SPAN 80. In comparison, the average mesopore size in the case of cationic trimethyl-stearylammonium surfactant (C18) was larger than the case of SPAN80 because water concentration in the disperse phase also significantly affected the pore size distribution. However, the concentration of nonionic polymeric fluorinated (FC4430) surfactant was found to affect the porous property more than the C/W ratio of the starting RF solution. In fact, a drastic effect of FC4430 concentration can be seen when R/C = 100[22].

*Improvement of mesoporosity by ultrasonic irradiation (USI)*

Carbon *sonogels* (CSG) were prepared with the aid of applying USI to the RF solution. The values of $S_{BET}$, $V_{mes}$ and $r_{peak}$ of CSGs prepared at C/W = 20 and 80 mol/dm$^3$ were significantly larger than their carbon cryogel counterparts. Unfortunately, the mesopore structure of carbon cryogels prepared at C/W = 80 mol/dm$^3$ could not be observed because these samples collapsed during freeze drying and pyrolysis[23]. Interestingly, when microwave drying was used in place of freeze-drying, USI could yield gels at a higher catalyst concentration (e.g. C/W = 80 mol/dm$^3$ with R/C > 100 mol/mol), which led to a decrease in the $V_{mes}$ as the C/W ratio increased. In fact, good mesoporous structure of RF gels was obtained at 22-Watt ultrasonic irradiation during gelation[23]. Thus microwave drying turned out to be a simple way to prepare 3D interconnected macroporous carbon monoliths[24].

*Preparation of mesoporous carbon spheres by ultrasonic emulsification (USE)*

The average size of RF carbon microspheres prepared by the homogenizer method depends on the applicable shear rate. Alternatively, (sub)micron carbon spheres could be produced with USE. In fact USE produced a larger fraction of significantly smaller submicron particles in the range of 6–20 nm than the homogenizer method. In addition, unique surface texture of carbon spheres could be obtained[25].

Synthesis of carbon monoliths by ultrasonic irradiation (USI)

It was found that the gelation time of RF gel monoliths could drastically be decreased from 48 h to 8 h by applying USI. The resulting sonogel monoliths showed 3D interconnected macroporous structures, experienced insignificant shrinkage after freeze drying, and maintained their macroporous structures after pyrolysis. In addition, their macropore size distributions were relatively narrow[26].

*Preparation of carbon monoliths (CM) with hierarchical pore structure*

First, the effect of sol–gel polycondensation temperature during USI ($T_{US}$) on the interconnected macroporous structures of CMs was investigated and the suitable $T_{US}$ was found to not exceed 45 °C. Interestingly, their macropore size distributions became narrower as $T_{US}$ increased. An increase in the molar ratio C/W was shown to result in size reduction of the interconnected macropores, whereas their size distribution was found to become broader as ultrasonic power ($P_{US}$) was reduced[27].

To see the effect of $Ca(NO_3)_2$, CMs prepared at C/W 10 mol/dm$^3$, $T_{US}$ 303 K and 22 Watt *without* (sample C30) and *with* the addition of calcium nitrate (sample C30-Ca850) were compared. Both samples were found to maintain their interconnected macroporous structures while their macropore size distributions were slightly different. Next C30 was activated with $N_2$ and C30-Ca850, with $CO_2$. The results reveal that *micropores* could be created on the interconnected macropores by either activation method at 1123 K (850°C). However, the use of $Ca(NO_3)_2$ was necessary to generate *mesopores* on the macroporores. To achieve this hierarchical pore structure, the appropriate activation temperature should be around 1123 K -- not significantly lower or higher[27].

Synthesis and applications of CNT foams

A typical sample of the bulk CNT foams is displayed in Figure 1. All foam samples exhibit sponge-like elastic nature and repeatedly recover their original shapes after being gently squeezed. In fact, the samples were highly macroporous (49-74 μm pore size) with porosity of ca. 97% and bulk density of 0.05 to 0.06 g cm$^{-3}$. Fig.ure 2 (a) shows typical macroporous networks observed on the cross sections of CNT foams. As expected, the foam strength depended on the surfactant concentration and MWCNT content, which constitute the foam wall[30]. As these two values increased, the average macropore size decreased[31-34]. Interestingly, it was found that, with a fixed initial formulation, the macropore diameter of CNT foams could be manipulated via the control of the freezing condition, more specifically, the ice crystal growth rate and temperature distribution in the freezing system[35-38].

*CNT foam as potential gas detector*

The CNT foams showed electrical resistivity in the range of 1 to 10 kΩ m$^{-1}$, one evidence they retain part of the MWCNT characteristics. Interestingly, the electric resistance responded rapidly and sensitively to a step change in the surrounding gas pressure (Figure 2 (b)) due to the adsorption/ desorption of dry air on the strands of CNTs constituting the foam walls. In fact, the normalized uptake rate of resistivity as the gas pressure rose with time was positively affected by the macropore size of CF. Therefore, properly designed CNT foams show potential as gas detector/ partial pressure sensor.

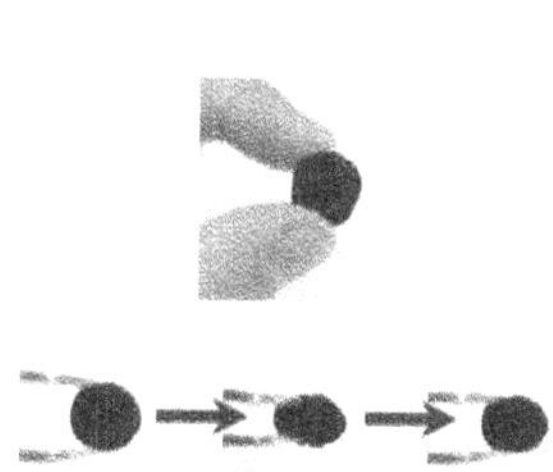

Figure 1. Elastic nature of the CNT foam.

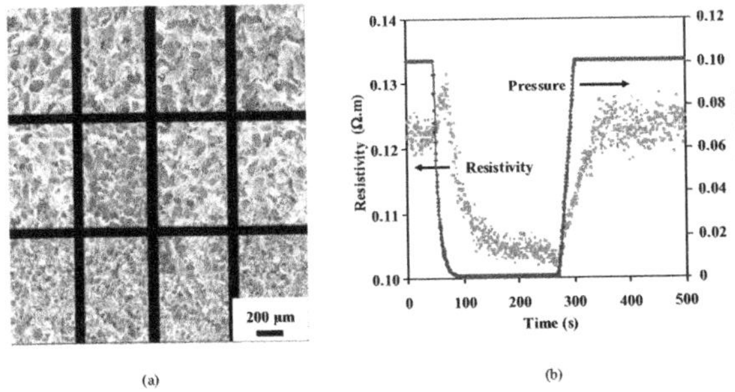


Figure 2. SEM images of sample cross sections (a) and relationship between electric resistivity and ambient pressure (b) of freeze-dried sample made of 1.0 wt.% of MWCNTs in 1.0 wt.% surfactant.

## CONCLUDING REMARKS

Research contributed by Thai researchers on the synthesis and applications of porous carbon materials has been reviewed. Spent ACs were favorably regenerated by $H_2O_2$-enhanced SWT. CGs with hierarchical pores and controlled pore size distributions were prepared by employing suitable surfactants and USI. Novel macroporous CNT foams were prepared by freeze-drying.

## ACKNOWLEDGEMENTS

Financial support from the Centennial Fund of Chulalongkorn University to CEPT is acknowledged.

## REFERENCES

[1] Y. Ngemyen, C. Tangsathitkulchai and M. Tangsathitkulchai, "Porous properties of activated carbon produced from Eucalyptus and wattle wood by carbon dioxide activation", *Korean J. Chem. Eng*, **23**, 1046-1054 (2006).

[2] A. Aworn, P. Thiravetyan, and W. Nakbanpote, "Preparation of $CO_2$ activated carbon from corncob for monoethylene glycol adsorption", *Colloids and Surfaces A: Physicochem. Eng. Aspects*, **333**, 19-25 (2009).

[3] A. Aworn, P. Thiravetyan, and W. Nakbanpote, "Preparation and characteristics of agricultural waste activated carbon by physical activation having micro- and mesopores", *J. Anal. Appl. Pyrolysis*, **82**, 279-285 (2008).

[4] P. Patnukao and P. Pavasant, "Activated carbon from Eucalyptus Camaldulensis Dehn bark using phosphoric acid activation", *Bioresource Technology*, **99**, 8540-8543, (2008).

[5] V. Sricharoenchaikul, C. Pechyen, D. Aht-ong and D. Atong, "Preparation and characterization of activated carbon from the pyrolysis of physic nut (*Jatropha curcas* L.)", *Waste Energy & Fuels*, **22**, 31-37 (2008).

[6] P. Patnukao, A. Kongsuwan and P. Pavasant, "Batch studies of adsorption of copper and lead on activated carbon from Eucalyptus Camaldulensis Dehn bark", *Environmental Sciences*, **20**, 1028-1034 (2008).

[7] S. Jumpirom, C. Tangsathitkulchai, M. Tangsthitkulchai and Y. Ngemyen, "Water adsorption in activated carbons with different burn-offs and its analysis using a cluster model", *Korean J. Chem. Eng*, **25**, 825-832 (2008).

[8] T. Banjonglaiad, E. Croiset, P. Silveston, P. L. Douglas, S. Douglas, W. Teppaitoon and S. Pongamphai, "Dynamic measurement of copper (II) ion adsorption in activated carbon fixed bed columns", *American Journal of Environmental Sciences*, **4**, 412-419 (2008).

[9] P. Thamtharai, P. Rangsunvigit and P. Malakul, "Surfactant-enhanced carbon regeneration in a vapor-phase application", *Separation Science and Technology*, **42**, 3333-3349 (2007).

[10] C. Ratasuk, C. Kositanont and C. Ratanatamskul, "Removal of haloacetic acids by ozone and biologically active carbon", *ScienceAsia*, **34**, 293-298 (2008).

[11] S. Sirianuntapiboon and P. Srisomsak, "Removal of disperse dyes from textile wastewater using bio-sludge", *Bioresource Technology*, **98**, 1057-1066 (2007).

[12] W. Nakbanpote, P. Thiravetyan and C. Kalambaheti, "Comparison of gold adsorption by Chlorella Vulgaris, rice husk and activated carbon", *Minerals Engineering*, **15**, 549-552 (2002).

[13] R. Ha and S. Vinitnantharat, "Competitive removal of phenol and 2,4-dichlorophenol in biological activated carbon system", *Enviromental Technology*, **21**, 387-396 (2000).

[14] K. Kraiwattanawong, S. R. Mukai, H. Tamon and A. W. Lothongkum, "Improvement of mesoporosity of carbon cryogels by acid treatment of hydrogels", *Microporous and Mesoporous Materials*, **115**, 432-439 (2008).

[15] K. Kriangsak, S. R. Mukai, H. Tamon and A. W. Lothongkum, "Control of mesoporous properties of carbon cryogels prepared from wattle tannin and furfural", *J Porous Mater*, **15**, 695-703 (2008).

[16] T. Chaijitrsakool, N. Tonanon, W. Tanthapanichakoon, H. Tamon and S. Prichanont, "Effects of pore characters of mesoporous resorcinol–formaldehyde carbon gels on enzyme immobilization", *Journal of Molecular Catalysis B: Enzymatic*, **55**, 137-141 (2008).

[17] N. Tonanon, W. Tanthapanichakoon, T. Yamamoto, H. Nishihara, S. R. Mukai and H. Tamon, "Formation of unique nanowhiskers on carbon gels", *Letters to the Editior/ Carbon*, **42**, 2113-2130 (2004).

[18] P. Ariyadejwanich, W. Tanthapanichakoon, K. Nakagawa, S. R. Mukai and H. Tamon, "Preparation and characterization of mesoporous activated carbon from waste tires", *Carbon*, **41**, 157-164 (2003).

[19] H. Tamon, K. Nakagawa, T. Suzuki, and S. Nagano, "Improvement of mesoporosity of activated carbons from PET by novel pre-treatment for steam activation", *Carbon*, **37**, 1643-1645 (1999).

[20] W. Tanthapanichakoon, P. Ariyadejwanich, P. Japthong, K. Nakagawa, S. R. Mukai, and H. Tamon, "Adsorption–desorption characteristics of phenol and reactive dyes from aqueous solution on mesoporous activated carbon prepared from waste tires", *Water Research*, **39**, 1347-1353 (2005).

[21] V. Boonamnuayvitaya, S. Sae-ung, and W. Tanthapanichakoon, "Preparation of activated carbons from coffee residue for the adsorption of formaldehyde", *Separation and Purification Technology*, **42**, 159-168 (2005).

[22] N. Tonanon, W. Tanthapanichakoon, T. Yamamoto, H. Nishihara, S. R. Mukai and H. Tamon, "Influence of surfactants on porous properties of carbon cryogels prepared by sol-gel polycondensation of resorcinol and formaldehyde", *Carbon*, **41**, 2981-2990 (2003).

[23] N. Tonanon, A. Siyasuk, W. Tanthapanichakoon, H. Nishihara, S. R. Mukai and H. Tamon, "Improvement of mesoporosity of carbon cryogels by ultrasonic irradiation", *Carbon*, **43**, 525-531 (2005).

[24] N. Tonanon, Y. Wareenin, A. Siyasuk, W. Tanthapanichakoon, H. Nishihara, S. R. Mukai and H. "Tamon, Preparation of resorcinol formaldehyde (RF) carbon gels: Use of ultrasonic irradiation followed by microwave drying", *Journal of Non-Crystalline Solids*, **352**, 5683-5686, (2006).

[25] N. Tonanon, W. Intarapanya, W. Tanthapanichakoon, H. Nishihara, S. R. Mukai and H. Tamon, "Submicron mesoporous carbon spheres by ultrasonic emulsification", *J. Porous Mater*, **15**, 265-270 (2008).

[26] N. Tonanon, A. Siyasukh, Y. Wareenin, T. Charinpanitkul, W. Tanthapanichakoon, H. Nishihara, S. R. Mukai and H. Tamon, "3D interconnected macroporous carbon monoliths prepared by ultrasonic irradiation", *Carbon*, **43**, 2808-2811 (2005).

[27]A. Siyasuk, P. Maneeprom, S. Larpkiattaworn, N. Tonanon, W. Tanthapanichakoon, H. Tamon and T. Charinpanitkul, "Preparation of a carbon monolith with hierarchical porous structure by ultrasonic irradiation followed by carbonization, physical and chemical activation", *Carbon*, **46**, 1309-1315 (2008).
[28]N. Thongprachan, K. Nakagawa, N. Sano, T. Charinpanitkul and W. Tanthapanichakoon, "Preparation of macroporous solid foam from multi-walled carbon nanotubes by freeze-drying technique", *Materials Chemistry and Physics*, **112**, 262-269 (2008).
[29]J. Rivera-Utrilla, M.A. Ferro-Garcia, I. Bautista-Toledo, C. Sanchez-Jimenez, F. Salvador, and M.D. Merchan, "Regeneration of ortho-chlorophenol-exhausted activated carbons with liquid water at high pressure and temperature", *Water Research*, **37** (2003): 1905–1911.
[30]S. R. Mukai, H. Nishihara and H. Tamon, "Formation of monolithic silica gel microhoneycombs (SMHs) using pseudosteady state growth of microstructural ice crystals", *Chemical Communication* ,**7**, 874-875 (2004).
[31]T. Fukasawa, Z. Y. Deng, M. Ando, T. Ohji, and S. Kanzaki, "Synthesis of porous silicon nitride with unidirectionally aligned channels using freeze-drying process", *Journal of American Ceramic Society*, **85**, 2151-2155 (2002).
[32]S. Deville, E. Saiz and A. P. Tomsia, "Ice-templated porous alumina structures",*Acta Materialia*, **55**, 1965-74 (2007).
[33]E. Landi, F. Valentini and A. Tampieri, "Porous hydroxyapatite/ gelatine scaffolds with ice-designed channel-like porosity for biomedical applications", *Acta Biomaterialia*, **4**, 1620-1626 (2008).
[34]C. Lau, M.J. Cooney and P. Atanassoy, "Conductive macroporous composite of chitosan-carbon nanotube scaffolds", *Langmuir*, **24**, 7004-7010 (2008).
[35]K. Nakagawa, A. Hottot, S. Vessot and J. Andrieu, "Influence of controlled nucleation by ultrasounds on ice morphology of frozen formulations for pharmaceutical proteins freeze-drying", *Chemical Engineering and Processing*, **45**, 783-791 (2006).
[36]A. Hottot, S. Vessot and J. Andrieu, "Freeze drying of pharmaceuticals in vials: Influence of freezing protocol and sample configuration on ice morphology and freeze-dried cake texture", *Chemical Engineering and Processing*, **46**, 666-674 (2007).
[37]A. Hazra, S. Paul, U. K. De, S. Bhar and K. Goswami, "Investigation on ice nucleation/ hydrate crystallization by aqueous solution of ammonium sulfate", *Progress in Crystal Growth and Characterization of Materials*, **47**, 45-61 (2003).
[38]K. Nakagawa, "Modeling of freezing step during freeze-drying of drugs in vials", *AIChE*, **53**, 1362-1372 (2007).

# MODERN REAL-TIME CHARACTERIZATION OF AIRBORNE NANOPARTICLES

Wladyslaw W. Szymanski* and Günter Allmaier**
* Faculty of Physics, Aerosol Physics and Environmental Physics Division, University of Vienna, Austria
**Institute for Chemical Technologies and Analytics, Technical University of Vienna, Austria

## ABSTRACT

Nanoparticles are ubiquitous in natural and technological environments. In the airborne state, they are called nanoaerosols. They are of substantial interest due to a variety of potential technological applications; particularly, they are important as precursors for the formation of larger particles and structures. Their properties are different from their bulk counterparts due to their much larger surface-to-volume ratio. Even with the same chemical composition, nanoparticles can show considerably different physicochemical characteristics. Experimental characterisation techniques are among the first measures towards understanding the behaviour of nanoparticles and of nanoaerosol systems. The application of an appropriate technique to measure the relevant parameters to obtain information and insight into processes determining the behaviour and fate of nanoaerosols is essential. Various moments of size distributions can be nowadays measured. In this contribution, selected real-time electrical methods for measuring aerosol nanoparticles are presented, and, based on experimental findings, their shortcomings and benefits are discussed.

## INTRODUCTION

Nanoparticles are considered the building stones of nanotechnology[1]. Nanotechnology has a huge potential with regard to industrial applications, but it also has real prospects to improve the environment - playing a role in energy production and storage, as well as the development of new technologies. The increasing drive to develop nanoscale materials and functional structures utilizing the advantageous physical, chemical, and biological properties brings with itself the obvious need to characterize nanoparticles; not only in technological processes but also in the ambient environment due to risks associated with their presence[2]. However, so far no standard procedures for measuring aerosol nanoparticles and their properties have been attempted to be invoked. Moreover, there is even no standard definition regarding nanoparticles. Definitions, as well as the upper limit in terms of particle size may vary depending on the source of information and application. The Publicly Available Specification (PAS) in the UK proposed that a nanoparticle is a particle "having one or more dimensions of the order of 100 nm or less". This seems to be a useful starting definition regarding the size of ultrafine particles (UFP) in the emerging discipline of nanotoxicology[2]. Due to increasing interest in aerosol nanoparticles and the necessity for a real-time measurement capability, measuring techniques need continuous development. High efficiency detection and characterisation for nanoaerosols even below 1 nm in size, along with the invention of new and/or improved techniques and the standardization of existing methods (determination of particle size, concentration, morphology, structure and, if possible chemical characterization and biological effect characterisation) are a constant challenge. The posed questions have interdisciplinary character leading to an increasing interdependence of research fields with decreasing particle dimension. During the past 60 years, or so, the drive to characterize airborne particles in real time resulted not only in a better definition and understanding of aerosols, but also rapid implementation of industrial applications utilizing nanoparticles (Figure 1).

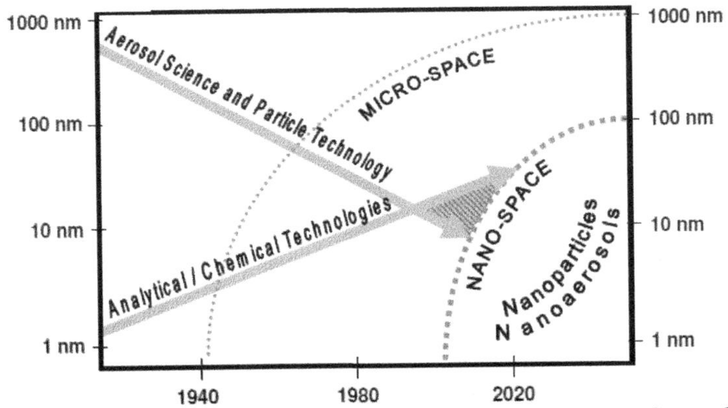


Figure 1. Development in time within past decades and into the near future of methods for characterizing particles of a given size.

The parameterization of nanoparticle dimension in terms of size (diameter) only is actually not the best approach. Particles with sizes on the order of few nanometers, or less, contain several hundreds of atoms only. This brings with itself a fact of the increasingly larger surface to volume ratios, which in consequence, influences the physical and/or chemical properties of nanoaerosols without changes in the chemical composition of the particle itself. Quantum size effects, material stability, conductivity, magnetic, or optical properties can be varied through changes in dimension or structure[1].

## REAL-TIME MEASURING TECHNIQUES

For real-time measurement of micro- and nanoparticles electrostatic techniques seems to be among the best and convenient approaches currently available. The operating principle is based on a two-step technique: conditioning of particle charge usually with ions[3] generated by means of a weak radioactive source, corona discharge, intense light source or soft X-rays and a consecutive manipulation of the charged aerosol by electrical field and particle detection.

In recent years the rise of aerosol mass spectrometry provided instruments delivering chemical information in real-time from the surface of single aerosol particles. The combination of aerosol measuring techniques and mass spectrometric techniques into a real-time measurement system offers unique possibility of physical and chemical characterization of nanoparticles[4-7].

The *conditio sine qua non* for utilization of electrostatic techniques for nanoparticle measurement is a known, well-defined and experimentally repeatable charge condition of investigated particles. Ideally, a highly efficient charging process resulting in a single charge per particle and independent of its size would be desirable. Unfortunately the currently known particle charging methods do not deliver such a perfect result. There are however various means to condition the charging status of aerosol particles. Likely the most widely spread method is the diffusion charging in a bipolar ion atmosphere, which is obtained by a weak radioactive source, typically Kr-85 (β radiation), Am-241, or Po-210 (both α radiation). The simplicity of use

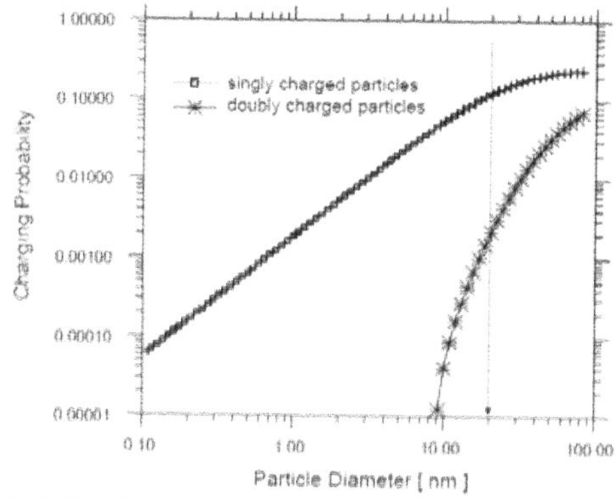


Figure 2. Diffusion charging probability of particles calculated with the Fuchs-Boltzmann charge equilibrium approach[8].

is offset by the trouble of handling radioactive materials and by the extremely low charging probability for particles with sizes below about 20 nm (Figure 2).

As can be seen in Figure 2 only about 10% of particles with 20 nm in size are charged, showing clearly the inefficiency of this charging process for nanoparticles. It should be also noted that particles with sizes below 100 nm, which were successfully charged, carry, from practical point of view, only a single charge due to the very low double charging probability.

Developments are under way to improve the charging probability with a corona discharge[9-11] or soft X-ray radiation[12], however a real breakthrough is yet to be achieved.

In past years the electrical approaches to measure aerosol particles was continuously developed and refined. The so-called electrical mobility of a charged particle in a defined electrostatic field is utilized in Differential Mobility Analyzers (DMA). The working principle of a DMA is based on the steady-state electrical drift velocity of a charged particle in an electrostatic field between two electrodes. Orthogonally to the electrostatic field there is a constant, laminar flow of a particle-free air. When a voltage applied to an electrode is varied, particles with certain electrical mobility which correlates with particle size can be extracted from such a DMA system can be consequently monitored and analyzed[13-16].

A specific further development of this method is the Parallel Differential Mobility Analyzer (PDMA) (Figure 3). This system allows measurement of nanoparticle size spectra and simultaneously a particle size fraction can be either collected or enriched for further handling. Moreover such a unit can be combined with complementing analytical tools such as mass spectrometry (MS). This combination of techniques delivers information even on the density of nanoparticles. The PDMA system is equipped with an electrical aerosol detection device working on the Faraday cup principle thus allowing in principle measurement of particles even below 1nm in size. Parallel with the scanning nano-DMA which scans and delivers the complete size spectrum of the generated particles, an identical (dimensions, flows, electric field) separation nano-DMA is in operation at one fixed but selectable voltage setting and used for sampling or enrichment of the selected size class of nano-particles, therefore we use the acronym PDMA[17].

The prerequisite for this detection method is the necessity of a certain minimum number of charged particles (concentration) needed to achieve measurable electrical signal. The detection limit of this system is currently in the order of $10^2$ charged particles per $cm^3$.

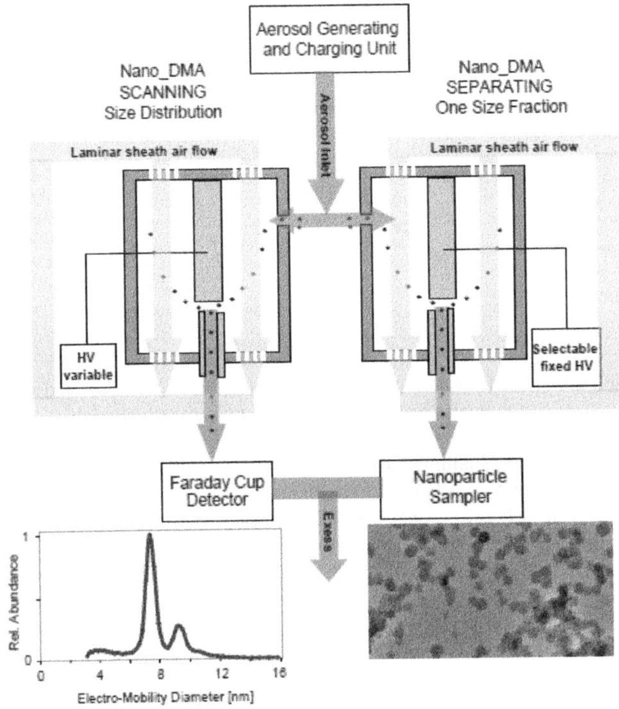


Figure 3. Diagram of the PDMA system containing two nanoDMAs indicating the function of each device: scanning (spectrum) and sampling (collected particles).

For sampling of a selected size fraction, an electrostatic nanoparticle sampler was constructed. A high voltage source is electrically connected with a TEM support grid imbedded into an insulator. The HV polarity between the housing and the TEM support grid was matched to the polarity of the central electrode of the PDMA. This leads to a deflection of the charged particles perpendicular to the aerosol flow in direction of the TEM-grid. After sampling of the desired particle fraction the sample can be investigated with a TEM providing e.g. an alternative way to determine the collected particle size distribution[17-18].

Nanoaerosol particles may form depending on their number concentration agglomerated, non-spherical nanoparticles. In such a case their mass or density can not be easy derived from differential mobility analysis. However, an innovative approach published recently[18-20] opens new avenues for more complete nanoaerosol characterisation as in the past, providing not only mass but also parameters such as fractal dimension, characteristic density and shape factors of particles in question. The principle of operation of this measuring method is shown in Fig. 4.

Balancing the centrifugal ($\boldsymbol{F_Z}$) and the electrostatic force ($\boldsymbol{F_{EL}}$) in a laminar flow field allows manoeuvring of particles through the gap between the rotating inner and outer electrode which have a certain specific mass-to-charge ratio. Certainly, prior to entering the instrument

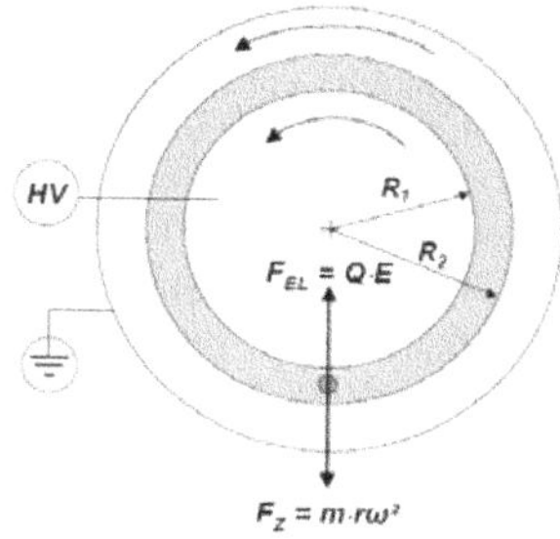


Figure 4. Principle of operation of the APMA. The electric field in the gap between the inner electrode (radius R1) and outer electrode (radius R2) is determined by the applied voltage (HV).

particle charge must be conditioned. Exiting the instrument they can be detected. Most commonly used methods are condensational particle counters or aerosol electrometers[22].

The ultimate objective for a nanoparticle characterisation would be measurement of each single particle over the entire size range of interest with size-resolved particle chemical composition and physical properties. Unfortunately, there is no single instrument capable of that task, but combination of measuring techniques opens new avenues for particle characterisation. Combining DMA and APMA provides real-time information regarding nanoparticle size, mass, density and fractal structure. Evidently for spherical particles simultaneous measurement of size (DMA) and mass (APMA) delivers instantaneously density of each detected particle[17].

The measurement of non-spherical particulate aggregates poses much more challenging task. However, the combination of these two measuring systems (DMA and APMA) allows approaching this problem as can be seen in Figure 5. The assessment of an effective density $\rho_{eff}$ of measured soot agglomerates is possible knowing both $D_{DMA}$ and $D_{APMA}$ which is a mobility diameter determined by the DMA and diameter of the average mass determined by the APMA, respectively.

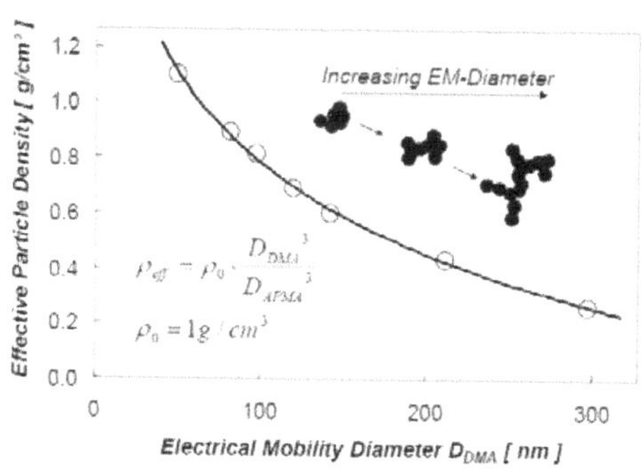


Figure 5. Effective density of soot agglomerates decreases with increasing complexity of the aggregate corresponding to the increasing electrical mobility diameter (from [20]).

The continuously increasing necessity for comprehensive characterisation of nanoaerosols and better understanding of processes regarding particles on the nanosize level caused that mass spectrometry obtains rather central position among various aerosol measuring techniques[4-6].

High sensitivity and diversity of applications of MS techniques combined with aerosol methods opens new ways of description of airborne nanoparticles including such species as macromolecules and viruses[6, 16, 23-24]. Particularly inviting and relatively straightforward is the combination of a DMA delivering size information with MS delivering mass information for globular nanoparticles such as many proteins, or noncovalent biospecific protein complexes or polymers like polystyrol, or dendrimeres[25]. Figure 6 shows electrical and mass spectra of dendimeres (Generation 6 (G6) Diameter = 6.7 nm; Generation 9 (G9) Diameter = 11.4 nm; values based on manufacturer's information – Dendritech, Inc., USA)

The dendrimeres (also called dendritic molecules) represent a new class of polymer-based molecular architecture, having globular shape, various sizes and clear distinction between core and surface making them suitable for many nanotechnological applications. To be able to apply dendrimers for specific use they have to be adequately characterized in terms of size, mass, density, or hydrophobic/hydrophilic properties. Using combined information regarding the size (DMA) and mass (MS) of nanoparticles provides means to an assessment of their density, as can be seen in Figure 7. Ideal dendritic growth assumes exponential molecular mass increase leading to steady increase of the density. As shown here it is obviously not the case, resulting probably from the deficient saturation of branching during the synthesis step from one dendrimer generation to the next one and leading in consequence to practically constant density.

It is evident that for a complete characterisation of nanoaerosol particles combinations of instruments measuring different particle properties lead to a better understanding and a more complete characterisation of nanoparticles.

CONCLUSION

The advantages in science, technology and the industrial utilization of nanoparticles are obvious. The risk to humans and environment of exposure to nanoaerosols is not clearly understood, but estimated to be rather high. The further development of nanotechnology is inescapable, but at the same time the presence of ambient nanoparticles is a growing concern. Continuous interdisciplinary efforts are needed for better, faster and affordable real-time

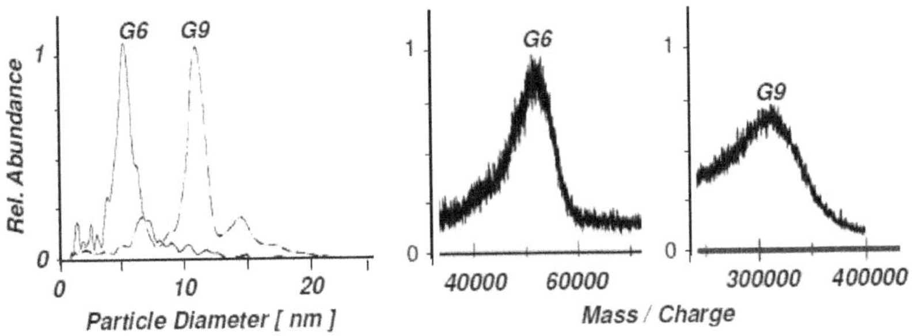


Figure 6. Spectra of dendrimers (G6 and G9) measured by means of a differential mobility analysis and positive ion MALDI-TOF mass spectrometry (adapted from [6])

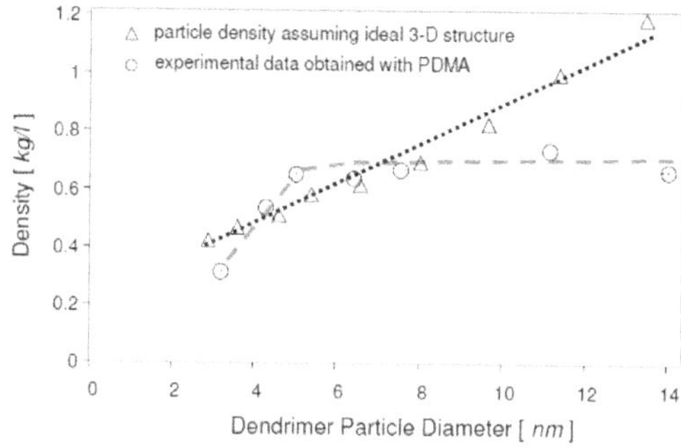


Figure 7. Measured and theoretical values of dendrimer densities vs. size obtained from data using differential mobility analysis and MALDI-TOF-MS (adapted from [6]).

characterisation of aerosol nanoparticles. Only the permanent cooperation between various areas of science and technology will result in new ways to characterize and improve the understanding of the nature and impact of nanoaerosol particles.

REFERENCES


[1] G. Cao, Nanostructures & Nanomaterials – Synthesis, Properties & Applications, *Imperial College Press* (2004).

[2] G. Oberdorster, E. Oberdörster, and J. Oberdörster, "Nanotoxicology: An Emerging Discipline Evolving from Studies of Ultrafine Particles". *Environ. Health Perspectives* 113 (7): 823–839 (2005).

[3] H. A. Erikson, "On the nature of the negative and positive ions in air, oxygen and nitrogen", *Phys. Review*, *20*, 117-126 (1922).

[4] J. Marijnissen, B. Scarlett, P. Verheijen, "Proposed on-line aerosol analysis combining size determination, laser-induced fragmentation and time-of-flight mass spectroscopy, *J. Aerosol Sci.* 19(7):1307-1310 (1988).

[5] C.A. Noble, K.A. Prather, "Aerosol time-of-flight mass spectrometry: a new method for performing real-time characterization of aerosol particles", *Appl. Occup. Environ. Hygiene* 13:439-443 (1998).

[6] R. Müller, C. Laschober, W. W. Szymanski and G. Allmaier, "Determination of molecular weight, particle size and density of high number generation", PAMAM dendrimers using MALDI-TOF-MS and nES-GEMMA, *Macromolecules*, 40:5599-5605 (2007).

[7] J.T. Jayne, D.C. Leard, X. Zhang, P. Davidovits, K.A. Smith, C.E. Kolb, and D.R. Worsnop, "Development of an aerosol mass spectrometer for size and composition analysis of submicron particles", *Aerosol Sci. Technol.*, 33, 49-70 (2000).

[8] N.A. Fuchs, "On the Stationary Charge Distribution on Aerosol Particles in a Bipolar Ionic Atmosphere", Geofis. Pura Appl. 56:185-193 (1963).

[9] Q. Chaolong, D-R. Chen and D.Y.H. Pui, "Experimental study of a new corona-based unipolar aerosol charger", *J. Aerosol Sci.*, 38: 775-792 (2007).

[10] C. Laschober, S. L. Kaufman, G. Reischl, G. Allmaier and W.W. Szymanski, "Comparison between an Unipolar Corona Charger and a Polonium-based Bipolar Neutralizer for the Analysis of Nanosized Particles and Biopolymers", *J. Nanosci. Nanotechnol.* 6, 1474–1481 (2006).
[11] X. Dexuan, S. Lianxi, Z. Jingsheng and Z. Jianwei, "Positive Short-Pulse Corona Discharge Charging of Aerosol Particles", *Jpn. J. Appl. Phys.* 42: 1766-1769 (2003).
[12] Kulkarni, P., Namiki, P.,Otani, Y. and Biswas, P., "Charging of Particles in Unipolar Coronas Irradiated by in-situ Soft X-Rays", *J. Aerosol Sci.* 33:1279-1296 (2002).
[13] E.O. Knutson and K. T. Whitby, "Aerosol classification by electric mobility: apparatus, theory, and applications", *J. Aerosol Sci.*, *6*, 443-451 (1975).
[14] D.R. Collins, A. Nenes, R.C. Flagan, and J. H. Seinfeld, "The scanning flow DMA", J. Aerosol Sci., 31:1129 – 1144 (2000).
[15] B.Y.H. Liu and D. Y. H. Pui, "On the performance of the electrical aerosol analyzer", *J. Aerosol Sci.*, *6*, 249-264 (1975).
[16] D.R. Chen and D. Y. H. Pui, "Numerical modeling of the performance of differential mobility analyzers for nanometer aerosol measurement", *J. Aerosol Sci.*, *28*, 985-1004 (1997)
[17] G. Allmaier, C. Laschober, and W.W. Szymanski, "Nano ES-GEMMA and PDMA, new tools for the analysis of nanobioparticles – protein complexes, lipoparticles and viruses", *J. Am. Soc. Mass Spectrom.* 19: 1062-1068 (2008).
[18] G. Allmaier, W.W. Szymanski, G. Reischl, and C. Laschober, "Method for classifying and separating particles and a device for carrying out said method", pat.*WO2007/016711A1* (2007).
[19] K. Ehara, C. Hagwood and K.J. Coakley, "Novel method to classify aerosol particles according to their mass-to-charge ratio—aerosol particle mass analyser", *J. Aerosol Sci.* 27: 217–234 (1996).
[20] P. H. McMurry, X. Wang, K. Park, and K. Ehara, "The Relationship Between Mass and Mobility for Atmospheric Particles: A New Technique for Measuring Particle Density", *Aerosol Sci.Technol.* 36: 227-238 (2002).
[21] K. Park, C. Feng, D. B. Kittelson and P. H. McMurry, "Relationship between particle mass and mobility for diesel exhaust particles". *Environ. Sci. Technol.* 37: 577-583 (2003).
[22] A. Ankilow, A. Baklanov, M. Colhoun, K.-H. Enderle, D. Filipovicova, A. Julanov, A. Lushnikov, A.Vrtala, R. Mavliev, F. McGovern, A. Mirme, T .C. O`Connor, J. Podzimek, O. Preining, G. Reischl, R. Rudolf, G. Sem, W.W. Szymanski, E. Tamm, P. Wagner and A. Zagaynov, "Particle size dependent response of aerosol counters", *Atmos. Res.* 62: 209-237 (2002).
[23] G. Bacher, W.W. Szymanski, S.L. Kaufman, P. Zöllner, D. Blaas and G. Allmaier, "Charge Reduced Nano-electrospray Combined with Differential Mobility Analysis of Peptides, Proteins, Glycoproteins, Noncovalent Protein Complexes and Viruses", *J. Mass Spectrom.* 36:1038-1052 (2001).
[24] C.J. Hogan, E.M. Kettelson, B. Ramaswami, D.-R. Chen, and P. Biswas, "Charge Reduced Electrospray Size Spectrometry of Mega- and Gigadalton Complexes: Whole Viruses and Virus Fragments", Anal. Chem. A. *78* (3), 844–852 (2006).
[25] D.A. Tomalia, and J.M.J. Frechet, "Discovery of dendrimers and dendritic polymers: A brief historical perspective", *J. Polymer Sci. A.* 40: 2719–2728 (2002).

# NOVEL RECYCLING PROCESS OF WASTE FRP FOR ADVANCED MATERIALS

Akira Kondo[1], Hiroya Abe[1], Makio Naito[1], Yasuko Okuni[2], Masashi Miura[2], and Norifumi Isu[2]
[1]Joining and Welding Research Institute, Osaka University, 11-1, Mihogaoka, Ibaraki, Osaka, 567-0047, Japan
[2]INAX Corporation, 3-77, Minatomachi, Tokoname, Aichi, 479-8588, Japan

## ABSTRACT

The volume of waste Fiber Reinforced Plastic, FRP, has increased every year. Almost all of the waste FRP is incinerated or disposed of in a landfill. In this study, a new recycling system involving the use of waste FRP to produce advanced nanoporous materials is proposed. An attrition type mill was used to separate the glass fibers in FRP from the other fine particles of matrix (resin) and fillers ($CaCO_3$, etc.) The separation rate was affected by the mechanical processing conditions, and it was found that the roughness of the inside wall of the chamber is an important factor for increasing the separation rate. The milled powder was subsequently coated with waste $Al(OH)_3$ nanoparticles, from which nanoporous composites compacts were formed by dry pressing. The flexural strength of these compacts was increased with a steam curing treatment.

## INTRODUCTION

Fiber Reinforced Plastic, FRP has been used in various applications including boats and building materials, because it is lightweight, has high strength, and has good workability. According to an estimate by the Japan Reinforced Plastics Society (JRPS), the waste FRP volume is around 400,000 tons/year in Japan. In the present situation, almost all of the waste FRP is burned or disposed in a landfill. Only 1-2 % of the waste FRP is recycled as cement raw material or as additives in concrete. JRPS started to produce cement including recycled FRP in 2002.[1] Incineration of FRP has the problems of lower calorific values for burning, and the residue generated by incineration also needs a waste disposal site.

To recycle waste FRP, some chemical recycling approaches have been studied, such as dissolution using solvent[2] and using a supercritical fluid. However, they have not been used practically. Those chemical methods still have serious problems. Both result in a high processing cost for recycling FRP due to the treatment with solvents or processing under high temperature and high pressure conditions. Another problem is the generation of by-products. Additionally, the recycled materials do not have better quality than that of the starting materials.

On the contrary, the recycling concept proposed by the authors does not aim to obtain original elements, but to develop further advanced materials using the disassembled blocks of the waste composite materials. According to the concept, waste FRP should be separated into its intermediate structure, and then, it should be assembled with another material to make further advanced materials. FRP usually contains 40-50% of calcium carbonate as filler, and 20-30% of glass fibers. To realize their concept, the authors developed a process to separate the glass fiber from the FRP using an attrition type mill. The separated powders are subsequently coated with additive nanoparticles using particle bonding. The composites obtained are compacted to make nanoporous materials by dry pressing [3]. In this study, the effect of the roughness of the inside wall of the milling chamber on the separation rate of glass fibers from the waste FRP was investigated. Furthermore, steam curing was applied to increase the flexural strength of the porous compacts obtained by dry pressing the composites.

EXPERIMENT

The attrition type mill employed for the separation of waste FRP is shown in Figure 1. This apparatus has a fixed chamber (inner diameter 80mm) and an oval rotor set with a certain clearance against the inside wall of the chamber. Both chamber and rotor were made from stainless steel. When the rotor rotates, the powder mixture is compressed into the clearance (4 mm in gap) and receives various kinds of mechanical forces including compression and shearing. To date, only a smooth surface chamber wall has been employed. In this study, a rough surface wall chamber (shown in Figure 1) was used to examine its effect on the separation rate of waste FRP.

Waste FRP chips crushed down to about 10mm were used as the starting material (shown in Figure 2). The constituent composition of the waste FRP is listed in Table I. The waste FRP has 47.6 mass% of $CaCO_3$ and $Al(OH)_3$ as filler.

As nanoparticles for coating the surface of glass fibers, waste $Al(OH)_3$ was used. The waste $Al(OH)_3$ was produced as a by-product from the aluminum sash process. Although the primary particle size of $Al(OH)_3$ was about 50 nm, most of the material was coarse aggregates.

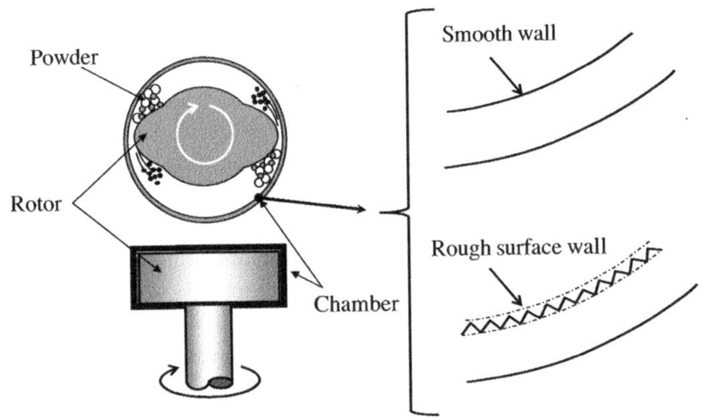


Figure 1. Illustration of the attrition type mill.

The waste FRP chips (20g) were processed with the attrition type mill (Figure 1). Then, the $Al(OH)_3$ nanoparticles (8g, 40% at mass base) were added to the processed powder, followed by an additional 30 minutes of processing in the attrition type mill with the smooth wall. The surface of the glass fiber and the other ground particles were uniformly coated with the nanoparticles. During the attrition type milling, the electric load of the motor was set about 0.45kW by adjusting the rotating speed of the rotor. As a result, it was adjusted in the range from 800rpm to 1800rpm.

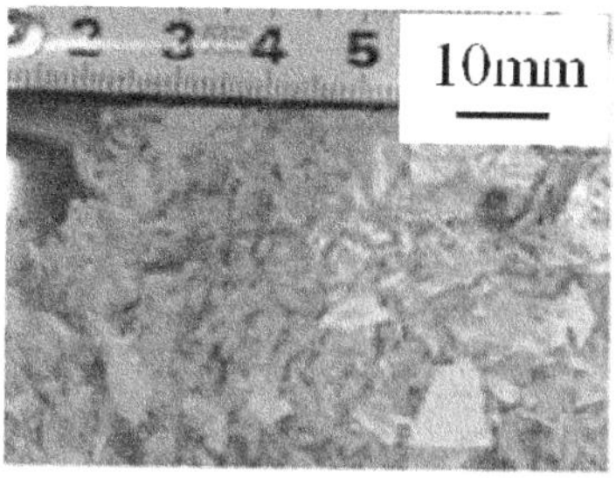


Figure 2. Photograph of the waste FRP chips.

Table I. The composition of the Waste FRP.

| Components | Contents (mass%) |
|---|---|
| Glass fiber | 25.6 |
| Resin | 24.0 |
| $CaCO_3$ | 29.3 |
| $Al(OH)_3$ | 18.3 |
| Others | 2.8 |

The composite powders were processed through a 355μm mesh size sieve. The powder processed through the sieve was easily processed into bulk compacts by dry pressing at 100 MPa, followed by the steam curing at 80 °C. The specimen size was 12 mm × 80 mm × 5mm. A vibration type mill was also applied to grind FRP into glass fibers and other elements.

## RESULTS AND DISCUSSION

Figure 3(a) shows a SEM image of the waste FRP after attrition milling in the smooth surface chamber. It is obvious that the glass fiber in FRP has its own shape. The length of the glass fibers ranges from about 100 μm to over several 100 μms. The glass fibers were also ground using the vibration type mill (Figure 3(b)). In vibration milling, impact and compression forces are mainly applied to the material, resulting in ground glass fiber powders. These results show that the proposed method using the attrition type milling apparatus is very effective in selective separation of glass fibers from the matrix components.

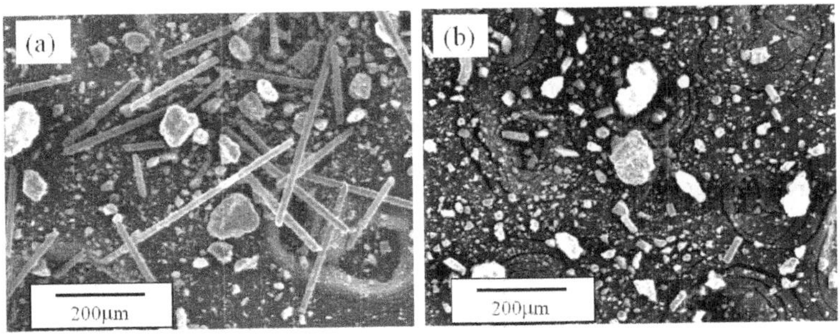


Figure 3. SEM images of the waste FRP powder milled using:
(a) an attrition type mill (passed through 355μm sieve), and (b) a vibration type mill.

Figure 4 shows the separation rate for the waste FRP by using the attrition type mill. First, for the case of the smooth wall, the percentage of the fine powder (under 355μm) increases with the milling time, and reaches about 50 mass% after 90 min. of processing. On the other hand, using the rough surface wall, the production rate of fine powder increased remarkably. In only

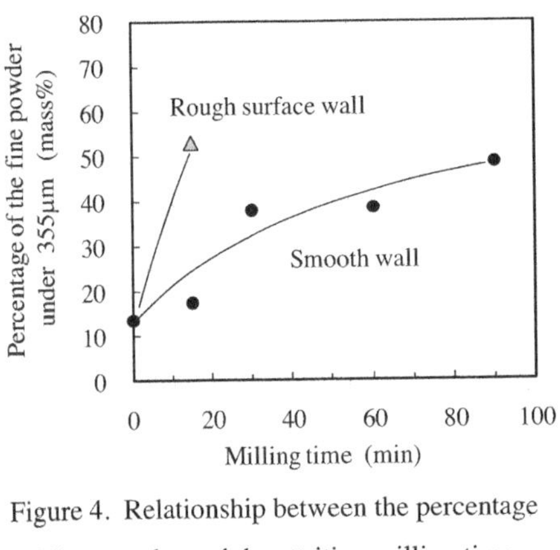


Figure 4. Relationship between the percentage of fine powder and the attrition milling time.

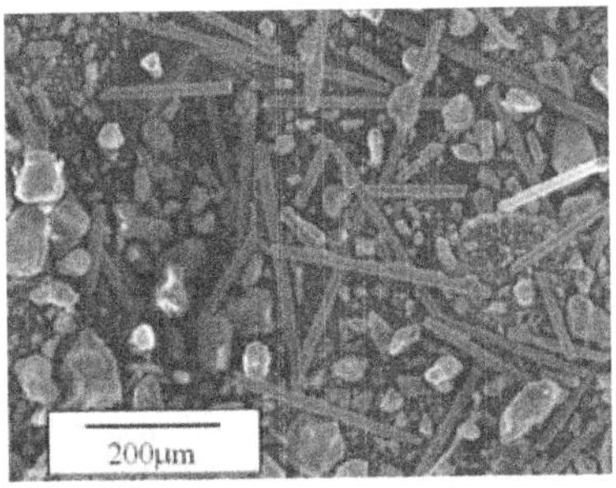


Figure 5. SEM image of the waste FRP powder milled using the attrition type mill with the rough surface wall (passed through 355μm sieve)

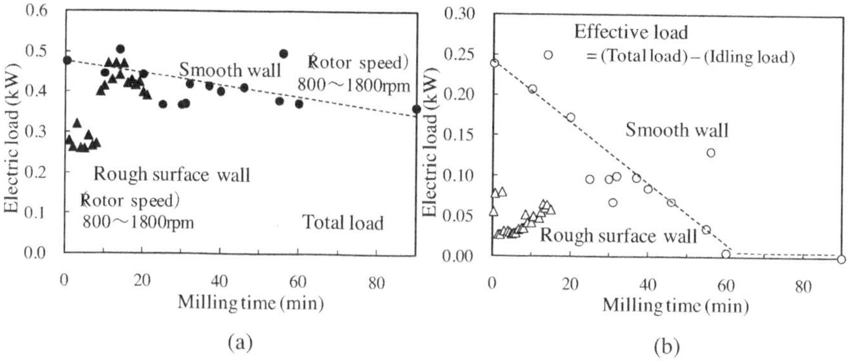


Figure 6. Energy consumption during attrition milling:

(a) Total electric load, (b) Effective load (Total load – Idling load)

15 min. of processing, the percentage of the fine powder (under 355μm) reached 50 mass% or more. Even in this case, the glass fibers retained their shape as shown in Figure 5.

Figure 6 shows the electric load chart against the milling time. Figure 6(a) shows the chart of total load given to the mill during its operation. Figure 6(b) shows the effective load which eliminated the idling load. As for the total load, the electric load of the milling with the rough surface wall was about half of that of the milling with the smooth wall at the initial milling time (Figure 6(a)). Furthermore, the effective electric load of the milling with the rough surface wall was only one tenth that of the milling with the smooth wall at the initial milling time as shown in Figure 6(b). Thus, the rough surface wall had a remarkable advantage on energy reduction compared to the smooth surface wall. In the case of the smooth surface wall, it was supposed that the specimens in the chamber partially slipped at the surface of the wall, thus leading to the surplus energy to process FRP chips. On the other hand, it was supposed that the rough surface wall was effective to prevent the slip of specimens on the inside wall of the chamber, thus leading to the effective separation of glass fibers from FRP.

The Al mapping of the surface of the glass fiber processed by nanoparticle bonding is shown in Figure 7. The bright signals in the Al-map observed on the surface of the separated glass fiber show that the glass fibers are coated with $Al(OH)_3$ nanoparticles homogenously

The flexural strength of the pressed compact is shown in Figure 8. For both the smooth and the rough surface wall, the flexural strength of the compacted specimens is about 3MPa before steam curing, and it increases to over 4MPa after steam curing at 80 °C. The increase in the strength after steam curing may be affected by the mechanochemical effect of the added $Al(OH)_3$ nanoparticles and the $CaCO_3$ filler particles separated from FRP during the attrition type milling. Further analysis should be conducted in the future.

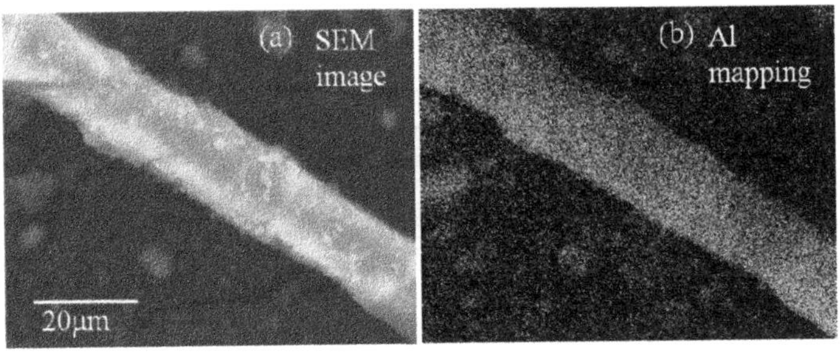


Figure 7. Glass fiber coated with $Al(OH)_3$ nanoparticles:
(a) SEM image, (b) Al-mapping

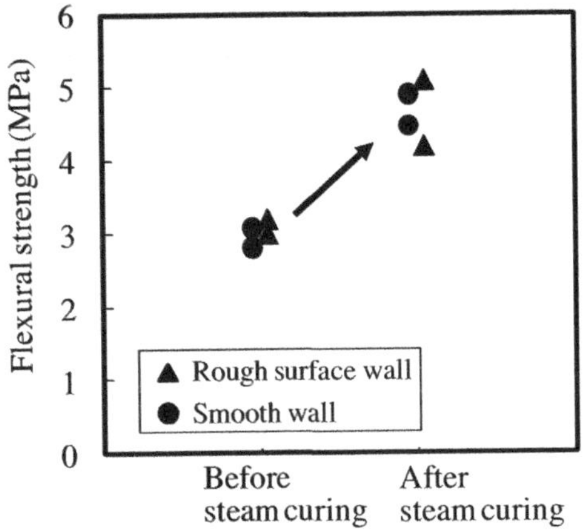


Figure 8. Flexural strength of the pressed compacts of the composites processed by the attrition type mill.

A large amount of processing of waste FRP will be necessary in future. In this case, one possibility is to design a long-gap mill for the horizontal direction. Then, the separation process and the coating process will be achieved continuously, thus leads to large amount processing. Further scale-up discussion should be conducted in future.

## CONCLUSION

Using an attrition type mill, waste FRP was effectively separated into glass fibers that retained their shape, along with fine powders of the matrix resins and fillers. The separation rate of glass fibers was affected by the configuration of the inside wall of the chamber of the attrition type mill. The rough surface wall was more effective for separating the glass fibers from the resins and fillers. The separated glass fibers were coated with $Al(OH)_3$ nanoparticles homogeneously. The flexural strength of the pressed compact of the nanoparticle-coated powder was increased by the steam curing.

## ACKNOWLEDGMENT

This work was supported by Research & Technology Development on Waste Management Research Grant of Ministry of the Environment

## REFERENCES

[1] Y. Shoji and K. Miyashiro, *Focus NEDO*, **2**, 5-6 (2002)

[2] T. Iwaya, S. Tokuno, M. Sasaki, M. Goto and K. Shibata, "Recycling of Fiber Reinforced Plastics Using Depolymerization by Solvothermal Reaction with Catalyst", *J. Mater. Sci.*, **43**, 2452-56 (2008).

[3] Y. Okuni, M. Miura, N. Isu, A. Kondo, H. Abe and M. Naito, "New Recycling Method to Make Advanced Materials from Waste FRP by using Attrition Type Milling Apparatus", Mater. Res. Bull., in press

# NUMERICAL SIMULATION OF FLOW AND NOx REDUCTION IN A DE-$NO_x$ CATALYST HONEYCOMB

Kenji Tanno[1], Ryoichi Kurose[2], Hisao Makino[3], and Satoru Komori[4]

[1] Energy Engineering Laboratory, Central Research Institute of Electric Power Industry (CRIEPI), 2-6-1 Nagasaka, Yokosuka, Kanagawa, 240-0196, Japan

[2] Department of Mechanical Engineering and Science, Kyoto University, Yoshida-honmachi, Sakyo-ku, Kyoto, 606-8501, Japan

[3] Energy Engineering Laboratory, Central Research Institute of Electric Power Industry (CRIEPI), 2-6-1 Nagasaka, Yokosuka, Kanagawa, 240-0196, Japan

[4] Department of Mechanical Engineering and Science, Kyoto University, Yoshida-honmachi, Sakyo-ku, Kyoto, 606-8501, Japan

## ABSTRACT

A selective catalytic de-NOx system, operated around 350°C, offers a very efficient method to reduce the NOx concentration in flue gas, and is therefore widely used in thermal power plants. In order to precisely predict the lifetime and efficiency of the system, it is of great importance to understand the NOx reduction mechanism in a honeycomb rectangular channel of a de-NOx catalyst. In this study, the effects of turbulent and laminar flow behavior in a honeycomb rectangular channel on the NOx reduction reaction is investigated using large-eddy simulations (LES). The results show that the NOx reduction reaction is strongly affected by inner flow behavior. The NOx reduction rate on the catalyst surface for the turbulent flow is higher than that for laminar flow. This is due to the fact that the diffusion of NOx in the vicinity of the wall is dominated only by molecular diffusion during laminar flow, whereas it is enhanced by turbulent motion during turbulent flow.

## INTRODUCTION

From an environmental point of view, a reduction in NOx emissions in the combustion process in thermal power plants is very important. Although selective catalytic de-NOx systems have been widely used to reduce NOx emission combined with low-NOx combustion technologies in thermal power plants, the de-NOx catalyst is degraded by long-term operation and hence needs to be periodically exchanged or regenerated. In thermal power plants, flue gas flows into the de-NOx catalyst honeycomb, which has a small hydraulic diameter, from the large-diameter duct, hence the inner flow and transition from laminar to turbulent flow needs to be considered. The authors recently showed that the flow in the de-NOx catalyst actually transitions from turbulent flow to laminar flow, and this turbulent-laminar flow transition causes local degradation of the de-NOx catalyst due to ash particle adhesion[1-2]. However, in order to precisely predict the efficiency of the de-NOx catalyst, it is of great importance to determine the relationship between the inner flow behavior and selective catalytic reduction process, namely NOx reduction reaction.

The purpose of this study is therefore to investigate the effect of the inner flow behavior on the NOx reduction reaction on the catalyst surface by applying a three-dimensional large-eddy simulation (LES) to a reacting flow in a honeycomb rectangular channel of a de-NOx catalyst.

## NUMERICAL SIMULATION

### Governing equations

In LES modeling, the large, energy-containing scales of motion are simulated numerically, while the small, unsolved subgrid scale (SGS) and their interactions with the large scales are modeled. The LES equations, which can be derived by the spatial filtering of the Navier-Stokes equation with a filter of around each cell width, are given as,

$$\frac{\partial \bar{\rho}}{\partial t} + \frac{\partial}{\partial x_j}\left(\bar{\rho}\tilde{u}_j\right) = S_m \tag{1}$$

$$\frac{\partial(\bar{\rho}\tilde{u}_i)}{\partial t} + \frac{\partial}{\partial x_j}\left(\bar{\rho}\tilde{u}_i\tilde{u}_j + \bar{p}\delta_{ij} + \tilde{\tau}_{ij} - \tau_{u_iu_j}\right) = S_{u_i} \tag{2}$$

$$\frac{\partial(\bar{\rho}\tilde{Y}_k)}{\partial t} + \frac{\partial}{\partial x_j}\left(\bar{\rho}\tilde{u}_j\tilde{Y}_k + \tau_{u_jY_k}\right) = \bar{\omega}_k + S_{Y_k} \tag{3}$$

here, $\bar{}$ and $\tilde{}$ denote the spatial and Favre filtering respectively. $\tau_{u_iu_j}$ and $\tau_{u_jY_k}$ appear in Eqs. (2) and (3) are the effects of SGS. In this study, the effects of SGS are taken into account by employing the dynamic Smagorinsky SGS turbulence model[3]

### Selective catalytic reduction reaction

The overall chemical reaction for NO reduction in selective catalytic reduction process can be expressed as follows:

$$4NO + 3NH_3 + O_2 \rightarrow 4N_2 + 6H_2O \tag{4}$$

Selective catalytic reduction rate is calculated by employing a Langmuir-Hinshelwood type rate expression which was proposed by Beeckman and Hegedus[4].

$$\bar{\omega} = Sk_0 \exp\left(-\frac{E}{RT}\right) X_{NO}\left[\frac{K_{NH_3} X_{NH_3}}{1+K_{NH_3} X_{NH_3}}\right] \tag{5}$$

where $\bar{\omega}$, $S$, $k_0$, $E$, $R$, $T$, $X_{NO}$, $X_{NH_3}$ and $K_{NH_3}$ are the NO reduction rate, catalyst volume specific surface area, pre-exponential factor, activation energy, gas constant, absolute temperature, NO and $NH_3$ mole fraction, and $NH_3$ absorption constant, respectively. $S$, $k_0$, $E$, $K_{NH_3}$ are set to $1.25 \times 10^4$ ($m^2/m^3$), $6.864 \times 10^8$ (m/s), $1.343 \times 10^5$ (J/mol), and 3.964 ($m^3$/mol), respectively[4-5].

Computational details

Figure 1 shows schematic diagram of the computational domain. The computational domain (main part) is 100mm in length and 7mm x 7mm in cross-section. The number of computational cells is about 170,000, and the grid spaces near the wall are compressed. The origin of this coordinate system is located at the inlet ($x$-axis) and the centre of the cross section ($y$- and $z$-axis). Inlet boundary condition is varied in order to investigate the effect of inner flow behavior on the NOx reduction reaction on the catalyst surface. One case is that fully developed turbulent flow, which is generated in the driver part, is used for the inlet boundary condition, and the other case is that laminar flow is used (referred to as "turbulent case" and "laminar case", respectively). For numerical approximation, discretization of nonlinear terms of the momentum equations is derived from a second-order central difference schemes, while the mass fractions is computed by a second-order finite difference scheme. An Adams-Moulton method and fully implicit scheme are used for the time advancement of momentum and mass fractions.

The cross-sectional mean velocity, $U_{ave}$, temperature and kinetic viscosity are set at 6.0 m/s, 350°C, and $5.67 \times 10^{-5}$ $m^2s^{-1}$, respectively. Reynolds number based on $U_{ave}$, the height of the square channel, $d_h$, Re, is 740. For the turbulent case, Re are too small to develop turbulent flow, hence value of Re in the driver part is set to 5 times larger value than the original Re value.

In order to compare numerical results with author's previous experimental results[1], gas compositions are set to the almost same as that in the experiment, which is simulated actual flue gas. Mole fractions of inflow gas are set at $N_2$=88%, $O_2$=4%, $H_2O$=8%, NO=300 ppm, and $NH_3$=246 ppm. Molecular ratio of $NH_3$/NO is set to 0.82.

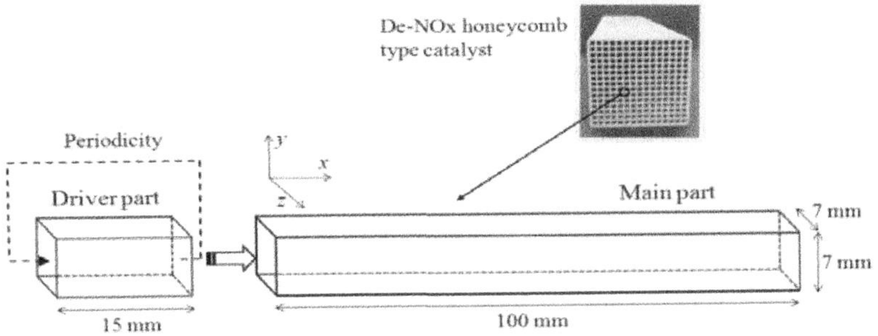


Figure 1. Diagrams of de- NOx catalyst and computational domain.

## RESULTS AND DISCUSSION

Figure 2 shows the contour distributions of instantaneous streamwise velocity for the turbulent and laminar cases. It is observed that the turbulent flow issued from the inlet gradually transients to laminar flow towards downstream[1] for the turbulent case, whereas the laminar flow keeps its state for the laminar case. Figure 3 shows the cross-sectional flow fields at $x$=20mm. The vector size is function of the flow magnitude. For the turbulent case, strong fluid motion from central region to wall is frequently observed, whereas this kind motion is quite weak for the laminar case. It is considered that this fluid motion caused by turbulent motion enhances the diffusion of NOx and affects the NOx reduction reaction on the catalyst surface (this mechanism will be discussed later in detail).

Figure 4 shows the streamwise variations of the NOx reduction rate $-\bar{\omega}$ on the centerline of the catalyst wall. For both cases, $-\bar{\omega}$ has the same trend that $-\bar{\omega}$ takes the maximum value in the inlet of the De-NOx catalyst and decreases towards downstream. This is due to the decrease of NO concentration in the vicinity of the wall towards downstream. Furthermore, $-\bar{\omega}$ for the turbulent case is higher than that for the laminar case, which means that the NOx reduction reaction is enhanced by turbulent motion. In order to evaluate the effect of this inner flow behavior on the total efficiency of NOx reduction, de-NOx efficiency is evaluated from the NO concentration of both the inlet and outlet. The de-NOx efficiency for the turbulent and laminar cases are 22.2% and 17.9 %, respectively, which indicates that turbulent flow promotes the total de-NOx efficiency. In author's previous experiment de-NOx efficiency was 68.3% using 600 mm length catalyst[1], and hence the values evaluated in this study using 100 mm length catalyst are considered to be reasonable.

In order to further investigate the effect of turbulent motion on the NOx reduction reaction, the contour distributions of instantaneous NO concentration and vector fields of instantaneous cross-sectional flow field at $x$=20mm are shown together in Figure 5. It is observed

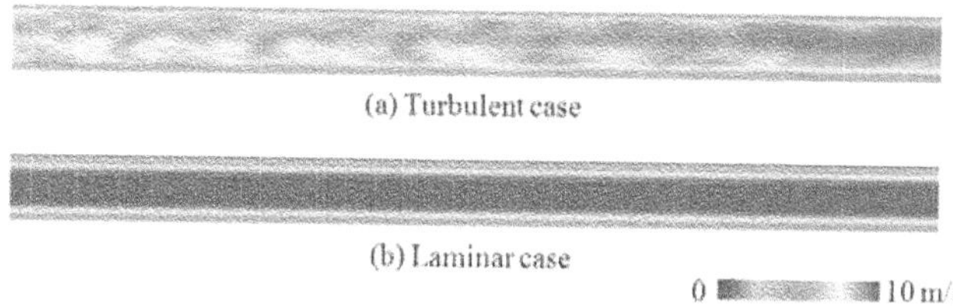


Figure 2. Contour distributions of instantaneous streamwise velocity (*x*-y plane, *z*=0mm).

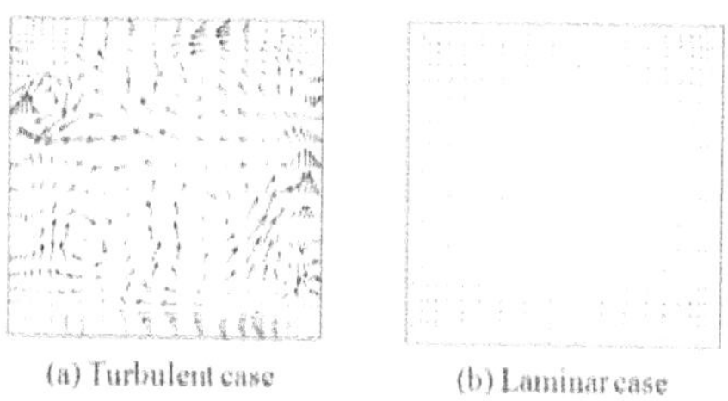


Figure 3. Instantaneous cross-sectional flow vector fields at *x*=20mm.

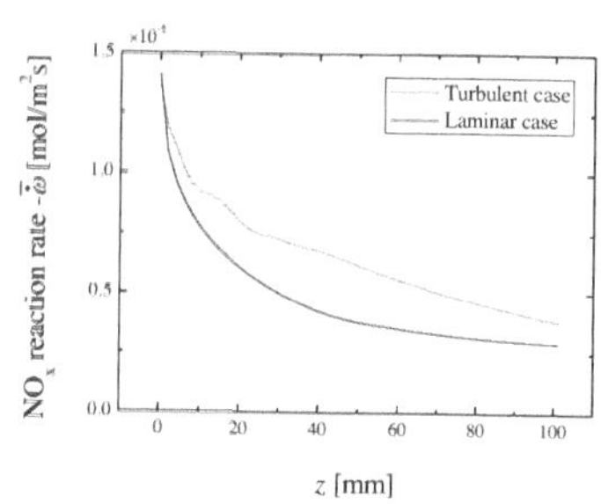


Figure 4. Streamwise variations of NO reduction rate on the centerline of the wall.

that the NO concentration distribution for the laminar case is symmetric with respect to the duct center, whereas that for the turbulent case is skewed in the vicinity of the wall. This means that NO transportation for the laminar case is dominated only by molecular diffusion, but that for the turbulent case is strongly affected by the turbulent motion. In fact, for the turbulent case, high-speed fluid motion towards the wall transports high NO concentration gas to the wall surface and low-speed fluid motion towards the central region transports low NO concentration gas. Therefore, it can be concluded that since this turbulent motion effectively enhances the diffusion of NOx in the vicinity of the wall, the NOx reduction reaction on the catalyst surface is also enhanced for the turbulent flow, compared to the laminar flow. This result suggests that there is a possibility to increase the de-NOx efficiency and reduce the costs of selective catalytic de-NOx systems by taking the turbulent motion into account in developing the new systems.

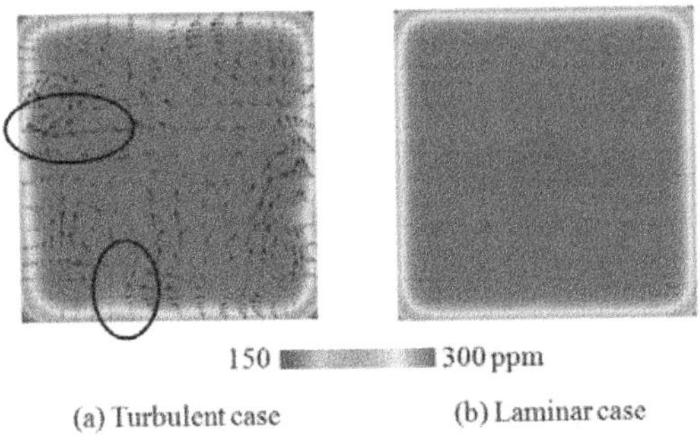


Figure 5. Contour distributions of instantaneous NO concentration and vector fields of instantaneous cross-sectional flow (*y*-*z* plane, *x*=20mm).

CONCLUSION

A three-dimensional large-eddy simulation (LES) was applied to a reacting flow in a de-NOx catalyst honeycomb, and the effect of flow behavior in the honeycomb rectangular channel on the NOx reduction reaction was investigated. It was found that the NOx reduction reaction is strongly affected by inner flow behavior. That is, the NOx reduction rate on the catalyst surface for the turbulent flow is higher than that for the laminar flow, since the diffusion of NOx in the vicinity of the wall is dominated only by molecular diffusion for the laminar flow, whereas it is enhanced by turbulent motions during turbulent flow.

REFERENCES

[1]K. Tanno, H. Makino, R. Kurose, S. Komori, H. Shimada, and S. M. Hwamg, "Effect of Turbulent-Laminar Flow Transition on Degradation of De-NOx Catalyst", *Fuel*, in press.

[2]R. Kurose, T. Michioka, H. Makino, and S. Komori, "Effect of Flow Behavior in De-NOx Catalyst Honeycomb on Adhesion of Particle to the Wall", *J. Soc. Powder Technol., Japan*, **44**, 107-112 (2007).

[3]M. Germano, U. Piomelli, P. Moin, and W.H. Cabot, "A dynamic subgrid-scale eddy viscosity model", *Phys.F luids A*, **3**, No.7, 1760-1765 (1991).

[4]J.A. Beeckman and L.L. Hegedus, "Design of Monolith Catalysts from Power Plant NOx Emission Control", *Ind. Eng. Chem. Res.*, **30**, 969-978 (1991)

[5]J.O.L. Wendt, W.P. Linak, P. W. Groff, and R. K. Srivastava, "Hybrid SNCR-SCR Technologies of NOx Control: Modeling and Experiment", *AIChE Journal*, **47**, 2603-2617 (2001).

# CREATION OF TRIPLE-PHASE-BOUNDARY IN A SOLID OXIDE FUEL CELL USING A THREE-DIMENSIONALLY ORDERED STRUCTURE

Hirokazu Munakata, Masashi Otani, Yuki Katsuki, Kiyoshi Kanamura
Department of Applied Chemistry, Graduate School of Urban Environmental Sciences, Tokyo Metropolitan University, 1-1 Minami-Ohsawa, Hachioji, Tokyo, Japan, 192-0397

## ABSTRACT

The surface modification of a three-dimensionally ordered macroporous (3DOM) nickel-yttria stabilized zirconia (Ni-YSZ) with samarium doped ceria (SDC) was examined to increase the triple phase boundary (TPB). The pore surface of the 3DOM Ni-YSZ was coated with SDC using the sol-gel method, and the coating uniformity was confirmed by scanning electron microscopy (SEM) and energy dispersive X-ray (EDX) analysis. The amount of TPB in the 3DOM Ni-YSZ was increased by the oxygen-ion and electron conductivities added by the SDC coating, resulting in 1.3 times higher cell performance (maximum power density: 306 mW $cm^{-2}$) at 700 °C than the cell without a SDC coating. The effect of the SDC coating on cell performance became even more remarkable at lower operating temperatures due to the high oxygen-ion conductivity of SDC at intermediate temperatures.

## INTRODUCTION

Many approaches have been examined to reduce the operating temperature of solid oxide fuel cells (SOFCs) to lower than 800°C, the so-called intermediate temperature (IT) range[1]. Low temperature operation enables a wider selection of materials to be used as cell components. Particularly, the availability of metallic materials for interconnects is a great advantage for low cost production of SOFC systems. However, lowering the operating temperature leads to large polarization resistances of the electrode reactions and to low oxygen-ion conductivity of solid electrolytes. The resistance of the electrolyte is directly proportional to the thickness of electrolyte. Therefore, cell configurations considered for IT-SOFCs are mainly electrode-supported ones, in which the electrolyte layer formed is less than 20 μm by using various thin film techniques[2-6]. On the other hand, large polarization resistances of the electrode reactions at low temperatures still remain in these electrode-supported configurations. The electrode reactions occur at spatially limited sites called triple phase boundaries (TPB)[7], where the electrolyte, reactant (fuel or oxygen), and catalyst easily contact each other. As such, the amount of TPB in an electrode has to be increased to reduce the polarization resistance.

In our previous work, we successfully prepared three-dimensionally ordered macroporous (3DOM) nickel-yttria stabilized zirconia (Ni-YSZ) with a colloidal crystal templating method using suction filtration[8]. Ni-YSZ is a standard material used as a SOFC anode. The 3DOM structure has high porosity and uniformity, which are favorable for reducing the concentration of polarization resistances at the SOFC anode. Actually, the cell with the 3DOM Ni-YSZ exhibited higher performance than the cell with a conventional anode structure[9].

Samarium doped ceria (SDC) has attracted much attention as a new electrolyte material due to its high oxygen-ion conductivity at intermediate temperatures. In addition, it shows electron conductivity under reducing atmosphere, so that it can work as a current collector at the anode. However, this material is difficult to use as a cell-supporting material due to its fragility. Therefore, in this study, we examined the application of SDC as a coating material on the pore surface of the 3DOM Ni-YSZ to increase the TPB, and evaluated the performance of the SDC-modified 3DOM Ni-YSZ as an anode for IT-SOFCs.

EXPERIMENTAL

An aqueous suspension containing NiO-YSZ particles ($\phi$ = 410 nm, 34.4 wt% $(Y_2O_3)_{0.08}(ZrO_2)_{0.92}$ - 65.6 wt% NiO, Hosokawa Powder Technology Research Institute, Japan) and mono-disperse polystyrene (PS) beads ($\phi$ = 2.5 μm) was prepared. The pH of suspension was adjusted using hydrochloric acid to 1.8 according to our previous result.[8] The PS beads were synthesized by the soap-free emulsion method from styrene, 2,2'-azobis [N-(2-carboxylethyl)-2-2-methylpropionamidine] hydrate, ammonia aqueous solution, and ammonium chloride. These chemicals were purchased from Wako Pure Chemical Industries, Ltd. The mixing ratio of NiO-YSZ particles and PS beads in the suspension was set to 4:6 in volume to obtain 3DOM Ni-YSZ anode with ≈65% porosity. 3DOM Ni-YSZ with an ideal porosity of 74% is favorable, but is a little fragile. The suspension was filtered through a nitrocellulose membrane filter (0.22 μm pore, Millipore GSWP02500), and then another aqueous suspension containing YSZ particles ($\phi$ = 90 nm, TZ-8YS, Tosoh Corp., Japan) was gently added. The sediment (NiO-YSZ/PS hybrid layer with a thin YSZ skin) on the membrane filter was peeled off and placed on a flat ceramic plate, and heated at 450 °C for 1 h and then 1350°C for 5 h in air. The heating rate was set to 10 °C $min^{-1}$ for each heating step. During this heat treatment, the PS beads were burned away. On the contrary, the NiO-YSZ and YSZ particles were sintered to produce 3DOM NiO-YSZ with YSZ thin electrolyte layer. The pore surface of 3DOM NiO-YSZ was then coated with SDC by sol-gel method as shown in Figure 1. The precursor sol of SDC was prepared by dissolving of $8.0 \times 10^{-3}$ mol of $Ce(NO_3)_3 \cdot 6H_2O$ and $2.0 \times 10^{-3}$ mol of $Sm(NO_3)_3 \cdot 6H_2O$ ($2.0 \times 10^{-3}$ mol) in 10 ml of pure water with 2.1 g of citric acid monohydrate and 0.9 g of polyethylene glycol (PEG #600) [10]. All the reagents used for the sol preparation were purchased from Kanto Chemical Co., Inc. The pores of 3DOM NiO-YSZ were impregnated with the precursor sol, and then heated in air at 300°C for 30 min. This process was repeated three times to ensure adequate coating of the pore surface of 3DOM NiO-YSZ with SDC precursor. Finally, the sample was annealed in air at 1200°C for 2h. The cathode composed of $La_{0.85}Sr_{0.15}MnO_3$ (LSM, $\phi$ = 400 nm, AGC Seimi Chemical CO., LTD., Japan)-YSZ in a weight ratio of 1:1 was formed on the YSZ electrolyte layer by casting a glycerol paste of LSM and YSZ particles. After the heat treatment at 1200°C for 2 h, the cell with LSM-YSZ / YSZ / SDC-modified 3DOM NiO-YSZ configuration was obtained.

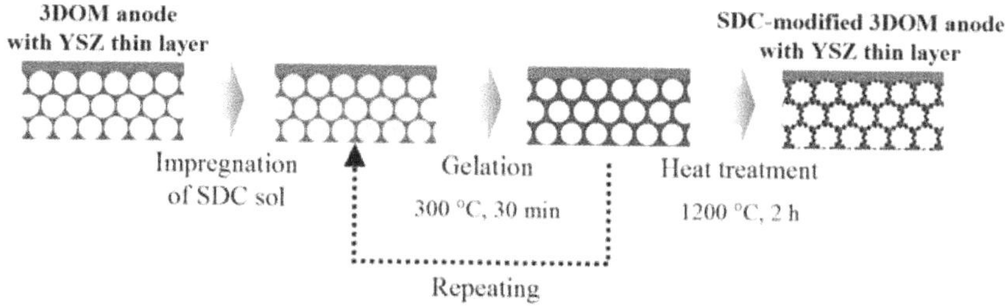


Figure 1. Schematic of SDC coated on the pore surface of the 3DOM anode using sol-gel.

The crystalline phases of cell components were analyzed by X-ray diffraction (XRD) method with RINT2000 (Rigaku Corp., Japan). The microstructure of cell was observed with a scanning electron microscope (SEM, JSM-5310, JEOL Ltd., Japan). Energy dispersive X-ray (EDX) spectrum imaging was also performed in SEM observation to characterize the cerium distribution in the cross-section of SDC-modified 3DOM anode. The cell performance was measured at temperatures between 600°C and 700°C with an electrochemical analyzer (PEMTest8900, TOYO Corp., Japan). The cell was sealed with a glass ring to alumina tube for preventing gas leakage. Hydrogen gas moisturized at room temperature ($H_2$ + 3% $H_2O$) was fed to the anode as a fuel at a flow rate of 150 ml $min^{-1}$, while dry oxygen gas was fed to the cathode at a flow rate of 300 ml $min^{-1}$. Before polarization measurements, the cell was kept at 800°C for 2 h in order to reduce NiO to Ni in the anode.

## RESULTS AND DISCUSSION

The suspension filtration process used for SOFC fabrication enables one to precisely control the thickness of the deposited layer by the amount of suspension. In this study, we fixed the thicknesses of YSZ electrolyte and the 3DOM NiO-YSZ anode to 10 μm and 500 μm, respectively. On the other hand, the LSM-YSZ cathode was adjusted to about 10 μm using a paste coating. Both the cathode and anode had good adhesion to the dense YSZ electrolyte layer. Figure 2 shows the XRD pattern of the SDC powder prepared by the sol-gel method. All the diffraction peaks observed were assigned to a cubic fluorite structure of $Ce_{0.8}Sm_{0.2}O_{1.9}$. Also in the case of the SDC-modified 3DOM NiO-YSZ, no peaks for impurities such as $Sm_2O_3$ were observed. After the heat-reduction of NiO to Ni at 800 °C, the XRD pattern of the SDC-modified anode was maintained, with the exception that the NiO diffraction peaks disappeared and Ni diffraction peaks appeared.

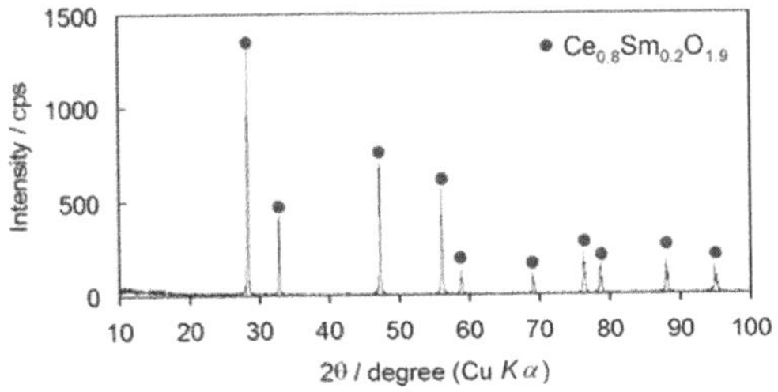


Figure 2. X-ray diffraction pattern of the SDC prepared by the sol-gel method.

Figure 3 shows the cross-sectional SEM and EDX images of the SDC-modified anode. The uniformity of SDC coating was confirmed from the SEM image, in which no segregation of SDC was observed both in and on the anode. It was also found from EDX images that the cerium is distributed uniformly in the 3DOM anode, and increases with the number of impregnations.

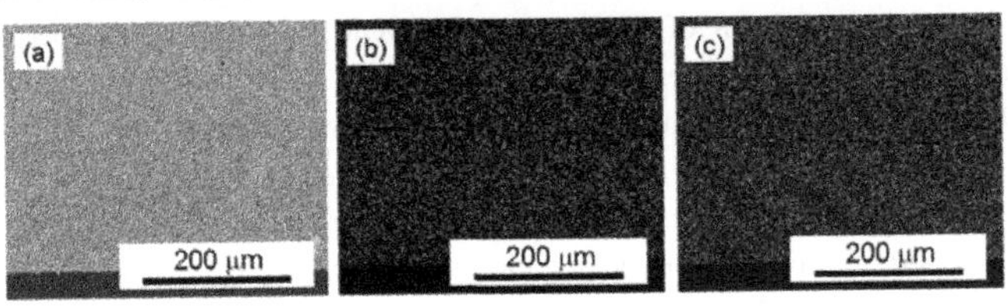


Figure 3. Cross-sectional SEM (a) and EDX (b, c) images for Ce mapping of the SDC-modified anode (the number of SDC impregnations: (a, c) three times, (b) one time).

Figure 4 shows typical fuel cell performance curves for the SOFC without (a) and with the (b) SDC coating on the pore surface of the 3DOM anode, measured at temperatures between 600°C and 700°C. The open circuit voltages (OCVs) of the SDC-modified and unmodified cells were similar and about 1.1 V at 600~700°C, indicating that both YSZ electrolyte layers were fairly dense without any pinholes. The SDC-modified cell showed the maximum power density of 306 mW $cm^{-2}$ at 700 °C. This power density is 1.3 times higher than that of the unmodified cell, even though the unmodified cell still showed better performance than previously reported cells. For example, Besra et al. reported a maximum power density of 144 mW $cm^{-2}$ for LSM-YSZ / YSZ / Ni-YSZ cell at 700 °C[11]. The superiority of the SDC coating on cell performance became more remarkable with decreasing operating temperature. This result can be understood by considering the high oxygen-ion conductivity of SDC at intermediate temperatures. Actually, the decrement of resistance was confirmed for the SDC-modified cell by alternating current (AC) impedance measurement (Figure 5). It is also known that SDC shows electron conductivity under reducing atmosphere by the reduction of $Ce^{4+}$ to $Ce^{3+}$. Therefore, the SDC on the pore surface of 3DOM anode, being fed with hydrogen gas, is expected to exhibit not only oxygen ion conductivity but also electron conductivity. This situation was confirmed as the increment of TPB in the AC impedance spectra shown in Figure 5, in which the semi-circle assigned to anode polarization resistance was shrunk by the SDC coating. By increasing the amount of SDC in the 3DOM anode, both oxygen-ion and electron conductivities of the 3DOM anode may be increased. However, adequate mass transport should also be maintained for the additional formation of TPB. A more precise design for the 3DOM anode by the SDC coating is now underway for further improvement of cell performance.

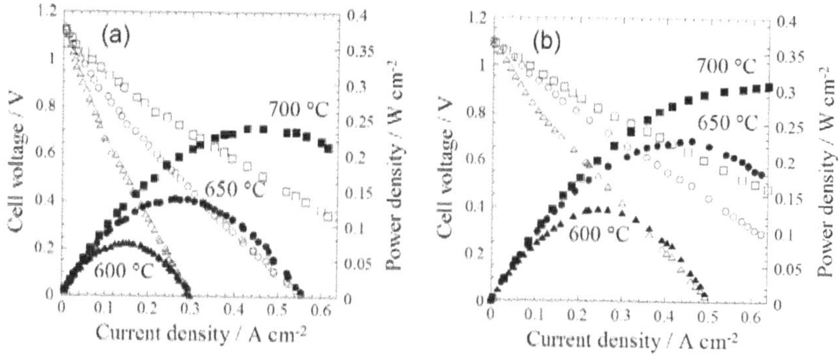


Figure 4. Typical fuel cell performance curves for SOFC without (a) and with the (b) SDC coating on the pore surface of the 3DOM anode, measured in the temperature range between 600°C and 700°C.

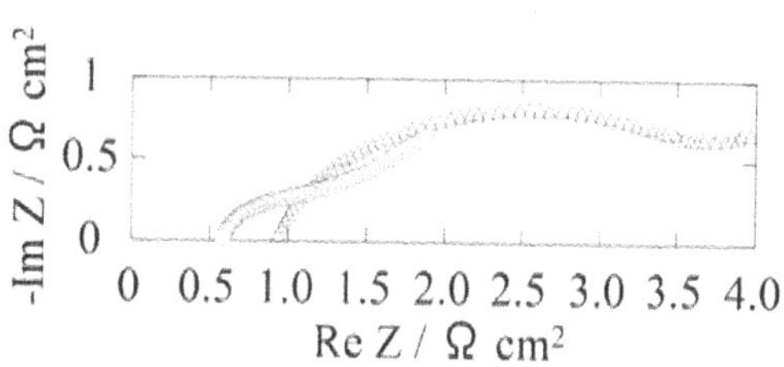


Figure 5. Typical impedance spectra of the SOFC without (Δ) and with the (○) SDC coating on the pore surface of the 3DOM anode, measured at OCV and 600°C.

CONCLUSION

The pore surface of the 3DOM Ni-YSZ was uniformly coated with SDC using the sol-gel method. The amount of TPB in the 3DOM Ni-YSZ was successfully increased by the high oxygen-ion conductivity of the SDC at intermediate temperatures, as well as its electron conductivity generated under reducing atmosphere. Accordingly, the cell with the LSM-YSZ / YSZ / SDC-modified 3DOM Ni-YSZ configuration showed 306 mW $cm^{-2}$ at 700°C as the maximum power density, corresponding to 1.3 time that of the unmodified cell. With decreasing operating temperature, the output of the unmodified cell drastically decreased. On the contrary, the SDC-modified cell maintained relatively high performance, i.e., 277 mW $cm^{-2}$ at 650 °C and 131 mW $cm^{-2}$ at 600 °C, due to the high oxygen-ion conductivity added by the SDC.

REFERENCES

[1]D. J. L. Brett, A. Atkinson, N. P. Brandon, and S. J. Skinner, "Intermediate temperature solid oxide fuel cells", *Chem. Soc. Rev.*, **37**, 1568-1578 (2008).

[2]K. W. Chour, J. Chen, and R. Xu, "Metal-organic vapor deposition of YSZ electrolyte layers for solid oxide fuel cell applications", *Thin Solid Films*, **304**, 106-112 (1997).

[3]H. Ruiz, H. Vesteghem, A. R. Di Giampaolo, and J. Lira, "Zirconia coatings by spray pyrolysis", *Surf. Coat. Technol.*, **89**, 77-81 (1997).

[4]D. W. Dees, T. D. Claar, T. E. Easler, D. C. Fee, and F. C. Mrazek, "Conductivity of Porous Ni/$ZrO_2$-$Y_2O_3$ Cermets", *J. Electrochem. Soc.*, **134**, 2141-2146 (1987).

[5]W. F. Chu, "Thin- and thick-film solid ionic devices", *Solid State Ionics*, **52**, 243-248 (1992).

[6]T. Ishihara, K. Sato, and Y. Takita, "Electrophoretic Deposition of $Y_2O_3$-Stabilized $ZrO_2$ Electrolyte Films in Solid Oxide Fuel Cells", *J. Am. Ceram. Soc.*, **79**, 913-919 (1996).

[7]R. E. Williford, L. A. Chick, G. D. Maupin, S. P. Simner, and J. W. Stevenson, "Diffusion limitations in the porous anodes of SOFCs", *J. Electrochem. Soc.*, **150**, A1067-A1072 (2003).

[8]H. Munakata, M. Otani, and K. Kanamura, "Preparation of a 3DOM Ni-YSZ anode with a dense YSZ thin electrolyte layer for solid oxide fuel cells", *J. Fuel Cell Sci. Technol.*, **5**, 031206 (2008).

[9]H. Munakata, M. Otani, Y. Katsuki, and K. Kanamura, "3-Dimensionally Ordered Macroporous Ni-YSZ for Low Temperature SOFC", *ECS Trans.*, accepted for publication.

[10]H. Lin, C. Ding, K. Sato, Y. Tsutai, H. Ohtaki, M. Iguchi, C. Wada, and T. Hashida, "Preparation of SDC electrolyte thin films on dense and porous substrates by modified sol–gel route", *Mater. Sci. Eng. B*, **148**, 73-76 (2008).

[11]L. Besra, S. Zha, and M. Liu, "Preparation of NiO-YSZ/YSZ bi-layers for solid oxide fuel cells by electrophoretic deposition", *J. Power Sources*, **160**, 207-214 (2006).

# CHEMICAL REDUCTIVE PREPARATION OF NI DECORATED CONDUCTIVE POROUS ALUMINA COMPOSITE AND ITS ELECTRO-PERFORMANCE IN ALKALINE SOLUTION

Chunxi Hai[a,b], Hideo Watanabe [a], Takashi Shirai [a], Masayoshi Fuji [a], Feng Wang [b], Jingjun Liu[a] and Minoru Takahashi [b]

[a]*Ceramics Research Laboratory, Nagoya Institute of Technology, Crystal Plaza 4th Floor, Hommachi 3-101-1, Tajimi, Gifu 507-0033, Japan*

[b]*College of Materials Science and Engineering, Beijing University of Chemistry and Technology, P.R.China*

**Corresponding Author: fuji@nitech.ac.jp*

## ABSTRACT

In this work, electrically conductive porous alumina (CPA) was decorated with uniformly dispersed Ni nanopartices by a chemical reductive reaction, thanks to the unique bone structure of the CPA. Surface pre-treatment plays an important role in achieving a uniform distribution of Ni nanoparticles without destroying the bone structure of the CPA, as confirmed by the characterization results. An electrochemical investigation of the CPA-based electrode indicates that increased performance can be expected because of the existence of the porous structure and the carbon networks in the CPA. Moreover, the reductive peak around +210 mV is mainly attributed to the uniform dispersion of the Ni nanoparticles in the Ni/CPA.

## INTRODUCTION

Although various technologies are available to prepare electrically conductive porous ceramics[1-3], some unavoidable defects still restrict the further applications of these ceramics for a number of fields. These defects include low mechanical strength, anisotropy in electrical conductivity, non-uniform density, and unwanted crystal growth. Therefore it is reasonable that there has been a surge of interest in electrically conductive porous alumina (CPA)[4-6]. The CPA used in this work was fabricated by a combination of gelcasting and reductive sintering in inert gas, where the gel monomer added was totally converted into uniformly inter-connected conductive pathways in the CPA. The crystal structure of the resultant three-dimensional nano-carbon networks is graphitic with a six-member regular ring. The alumina grains bond to each other forming a porous bone structure that provides perfect mechanical strength for the as-fabricated CPA. In general, a porous body is weaker than a dense one, while a lower amount of carbon does not reduce the mechanical strength of the CPA appreciably. The unique properties of CPA such as its ideal electrically conductivity, high porosity, and mechanical strength, enable all sorts of potential applications in many fields, especially as an electrode. It is possible that a series of wider applications can be developed with suitable surface modification.

In our previous work[7-9], it was discovered that enhanced electro-performance could be expected due to the inter-connected nano-carbon networks and the co-performance of the alumina grains compared with the classical electrodes. We have also effectively investigated the influences of a mixed-acid pre-treatment and deposition with Ni nanoparticles, considering the catalytic merits of Ni[10-11]. However, it was discovered that some of the Ni nanoparticles deposited had a little coagulation, and the particle size is large. In this paper, we are mainly devoted to finding a simple method to modify the surface of CPA with uniformly dispersed Ni nanoparticles, avoiding the problems we encountered before. The resultant composites were characterized by FE-SEM and Raman spectroscopy. The electro-performance of the composite was examined while operating in an alkaline electrolyte at room temperature.

## EXPERIMENTAL

In our group, a novel fabricating method involving the gelcasting and reductive sintering was employed to prepare the CPA and the detail of preparing procedures were introduced in references [4-9]. For the CPA used in this work, the mechanical strength is about 22.07 MPa, carbon content measured by TG/DTA methods is about 0.54 % and porosity is about 67.50 %. To investigate the influence of surface modification, as-fabricated CPA was cut into small pieces (20 mm × 10 mm × 2 mm) and was subjected to the following procedures:

Firstly, the 18.208 g of CPA plates were immersed into the mixed acids for 1 h, followed by thorough washing and drying at 90°C for 12 h. The sample can be named as acid-CPA. Then 3.094 g acid-CPA was soaked into 10 ml of 0.027 mol/L $Ni(AC)_2$ solution and evaporated at 60°C for 24 h. Finally, as-resulted sample was refluxed in 20 ml 0.4 M $N_2H_4$ at 75°C for 3 h, followed by thorough washing and drying, the obtained sample was named as Ni/CPA.

In this work, Raman spectroscopy (JASCO, NRS-3100, Corp.) was used to characterize the nano-carbon in CPA. FE-SEM (JSM7000F, JEOL, Ltd.) was utilized to investigate the changes of morphology of as-resulted composite.

The electro-performance of as-resulted sample was operated in 1.0 M NaOH aqueous solution using saturated calomel electrode (SCE) as a reference electrode. The counter and working electrodes used in this work were Pt gauze and as-resulted Ni/CPA. At the same time, the classical graphite electrode was used for the comparison. All of the experiments were operated at room temperature.

## RESULTS AND DISCUSSION

Figure 1 shows the Raman spectrum for the as-prepared CPA, acid-CPA, and Ni/CPA, employing graphite as a reference. According to the appearance of typical peaks around 1575 $cm^{-1}$ and 1350 $cm^{-1}$ for the CPA compared with graphite[12-13], it is no doubt that the crystal structure of nano-carbon in CPA belongs to graphite, and has a carbon crystal structure as good as graphite. After acid-treatment, the precious nano-carbon is not destroyed and according to the increased ratio of D band to G band ($I_D/I_G$, as shown in Table I), it is facilitated to arrive a conclusion that the surface defects have been greatly increased by pre-treatment. Moreover, the highest intensity ratio for Ni/CPA can be observed, which may be mainly resulted from the deposition of Ni nanoparticles resulting increased specific surface.

The changes in morphology in this work are shown in Figure 2. The nano-carbon inter-connects and forms conductive networks in the CPA as shown in networks along the bone

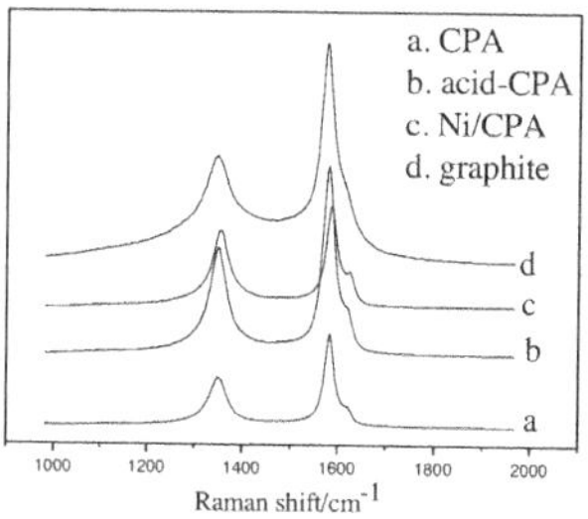


Figure 1. Raman spectrum of (a) CPA, (b) acid-CPA, (c) Ni/CPA, and (d) graphite

Table I. The intensity ratio of the D and G bands ($I_D/I_G$) in the Raman spectrum.

| Samples | $[I_D/I_G]$ |
|---|---|
| CPA | 0.53 |
| Acid-CPA | 0.68 |
| Ni/CPA | 0.79 |
| graphite | 0.58 |

structure of the CPA as shown in Figure 2 (a), and the bone structure of CPA has not been totally destroyed by acid-treatment (Figure 2 (b)). By comparing with (a) and (b) images, as shown in Figure 2 (c), it is obvious that the Ni nanoparticle have been successfully deposited onto the surface of CPA with smaller particles size and little coagulation.

The role of acid-treatment for particles dispersion was interested, so Ni particles were also deposited onto CPA plate without acid-treatment by the same experimental steps. According to the FE-SEM image in Figure 2 (d), much more agglomeration and bigger deposited Ni particles are observed when the un-treated CPA is directly used as substrate materials. Thus, acid-treatment plays an important role for decreasing particles size and uniform distribution. According to the introduction of references[14-15], various functional groups can be introduced on to the surface of CPA by acid pre-treatment, which can serve as the active sites for particles growth.

The electro-performances of graphite, CPA, acid-CPA and Ni/CPA in 1 M NaOH solution were investigated at room temperature. As shown in Figure 3, all of the detected electrodes displays an electro-activity toward oxygen reductive reaction in alkaline because of the appearance of oxygen reductive peaks around -430 mV and -820 mV [8-9]. An enhanced elecrtro-performance can be obtained for various CPA-based electrodes because of the existence

Figure 2. FE-SEM images of (a) CPA; (b) acid-CPA; (c) Ni/CPA; and (d) Ni/CPA without acid-pretreatment.

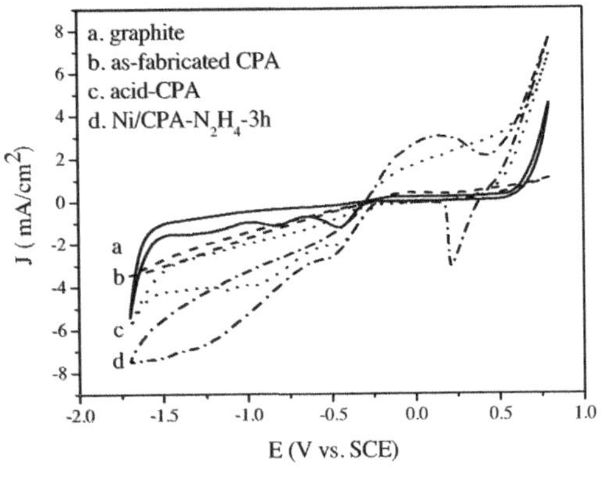


Figure 3. Cyclic voltammetric curves of different electrodes in 1 M NaOH solution at a scan rate of 2 mV/s: (a) graphite; (b) as-fabricated CPA; (c) acid-CPA; and (d) Ni/CPA-$N_2H_4$

of nano-carbon networks in CPA and porous structure. Moreover, the appearance of reductive peak around +210 mV for Ni/CPA implies the typical reductive ability toward oxygen in alkaline solution, which is related to the deposition of uniformly dispersed Ni nanoparticles.

## CONCLUSIONS

Uniformly sized and dispersed Ni nano-particles have been successfully deposited onto the surface of acid-CPA. Compared to former work, it is obvious that the morphology of the Ni particles deposited has been improved in this work, as smaller particles can be realized. The surface defects have also been increased, as confirmed by Raman spectroscopy, because of the deposition with the uniformly dispersed Ni nanoparticles. Moreover, as shown in cyclic voltammetric curves, it is logical that the performance of the as-fabricated nano-composite has been largely improved, which indicates the potential application of Ni/CPA as an electro-catalyst for ORR.

## ACKNOWLEDGEMENT

This research was supported by the Ministry of Education, Culture, Sports, Science and Technology (MEXT), Project "Cooperation for Innovative Technology and Advanced Research in Evolution Area". We also acknowledge the financial support from the program of Introducing Talents of Discipline to Universities (B08003).

## REFERENCES

1. J. R. Howell, M. J. Hall and J. L. Ellzey, "Combustion of hydrocarbon fuels within porous inert media", *Energy Combust. Sci.*, **22**, 121-127 (1996).
2. K. R. Mikeska, "Electrically conductive ceramics with oxides of Al, Cr, and Mg", *US Patent 5 656 203* (1997).
3. Y. Yasutomi, T. Miyoshi, M. Sobue, N. Yamashita, H. Nagase, K. Tanno, S. Arimoto and F. Jooraku, "Conductive material and process for preparing the same", *US Patent 5403674* (1995).
4. R. L. Menchavez, M. Fuji, T. Yamakawa, T. Endo and M. Takahashi, "Investigation of phase Composition in Dense and Porous Gelcast Alumina Sintered under Argon Atmosphere", *Mater. Sci. Forum*, **561-565**, 2123-2126 (2007).
5. R. L. Menchavez, M. Fuji, M. Takahashi, "Electrically Conductive Dense and Porous Alumin with In-Situ-Synthesized Nanoscale Carbon Networks", *Adv. Mater.* **20**, 2345-2351 (2008).
6. R. L. Menchavez, M. Fuji, H. Taroaki, M. Takahashi, "Electrically conductive gelcast porous alumina sintered under argon atmosphere", *Mater. Lett.* **61 (3)**, 754 -756 (2007).
7. J. Liu, H. Watanabe, M. Fuji, M.u Takahashi, "Electrochemical reduction of oxygen on alumina/gelcasting-derived nano-carbon network composite in alkaline solution", *Electrochem. Commun.*, **10**, 922-925 (2008).
8. J. Liu, R. L. Menchavez, H. Watanabe, M. Fuji and M. Takahashi, "Highly Conductive Alumina/NCN Composites Electrodes Fabricated by gelcasting and Reduction Sintering—An Electrochemical behavior Study in Aggressive Environments", *Electrochim.*

*Acta*, **53(24)**, 7191-7197 (2008).
9. J. Liu, R. L. Menchavez, H. Watanabe, M. Fuji, M. Takahashi, "Electrochemical Behaviors for Conductive Porous Alumina Composites with Gelcast-Derived Nanocarbon Networks in Aqueous Solutions", *J. Electrochem. Soc.*, **155(11)**, H878-H884 (2008).
10. C. Hai, H. Watanabe, T. Shirai, M. Fuji, M. Takahashi, F. Wang, "Modifying the Surface of Electrically Conductive Porous Alumina", *Mater. Lett.*, **63**, 1320-1322 (2009).
11. C. Hai, J. Liu, H. Watanabe, M. Fuji, F. Wang and M. Takahashi, "Surface activation of Conductive Porous Alumina by Depositing Nickel Particles", *J. Am. Ceram. Soc.*, **92 (S1)**, S38-S41 (2009).
12. S. Banerjee, T. Hemraj-Benny, S. S. Wong, "Covalent Surface Chemistry of single-Walled Carbon Nanotubes", *Adv. Mater.*, **17(1)**, 17-29 (2005).
13. Z. Tang, N. A. Kotov, "One-Dimensional Assembilies of Nanoparticles: Preparation, Properties, and Promise", *Adv. Mater.*, **17(8)**, 951-962 (2005).
14. R. Loscutova, A. R. Barron, Coating single-walled carbon nanotubes with cadmium chalcogenides, *J. Mater. Chem.*, **15**, 4346-4353 (2005).
15. M. Olek, T. Büegeb, M. Hilgendorff, and M. Giersig, "Quantum Dot Modified Multiwall Carbon Nanotubes", *J. Phys. Chem. B*, **110**, 12901-12904 (2006).

# INFLUENCE OF THE FUNCTIONAL GROUPS ON CARBON BLACK PARTICLES ON THE ZETA-POTENTIAL

Masahide Kawaraya*, Sumiyo Shimizu*, Hideyuki Hisashi*, and Hidehiro Kamiya**
* Research and Development Laboratory, Mikuni Color Ltd., 101 Kokubunji Mikunino-cho, Himeji, Hyogo, Japan, 671-0234
** Tokyo University of Agriculture and Technology, 2-24-16 Naka-cho, Koganei, Tokyo 184-8588

## ABSTRACT

Oxidation states of four kinds of carbon black (CB) powders, which varied with the length of ozone oxidizing time, were determined using vacuum pyrolysis. The zeta-potential and dispersibility were determined from the median particle size, $d_{50}$, measured in an aqueous suspension prepared from each oxidized CB powder. The relationship between the oxidation state of the CB and the dispersion behavior is discussed. Additionally, the effect of adding alkali species to neutralize each suspension on the zeta-potential was evaluated. The amounts of carboxyl and quinine groups of the CB increased concomitantly with ozone oxidation time. The absolute value of the zeta-potential of the CB suspension increased with the amount of oxidized groups, and the maximum value of the zeta-potential reached -71.3 mV. The high absolute value of the zeta-potential indicates good dispersibility of the CB particles in the aqueous suspension. The alkali addition influences the dispersibility, the zeta-potential, and dispersibility of the oxidized CB.

## INTRODUCTION

Carbon black (CB) has been a useful and important material in the reinforcement of tires and other organic materials, and for the development of functional properties of ink, paint, and toner. In the 21st century, CB has become important for electrical devices such as Li-ion batteries, electric double-layer capacitors, and new solar cells. Despite the use of CB in various industrial fields, the basic understanding of the characteristics of CB has not advanced in recent decades. Because CB pastes in a liquid state are generally applied to electrodes of secondary batteries or dye-sensitized solar cells, and the condition of the electrodes in those devices influences their performance, research into dispersibility, stability, and viscoelasticity of CB pastes is crucial for the performance of those cells.

Because CB pastes, as a coloring material, must be controlled in terms of their physical properties to support good applications, some studies[1] have specifically investigated the rheological behavior of the CB pastes to elucidate the dispersion behavior of CB particles. Alternatively, to generate better plastic or rubber products such as tires with mixed CB, some investigators[2] have studied chemical characteristics and reactivity of the CB or the CB surface using complicated methods, pyrolytic volatile analysis[3] and selective neutralization methods[4]. Nevertheless, few works have specifically examined the relation between surface oxidation behavior and the aqueous suspension behavior of CB.

For this study, we examined the surface oxidation states of ozone treated CB powders using a vacuum pyrolysis method. The dispersibility of CB in aqueous suspension was characterized using mean aggregate size measured using laser diffractive scattering systems and zeta-potential with alkali addition. Based upon the results described above, a relationship between the CB's oxidation states and the dispersion behaviors of the CB in aqueous suspension is discussed.

## EXPERIMENT

Surface treatments with an ozone reaction using the following equipment and processes were employed to oxidize CB powder. First, 10 kg of CB power dried at 100°C in advance was put into a SUS drum (0.6 m diameter, 1.5 m length). Then ozone air attenuated with dry air to 5000 ppm was introduced at 20 L/min into the rotating drum at 40 rpm to oxidize the CB powder. Oxidation times were 0, 6, 12, 24, and 42 hr to produce five different surface treatment states of CB. The amounts of $CO_2$ and CO, indicating oxidation states of each treated CB, were evaluated using the vacuum pyrolysis method[4]. The generalized characteristics of the oxidized CB are presented in Table I. Dibutyl phthalate (DBP) absorption, pH, and specific surface area were estimated, respectively, in accordance with JIS K6217, JIS K6219, and JIS K6217. The geometry of each CB was observed using STEM (S-5200; Hitachi High Technologies Corp.).

The CB powder (10 g) and deionized water (90 g) with either alkali, triethanolamine (TEOA) or potassium hydroxide (KOH) were put into plastic bottles (250 ml). The bottle contents were dispersed for 60 min using a shaking apparatus with ceramic beads. The quantities of mixed alkali in each equivalent ratio, 0, 0.5, 1.0, 1.5, 2.0, and 2.5 were determined by the amount of $CO_2$ measured using vacuum pyrolysis[3]. A laser diffractive scattering system (Zetasizer-Nano; Sysmex Corp.) was used to measure the sample pastes' median particle size ($d_{50}$) and the zeta-potential. Concentrations of the solutions in this measurement were diluted to $10^{-5}$wt%.

## RESULTS and DISCUSSION

Table II presents the amounts of desorption CO or $CO_2$ measured using the vacuum pyrolysis method. The longer ozone treatment time produced the larger amounts of CO or $CO_2$ of CB. A close relation exists between the increase in CO or $CO_2$ and the amount of volatile portion. Therefore, the adsorbed groups on CB in the treatment are CO and $CO_2$.

The FE-SEM images of the raw CB without treatment and sample D with the strongest treatments in Table 2 are depicted in Figure 1. In these FE-SEM images, no marked differences are apparent. Data of the specific surface area also clarified that the conditions of aggregation or flocculation of these CB powders are equivalent.

Dispersion behaviors of each surface treated CB in aqueous suspension are portrayed in Figures 2, 3, and 4. The figures respectively depict the measured mean aggregate size in aqueous suspension without alkali (Figure 2), with triethanolamine (Figure 3), and with potassium hydroxide (Figure 4). In the suspension without alkali, the higher oxidation CB, samples C and D show better dispersibility because of the fine $d_{50}$ value. Samples B, C, and D in the pastes with TEOA in Figure 3, and samples A, B, C, and D in the pastes with KOH in Figure 4 show good dispersibility.

Table I. CB samples, raw materials, and ozone oxidation treated.

| | Raw Material | Treated Sample | | | |
|---|---|---|---|---|---|
| | | Sample A | Sample B | Sample C | Sample D |
| Ozone oxidation time (hr) | 0 | 6 | 12 | 24 | 42 |
| DBP absorbed amount (ml/100 g) | 46 | 48 | 52 | 53 | 58 |
| Specific surface area ($m^2/g$) | 55 | 55 | 56 | 57 | 57 |
| pH | 9.3 | 3.7 | 2.6 | 2.2 | 2.1 |
| Volume resistivity ($\Omega \cdot cm$) | 0.4 | 0.7 | 5.5 | 23.9 | 65.5 |

Table II. Desorption gases of treated CB measured using vacuum pyrolysis.

| | | R.M. | Sample A | Sample B | Sample C | Sample D |
|---|---|---|---|---|---|---|
| CO | (mg/g) | 2.80 | 4.81 | 10.93 | 16.35 | 24.98 |
| $CO_2$ | (mg/g) | 0.70 | 2.45 | 8.71 | 15.40 | 27.34 |
| Volatile portions at 1000°C (%) | | 0.36 | 0.74 | 1.97 | 3.18 | 5.24 |

The relation between the surface oxidation level and zeta potential in aqueous suspension are depicted in Figure 5. It is noted that higher oxidized CB has a larger absolute value of the zeta-potential. This tendency is the same as that of each suspension with alkali, TEOA, and KOH in equivalent points. Furthermore, the addition of alkali is more effective in pastes with the medium oxidation conditions (Samples B and C) to increase the zeta-potential. As described herein, the amount of $CO_2$ generation during the oxidation treatment of the CB is used for these evaluations because Murata[5] suggested that the quinine groups (CO) have a lower effect on the zeta-potential of the CB aqueous solution than the carboxyl groups ($CO_2$); the dissociation constant (pKa) of the carboxyl group is lower than those of the quinine groups.

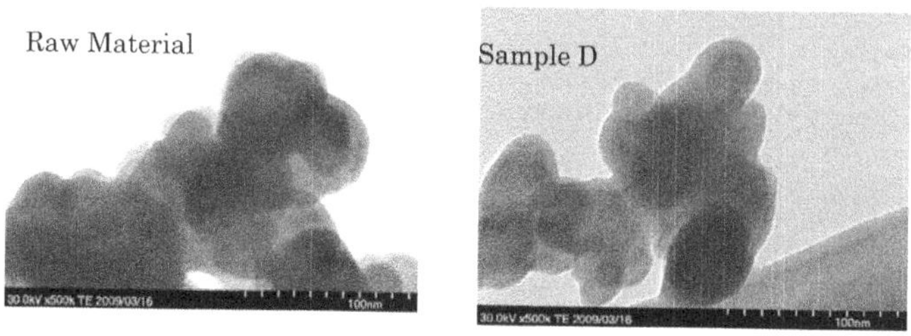


Figure 1. FE-SEM images of CB samples: raw CB without treatment and sample D.

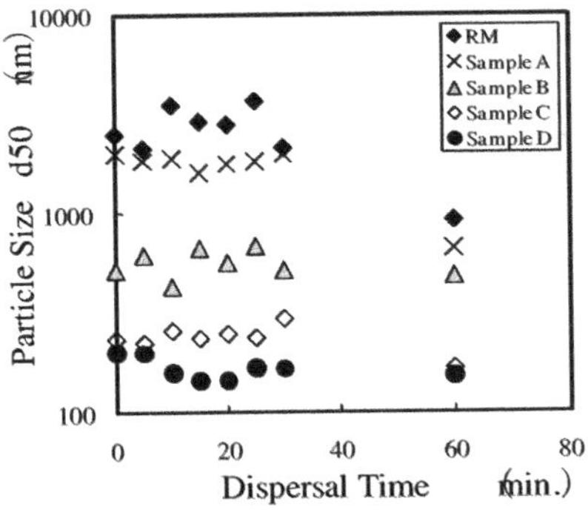


Figure 2. Dispersal behaviors of the CB pastes without alkali pastes.

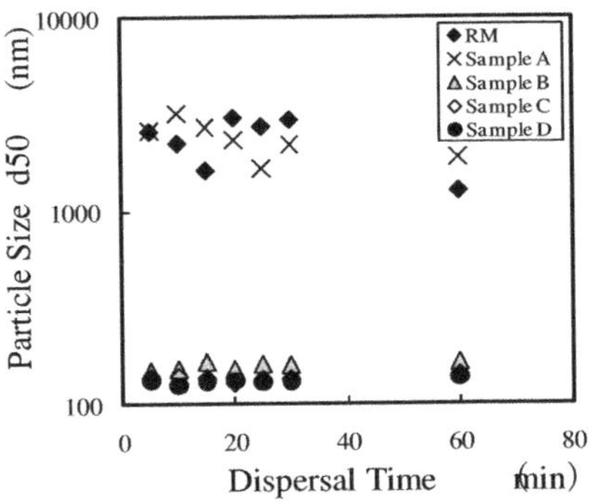


Figure 3. Dispersal behaviors of the CB pastes with triethanolamine.

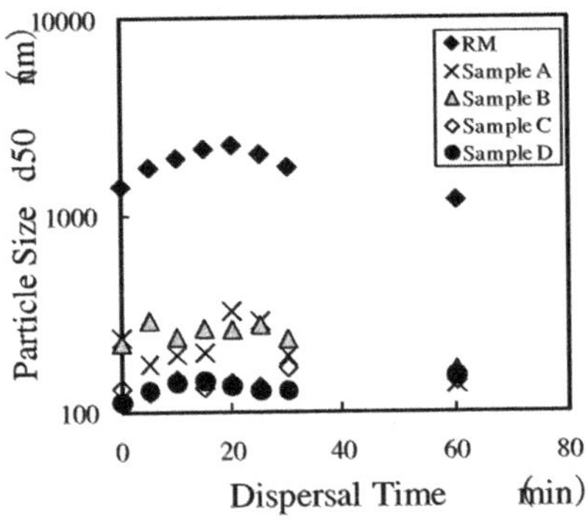


Figure 4. Dispersal behaviors of the pastes of CB with potassium hydroxide.

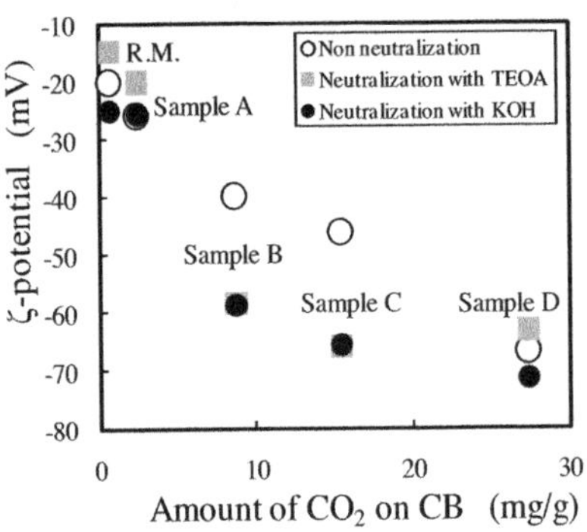


Figure 5. Relation between the amounts of $CO_2$ on the samples and the zeta-potential of the pastes.

Good correlation was found between the particle sizes and the zeta-potentials in all samples shown in Figures 6 and 7, except for sample A in KOH neutralization (Figure 7). To gain good dispersibility in a suspension, a higher absolute value of the zeta-potential is necessary: much more oxidation of CB is needed. The particle sizes of sample B with TEOA (Figure 6) and samples A and B with KOH (Figure 7) are decreased dramatically. Results show that it is also possible that CB oxidized less produces a good dispersion by adding alkali. Comparison of sample A in Figure 6 to sample A in Figure 7 shows that the zeta-potential in the KOH solution (Figure 7) becomes slightly lower: less than -30 mV. The particle size of sample A with KOH becomes less than with TEOA. Therefore, the difference between sample A's zeta-potential in TEOA and KOH solution might be a factor to decide dispersibility. In other words, suspensions with the zeta-potential controlled below -30 mV show good dispersibility. In earlier studies, it was considered that -30 mV was the threshold value to coagulate particles in a colloid system[6].

The pH changes in alkali ion concentrations are portrayed in Figures 8 and 9. As both figures show, the achieved pH of TEOA is less than that of KOH. It is agreed that activated hydrogen on the oxidation CB reacts with $OH^-$ of water[7], and that the pH in the suspension with TEOA in which there are few $OH^-$ might be lower by taking $OH^-$ onto CB, whereas KOH solutions have excessive $OH^-$ and that pH might be higher. For low concentrations of KOH for sample A, it is sufficient to obtain the absolute value of the zeta-potential as slightly larger. That is sufficient to give good dispersibility of suspension.

For high concentrations of alkali ion— especially KOH, which has higher pKa—pH still increases slightly. Results suggest that the electrical double layer or the slipping plane would be compressed by the cation components, $K^+$ or $H^+$, and that the surface potential would be raised because of neutralization of the functional groups on the CB (Figure 10). Results show that the ion concentration was insufficient or that the alkali type was inappropriate in this study to neutralize all functional groups on CB. It will be necessary to produce pastes that are more stable by controlling the zeta-potential, with accurate estimation of the functional group amounts and types on CB surface, and selection of the appropriate neutralization alkali for those functional groups.

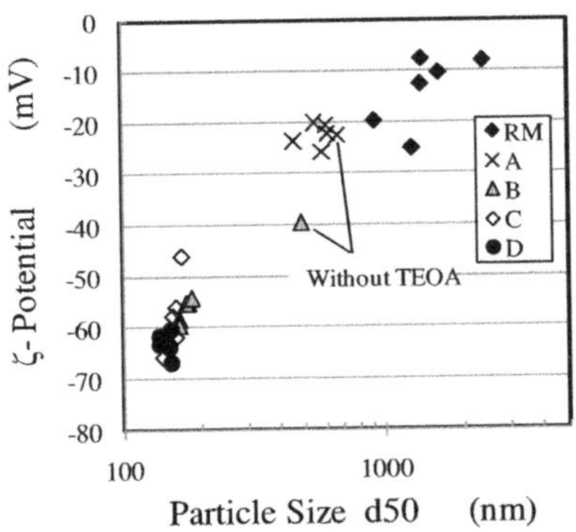


Figure 6. Particle size and the zeta-potential of pastes in TEOA neutralization.

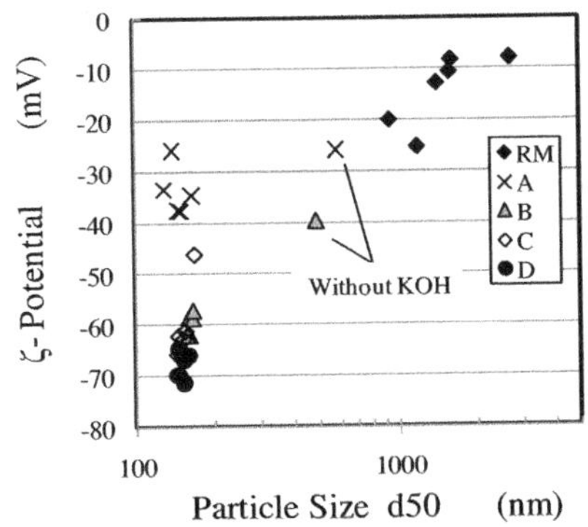


Figure 7. Particle size and the zeta-potential of pastes in KOH neutralization.

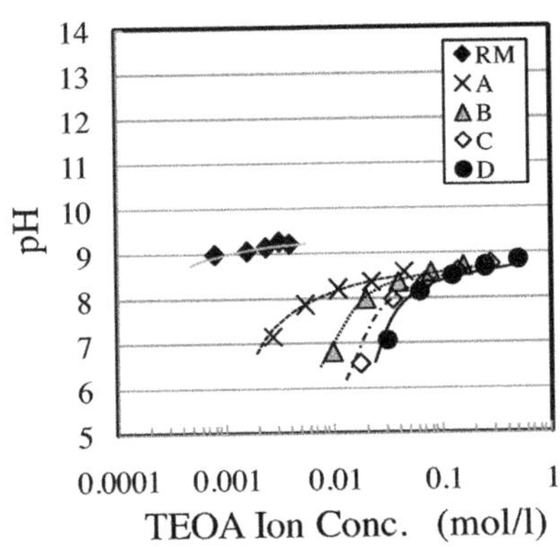


Figure 8. pH of the pastes at each TEOA concentration.

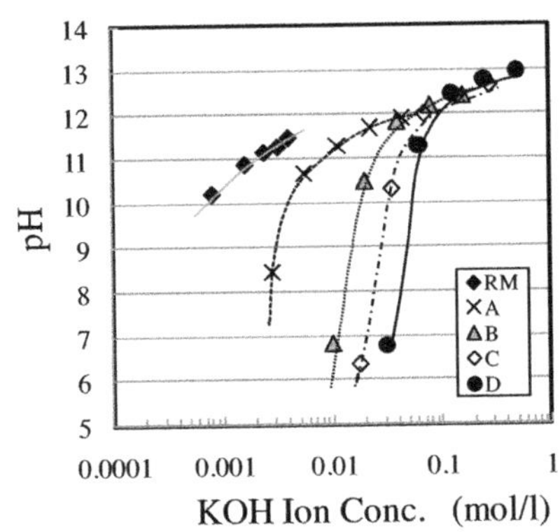


Figure 9. pH of the pastes at each KOH concentration.

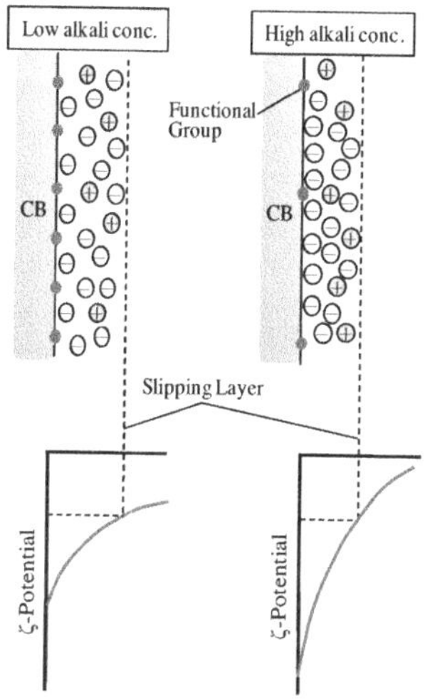


Figure 10. Scheme of the electrical double layer, the potential distribution, and the zeta-potential on CB in low and high alkali ion concentrations.

## CONCLUSION

CB oxidation states, graded in stages, were measured exactly using vacuum pyrolysis. Good dispersibility was obtained in pastes made of highly oxidized CB. A correlation was found between the amount of $CO_2$ generated during the CB oxidization and the zeta-potential in suspensions; results show that the zeta-potential produced an effect on the particle size ($d_{50}$) of the paste. Adding alkali, especially KOH, increased the absolute value of the zeta-potential for poorly oxidized samples. Good dispersion was possible, even with slightly treated CB, using an appropriate amount of a suitable alkali.

## REFERENCES

[1] T. Amari, "Non-linear viscoelastic properties of concentrated suspensions", Progr. Org. Coatings, **31**, 11-19(1997)
[2] J. B. Donnet, "The chemical reactivity of carbons", Carbon, **6**, 161-176(1968)
[3] D. G. Barton and B. H. Harrison, "Surface studies of carbon: Acidic oxides on spheron 6", Carbon, **11**, 649(1973)
[4] D. Rivin, "Surface Properties of Carbon", Rubber Chem.Technol., 44, 307-343(1971)
[5] T. Murata, Y. Matsuda and H. Imagawa, Effects of Alkali Metal Cations on the ζ-Potential of Carbon Black Particles, Electrochemistry, **47**, 334(1979)
[6] K. Furusawa, "Zeta-denni no Sokutei", Bunseki, **5**, 247(2004)
[7] S. Hagiwara, H. Takahashi, K. Tsutsumi, "Interaction between active hydrogen of carbon surface and water", Carbon, **9**, 693(1971)

# Smart Processing Technology

# EUDRAGIT POLYMER ENCAPSULATED HYDROXYAPATITE MICROSPHERES AS A CARRIER FOR THE SUSTAINED RELEASE OF SMALL MOLECULAR DRUGS

Jianbo GUO,[1] Hui XU,[1,*] Shaoning WANG,[2] Sijie ZHANG,[1] Hiroya ABE,[3] Xiaoyun WANG,[1] Yanqiu ZHAO,[1] Makio NAITO[3]

* School of Pharmacy, Shenyang Pharmaceutical University, 103, Wenhua Road, Shenyang, 110016, Liaoning, China

** School of Pharmaceutical Engineering, Shenyang Pharmaceutical University, 103, Wenhua Road, Shenyang 110016, Liaoning, China

*** Joining and Welding Research Institute, Osaka University, 11-1, Mihogaoka, Ibaraki, Osaka 567-0047, Japan

## ABSTRACT

The adsorption/desorption of oleanolic acid hemiphthalate disodium (OPS) with hydroxyapatite microspheres (HAPm) was investigated. XPRD and FT-IR analysis indicated a highly dispersing state of OPS on HAPm, and weak physical interaction between OPS and HAPm. OPS loaded HAPm (OPS-HAPm) was encapsulated with Eudragit RS using the S/O/W emulsion-solvent evaporation method. Both the desorption of OPS from HAPm and the release of OPS from the Eudragit encapsulated HAPm were greatly affected by factors including the volume, pH, types, and ionic strength of the release medium. The polymer encapsulated OPS-HAPm showed sustained drug release profiles with no obvious initial burst release in various release medium, the drug release rate can also be tailored by utilizing various polymer/HAPm ratios in the encapsulation of OPS-HAPm. These results indicate great potential of polymer encapsulated HAPm as the carrier for the modified release of small, molecular, water soluble drugs.

## INTRODUCTION

Hydroxyapatite (HAP) is a natural form of calcium apatite with the formula $Ca_5(PO_4)_3(OH)$, which is usually written as $Ca_{10}(PO_4)_6(OH)_2$ to denote that the crystal unit cell comprises two molecules[1]. Due to its excellent biocompatibility and bioactivity, and binding affinity to various molecules, HAP has found success in hard-tissue replacement, purification and separation of proteins, peptide and antibodies, and waste water treatment [2-5]. In recent years, HAP and HAP microspheres (HAPm) fabricated by spray-drying aqueous suspension of nano-sized HAP grains have aroused great interest, focusing on it as a promising carrier material for drug delivery, such as bone cement, scaffolds of growth factors and antibiotics, and injectable carriers for the sustained release of various active substances[6-10]. But the initial burst and rapid release associated with the drug loading mechanism of HAP limited its wide application in sustained drug delivery, especially for small molecular and water soluble drugs. To use HAP as the carrier for a sustained-release preparation, Mizushima *et al*[4] suggested it might be necessary to not only modify the manufacturing method to achieve slower release, but also to add other substances that delay drug release. Polymer encapsulation and polymer-HAP hybrid gels were reported to control or sustain drug release[6-7,10].

In this study, oleanolic acid hemiphthalate disodium (OPS, see Figure 1), an esterified prodrug of oleanolic acid, which exhibits hepato-protection, anti-inflammatory and antiulcer effects, was selected as a small molecular, water soluble model drug. OPS loaded HAPm (OPS-HAPm) was encapsulated with water permeable Eudragit RS polymer, a widely used coating material for oral dosage forms, to sustain drug release. Here, the adsorption/desorption, and the *in vitro* release properties of OPS-HAPm and encapsulated OPS-HAPm are reported.

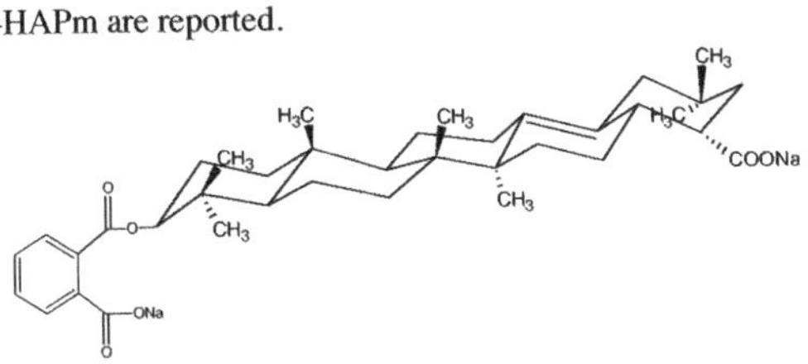


Figure 1. The chemical structure of OPS.

## MATERIALS AND METHODS

### Materials

Spherical hydroxyapatite (HAPm) with mean particle size of 11.4μm was obtained from Taihei Chemical (Osaka, Japan) and used as received. OPS (>98%) was synthesized in our laboratory. Eudragit RS was kindly donated by Degussa (Germany). Polyvinyl alcohol (PVA) (MW: 88,000-97,000, degree of alcoholysis: 87-89%) was obtained from Alfa Aesar (USA). All other chemicals used were analytical grade.

### Adsorption of OPS on HAPm

OPS was dissolved in methanol to make solutions of 1, 5, 10, and 12 mg/ml, respectively. 1.0g of HAPm was added to a 100 ml drug solution, while shaking the mixture in a water bath kept at 25°C. At predetermined intervals, 1 ml of suspension was withdrawn and centrifuged at 10,000 r/min for 5 min. The amount of OPS adsorbed was determined by measuring the difference between the OPS concentrations of the initial stock solution and the supernatant after adsorption.

### Desorption of OPS from OPS-HAPm

The HAPm suspensions soaked in OPS solutions for 24 h were filtrated; the obtained OPS-HAPm was vacuum-dried at 50°C overnight. Then the OPS-HAPm (drug loading: 96.5mg/g) was resuspended in different media and shaken in a water bath kept at 25°C. A 1 ml sample was withdrawn from the suspension at different time intervals. The samples were centrifuged at 10,000r/min for 5 min, and the amount of OPS released into the supernatants were determined by HPLC.

### Preparation of encapsulated OPS-HAPm

Polymer encapsulated microspheres were prepared by the S/O/W method. In brief, 0.3 g of Eudragit RS was dissolved in a 3 ml of the mixture of dichloromethane-ethyl ether (1:1) to make the oil phase. The drug loaded HAPm was then suspended in the oil phase uniformly by vortexing (about 30s). The suspension was then dropped into 10 ml of 2% PVA aqueous solution, stirred at 1,000 r/min for 5 min to form the S/O/W emulsion. And then, the emulsion was poured into 50 ml of 1% PVA aqueous solution, and stirred gently to remove the organic solvent at 37°C for 4 h. The encapsulated microspheres were collected by filtration, washed with distilled water, and vacuum dried at 50°C over night.

In vitro release of OPS from encapsulated microspheres

Microspheres of about 30 mg were placed into different release media kept at 37±0.5°C and stirred at 100 r/min in a water bath. At predetermined intervals, 1 ml of sample was withdrawn and passed through filter membrane (0.45μm pore size), the filtrate was assayed for OPS using HPLC.

Assay of OPS

A HPLC system (LC-10AD pump, SPD-10A UV-vis detector, and CTO-10A column oven, Shimadzu, Japan) with a column (4.6 mm×150 mm, Hanbang, China) packed with Hypersil ODS C18 filler (5 μm) was used for the assay of OPS. The mixture of methanol-water-phosphoric acid (90:10:0.1, v/v) adjusted to pH 4.0 with triethylamine was used as the mobile phase at a flow rate of 1.0 ml/min. The HPLC spectrum was recorded at 225 nm.

X-ray powder diffraction (XRPD)

X-ray powder diffraction patterns were obtained using a LabX XRD-6000 X-ray diffratometer (Shimadzu, Japan) at 40 kV, 30 mA and scanning rate of 5°/min over the range $2\theta$=0~40°, using CuKα1 radiation of wavelength 1.5405 Å.

Fourier transform infrared (FT-IR) spectroscopy

Transmission FT-IR spectra were recorded on a FT-IR spectrophotometer (Bruker IF55, Germany) at a resolution of 2 $cm^{-1}$.

Optical microscopy and scanning electron microscopy (SEM) observation

The morphology of encapsulated microspheres was observed under optical microscope and SEM (Superscan SSX-550, Shimadzu, Japan). Samples were sputter-coated with carbon before SEM observation.

RESULTS AND DISCUSSION

Adsorption properties of OPS on HAPm

The adsorption kinetics of OPS on HAPm in drug solutions was shown in Figure 2. The amount of OPS adsorbed increased with the increase of drug concentrations, but the adsorption equilibrium was not established even after soaking the HAPm in drug solutions for over 12 h.

The overlaid XRPD patterns of OPS, HAPm, physical mixture of OPS and HAPm, and OPS-HAPm were given in Figure 3. The XRPD patterns showed no obvious drug crystal peaks for OPS-HAPm. There were no significant differences between the FT-IR spectrum (Figure 4) of HAPm and OPS-HAPm, except that a C=O stretching vibration of OPS-HAPm occurred at 1560.1 $cm^{-1}$ (weaker than that of OPS at 1568.3 $cm^{-1}$). These results indicated highly dispersing state of OPS on HAPm and weak interaction between the C=O groups of OPS and HAPm.

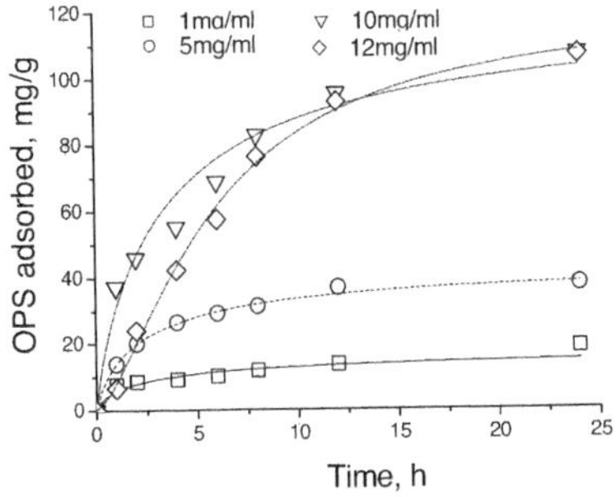


Figure 2. The amount of OPS adsorbed versus time.

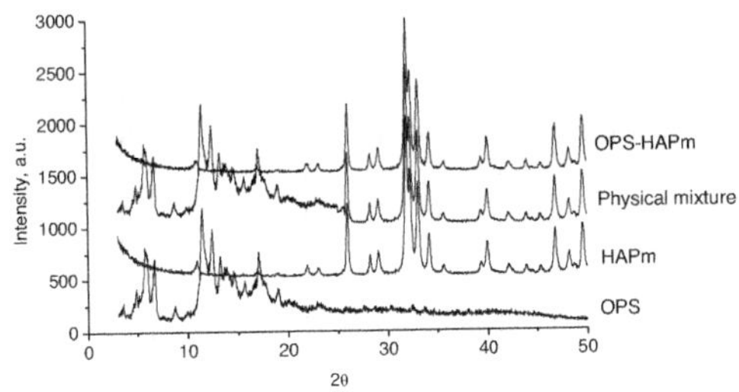


Figure 3. The X-ray powder diffraction patterns of OPS, HAPm, their physical mixture, and OPS-HAPm.

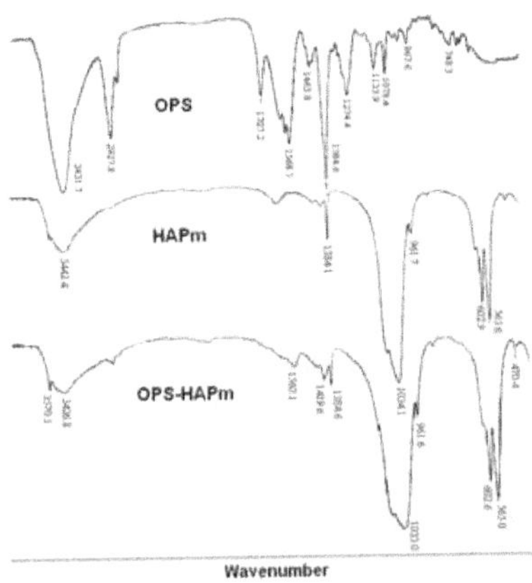


Figure 4. The FT-IR spectrum of OPS, HAPm and OPS-HAPm.

Desorption (release) of OPS from OPS-HAPm

Different to drug adsorption, the desorption equilibrium of OPS from HAPm was rapidly established (normally within 1.0h) in various media (Figure 5 A and B). The equilibrium desorption percents increased with the increase of the volume (Figure 5 A) and pH (Figure 5 C), and with the decrease of ionic strength of release media (Figure 5 D). The species presented in the buffer solutions also affected the desorption percents due to the common ion effect. Figure 5 B showed that the desorption percent in citric acid-citrate sodium buffer solution was the highest compared with other buffer solutions containing phosphate anion, but significantly lower than in purified water. The sharp increase of desorption amount at higher pH might be attributed to the increase of solubility of OPS with pH.

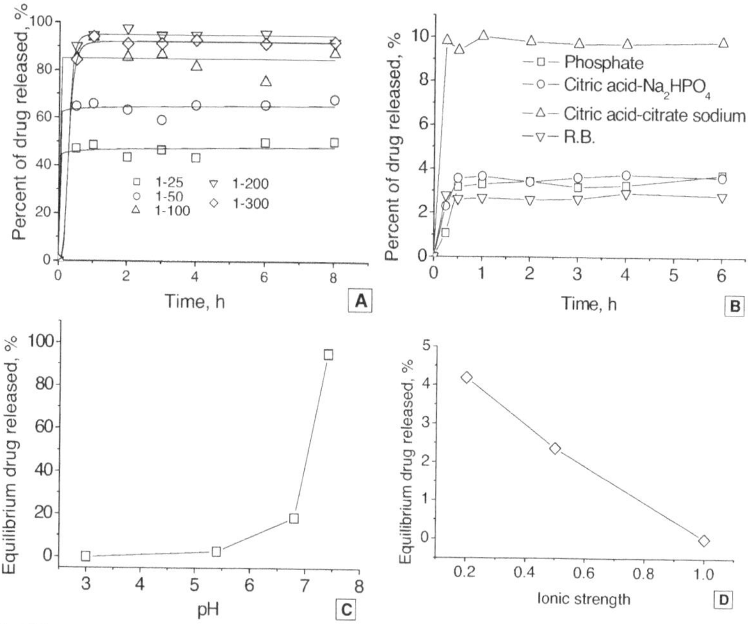


Figure 5. Effect of various factors on desorption of OPS from HAPm. (A) Volume of release medium (1g in milliliters of purified water), (B) types of buffer solutions (pH5.4), (C) pH (R.B. buffer solutions), and (D) ionic strength (R.B. buffer solutions). B, C, and D: OPS-HAPm/release media=1/200 g/ml.

In vitro release of OPS from encapsulated microspheres

The optical microscope image of OPS-HAPm showed spherical shape and multi-particulates structure inside the encapsulated microspheres, while the SEM images showed uneven or subsided surface, which might due to the shrinkage of encapsulation polymer during solvent-evaporation process (Figure 6).

The release profiles of encapsulated OPS-HAPm characterized with initial sustained drug release phase, followed by an equilibrium release percents dependent on the release medium used (Figure 7). It can be concluded that it was the polymer encapsulation that sustained drug release, while the HAPm that determined the equilibrium release percent that dependent on the adsorption/desorption balance between the drug and HAPm. An interesting finding was that the release percents of OPS increased with the pH of the medium, except that at pH 1.0 which was higher than that at pH 3.2, 5.4 and 6.8. The possible explanation was that HAPm readily dissolved in the strong acidic medium (0.1mol/L HCl) penetrated into the coated polymer membrane, thus the release percent of OPS increased.

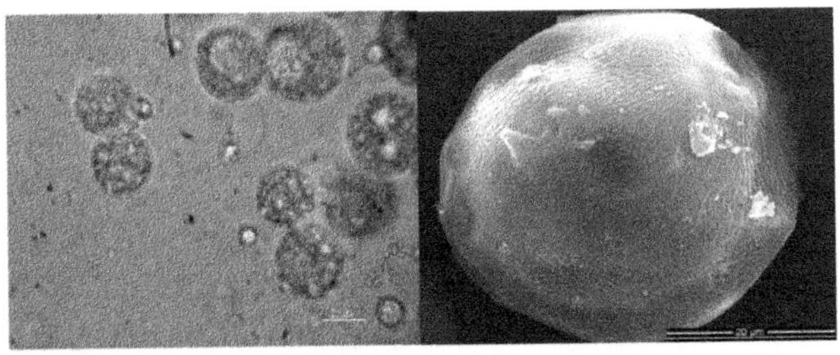

Figure 6. The optical microscope photo (Left) and SEM image (Right) of Eudragit encapsulated OPS-HAPm.

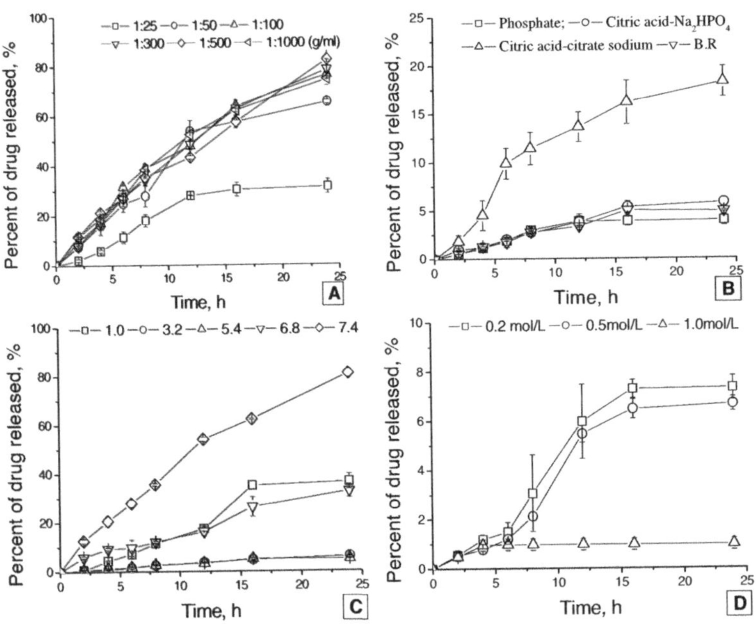


Figure 7. Effect of various factors on the release profiles of encapsulated OPS-HAPm (Eudragit RS/OPS-HAPm=1/1). (A) Volume of release medium (1g in milliliters of purified water), (B) types of

buffered solutions (pH5.4), (C) pH (R.B. buffer solutions), and (D) ionic strength (R.B. buffer solutions). B, C, and D: OPS-HAPm/release media=1/100 g/ml.

It is obvious that the thickness of the encapsulated polymer, which was directly related to the ratio of encapsulating polymer to HAPm, also affect the drug release rate. As shown in Figure 8, the release rates decreased with the increase of the ratio of Eudragit RS to OPS-HAPm.

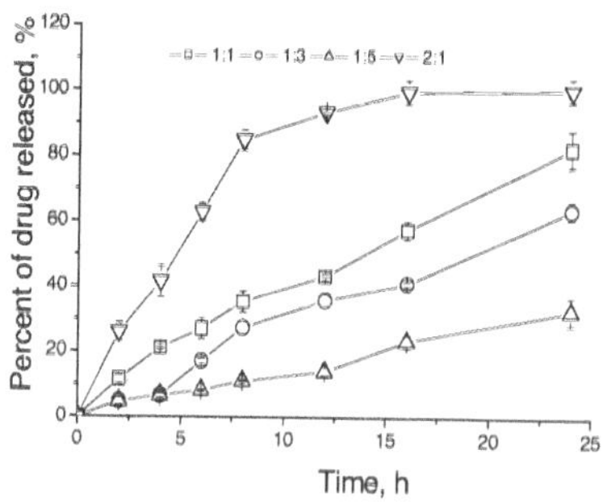


Figure 8. The drug release profiles of encapsulated microspheres of different ratios of Eudragit RS to OPS-HAPm. Encapsulated OPS-HAPm/release media (purified water)=1/200g/ml.

## CONCLUSION

The adsorption/desorption properties of OPS with HAPm was investigated. OPS was adsorbed on HAPm by soaking in drug solutions in methanol, and OPS loaded HAPm was encapsulated with Eudragit RS in order to obtain sustained drug release profiles. The polymer encapsulated OPS-HAPm showed sustained drug release profiles with negligible initial burst release in various release media. It is believed that both the polymer encapsulation and adsorption/desorption equilibrium of OPS with HAPm contribute to the modified drug release mechanism, and the drug release rate can be tailored by utilizing various polymer/HAPm ratios. These results indicated great potential of polymer encapsulated HAPm as a carrier for the modified release of small molecular, water soluble drugs.

## ACKNOWLEDGEMENT

The authors would like to thank Miss Yanli Yang for her technical support on XRPD. This work is partly supported by the Scientific Research Foundation for Returned Scholars of Shenyang Pharmaceutical University, China.

## FOOTNOTES

* Corresponding author: Hui Xu; E-mail: xuhui-spu@163.com

## REFERENCES

[1] D. L. Wise, Biomaterials and bioengineering handbook, CRC Press, 433-457 (2000).

[2] S. Schubert, R. Freitag, "Comparison of ceramic hydroxyl-and fluoroapatite versus protein A/G-based resins in the isolation of a recombinant human antibody from cell culture supernatant", *J. Chromatogr. A*, **1142**, 106-113 (2007).

[3] Q. J. Xu, J. T. Czernuszka, "Controlled release of amoxicillin from hydroxyapatite-coated poly (lactic-co-glycolic acid) microspheres", *J. Control. Release.*, **127**, 146- 153 (2008).
[4] Y. Mizushima, T. Ikoma, J. Tanaka, K. Hoshi, T. Ishihara, Y. Ogawa and A. Ueno, "Injectable porous hydroxyapatite microparticles as a new carrier for protein and lipophilic drugs", *J. Control. Release.*, **110**, 260–265 (2006).
[5] T. Kawasaki, "Hydroxyapatite as a liquid chromatography packing", *J. Chromatogr.*, **544**, 147-184 (1991).
[6] Y. Boonsongrit, K. Sato, A. Kondo, H. Abe, M. Naito, M. Yoshimura, H. Ichikawa, Y. Fukumori, "Controlled release of bovine serum albumin from hydroxyapatite microspheres for protein delivery system", *Mat. Sci. Engineer B.*, **148**, 162-165 (2008).
[7] M. P. Ginebra, T. Traykova, "Calcium phosphate cements as bone drug delivery systems: A review", *J. Control. Release.*, **113**, 102–110 (2006).
[8] C.C. Ribeiro, C.C. Barrias, M.A. Barbosa, "Calcium phosphate-alginate microspheres as enzyme delivery matrices", *Biomaterials*, **25**, 4363-4373 (2004).
[9] H. H. Pham, P. Luo, F. Génin, A. K. Dash, "Synthesis and characterization of hydroxyapatite-ciprofloxacin delivery systems by precipitation and spray drying technique", *AAPS Pharm. Sci. Tech.*, **3**, 1-9 (2002).
[10] T. Matsumoto, M. Okazaki, M. Inoue, S. Yamaguchi, T. Kusunose, T. Toyonaga, Y. Hamada, J. Takahashi, "Hydroxyapatite particles as a controlled release carrier of protein", *Biomaterials*, **25**, 3807-3812 (2004).

# PREPARATION OF ESTRADIOL SUBMICRON EMULSIONS TO IMPROVE BRAIN TARGETING THROUGH THE NASAL ROUTE IN RATS

Chaoqun Yu, Xiaomei Wang, Xing Tang*
Department of Pharmaceutics, Shenyang Pharmaceutical University, No. 103, Wenhua Road, Shenyang 110016, China. Tel.: +86 24 23986343; fax: +86 24 23911736.
E-mail address: tangpharma@126.com (X. Tang).

## ABSTRACT

Estradiol ($E_2$)-loaded submicron emulsions (SE) were prepared with the aim of improving brain targeting of $E_2$ through the nasal route. $E_2$-SE was administered to male Wistar rats, either intranasally or intravenously, at a dose of 0.48 mg $kg^{-1}$. The levels of $E_2$ in the blood were investigated and the levels of $E_2$ in the cerebrospinal fluid (CSF) in rats were evaluated using the microdialysis method. The plasma levels achieved following intranasal administration ($C_{max}$ 34.2 ± 8.8 ng·$ml^{-1}$; $AUC_{plasma}$ 4235.7 ± 1073.2 ng·min·$ml^{-1}$) were significantly lower than those after intravenous administration ($C_{max}$ 123.4 ± 24.5 ng·$ml^{-1}$; $AUC_{plasma}$ 11632.2 ± 3902.3 ng·min·$ml^{-1}$), while the CSF concentrations achieved after intranasal administration ($C_{max}$ 97.3 ± 20.4 ng·$ml^{-1}$; $AUC_{csf}$ 17387.1±2294.5 ng·min·$ml^{-1}$) were significantly higher than those after intravenous administration ($C_{max}$ 42.8 ± 18.6 ng·$ml^{-1}$; $AUC_{csf}$ 9505.1±4452.8 ng·min·$ml^{-1}$). The drug targeting index (DTI) for the nasal route was 5.0, and the drug targeting percentage (DTP%) was 80.1%. These results showed that the $E_2$ must be directly transported from the nasal cavity into the CSF in rats. In comparison with the $E_2$ inclusion complex in our previous study, $E_2$-SE significantly improved the transport of $E_2$ into the CSF, and exhibited greater brain targeting efficiency.

## INTRODUCTION

Alzheimer's disease (AD) is a progressive neurodegenerative disorder associated with neurofibrillary tangle formation and deposition of amyloid plaques in susceptible brain regions.[1] Enhancement of the amount of estradiol (17β-estradiol) transported into brain could improve the therapeutic effect of AD treatment. Nasal delivery has been explored as an alternative administration route to target drugs directly to the brain via the olfactory neurons[2-3]. Estradiol ($E_2$), as a model lipophilic compound with poorly water soluble, can be prepared as submicron emulsions with vegetable oils and lecithin aimed at enhancing nasal absorption. On the basis of our earlier study[4], the microdialysis method was used again to measure $E_2$ levels in the CSF. The bioavailability and brain-targeting of $E_2$ were evaluated after intranasal and intravenous administration of SE to rats.

## MATERIALS AND METHODS

### Materials

Estradiol (17β-estradiol) was purchased from Xianju Pharmaceutical Factory, China. Poloxamer 188 (Pluronic F68®) was obtained from BASF Corporation, Germany. Egg lecithin

(Lipoid E80®) and medium-chain triglyceride (MCT) were purchased from Lipoid KG (Ludwigshafen, Germany). Soybean oil (LCT) was obtained from Tieling BeiYa Medicated Oil Co. (Tieling, China); Glycerol was supplied by Zhejiang Suichang Glycerol Plant, (Zhejiang, China). All other reagents were of analytical grade or the highest grade commercially available. Microdialysis probes and Artificial CSF were the same as described in an earlier study[4].

Preparation of estradiol ($E_2$)-loaded submicron emulsions (SE) for nasal administration

MCT (10%, w/v), LCT (10%, w/v), egg lecithin (2%, w/v), oleic acid(0.06%, w/v) and $E_2$ (0.2%, w/v) were mixed by stirring at 80°C to prepare the oil phase, while the glycerol (2.5%, w/v) and F68 (0.5%, w/v) were dispersed in water and stirred at 70°C to obtain the aqueous phase. The oil phase was added to the aqueous phase with continuous stirring using a high-shear mixer (ULTRA TURRAX® T18 basic, IKA® WORKS Guangzhou, Germany) at 14000 rpm for 15 min to obtain the primary emulsion. After adjusting the pH to about 5.5 with 0.1 mol/L NaOH solution, the primary emulsion was passed through a high pressure homogenizer (Niro Soavi NS10012k, Niro Soavi S.p.A., Via M. Da Erba, Italy) at 90 MPa for eight cycles. The temperature of the entire homogenization process was maintained at 40°C using an ice-water bath. Finally, the preparation was transferred to vials after adding nitrogen gas, and sterilized in a 100°C rotating water bath for 30 min.

Physicochemical characterization of $E_2$-SE[5].

The particle size and zeta potential were measured using a NICOMPTM 380 Zeta Potential/Particle Sizer (Particle Sizing Systems, Santa Barbara, USA). The zeta potential was determined using the electrophoretic light scattering (ELS) technique.[5]

High-performance liquid chromatographic analysis in vitro.

The $E_2$ in vitro was determined using a high performance liquid chromatographic method at a wavelength of 205nm. The HPLC equipment consisted of a HITACHI L-7110 Intelligent HPLC pump, and a HITACHI L-7200 Intelligent HPLC Autosampler, a HITACHI L-7420 Intelligent HPLC Detector, and an ANASTAR Chromatography Data System. A Kromasil C18 column (200 mm × 4.6 mm, particle size 5μm, Zirchrom company) was used for the separation of $E_2$ and a mixture of acetonitrile-water (50:50, v/v) was used as the mobile phase. The injection volume was 20μl. $E_2$ could be completely separated from the other components by this HPLC method. The linear range of the $E_2$ determination was 1.0~40.0μg/ml (A=41596C+2698.1, R=0.9999) and this HPLC method was very precise and accurate.

Evaluation of $E_2$ loading capacity and entrapment efficiency (EE%)[6]

For determining $E_2$ loading capacity, the standard solution (about 10μg $ml^{-1}$) was prepared by dissolving $E_2$ in methanol, and the test solution was obtained as follows: 0.5 ml $E_2$-SE suspension of three batches was placed in a 50 ml volumetric flask, and then diluted with isopropanol, then damaged by ultrasound and, finally, filtered through a 0.22 ml microporous membrane. The ultracentrifugation method[5] was chosen to determinate the entrapment efficiency.

The entrapment efficiency of the emulsions was determined by measuring the amount of $E_2$ with the HPLC assay in the lowest aqueous layer obtained after ultracentrifugation. The EE observed by this method is the percentage of $E_2$ incorporated in the oil phase and located in the interfacial lecithin layer with respect to the total amount of $E_2$ located on the SE.

Animal experiments

The operations on rats before drug administration, microdialysis procedure, analytical method, data analysis were described clearly in our earlier study[4].

For intranasal administration, the $E_2$-SE suspension (2.0 mg $ml^{-1}$) was injected into the right nostril at a dose of 0.48 mg $kg^{-1}$ via a microsyringe, which was attached to a blunt needle. Blood samples (0.3 ml) were withdrawn from the tube in the jugular vein into heparin stabilized test-tubes at intervals: 0, 5, 10, 20, 30, 45, 60, 90, 120, 150, and 180 min. In addition, a cranial burr hole was made with a dental drill to a depth of 3.1 mm and the hole was 1.5 mm lateral and 0.9 mm posterior to the bregma. The microdialysis probe was stereotaxically inserted through this cranial burr hole, and was attached to the skull with dental cement. CSF perfusate samples were collected at intervals of 20 min for 240 min. Plasma was separated by centrifugation at 3000 rpm for 15 min, and kept frozen at -20°C together with CSF perfusate for subsequent analysis. Intravenous administration was carried out by injecting a dose of 0.48 mg $kg^{-1}$ $E_2$ -SE suspension (0.2 mg $ml^{-1}$) via the femoral vein. The method of sample collection was the same as above. Both nasal and intravenous administration should be performed following the successful implantation of a microdialysis probe and stabilization for 1 h with artificial CSF.

RESULTS

Physico-chemical properties of $E_2$-SE

Three batches of $E_2$-SE were prepared. The mean particle size of $E_2$-SE was approx 220nm, and the zeta potential was approximately -30mv. Egg lecithin was used as the major emulsifier and, with the help of oleic acid, the zeta potential of $E_2$-SE was greater than 20 mv which ensures the stability by preventing coalescence of droplets upon random collision[7]. The EE% of the three batches was higher than 85% and the loading capacity (LC) of E2 ranged from 1.8 to 2.0 mg $ml^{-1}$.

Pharmacokinetic study

The calculated concentrations of estradiol in blood after intranasal (i.n.) and intravenous (i.v.) administration are shown in Figure 1. The concentrations calculated by intranasal administration are significant lower than those calculated following intravenous administration. For intravenous delivery, the peak plasma concentration (Cmax 123.4±24.5) was reached at the start and the drug concentration-time curve, and then exhibited an exponential decline. Comparatively, 26 min after nasal administration, the $E_2$ plasma concentration reached a peak value of 34.2±8.8 ng $ml^{-1}$, far lower than the maximum plasma concentration after i.v. injection.

Selected pharmacokinetic parameters for intranasal and intravenous administration are shown in Table I, where the values of AUC, Cmax, absorption rate, half-life and $T_{max}$ can be

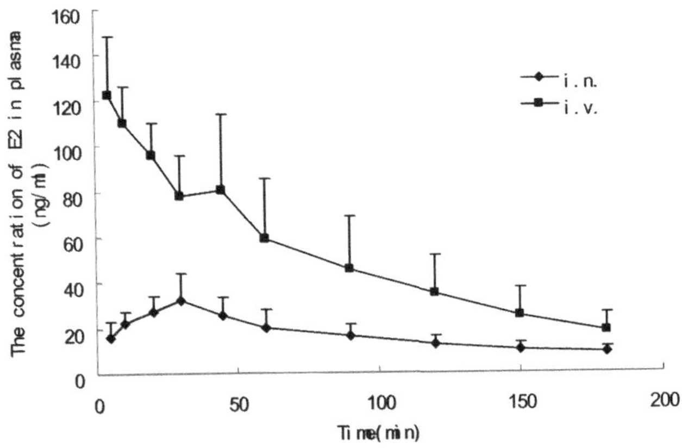


Figure. 1. The mean plasma concentration–time curves of estradiol in rats after intranasal and intravenous administration of $E_2$-SE at the dose of 0.48 mg kg-1. (n = 5, mean ± SD).

found. The absolute bioavailability for i.n. administration of $E_2$-SE was 36.41%. Significant differences ($P < 0.05$) between intranasal and intravenous delivery were found (t-test) for some of the calculated pharmacokinetic parameters, such as $AUC_{plasma}$ and $C_{max}$.

Brain-targeting study

To determine the brain-targeting of estradiol after preparation as a submicron emulsion, $E_2$-SE was administered intranasally and intravenously to rats. The calculated concentrations of estradiol in CSF after intranasal and intravenous administration are shown in Figure 2, and the important parameters are listed in Table II. In CSF, estradiol reached a Cmax (97.3±20.4ng $ml^{-1}$) at 24 min for the nasal route and Cmax (42.8±18.6 ng ml-1) was reached at 68 min after i.v. injection. In plasma, the concentrations after the nasal delivery were significant lower and the Cmax was reached later than those after i.v. administration while, in CSF, the concentrations after i.n. administration were significant higher and the Tmax was reached earlier than those after i.v. injection. This shows that $E_2$ reaches the brain tissue earlier following intranasal administration.

The degree of $E_2$ targeting to CSF after intranasal administration can be evaluated by the drug targeting index (DTI) and the higher the DTI, the greater the degree of $E_2$ targeting to the CSF can be expected after intranasal administration. As seen in Table III, the ratio of $AUC_{CSF}/AUC_{plasma}$ by the nasal route (4.10±0.54) was significant higher than that of i.v. injection (0.82±0.38). The DTI was 5.0, and the DTP was 80.1%. These results showed that a certain amount of E2 must be directly transported from the nasal cavity into the CSF in rats.

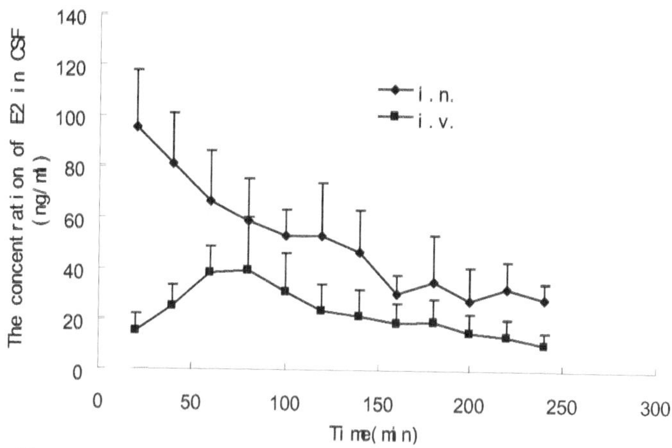


Figure. 2. The mean CSF concentration–time curves of estradiol in rats after intranasal and intravenous administration of estradiol $E_2$-SE at the dose of 0.48 mg kg-1 (n = 5, mean ± SD).

Table I. Pharamacokinetic parameters of estradiol E2-SE after intranasal and intravenous administration to rats at the dose of 0.48 mg $kg^{-1}$ in plasma (n = 5, mean ± SD)

| Pharmacokinetic parameters | $C_{max}$ ($ng \cdot ml^{-1}$) | Tmax (min) | $AUC_{plasma}$ ($ng \cdot ml^{-1} \cdot min$) | Ke (min-1) | $t_{1/2}$ (min) |
|---|---|---|---|---|---|
| E2-SE (i.v.) | 123.4±24.5 | 6.0±2.2 | 11632.2±3902.3 | 0.01±0.002 | 74.9±17.1 |
| E2-SE (i.n.) | 34.2±8.8 | 26.0±5.5 | 4235.7±1073.2 | 0.008±0.003 | 97.3±30.5 |

Table II. Pharamacokinetic parameters of estradiol E2-SE after intranasal and intravenous administration to rats at the dose of 0.48 mg $kg^{-1}$ in CSF (n = 5, mean ± SD)

| Parameter | $C_{max}$ ($ng \cdot ml^{-1}$) | Tmax (min) | $AUC_{CSF}$ ($ng \cdot ml^{-1} \cdot min$) |
|---|---|---|---|
| E2-SE (i.v.) | 42.8±18.6 | 68.0±11.0 | 9505.1±4452.8 |
| E2-SE (i.n.) | 97.3±20.4 | 24.0±8.9 | 17387.1±2294.5 |

Table III. The ratio of $AUC_{CSF}$ to $AUC_{plasma}$, DTP and DTI of E2 after i.v. or i.n. administration of $E_2$-SE to rats (n = 5, mean ± SD)

| Parameter | $AUC_{CSF}$ ($ng \cdot ml^{-1} \cdot min$) | $AUC_{plasma}$ ($ng \cdot ml^{-1} \cdot min$) | $AUC_{CSF}/AUC_{plasma}$ | DTI | DTP (%) |
|---|---|---|---|---|---|
| E2-SE (i.v.) | 9505.1±4452.8 | 11632.2±3902.3 | 0.82±0.38 | — | — |
| E2-SE (i.n.) | 17387.1±2294.5 | 4235.7±1073.2 | 4.10±0.54 | 5.0 | 80.1 |

DISCUSSION

Compared with the $E_2$ inclusion complex in my previous study[4], the value of $AUC_{CSF}/AUC_{plasma}$ (4.10±0.54) for $E_2$-SE was significantly higher than that of the $E_2$ inclusion complex (1.60±0.67), which proved that $E_2$-SE was a more suitable formulation for $E_2$ as far as transport into the CNS is concerned. In conclusion, the results obtained showed that $E_2$ must be directly transported from the nasal cavity into the CSF in rats after nasal administration of $E_2$-SE. In comparison with the $E_2$ inclusion complex in our previous study, $E_2$-SE significantly improved $E_2$ transport into the CSF and exhibited greater brain targeting efficiency. It was believed that the absorption enhancement mechanisms were probably due to the fact that the egg lecithin acted as an absorption enhancer, and the retention time in the nasal mucosa was prolonged because of the special character of the submicron emulsion.

REFERENCES


[1] J. Grutzendler, K. Helmin, J. Tsai, W.B. Gan, "Various dendritic abnormalities are associated with fibrillar amyloid deposits in Alzheimer's disease", Ann. N. Y. Acad. Sci., 1097, 30 - 9 (2007).

[2] L. Illum, "Transport of drugs from the nasal cavity to the central nervous system", Eur. J. Pharm. Sci., 11, 1-18 (2000).

[3] S. Mathison, R. Nagilla, U.B. Kompella, "Nasal route for direct delivery of solutes to the central nervous system: fact or fiction?", J. Drug Target., 5, 415- 41 (1998).

[4] X. Wang, H. He, W. Leng, X. Tang, "Evaluation of brain-targeting for the nasal delivery of estradiol by the microdialysis method", Int. J. Pharm., 317, 40-6 (2006).

[5] H. Komatsu, A. Kitajima, S. Okada, "Pharmaceutical characterization of commercially available intravenous fat emulsions: estimation of average particle size, size distribution and surface potential using photon correlation spectroscopy", Chem. Pharm. Bull, 43, 1412-5 (1995).

[6] X. Liu, Y. Zhang, X.Tang, H. Zhang, "Determination of entrapment efficiency and drug phase distribution of submicron emulsions loaded silybin, J. Microencapsul", 26, 180-6 (2009).

[7] M. Jumaa, B.W. Müller, "Parenteral emulsions stabilized with a mixture of phospholipids and PEG-660-12-hydroxy-stearate: evaluation of accelerated and long-term stability", Eur. J. Pharm. Biopharm., 54, 207-12 (2002).

# DEVELOPMENT OF A COMBINATORIAL PLASMA PROCESS ANALYZER FOR ADVANCED R&D OF NEXT GENERATION NANODEVICE FABRICATION

Kosuke Takenaka[1], Ken Cho[1], Yuichi Setsuhara[1], Masaharu Shiratani[2], Makoto Sekine[3], and Masaru Hori[3]
[1]Joining and Welding Research Institute, Osaka University, and JST, CREST, 11-1, Mihogaoka, Ibaraki, Osaka 567-0047, Japan,
[2]Department of Electronics, Kyushu University and JST, CREST, 744 Motooka, Nishi-ku, Fukuoka 819-0395, Japan
[3]Graduate School of Engineering, Nagoya University and JST, CREST, Furo-cho, Chikusa-ku, Nagoya 464-8603, Japan

## ABSTRACT

To develop a combinatorial plasma process analyzer in which the plasma distributions are finely controlled, we have carried out plasma-fluid simulations. The simulations show the feasibility of using low inductance antenna (LIA) modules to produce density-inclination plasmas. Numerically predicted plasma density profiles are generally in excellent agreement with experiments. A considerable inclination in plasma density was obtained using localized power deposition profiles. The results show that a variety of process results can be efficiently analyzed via the inclination distribution along the substrate for the process parameters.

## INTRODUCTION

Plasma processing is currently a key technology for microelectronics device fabrication, including ULSIs[1] and flat-panel displays[2]. However, to successfully fabricate next-generation advanced devices including MEMS[3-7], NEMS[8-9], and bio-chips[10-11], it is essential to develop plasma process technologies with nano-precision structure control. Tighter process control is required to successfully fabricate devices with nanometer-scale precision. This makes conventional trial and error methodologies for process development prohibitively expensive.

To overcome the constraints associated with developing next-generation plasma nano fabrication technologies, we have proposed a new approach for plasma nano fabrication technology in which we shift from characterizing the conventional process parameters (gas pressure, plasma sustaining power, and so on) to characterizing the basic elements of the processes (fluxes and energy distributions of ions, radicals and photons encountering the process).

To enable this breakthrough for next-generation plasma nano technologies, we have developed a plasma process analyzer, based on combinatorial methods, in which a variety of process results can be efficiently analyzed using the inclination of the plasma distribution along the substrate for the process parameters (ion flux, radical flux), where the distributions of particle fluxes are finely controlled and characterized via particle measurements.

In this study, a plasma process analyzer based on combinatorial methods has been developed in which the plasma profiles are finely controlled with a low-inductance antenna (LIA) module. The LIA module consists of a U-shaped internal antenna with dielectric isolation, each of which was coupled to an RF power system for independent power control to drive the inductively coupled plasma (ICP), and which also allowed low-voltage high density plasma production with active control of the power deposition profiles[12]. In our previous studies using LIA modules, we reported that this unique antenna for ICP generation allows for high-density

($10^{11}$ -$10^{12}$ $cm^{-3}$) plasma production with low plasma damage (sheath-edge ion energy as low as or less than 10V) via low-voltage operation of ICPs[12-19].

In this paper, we have carried out plasma-fluid simulations to show the feasibility of producing density-inclination plasmas using LIA modules.

## SIMULATION CODE

The model developed for this investigation of the plasma density profile is a three-dimensional (3D) fluid-simulation code. Before describing the details of the code, let us touch on the validity of the simulation model based on the fluid dynamics. In the present investigation, simulations have been carried out for argon plasmas sustained in 500 x 150 $mm^2$ rectangular chamber at pressures of 1.3 Pa. These conditions refer to the Knudsen number (defined as the ratio of the collision mean free path length to a representative physical length scale of the system) of the order of $10^{-2} << 1$, indicating that the collision mean free path is much less than a length scale of the problem. Therefore, the continuum assumption of fluid mechanics is regarded as a good approximation in the present investigation.

In the present model, the plasma density profile was predicted by numerically solving continuity and energy equations for neutral and charged species under the drift-diffusion approximation, in which the motions of electrons and ions were assumed to be governed by ambipolar diffusion, and the plasma was assumed to be quasi-neutral. Argon (Ar) plasmas sustained at an RF power (13.56 MHz) of 1000 W are simulated in the present study.

In this 3D model, the simulation space is divided into rectangular numerical meshes, whose typical size is set to be 44×15×30 $mm^3$ in the x-, y- (perpendicular to axis of discharge chamber) and z- (parallel to axis of discharge chamber) directions, respectively. The RF power absorption is modeled as the power deposition in the electron energy equation, in which the amount of the absorbed power delivered to the numerical mesh is in proportion to the fraction of the antenna included in the mesh; i.e., the absorbed power in the mesh is calculated as (total power to the antenna)×(volume fraction of the antenna included in the mesh)/ (total antenna volume).

For the cross sections of argon-electron collisions including excitation, ionization and elastic collisions, and the rates of electron neutral momentum transfer as a function of electron energy, we used the data reported by Bogaerts et al. [20] and Vahedi [21].

Plasma parameters (plasma density and electron temperature) were evaluated at a position of 15 mm (as the center of the mesh) above the substrate holder as those corresponding to the values at the neighbor mesh facing to the substrate.

## EXPERIMENTAL

Examinations of properties in inclination plasma were carried out in a cylindrical chamber installed with a set of LIA modules, as schematically shown in Figure 1. The LIA module consists of a U-shaped antenna conductor, which was fully covered with dielectric tubing for complete isolation from the plasma[12]. The LIA unit with a 70 mm width and a 160 mm height was mounted on the top flange of the discharge chamber and coupled to a 1000 W RF power generator at 13.56 MHz via a matching network. The LIA unit was connected in parallel to the matching network. This discharge chamber had a 500 mm inner diameter and a 200 mm height, which was connected to a diffusion chamber with a 500 mm inner diameter and a 400 mm height. The punched aluminum plates were located at the discharge chamber for which 500 x 150 $mm^2$ rectangular space was simulated as shown in Figure 1. The substrate holder was placed at a distance of 297 mm from the top flange. The plasmas were produced at a pressure of 0.67 - 2.6 Pa.

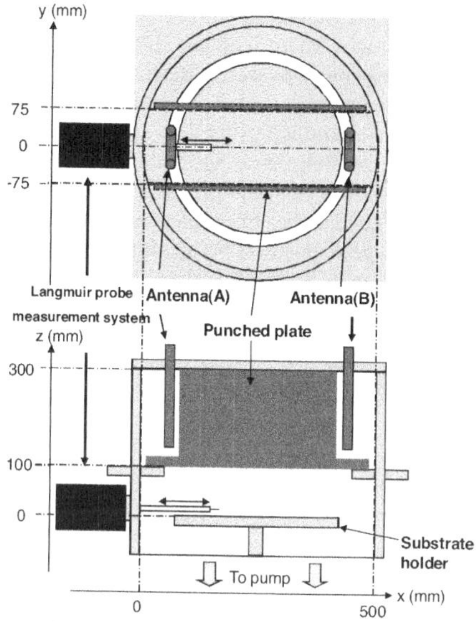


Figure 1. Schematic of experimental apparatus.

Plasma parameters were measured with a cylindrical Langmuir probe which was inserted radially at an axial position of 280 mm below the top flange as shown in Figure 1. For stroke restriction of the Langmuir probe measurement system, the measurement of the plasma distribution of near side as viewed from the center of the chamber was carried out using of antenna (A) and that of far side from the antenna was carried out using antenna (B)

RESULTS AND DISCUSSION

To examine an optimized configuration of LIA modules for obtaining inclination plasmas distributions, plasma density distributions for a variety of configurations with the LIA modules were simulated using plasma fluid-simulation. In this simulation, the variation of plasma distributions with the number and location of the LIA modules was investigated at 15 mm on the substrate holder. The simulated results showed that considerably inclined distribution of plasma density can be obtained by localized power deposition for sustaining plasmas as shown in Figure 2.

Based on the simulation results, using the experimental apparatus with LIA modules which were localized near wall as shown in Figure 1, generation of inclination plasma have been carried out and the experimental results in terms of plasma parameter were compared with the simulated results. Figure 2 shows the profiles of normalized plasma density measured at 15 mm on the substrate together with the profile numerically predicted results using the plasma fluid-simulation. The profiles of normalized plasma density showed excellent agreement with the distribution numerically predicted using the fluid-simulation code. The results showed the considerable inclination of plasma density was obtained by localized power deposition profiles. Figure 3 shows the distributions of electron temperature together with the distribution numerically predicted results using the fluid-simulation code. The distributions of electron temperature also showed excellent agreement with the distribution numerically predicted using the fluid-simulation code and the distributions were almost constant at 2.2-2.4 eV.

In order to investigate the properties of density-inclination plasmas, effect of plasma parameter on Ar pressure have been examined using the experimental apparatus. Figure 4 shows the variation of plasma density distributions at the substrate position as a parameter of Ar pressure. The plasma density distributions show peaks at around x = 150 mm, which is about 100 mm shifted toward the discharge chamber from the antenna position.

This feature of the density-peak position being shifted from the antenna position to the center of the chamber is considered to be attributed to the nature that the distribution of the

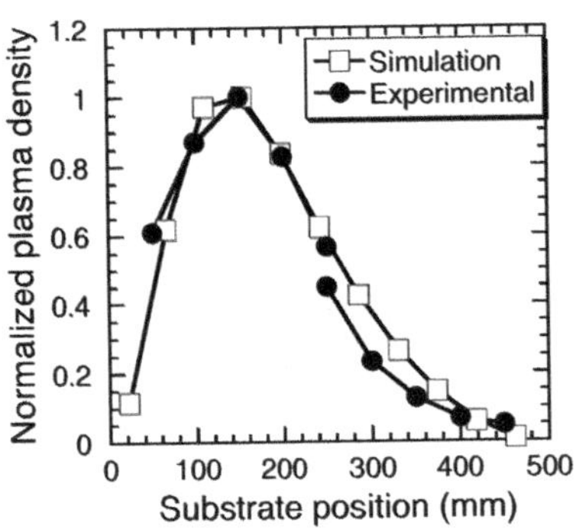


Figure 2. Simulated distribution of normalized plasma density with experimental measurements.

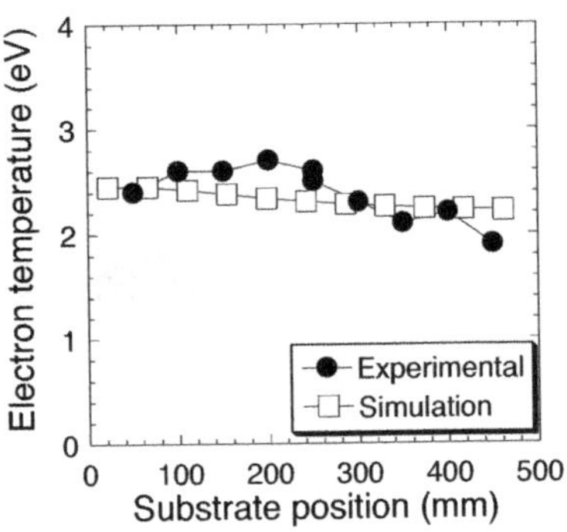


Figure 3. Simulated distribution of electron temperature with experimental measurements.

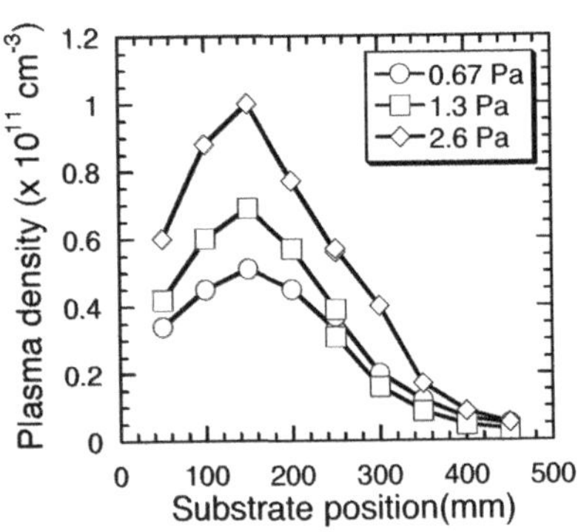


Figure 4. Plasma density distribution as a function of Ar pressure.

charged particle is primarily determined by the significant ambipolar-diffusion flux of the charged particles toward the walls of the chamber, on which charged particles are considerably lost due to recombination. In other words, the charged-particle density tends to peak at center of the chamber. Therefore for obtaining the charged-particle distributions so as to enhance the inclination length much longer than the half size of the discharge chamber, it is considered to be effective to employ the discharge method with a localized power-deposition profile and to locate it near the non-central position as schematically illustrated in Figure 1. In the example as shown in Figure 1, due to the installation of the LIA module being shifted from the center to the wall, the inclined profile of the charged-particle density as long as 350 mm was attained.

## SUMMARY

We have carried out plasma-fluid simulations to show the feasibility of producing density-inclination plasmas using LIA modules. Numerically predicted plasma density profiles are generally using localized power deposition profiles. The results show that a variety of process results can be efficiently analyzed via the plasma inclination distribution along the substrate for the process parameters used

## ACKNOWLEDGMENT

This work was supported partly by The Global COE Program "Center of Excellence for Advanced Structural and Functional Materials Design" from the Ministry of Education, Culture, Sports, Science and Technology (MEXT), Japan, and Grant-in-Aid for Cooperative Research Project of Nationwide Joint-Use Research Institutes on Development Base of Joining Technology for New Metallic Glasses and Inorganic Materials from MEXT, Japan.

## REFERENCES

[1] International technology roadmap for semiconductors, 2007 edition, Executive summary.

[2] L. Sansonnens, A. A. Howling and C. Hollenstein, "Electromagnetic field nonuniformities in large area, high-frequency capacitive plasma reactors, including electrode asymmetry effects", *Plasma Sources Sci. Technol.*, **15**, 302-313 (2006).

[3] H. Wada, D. Lee, U. Krishnamoorthy, S. Zappe and O. Solgaard, "Process for High Speed Micro Electro Mechanical Systems (MEMS) Scanning Mirrors with Vertical Comb Drives", *Jpn. J. Appl. Phys.*, **41**, L899-L901 (2002).

[4] C.-Y. Lee, G.-W. Wu and W.-J. Hsieh, "Fabrication of micro sensors on a flexible substrate", *Sen. Actuators A*, **147**, 173-176 (2008).

[5] C. Biasotto, J.A. Diniz, A.M. Daltrini, S.A. Moshkalev and M.J.R. Monteiro, "Silicon nitride thin films deposited by electron cyclotron resonance plasma enhanced chemical vapor deposition for micromechanical system applications", *Thin Solid Films*, **516,** 7777-7782 (2008).

[6] T. Ideno and T. Ichiki, "Maskless etching of microstructures using a scanning microplasma etcher", *Thin Solid Films*, **506-507**, 235-238 (2006).

[7] T. Ichiki, Y. Sugiyama, T. Ujiie and Y. Horiike, "Deep dry etching of borosilicate glass using fluorine-based high-density plasmas for microelectromechanical system fabrication", *J. Vac. Sci. Technol. B* ***21***, 2188-2192 (2003).

[8] W. W. Jang, J.-B. Yoon, M.-S. Kim, J.-M. Lee, S.-M. Kim, E.-J. Yoon, K. H. Cho, S.-Y. Lee, I.-H. Choi, D.-W. Kim and D. Park, "NEMS switch with 30 nm-thick beam and 20 nm-thick air-gap for high density non-volatile memory applications", *Solid-State Electronics*, **52**, 1578-1583 (2008).

[9] M. A. F. van den Boogaart, G. M. Kim, R. Pellens, J.-P. van den Heuvel, and J. Brugger, „Deep-ultraviolet–microelectromechanical systems stencils for high-throughput resistless patterning of mesoscopic structures", *J. Vac. Sci. Technol. B*, **22**, 3174-3177 (2004).

[10] E. Thiénot, F. Domingo, E. Cambril and C. Gosse, "Reactive ion etching of glass for biochip applications: Composition effects and surface damages", *Microelectron. Eng.*, **83**, 1155-1158 (2006).

[11] T. Yamada, T. Uezono, K. Okada, K. Masu, A. Oki and Y. Horiike, "In Vivo Batteryless Wireless Communication System for Bio-MEMS Sensors", *Jpn. J. Appl. Phys.*, **44**, 2879-2882 (2005).

[12] Y. Setsuhara, T. Shoji, A. Ebe, S, Baba, N. Yamamoto, K. Takahashi, K. Ono, and S. Miyake, "Development of internal-antenna-driven large-area RF plasma sources using multiple low-inductance antenna units", *Surf. Coat. Technol.*, **174 – 175**, 33-39 (2003).

[13] Y. Setsuhara, "Meter-Scale Large-Area Plasma Sources for Next-Generation Processes", *J. Plasma Fusion Res.*, **81**, 85-93 (2005).

[14] K. Takenaka, Y. Setsuhara, K. Nishisaka and A. Ebe, "Characterization of Inductively-Coupled RF Plasma Sources with Multiple Low-Inductance Antenna Units", *Jpn. J. Appl. Phys.*, **45**, 8046-8049 (2006).

[15] O. Tsuda, M. Ishihara, Y. Koga, S. Fujiwara, Y. Setsuhara, and N. Sato, "Characterization of Amorphous Hydrogenated Carbon Formed by Low-Pressure Inductively Coupled Plasma Enhanced Chemical Vapor Deposition Using Multiple Low-Inductance Antenna Units", *J. Phys. Chem. B*, **109**, 4917-4922 (2005).

[16] H. Deguchi, H. Yoneda, K. Kato, K. Kubota, T. Hayashi, K. Ogata, A. Ebe, K. Takenaka and Y. Setsuhara, "Effects of Antenna Size and Configurations in Large-Area RF Plasma Production with Internal Low-Inductance Antenna Units", *Jpn. J. Appl. Phys.*, **45**, 8042-8045 (2006).

[17] Y. Setsuhara, K. Takenaka, A. Ebe and K. Nishisaka, "Production and Control of Large-Area Plasmas for Meters-Scale Flat-Panel-Display Processing with Multiple Low-Inductance Antenna Modules", *Plasma Process. Polym.*, **4**, S628-S632 (2007).

[18] E. Takahashi, Y. Nishigami, A. Tomyo, M. Fujiwara, H. Kaki, K. Kubota, T. Hayashi, K. Ogata, A. Ebe, and Y. Setsuhara, "Large-Area and High-Speed Deposition of Microcrystalline Silicon Film by Inductive Coupled Plasma using Internal Low-Inductance Antenna", *Jpn. J. Appl. Phys.*, **46**, 1280-1285 (2007).

[19] K. Takenaka, Y. Setsuhara, K. Nishisaka and A. Ebe, "Characterization of Ion Energy Distribution in Inductively Coupled Argon Plasmas Sustained with Multiple Internal-Antenna Units", *Jpn. J. Appl. Phys.*, **47**, 6900-6902 (2008).

[20] A. Bogaerts, R. Gijbels and W. Goedheer, "Hybrid Modeling of a Capacitively Coupled Radio Frequency Glow Discharge in Argon: Combined Monte Carlo and Fluid Model", *Jpn. J. Appl. Phys.*, **38**, 4404-4415 (1999).

[21] V. Vahedi, Ph. D. Thesis, University of California, Berkeley, 1993.

# SPARK PLASMA SINTERING OF CARBON PARTICLES COATED WITH $Al_2O_3$ CERAMIC

Weiwu Chen[a], Yoshinari Miyamoto[a,b], Taihei Matsumoto[a,b] and Tstsuro Tojo[a,b]
[a]Joining and Welding Research Institute, Osaka University,
11-1, Mihogaoka, Ibaraki, Osaka, Japan, 567-0047
[b]Advanced Carbon Technology Center, Toyo Tanso. Co., Ltd.
5-7-12, Takeshima, Nishiyodogawa-Ku, Osaka, Japan, 555-0011

## ABSTRACT

To obtain super-light and multi-functional materials, new carbon-ceramic composites have been developed. Two kinds of carbon particles (meso graphite and nature graphite) coated with $Al_2O_3$ ceramic were prepared and then densified by spark plasma sintering. The results showed that carbon particles (around 15μm in size) could be coated homogenously with $Al_2O_3$ particles (170nm in size) by a centrifugal mixing method. The sintered samples had unique microstructures consisting of micro carbon grains with an $Al_2O_3$ grain boundary. The mechanical properties and thermal conductivities of sintered samples were analyzed, and the effects of the starting graphite powders were investigated as well.

## INTRODUCTION

With the development of high speed and miniaturized electronic systems, the thermal transport property of devices is becoming an emergent issue. So far, these devices are made from heavy metal materials and an insulating resin by a complex design and preparation process.[1] Graphite materials have many applications because they possess promising properties, such as easy machining, light weight (1.8~2.2 g/cm$^3$), low coefficient of thermal expansion, corrosion resistance, and high thermal and electrical conductivity. [2] However their lower strength limits their applications for electronic systems even though carbon materials show more advantages than metal materials.

Generally, graphite materials were prepared by using pulverized cokes as fillers and pitch as a binder. The baking and impregnation procedure has to be repeated for several times to achieve sufficient strength.[2] Conventional graphite products have a thermal conductivity around 50-140 W/m.K, and a bending strength of 30~100 MPa. The strength of graphite depends on the carbon/graphite network made from the pitch binder. This relatively weak network may limit the strength.

Metal carbides or boride additions have been used to improve the oxidation-resistance of carbon materials.[3] Some of them are also considered as reinforcement phases by simply mixing with carbon or coke powder. However, the graphitization steps are still necessary, and these additives were not homogenously distributed.[4]

Here we propose a new approach to prepare novel graphite-ceramic composites, as shown in Fig.1. In this new approach, graphite particles will be coated by nano ceramic powders, and then densified to a unique microstructure consisted of micro carbon grains with a ceramic grain boundary. The ceramic grain boundary is expected to bond graphite particles, and provide high strength, and other functional properties.

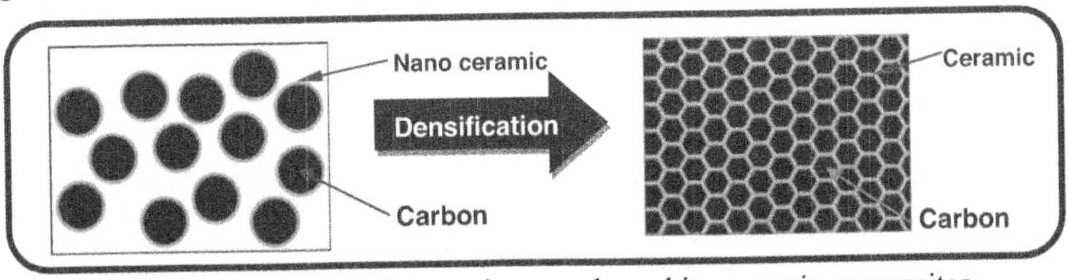


Figure 1. Concept of preparing novel graphite-ceramic composites

$Al_2O_3$ is extensively employed as an engineering ceramic with a bending strength in the range of 500-600 MPa. Spark plasma sintering (SPS) has been used successfully for fast densification of alloys and ceramics [5] at relatively lower temperatures. Rapid baking of graphite powders by SPS has also been reported.[6]

In this study, a basic attempt has been done to test this novel approach and concept. Two kinds of graphite particles, artificial and natural graphite were coated uniformly with nano $Al_2O_3$ particles, and then densified by SPS technique. The mechanical properties and thermal conductivities of sintered samples were analyzed, and the effects of the starting graphite powders on the structure and properties have been investigated.

## EXPERIMENTAL

The starting materials used were graphite A (artificial graphite made from meso phase), graphite B (nature graphite), $Al_2O_3$ powder (TM-DAR, Taimei Chemicals Co. Ltd., Japan), and Hydroxyethyl Cellulose (Extra pure, Nacalai)

Nano $Al_2O_3$ particles were coated onto graphite particles by a centrifugal mixing machine (AR-250, Thinky Co., Japan) with 15wt% hydroxyethyl cellulse water solution (2.5 wt%) as a binder and mixing agent. The coated powders were then dried and sintered in vacuum using a Dr. 1050 spark plasma sintering (SPS) apparatus (Sumitomo coal Co. Ltd.,

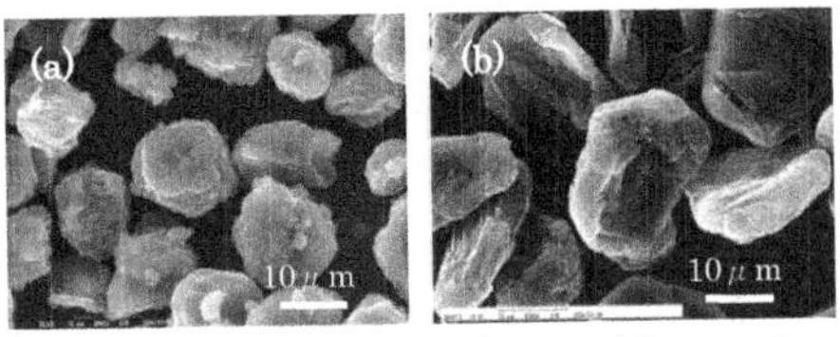


Figure 2. SEM images of starting graphite powders (a) GA, (b) GB

Japan) at 1600 ℃ with a heating rate of 100 ℃/min and a pressure of 30 MPa. The soaking time was 5 min.

Microstructure characterization was carried out using field emission scanning electron microscopy (ERA-8800, Elionix). X-ray diffraction (XRD, JDX-3530M, JEOL) was employed to characterize the degree of graphitization of powders and the degree of grain orientation in the sintered bodies. The density of the sintered pellets was measured by the Archimedes method. Rectangular bars, 4x2x20 $mm^3$, were prepared for three-point bending strength using a Table-Top Universal Tester (EZ-Test Type S, Shimadzu) with a span 15mm at room temperature. The thermal conductivity of sintered samples was measured by the laser-flashing method (TC-7000, ULVAC-RIKO).

## RESULTS AND DISCUSSION

Figure 2 shows the particle morphology of graphite A (GA) and B (GB). GA has a spherical shape with the particle size in a range of 5-20μm. GB also shows a spherical shape with the particle size of 15~20μm. The particles contain plate-like shape graphene several microns in size. Figure 3 shows the XRD patterns of two graphite powders. The (002) peaks with a slow scanning speed was inserted in order to identify the d-value of (002) peak. Only graphite crystals are noticed in both powders. However, the d-value of (002) peak is quite different, 0.3382 nm for GA and 0.3357 nm for GB. The degree of graphitization in the two powders was then calculated using the following equation [7]:

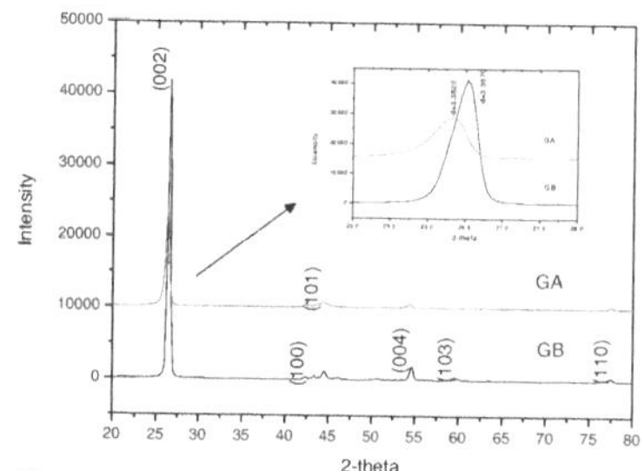


Figure 3. XRD pattern of starting graphite powders. Inserts are slowly scanned (002) peaks.

$$G = (0.344-d_{002})/(0.344-0.3354)\times 100\% \quad (1)$$

where G is the degree of graphitization and $d_{002}$ is the d value of (002) peak. The calculated degree of graphitization of GA and GB is 67% and 96%, respectively.

To coat graphite particles with $Al_2O_3$ fine powder, a centrifugal mixing method was used to avoid breaking graphite particles. Due to the large size ratio between graphite and $Al_2O_3$ particles, with the

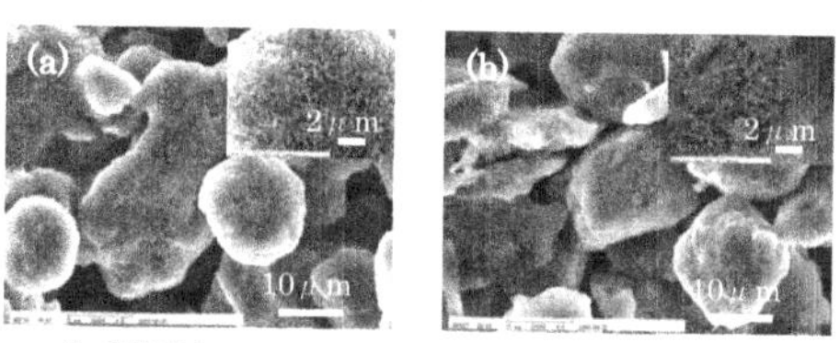


Figure 4. SEM images of $Al_2O_3$ coated graphite particles (a) GA, (b) GB. Inserts are large magnification images.

assistance of a binder solution, the nano meter sized $Al_2O_3$ particles were coated easily onto graphite particles. Figure 4 shows the SEM photos of mixed powders. In both compositions, nano meter sized $Al_2O_3$ are coated on graphite particles homogenously. During the centrifugal milling step, the spherical shape graphite particles are easily rotated in the $Al_2O_3$ powder matrix. A binder solution can smooth this kind of rotation and fix the $Al_2O_3$ particles onto the graphite surface.

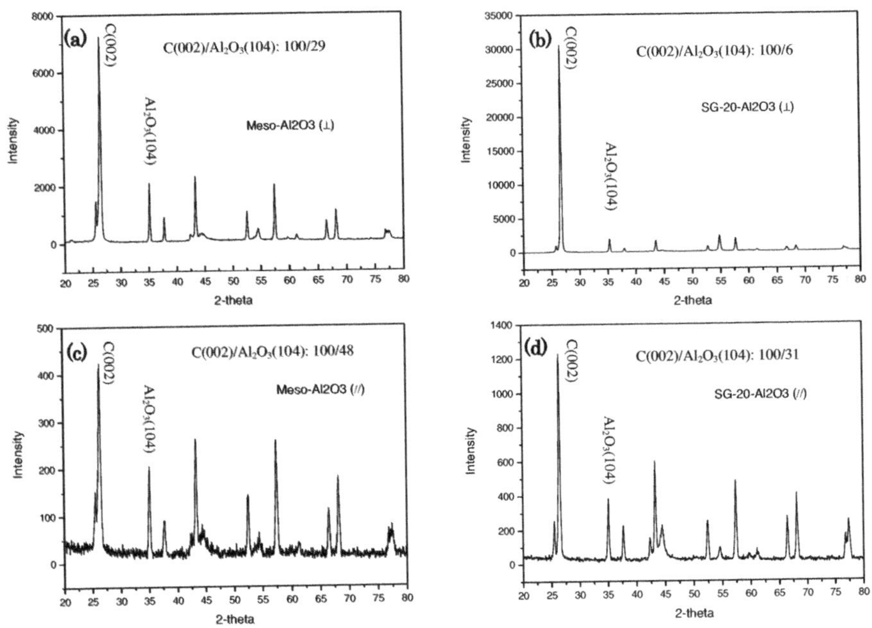


Figure 5. XRD of sintered graphite- $Al_2O_3$ composites
(a) GA ⊥, (b) GB ⊥, (c) GA //, (d)GB //.

SPS is a fast densification method for materials using direct pulsed current and high pressure. However, graphite grains may be oriented following to the pressure direction. Therefore, XRD patterns of the surface of perpendicular (⊥) and parallel (//) to the pressure direction were compared, as shown in Fig.5. Here, the intensity ratio between graphite peak (002) and $Al_2O_3$ peak (104) was selected for orientation degree comparison because the other graphite peaks are too weak and $Al_2O_3$ peak (104) can be considered as no grain orientation. In the GA sample, the peak intensity ratio is 100/29 and 100/48, respectively for surfaces of perpendicular and parallel to the pressure. However, in the GB sample, this ratio is 100/6 and 100/31, respectively. This indicates that the GB sample shows much higher grain orientation than the GA sample. The GB particle contains pieces of natural graphene, among which a lot of voids exist. Therefore, the GB particles are soft and weak, and easily deformed when pressure is applied.

The microstructures of the sintered samples are shown in Fig.6. In the surface perpendicular to pressure, the GA sample shows a unique microstructure as expected, consisting of micro graphite grains with an $Al_2O_3$ grain boundary. This unique microstructure can be also observed in the GB sample and even the GB particles deformed after SPS. In the surface parallel to pressure, the GA sample shows a loose structure. The GB sample shows dense packing due to the deformation of graphite grains with pressure.

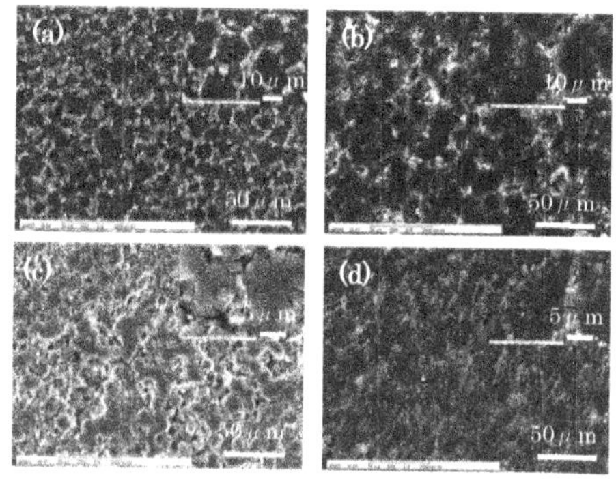


Figure 6. SEM images of sintered graphite-$Al_2O_3$ composites (a) GA, $\perp$, (b) GB, $\perp$, (c) GA, //, (d) GB, //. Inserts are large magnification images.

Table 1. Density, strength and thermal conductivity of sintered graphite- $Al_2O_3$ composites

| Samples | Density ($g/cm^3$) | Bending Strength (MPa) | Thermal conductivity(W/m•K) |
|---|---|---|---|
| GA- $Al_2O_3$ | 2.20 | 36.2(//) | 50.8(//) |
| | | 35.8($\perp$) | 40.3($\perp$) |
| GB- $Al_2O_3$ | 2.36 | 14.9(//) | 44.1(//) |
| | | 17.3($\perp$) | 118.5($\perp$) |

Table 1 shows the measured properties of the sintered composites. Considering the graphite grain orientation, the properties parallel and perpendicular to pressure direction are compared. For thermal conductivity, the GA sample shows a weak relationship with orientation, with a value around 40~50 W/m•K in both directions. But in the GB sample, a strong effect of orientation on thermal conductivity is noticed, 44 and 118 W/m•K, respectively in the two directions. For the bending strength, in both samples, no obvious orientation effect is observed. The GA sample has a value around 36 MPa and the GB sample shows a value around 16 MPa. It should be noticed that GB sample has a higher density because the GB particles are easily deformed. However, its bending strength is much lower (around 50%) than the GA sample.

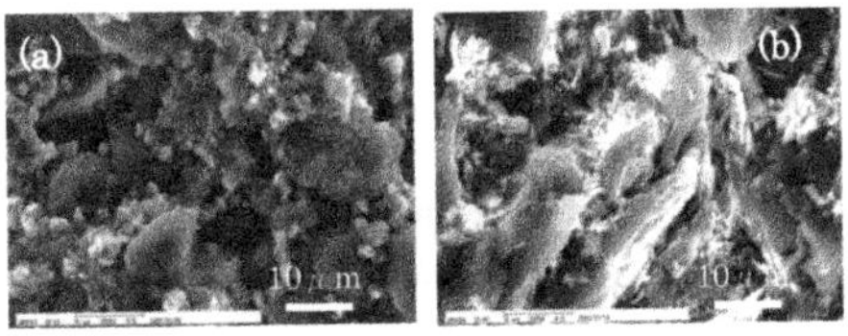

Figure 7. SEM images of fracture surfaces of composites (a) GA, (b) GB.

From the analysis of their fracture surfaces, as shown in Fig.7, the GA sample shows inter-granular fracture only, while the GB sample shows both inter-granular and trans-granular fracture. This means that the GB grains have lower strength (below 18 MPa) than the GA grains. In addition, in both samples, the $Al_2O_3$ grain boundaries do not show strong bonding with graphite particles, and themselves are not strong due to their non-continuous structure. This may be the reason for the low mechanical strength and thermal conductivity of both samples. The reason the GB-$Al_2O_3$ sample is lower may be mainly attributed to the lower strength of the GB particles.

## CONCLUSIONS

1. A concept for preparing novel graphite-ceramic composites was proposed and tested. Based on this concept, graphite particles were homogenously coated with nanometer size $Al_2O_3$ powders, and then densified by SPS. As expected, this produced a unique microstructure consisting of micro graphite grains with an $Al_2O_3$ grain boundary. The weak interface between graphite grain and ceramic grain boundary, and the non-continuous ceramic grain boundary limited the reinforcing function of the $Al_2O_3$ ceramic.

2. The effects of graphite particles on novel graphite-ceramic composites were compared. Natural graphite shows a high grain orientation, a higher thermal conductivity, and weak particle strength. Artificial graphite made from meso phase pitch has high mechanical strength and an isotropic thermal conductivity.

## REFERENCES

[1] T. Matsukawa, C.Neumeyer, T. Takaku S. Tsuji-Iio, R. Shimada, "Application of SiC-based Power Element to High Current AC/DC Converter System", *Fus. Engin. and Design*, **66-68**, 1155-58 ( 2003).

[2] E. Fitzer, The Future of Carbon-Carbon Composites, *Carbon,* **25**, 163-190 (1987).

[3] V. Raman, G. Bhatia, A. Mishra, P. Sengupta, M. Saha, Rashmi, "Development of carbon-ceramic composites", *Mater Sci. Eng. A.,* **412**, 31-6 (2005).

[4] C. Norfolk, A. Mukasyan, D. Hayes, P. McGinn, A. Varma, "Processing of meso carbon micro beads to high-performance materials", *Carbon,* **44**, 293-300 (2006).

[5] E. Olsvsky and L. Froyen, "Impact of Thermal Diffusion on Densification During SPS", *J. Am. Ceram. Soc.*, **92**, S122–S132 (2009)

[6]S. Hoshii, A. Kojima and M. Goto, "Rapid Baking of Graphite Powders by the Spark Plasma Sintering Method", *Carbon,* **38**, 1879-1902 (2000).
[7]L. Zou, B. Huang, Y. Huang, Q. Huang and C. Wang, "An Investigation of Heterogeneity of the Degree of Graphitization in Carbon-Carbon Composites", Mater. Chem. Phy., **82**, 654-662 (2003)

# PLASMA-ENHANCED NANOPARTICLES-BEAM DEPOSITION FOR HIGH-RATE FORMATION OF NANOCOMPOSITE FILMS

Kosuke Takenaka[1], Yuki Nakajima[1], Yuichi Setsuhara[1], Hiroya Abe[1], and Kiyoshi Nogi[2]
[1]Joining and Welding Research Institute, Osaka University, 11-1, Mihogaoka, Ibaraki, Osaka 567-0047, Japan,
[2]Hosokawa Powder Technology Research Institute, 1-9 Shoudai Tajika, Hirakata, Osaka 573-1132, Japan

## ABSTRACT

A plasma-enhanced nanoparticle-beam deposition method under vacuum has been developed. The deposition rate is 150 nm/min, which is 1-2 orders higher than that obtained using conventional reactive sputter deposition. Using this method, the surface smoothness of the resultant nanoparticle films is significantly enhanced.

## INTRODUCTION

Oxide-ceramics have attracted considerable attention as functional materials in a variety of applications, including as photo catalysts, gas sensors, microwave devices, and magneto-optical devices[1-5]. Oxide-ceramics nanocomposites formed by inclusion of nanoparticles are expected to be applicable to erasable holographic data-storage media, which can record huge amounts of data, as high as tera byte, on one disk[6-7]. Deposition techniques that can control phase structures and/or the chemical states of the surfaces of nanoparticles are necessary to fabricate oxide-ceramic nanocomposites[8]. The deposition of oxide-ceramic nanoparticles via conventional methods involving annealing processes have problems associated with nature of nanoparticles, including agglomeration, grain growth due to sintering, and/or the formation of complex oxide compounds. Therefore, it is highly desirable to develop new deposition methods that can avoid these problems during processing.

To address this problem, we adopted a nanoparticle-beam deposition method in an attempt to prevent nanoparticle agglomeration and grain growth. In this method, dispersed nanoparticles are generated, and then transferred through a tube and ejected from a nozzle in a vacuum chamber where they bombard a substrate. The most notable features of this method are that a dense ceramic film can be deposited at around room temperature similar to aerosol deposition[9] and the properties of the raw ceramic nanoparticles used as can be modified by exposure to the plasma. Therefore the composition and electrical properties can be controlled in the same way as in bulk ceramics without agglomeration of the nanoparticles.

In this paper, we report the preliminary experimental results regarding nanoparticle film deposition using the nanoparticle-beam deposition method, including the effects of plasma exposure on the surface morphology of the nanoparticle films.

## EXPERIMENTAL

A diagram of the chamber used for the deposition of nanoparticle film by plasma exposure is shown in Figure 1. A water cooled copper coil antenna (three turns) was looped around an alumina discharge tube of 50 mm in outer diameter and 42 mm in inner diameter. The antenna is coupled to a 1000 W RF power generator at 13.56 MHz via a matching network. Air was supplied from upstream of the stainless tube of 18 mm in outer diameter and 15 mm in inner diameter. In this study, $Y_3Fe_5O_{12}$ particles (Hosokawa Micron Corporation) having hexagonal phase and with an average size of 30 nm were used. The supply of nanoparticles was performed by the ejection of droplets of dispersed nanoparticles from a nozzle in the vacuum chamber. The total pressure was kept constant at 50-100 Pa. The Si and glass substrates were located in the downstream region at 300 mm from the copper coil antenna. The plasma parameters for the plasma were measured with a cylindrical Langmuir probe. The surface morphology of the nanoparticle films was studied with a scanning electron microscope (SEM). Surface roughness was observed using atomic force microscopy (AFM), and the measured area was 1 μm x 1 μm for all AFM images in this paper.

## RESULTS AND DISCUSSION

To investigate the characteristics of plasmas used in the deposition of the nanoparticle films, the ion saturation current for the plasma at a pressure of 50 Pa was measured at a distance of 50 mm from the antenna. Figure 2 shows the RF power dependence of the ion saturation current at the generation region. With increasing RF power, the plasma density slightly increased from 100 W to 600 W, and then drastically increased at 1000 W. The drastic increase of the ion saturation current for the plasma over 600 W is considered to be the transition from a weak capacitive discharge known as the E mode to an inductive discharge known as the H mode. From the viewpoint of the efficient heating and modification of the nanoparticles, the plasmas for deposition of nanoparticle films were operated at an RF power over 600 W.

Using the nanoparticles-beam deposition method, deposition of nanoparticle films has been performed. Figure 3 shows an SEM image of a nanoparticle film deposited with the nanoparticle-beam deposition method. The supplied RF power was 1000 W. The thickness of the nanoparticle films was about 100 nm. As shown in Figure 3, we can observe round nanoparticles of around 100 nm diameter. The deposition rate was 150 nm/min, which is 1-2

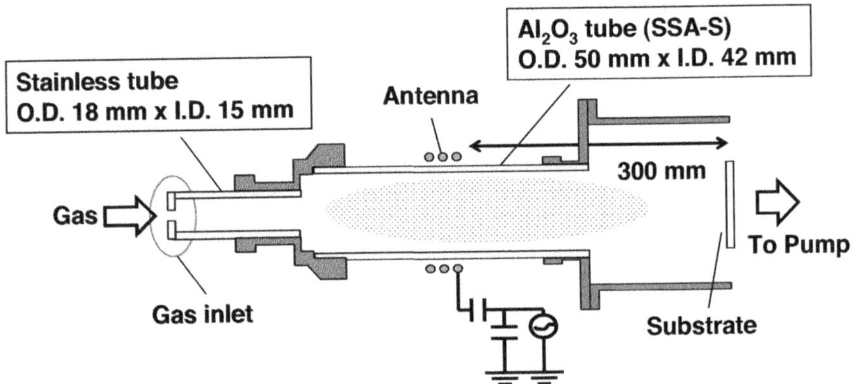


Figure 1. Schematic of the chamber for used for the nanoparticle film deposition.

orders higher than that obtained using conventional reactive sputter[10-13].

The effect of plasma exposure on the surface morphology of the nanoparticle films deposited using nanoparticle-beam deposition method was also investigated. Figure 4 (a) and (b) show the typical surface morphology of nanoparticle films deposited with and without plasma exposure of the droplets including the nanoparticles. The nanoparticle films deposited without plasma exposure were rough, while the nanoparticles films deposited with plasma exposure were comparatively flat. The plasmas used in this study have enough enthalpy to evaporate the liquid from the droplets including the nanoparticles since the plasma is high density as shown in Figure 2[14]. Therefore the nanoparticles in the droplets were isolated in the plasmas due to the evaporation of the liquid in droplets, and then the isolated nanoparticles arrived on the substrates

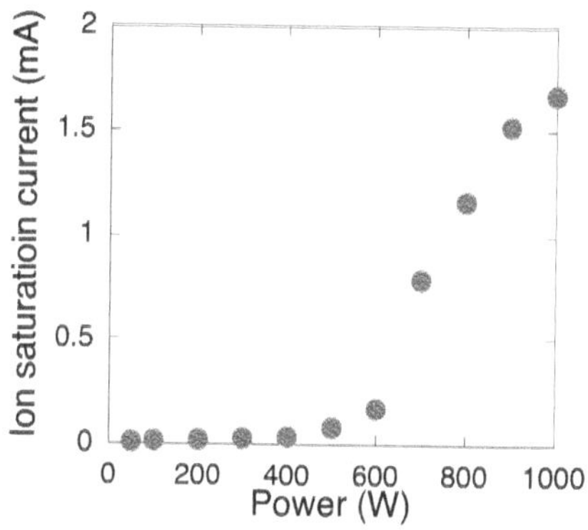


Figure 2. RF power dependence of ion saturation current at a distance of 50 mm at a pressure of 50 Pa.

Figure 3. Typical SEM image of nanoparticle films deposited by plasma-enhanced nanoparticle-beam deposition method.

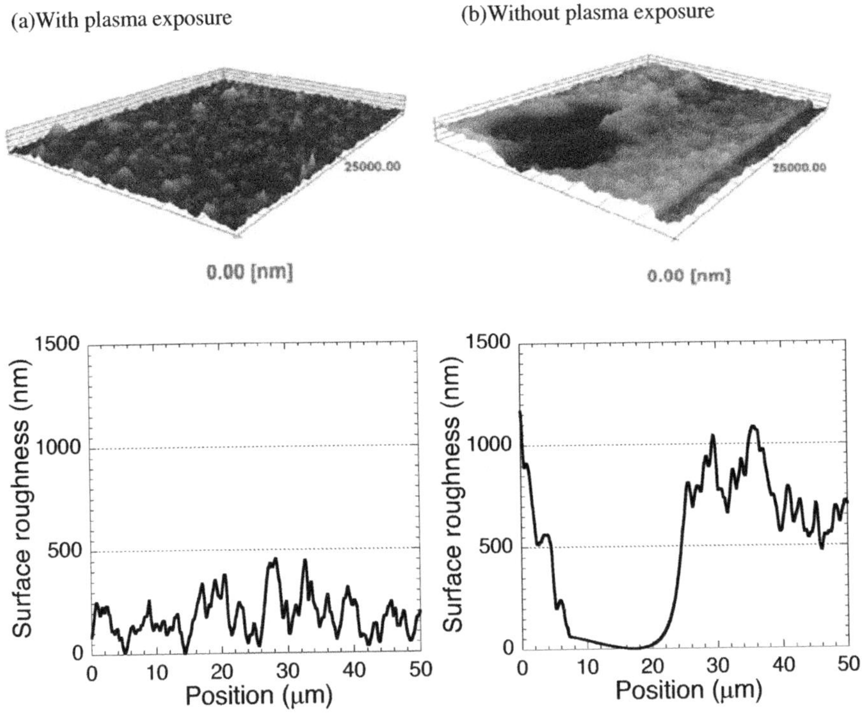


Figure 4. Surface morphology of nanoparticle films deposited (a) with and (b) without plasma exposure.

without agglomeration.

SUMMARY

The deposition of nanoparticle films has been performed using a nanoparticle-beam deposition method. Using this method, deposition of nanoparticle films without agglomeration was realized due to the effects of the thermal energy of the plasma. The deposition rate was 150 nm/min, which is 1-2 orders higher than that obtained using conventional reactive sputter deposition. The surface smoothness of the nanoparticle films was significantly enhanced by the use of this method. The results show that the plasma-enhanced nanoparticles-beam deposition method is quite attractive as a technique for fabricating high quality nanocomposites.

ACKNOWLEDGMENT

This work was supported partly by The Global COE Program "Center of Excellence for Advanced Structural and Functional Materials Design" from the Ministry of Education, Culture, Sports, Science and Technology (MEXT), Japan, and Grant-in-Aid for Cooperative Research

Project of Nationwide Joint-Use Research Institutes on Development Base of Joining Technology for New Metallic Glasses and Inorganic Materials from MEXT, Japan.

REFERENCES

[1] X. Z. Guo, B. G. Ravi, P. S. Devi, J.C. Hanson, J. Margolies, R. J. Gambino, J. B. Parise, and S. Sampath, "Synthesis of yttrium iron garnet (YIG) by citrate–nitrate gel combustion and precursor plasma spray processes", *J. Magn. Magn. Mater.*, **295**, 145-54 (2005).

[2] G. J. Chen, H. M. Lee, Y. S. Chang, Y. J. Lin and Y. L. Chai, "Preparation and properties of yttrium iron garnet microcrystal in $P_2O_5$–MgO glass", *J. Alloys Compounds*, **388**, 297-302 (2005).

[3] T. V. Dolgova, A. A. Fedyanin, O. A. Aktsipetrov, K. Nishimura, H. Uchida, M. Inoue, "Nonlinear magneto-optical Kerr effect in garnet magnetophotonic crystals", *J. Appl. Phys.*, **95**, 7330-7332 (2004)." Growth of YIG and BiGdIG single crystals for magnetooptical applications", *Mat. Chem. Phys.*, 45, 124-129 (1996).

[4] S. Higuchi, Y. Furukawa, S. Takekawa, O. Kamada, K. Kitamura, K. Uyeda, "Magnetooptical properties of cerium-substituted yttrium iron garnet single crystals for magnetic-field sensor", *Sens Actuators,* A **105**, 293-296 (2003).

[5] S. Taketomi, C. M. Sorensen and K. J. Klabunde, "Preparation of yttrium-iron-garnet nanocrystals dispersed in nanosize-pore glass", *J. Magn. Magn. Mater.*, **222,** 54-64 (2000).

[6] R. D. Sáncheza, C. A. Ramosa, J. Rivasb, P. Vaqueiroc and M. A. López-Quintela, "Ferromagnetic resonance and magnetic properties of singledomain particles of $Y_3Fe_5O_{12}$ prepared by sol–gel method", *Physica B*, **354** , 104-107 (2004).

[7] B.J.H. Stadler and A. Gopinath, "Magneto-optical garnet films made by reactive sputtering", *IEEE Transactions on Magnetics*, **36**, 3957-3961 (2000).

[8] J. Akedo, M. Lebedev, "Piezoelectric properties and poling effect of Pb(Zr, Ti)$O_3$ thick films prepared for microactuators by aerosol deposition", *App. Phys. Lett.*, **77**, 1710-1712 (2000).

[9] J. Qua, Y. Wanga, L. Xiea, J. Zhenga, Y. Liua, X. Lia, "Superior hydrogen absorption and desorption behavior of Mg thin films", *Journal of Power Sources*, **186**, 515–520 (2009).

[10] Y.C. Lin, H.A. Chuang, J.H. Shen, "PZT thin film preparation by pulsed DC magnetron sputtering", *Vacuum*, **83**, 921-926 (2009).

[11] Y.-S. Lin, Y.-L. Chiang, J.-Y. Lai, "Effects of oxygen addition to the electrochromic properties of $WO_{3-z}$ thin films sputtered on flexible PET/ITO substrates", *Solid State Ionics*, **180**, 99-105 (2009).

[12] Z. Lina, K. Liua, Y.-C. Zhanga, X.-J. Yuea, G.-Q. Songa, D.-C. Ba, "The microstructure and wettability of the TiOx films synthesized by reactive DC magnetron sputtering", *Materials Science and Engineering B*, **156**, 79-83 (2009).

[13] K. Takenaka , H. Nakayama, Y. Setsuhara, H. Abe, K. Nogi, "Modification of yttrium-iron-oxide nanoparticle films using inductively-coupled plasma annealing", *Surf. Coat. Technol.*, **202**, 5336-5338 (2008).

# MICROSTRUCTURES FORMED ON STAINLESS STEEL BY FEMTOSECOND LASER IRRADIATION

Hiroshi Honda*, Masahiro Tsukamoto**, Nobuyuki Abe**, Togo Shinonaga*** and Masayuki Fujita****

*Structural Metals Center, National Institute for Materials Science
1-2-1 Sengen, Tsukuba, Ibaraki 305-0047, Japan
**Joining and Welding Research Institute, Osaka University
11-1 Mihogaoka, Ibaraki, Osaka 567-0047, Japan
***Graduate School of Engineering, Osaka University,
2-1 Yamadaoka, Suita, Osaka 565-0871, Japan
****Institute for Laser Technology, 2-6 Yamadaoka, Suita, Osaka 565-0871, Japan

## ABSTRACT

Various types of microstructures are formed on material surfaces by femtosecond laser irradiation. The formation conditions depend on the laser irradiation conditions, the materials, and so on. In this study, microstructures formed on stainless steel by femtosecond laser irradiation were investigated. The laser fluence and the number of irradiated laser pulses were changed to explore the microstructure formation conditions. Different types of periodic structures were found to be formed depending on the laser fluence and the number of irradiated laser pulses.

## INTRODUCTION

Femtosecond lasers provide precise material processing, such as drilling, cutting, and grooving in metal[1]. It was reported that precisely-shaped drilled holes were obtained by using a femtosecond laser in comparison to a nanosecond laser[2]. The precise processing was based on very rapid heating to high temperature to create a vapor and plasma phases, and negligible heat conduction in femtosecond laser ablation.

In addition to these features of femtosecond laser material processing, periodic nanostuctures are formed in the laser irradiation area[3-10]. The grooves of the periodic nanostructures are perpendicular to the electric-field polarization vector of linearly polarized laser light. Similar periodic structure formation is also observed for long pulse lasers[11-14]. It is suggested that the periodic structures arise from the interaction of the incident and scattered wave with a surface wave created from any periodic modulation in the surface. However, it was found that the periodic nanostructures formed by femtosecond laser irradiation near the laser fluence of the ablation threshold have periods shorter than those derived from the interaction model. Furthermore, at higher laser fluences, periodic or cone-like protrusion microstructures are also formed by femtosecond laser irradiation[15,16]. Such various structures are interesting for surface modification as well as the features of precise material processing.

In this study, in order to determine the conditions for the formation of periodic structures on stainless steel by femtosecond laser irradiation, the laser fluence per pulse and the number of irradiated laser pulses were changed, and the structures formed were investigated.

## EXPERIMENTAL

A commercial femtosecond Ti:sapphire laser system was used. The laser wavelength was 775 nm and the pulse duration was 150 fs. The pulse duration was measured with a background-free second-order autocorrelator. The output energy was changed by using an attenuator composed of polarizing filter. The thickness and the surface roughness (Ra) of a SUS304 stainless steel plate were 2 mm and 0.02 μm, respectively. The stainless steel plate was attached to an X-Y stage controlled by a computer. The laser beam was focused on the stainless steel plate surface at normal incidence in air by a lens with a focal length of 100 mm. The focal spot diameter was 50 μm at $1/e^2$ in intensity. The laser fluence per pulse varied from 0.25 to 1.0 $J/cm^2$ by changing the output energy. The number of irradiated laser pulses was varied from 1 to 50 pulses. The surface morphology of the stainless steel plate was observed with a scanning electron microscope (SEM).

## RESULTS AND DISCUSSION

SEM images of samples exposed to a laser fluence of 0.25 $J/cm^2$ are shown in Figure 1 for (a) 5, (b) 10, and (c) 50 pulse irradiation. Two different periodic nanostructures were observed at this laser fluence. When the number of irradiated laser pulse is low, the periodic nanostructure whose grooves were parallel to the electric-field polarization vector $E$ of the laser light were found to be formed (Fig. 1 (a)). As the number of laser pulses increased, the periodic nanostructures whose grooves were perpendicular to $E$ appeared (Fig. 1 (b)). The spatial periods of the structures whose grooves were parallel to $E$ were smaller than those of the structures whose grooves were perpendicular to $E$. Finally, the periodic nanostructures whose grooves were perpendicular to $E$ became dominant (Fig. 1 (c)).

Two such different spatial periods of the periodic structures were also reported for stainless steel[10]. In that case, the direction of the grooves of the smaller periodic structures was, however, the same as the direction of the grooves of the larger periodic structures.

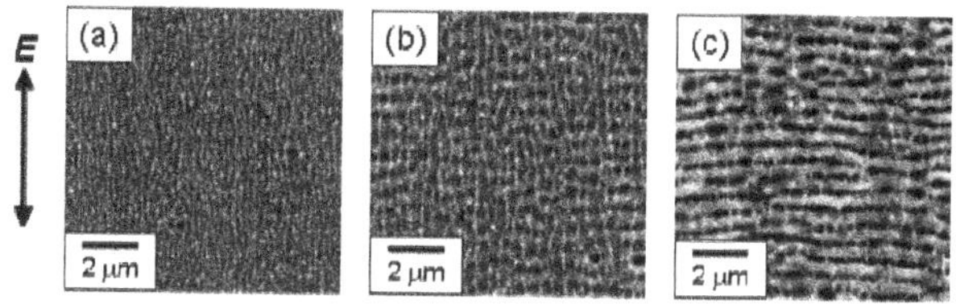


Figure 1. SEM images of structures formed in the center region of the irradiation area at 0.25 $J/cm^2$ for (a) 5, (b) 10, (c) 50 pulses.

SEM images of samples exposed to a laser fluence of 0.5 $J/cm^2$ are shown in Figure 2. The number of irradiated laser pulses was (a) 5, (b) 10, and (c) 50. At this laser fluence, the periodic nanostructures whose grooves were perpendicular to $E$ were produced in every image. The periodic nanostructures whose grooves were parallel to $E$ were also seen after 5 pulses as shown in Figure 2 (a). As the number of laser pulses increased, the formation of periodic nanostructures whose grooves were perpendicular to $E$ became gradually disturbed (Fig. 2 (c)).

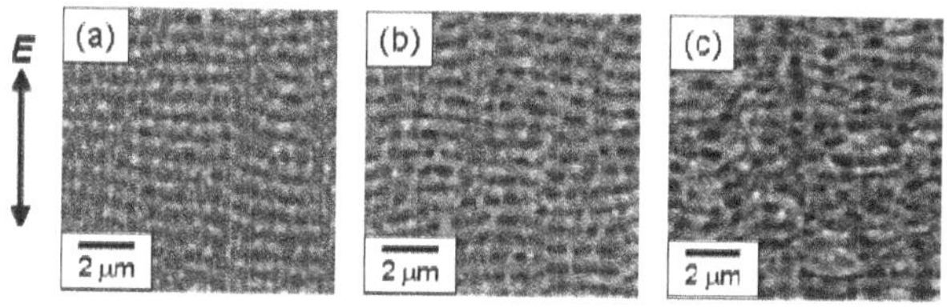


Figure 2. SEM images of structures formed in the center region of the irradiation area at 0.5 $J/cm^2$ for (a) 5, (b) 10, (c) 50 pulses.

SEM images of samples processed at a laser fluence of 1.0 $J/cm^2$ are shown in Figure 3 for (a) 5 (b) 10, (c) 50 pulse irradiation. In Figure 3 (a), the periodic nanostructures whose grooves were perpendicular to $E$ were also observed and the periodic microstructures whose grooves are parallel to $E$ started to appear on the periodic nanostructure whose grooves were perpendicular to $E$. As the number of laser pulses increased, the periodic microstructures were clearly observed (Fig. 3 (b)) and the formed structures came to be almost composed of the periodic microstructures (Fig. 3 (c)). A similar microstructure formation process was observed for titanium[15].

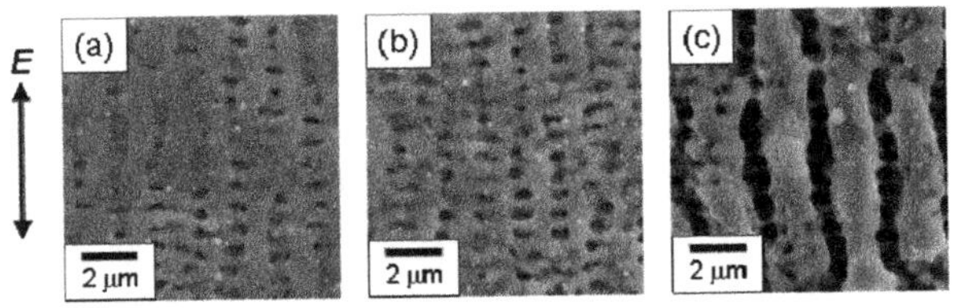


Figure 3. SEM images of structures formed in the center region of the irradiation area at 1.0 J/cm$^2$ for (a) 5, (b) 10, (c) 50 pulses.

SUMMARY

Microstructures formed by femtosecond laser irradiation of stainless steel were investigated. The laser fluence and the number of irradiated laser pulses were changed. At the lower laser fluence, two different types of periodic nanostructure were formed depending on the number of laser pulses. The first one had grooves parallel to the electric-field polarization vector. The other one had grooves perpendicular to the electric-field polarization vector. The periods of the former were smaller than the periods of the later. As the laser fluence was increased, the later periodic nanostructures were mainly formed. At the higher laser fluence, the periodic microstructures whose grooves were parallel to the electric-field polarization vector appeared on the later periodic nanostructure. As the number of laser pulses increased, the periodic microstructures became dominant.

REFERENCES

[1]P. S. Banks, B. C. Stuart, M. D. Perry, M. D. Feit, A. M. Rubenchik, J. P. Armstrong, H. Nguyen, F. Roeske, R. S. Lee, B. R. Myers and J. A. Sefcik, "Femtosecond laser machining", *Technical Digest of Conf. on Lasers and Electro-Optics*, **6**, 510 (1998).

[2]B.N. Chichkov, C. Momma, S. Nolte, F. von Alvensleben, A. Tünnermann, "Femtosecond, picosecond and nanosecond laser ablation of solids", *Appl. Phys. A*, **63**, 109-15 (1996).

[3]P. S. Banks, M. D. Feit, A. M. Rubenchik, B. C. Stuart, M. D. "Perry, Material effects in ultra-short pulse laser drilling of metals", *Appl. Phys. A*, **69**, S377-80 (1999).

[4]J. Bonse, J. M. Wrobel, J. Kruger, W. Kautek, "Ultrashort-pulse laser ablation of indium phosphide in air", *Appl. Phys. A*, **72**, 89-94 (2001).

[5]A. Borowiec, H. K. Haugen, "Subwavelength ripple formation on the surfaces of compound semiconductors irradiated with femtosecond laser pulses", *Appl. Phys. Lett.*, **82**, 4462-4 (2003).

[6]W. Kautek, P. Rudolph, G. Daminelli, J. Kruger, "Physico-chemical aspects of femtosecond-pulse-laser-induced surface nanostructures", *Appl. Phys. A*, **81**, 65-70 (2005).

[7]M. Hashida, M. Fujita, M. Tsukamoto, A. F. Semerok, O. Gobert, G. Petite, Y. Izawa, and J. F.

Wagner, "Femtosecond laser ablation of metals: precise measurement and analytical model for crater profiles", *Proc. of SPIE*, **4830**, 452-7 (2003).

[8]M. Hashida, K. Nagashima, M. Fujita, M. Tsukamoto, M. Katto, and Y. Izawa, "Femtosecond Laser Ablation of Metals: Characterization of New Processing Phenomenon and Formation on Nano-structures", *Proc. of the 9th Symposium on Microjoining and Assembly Technology in Electronics*, **9**, 517-22 (2003) (in Japanese).

[9]E. M. Hsu, T. H. R. Crawford, H. F. Tiedje, and H. K. Haugen, "Periodic surface structures on gallium phosphide after irradiation with 150 fs-7 ns laser pulses at 800 nm", *Appl. Phys. Lett.*, **91**, 111102 (2007).

[10]L. Qi, K. Nishii, and Y. Namba, "Regular subwavelength surface structures induced by femtosecond laser pulses on stainless steel", *Opt. Lett.*, **34**, 1846-8 (2009).

[11]D. C. Emmony, R. P. Howson, L. J. Willis, "Laser mirror damage in germanium at 10.6 μm", *Appl. Phys. Lett.*, **23**, 598-600 (1973).

[12]P. M. Fauchet, A. E. Siegman, "Surface ripples on silicon and gallium-arsenide under picosecond laser illumination", *Appl. Phys. Lett.*, **40**, 824-6 (1982)

[13]Z. Guosheng, P. M. Fauchet, A. E. Siegman, "Growth of spontaneous periodic surface structures on solids during laser illumination", *Phys. Rev. B*, **26**, 5366-81 (1982)

[14]J. F. Young, J. S. Preston, H. M. van Driel, J. E. Sipe, "Laser-induced periodic surface-structure. II. Experiments on Ge, Si, Al, and brass", *Phys. Rev. B*, **27**, 1155-72 (1983)

[15]M. Tsukamoto, K. Asuka, H. Nakano, M. Hashida, M. Katto, N. Abe, and M. Fujita, "Periodic microstructures produced by femtosecond laser irradiation on titanium plate", *VACUUM*, **80**, 1346-50 (2006).

[16]M. Tsukamoto, T. Kayahara, H. Nakano, M. Hashida, M. Fujita, M. Tanaka, N. Abe, "Microstructures formation on titanium plate by femtosecond laser ablation", *J. Phys. Conference* Seriese, **59**, 666-9 (2007).

FABRICATION AND DENSIFICATION OF SiC COATED CARBON MICROPARTICLES

Masaharu Nakamura, Yoshinari Miyamoto, Taihei Matsumoto, and Tetsuro Tojo
Toyo Tanso "Advanced Caron Design" Joint Research Chair
Joining and Welding Research Institute, Osaka University
11-1, Mihogaoka, Ibaraki, Osaka, Japan, 567-0047

ABSTRACT

A new SiC coating on carbon particles has been developed using SiO powders as the Si source, and CVD and CVR processing routes. Uniform SiC coatings 10 nm to 10 μm thickness were obtained on individual carbon particles. The oxidation resistance of carbon coated with SiC by CVR reached 750°C which is nearly 300°C higher than that of uncoated carbon. The SiC coated carbon particles sintered by SPS exhibited unique microstructures having graphite grains covered with SiC grain boundaries. The products sintered at 2000°C showed a high bending strength of 150 MPa.

INTRODUCTION

Advanced carbons and SiC materials are increasingly required because of their light weight and robust properties for various applications in electronics, mechanics, energy, aerospace, and other industries[1-5]. Formation of SiC/C composites can further extend their performances and applicable fields. SiC infiltrated carbon fiber composites are used for rocket engine nozzles with high temperature oxidation resistance and high toughness[6]. Graphite disks coated with SiC by CVD are used as disks for the epitaxial growth of compound semiconductors[7]. Other graphite parts coated with SiC by a reactive coating using silicon powders are used for wear and chipping resistant mechanical parts.

Recently there are some reports on new SiC coatings for multiwall carbon nanotubes, MWCNT[8], diamond[9-11], and pyro-carbon coated carbon fibers[12]. However, SiC coatings on carbon particles are not reported yet. Such SiC coated carbon particles are expected for oxidation, wear, electrically resistant carbon fillers, and for dense graphite composites with SiC grain boundaries.

We have developed a new simple method for coating SiC on carbon particles by controlling CVD (Chemical Vapor Deposition) and CVR (Chemical Vapor Reaction) processing routes. It is possible to form 10 nm to 10 μm thick SiC coatings. We found that the SiC/C composites sintered by SPS have a unique microstructure of graphite grains covered with SiC boundaries. The SiC coating on natural graphite particles and the sintering behavior will be reported.

EXPERIMENTAL

Carbon particles of natural graphite about 20 μm in diameter were used for SiC coating. They consist of spherical graphite granules as seen in Figure 1. SiO powders (purity 99.9%, Osaka Titanium Technologies Co., Ltd., Japan) were provided as the Si source. Carbon and SiO powders were placed separately in an alumina crucible as illustrated in Figure 2. The SiO powders were set on a carbon sheet at the

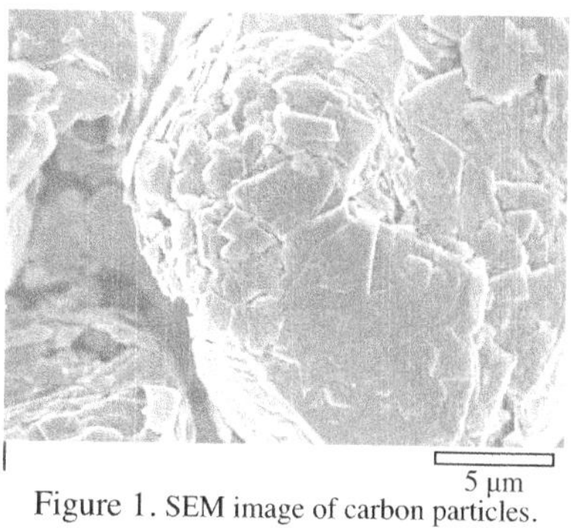

Figure 1. SEM image of carbon particles.

bottom of an alumina crucible and carbon particles were placed on SiO powders via a carbon felt. This assemble was covered with an alumina lid and a carbon sheet to keep the SiO gas pressure inside the crucible, and heated at temperatures between 1400°C to 1600°C under a vacuum of about 1 Pa or argon atmosphere for 2 h.

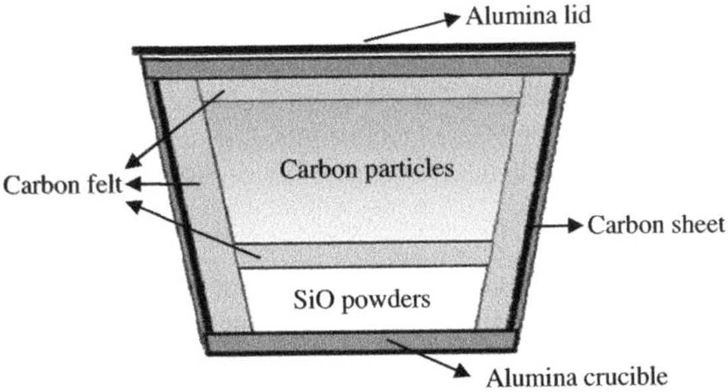


Figure 2. Assemble for SiC coating of carbon particles.

SiC coated carbon particles were filled in a graphite die and sintered at 2000°C under 40 MPa for 20 min by SPS (Spark Plasma Sintering). A carbon seat liner was rolled inside the die to prevent the reaction and adhesion of sample and die.

The SiC coated carbon particles were characterized by XRD (Rigaku, RINT-TTR III, 50 kV, 300 mA, Cu Kα). The surface morphology and microstructure were observed using FE-SEM (JEOL JSM-7500F). The oxidation resistance was evaluated by thermogravimetric analysis (Seiko Instruments Inc., TG/DTA6300). The three point bending strength of SiC/C compacts was measured by using a compact table-top universal tester (Shimadzu Co., Ltd., EZ-Test). The resistivity measurement was performed by means of a conventional four-point probe DC method (Kyowariken Co., Ltd., K-503RB).

## RESULTS AND DISCUSSION

### Synthesis of SiC Coated Carbon Particles

It is reported that the SiC formation and coating on diamond and MWCNTs using SiO sources is promoted by the following reactions[9]. We consider similar reactions take place on the carbon particles.

$$SiO(g) + C(s) \rightarrow SiC(s) + CO(g) \quad (1)$$
$$SiO(g) + 3CO(g) \rightarrow SiC(s) + 2CO_2(g) \quad (2)$$
$$CO_2(g) + C(s) \rightarrow 2CO(g) \quad (3)$$

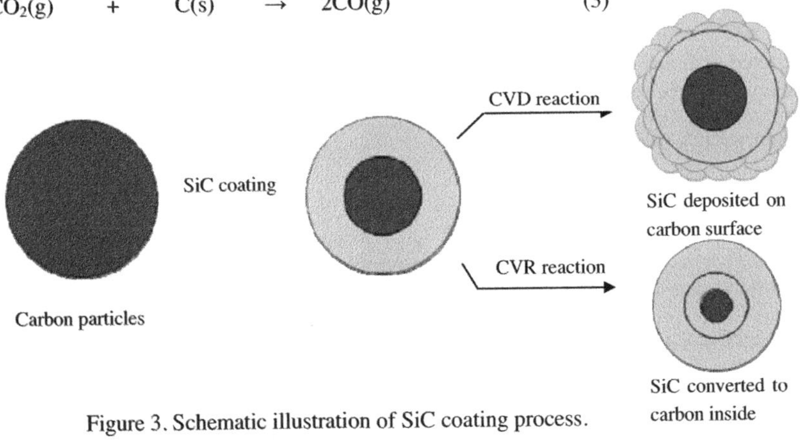


Figure 3. Schematic illustration of SiC coating process.

However, we found two different coating processes, depending on atmosphere, as illustrated in Figure3. One is the CVD route in which fine SiC crystals are deposited on carbon particles under an Ar gas atmosphere (105 kPa) mainly by the cyclic gas phase reactions of (2) and (3). The other route is CVR in vacuum in which carbon particles convert to SiC from the surface to the interior by the reaction (1). Figure 4 shows the change of SiC coating in Ar by CVD depending on the coating temperatures.

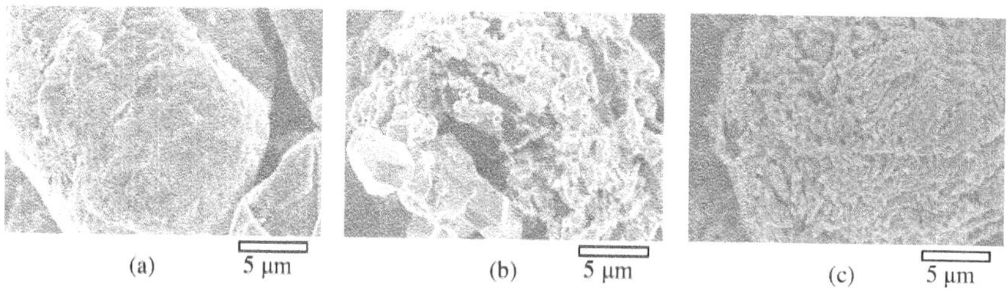


Figure 4. SiC coated carbon particles (a: 1200°C, b: 1500°C, c: 1800°C).

The solid SiO particles start to dissociate, but the coating is scarcely promoted at temperatures below 1300°C. The uniform SiC coating was obtained in a temperature range between 1400°C to 1550°C. It was porous, consisting of SiC microcrystals 50~200 nm in size. At over 1600°C, the coating was very rough. Large SiC crystals with several or ten μm sizes were produced. The SiC formation can be controlled by the coating temperature and the weight ratio of SiO/C.

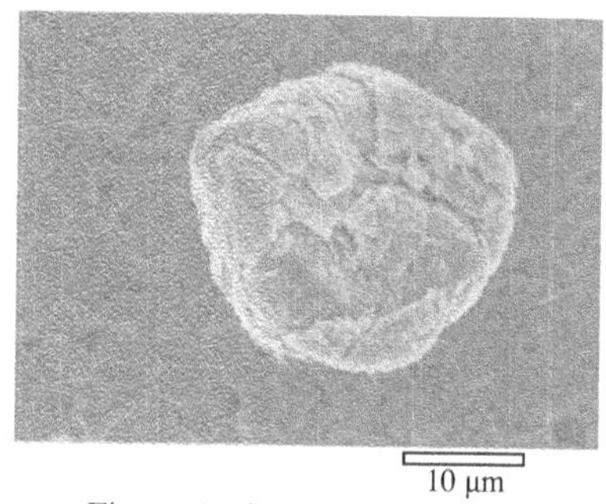


Figure 5. SiC coated carbon prepared at 1500°C by CVR.

In case of the CVR route, the dense and uniform SiC layer was formed as seen in Figure 5. The Gibbs energy of the formation of SiC coating through the gas phase reactions (2) and (3) was estimated to be about 100 kJ/mol[13]. In the solid/gas reaction (1), it was higher, about 500 kJ/mol[14]. The former energy is considered due to the control of gas phase reaction (2) for the deposition of SiC. The edges of the surface graphene layers would provide for growth sites of SiC deposition at temperatures of 1450°C ~ 1600°C. The latter process would be controlled by the diffusion of carbon through the SiC coating[14]. We can say therefore, the CVD route is useful to form thick and porous SiC coatings, while the CVR to form thin and dense coatings. Figure 6 compares the starting temperatures of oxidation for SiC coatings by CVD and CVR routes. The measured temperatures were 650°C, 750°C, and 400°C for SiC coated carbons by CVD, CVR, and for the raw carbon particles used as the reference, respectively. The carbon coated with SiC by CVR showed the highest starting temperature of oxidation reflecting the dense coating. The SiC/C ratio of SiC coated carbon particles was measured by X-ray peak intensity ratio of SiC(101)/Graphite(002) as shown in Fig.ure 7. We can calculate the mean thickness of SiC coatings using this SiC/C ratio.

These SiC coatings are bonded strongly on carbon particles. If the dense compacts of SiC coated carbon particles can be produced by sintering, it is expected that the products would have high strength, high electrical resistivity, and good oxidation resistance.

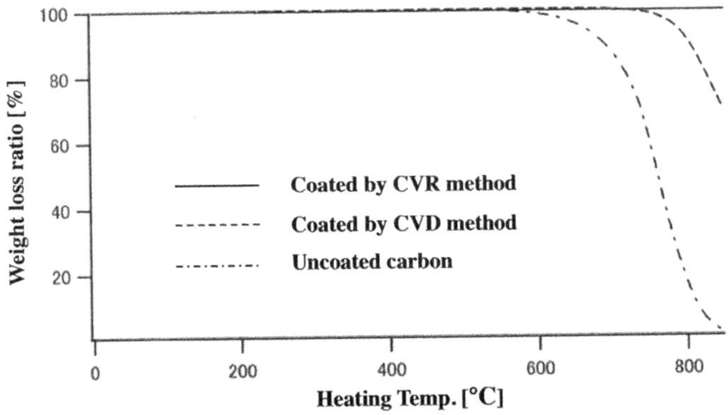


Figure 6. Comparison of starting temperatures of oxidation for carbons with and without SiC coating measures by TG under the heating rate of 10°C /min in air.

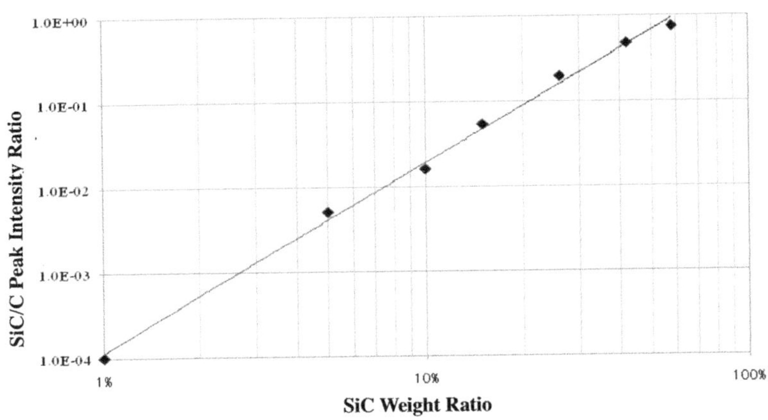


Figure 7. Relation of X-ray peak intensity ratio of SiC(111)/C(002) and SiC weight ratio on carbon particles.

Fabrication of Dense SiC/C Compacts

We tried to sinter two kinds of SiC coated carbon particles with different SiC contents of 22 wt.% (A) and 55 wt.% (B) at 2000°C, 40 MPa, for 20 min by SPS. The calculated thicknesses of SiC coating on carbon particles were about 0.6 μm, and 1.5 μm, respectively. They were densified well without cracks and deformation as seen in Figure 8. Figure 9 shows the fractured surface of a SiC/C sample (B). It is seen that a graphite grain is

Figure 8. SEM of a SiC/C compact

covered with SiC, but the transgranular fracture occurred due to weak graphite grains. The characteristics measured for density, strength, oxidation and electrical resistances are listed in Table I. The reference data were taken from a commercially available high strength graphite block. Sample A was almost fully densified with 99% density, but the strength is low about 57 MPa which is lower than that of the reference value. While sample B was not fully densified with the relative density of 95%, it shows a higher strength of 150 MPa. The strength of conventional graphite bulks is below 100 MPa. These results suggest that the three dimensional network of SiC grain boundary is basically formed in sample B, but not in sample A. The SiC boundaries of sample A are thin, weak, and broken. This tendency is seen in electrical resistivity which is higher in sample B than A. The resistivities of both samples remain in 30 to 60 μΩm in spite of the high resistance of SiC.

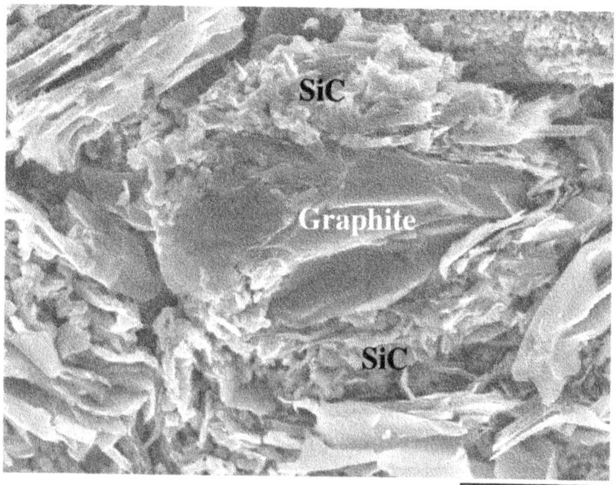

Figure 9. SEM image of sample B (cross section)

Table I. Characteristics of SiC/C compacts.

| Sample | Bulk Density [g/cm$^3$] | Relative Density [%] | Bending Strength [MPa] | Resistivity [μΩm] |
|---|---|---|---|---|
| A (SiC22 wt.%) | 2.20 | 99 | 57 | 33.9 |
| B (SiC55 wt.%) | 2.52 | 95 | 150 | 59.5 |
| High Strength Graphite | 1.90 | - | 93 | 15.5 |

The results show that many direct paths of graphite grains still exist. obtain the desired properties of SiC/C composites, it is necessary to further optimize the microstructures by selecting carbon particles, controlling SiC coating conditions, and the sintering conditions of SiC coated carbon particles.

CONCLUSIONS

We have developed new methods for coating SiC on carbon particles using SiO powders as the Si source, and by controlling the CVD and CVR processing routes. Porous and dense coatings can be produced by CVD and CVR, respectively. The oxidation resistance of the carbon coated with SiC by CVR reached 750°C, which was nearly 300°C higher than that of normal graphite. The SiC coated carbon particles densified well by SPS. They had unique microstructures having graphite grains covered with SiC grain boundaries. The products sintered at 2000°C showed a high bending strength of 150 MPa.

## REFERENCES

[1]W. Krenkel, "Carbon fiber reinforced CMC for high performance structures", *Int. J. Appl. Ceram. Tech.*, **1**, 188–200 (2004).

[2]Narottam Bansal, P. Kluwer, *Hand Book of Ceramic Composites*. Acadamic Publishers, 117–127 (2005).

[3]A. W. Weimer, K. J. Nilsen, G. A. Cochran, and R. P. Roach, "Kinetics of carbothermal reduction synthesis of beta silicon carbide reactors", *Kinet. Catal.*, **39**(3), 493–503 (1993).

[4]V. Raman, V. K. Parashar, S. Dhakate, O. P. Bahl, and U. Dhawan, "Synthesis of silicon carbide through the sol–gel process from rayon fibers", *J. Am. Ceram. Soc.*, **83**(4), 952–954 (2000).

[5]A. Leonhardt, H. Liepack, K. Biedermann, and J. Thomas, "Synthesis of SiC nanorods by chemical vapor deposition" *Fullerenes, Nanotubes, Carbon Nanostruct.*, **13**, 91–97 (2005).

[6]Q.S. Zhu, X.L. Qiu, C.W. Ma, "Oxidation resistant SiC coating for graphite materials", *Carbon* **37**, 1475 (1999).

[7]S.P. Vanka, G. Luo, N.G. Glumac, "Parametric effects on thin film growth and uniformity in an atmospheric pressure impinging jet CVD reactor", *J. Cryst. Growth*, **267**, 22–34 (2004).

[8]Y. Morisada, M. Maeda, T. Shibayanagi, and Y. Miyamoto, "Oxidation Resistance of Multiwalled Carbon Nanotubes Coated with Silicon Carbide", *J. Am. Ceram. Soc.*, **87** (5) 804–808 (2004).

[9]Y. Miyamoto, J. Lin, Y. Yamashita, T. Kashiwagi, O. Yamaguchi, H. Moriguchi, and A. Ikegaya, "Reactive Coating of SiC on Diamond Particles", *Ceram. Eng. Sci. Proc.*, **21**, 185–92 (2000).

[10]Y. Miyamoto, T. Kashiwagi, K. Hirota, O. Yamaguchi, H. Moriguchi, K. Tsuduki, and A. Ikegaya, "Fabrication of New Cemented Carbide Containing Diamond Coated with Nanometer Sized SiC Particles", *J. Am. Ceram. Soc.*, **86**, 73–76 (2003).

[11]Y. Morisada, Y. Miyamoto, H. Moriguchi, K. Tsuduki, and A. Ikegaya, "Growth Mechanism of Nanometer-Sized SiC and Oxidation Resistance of SiC-Coated Diamond Particles", *J. Am. Ceram. Soc.*, **87** (5), 809–13 (2004).

[12]H. Ouyang, H. Li, L. Qi, Z. Li, J. Wei, J. Wei, "Synthesis of a silicon carbide coating on carbon fibers by deposition of a layer of pyrolytic carbon and reacting it with silicon monoxide", *Carbon*, **46**, 1339-1344 (2008).

[13]Y. Miyamoto, et al. "Nano SiC Coating on Carbon Nanotubes and Its Applications", *Materials Integration*, **17** (10), 39-43 (2004).

[14]T. Shimoo, F. Mizutaki, S. Ando, and H. Kimura, "Mechanism of Formation of SiC by Reaction of SiO with Graphite and CO", *J. Japan Inst. Metals*, **52**, 279-287 (1988).

# EFFECT OF PARTICLE DISPERSION ON A GEL-CAST GREEN BODY

K. Yoshino, M. Fuji, T. Shirai and M. Takahashi
Ceramics Research Laboratory, Institute of Nagoya, Tajimi, Gifu, Japan

## ABSTRACT

In this study, the effect of the slurry particle dispersion state on a gel-casting green body was investigated. The viscosity, η, of an alumina slurry, green density, and strength were measured. The particle size distribution in the slurry was controlled by blending two kinds of commercial alumina powder with different particle size; "A" powder, D50: 0.4μm, and "B" powder, D50: 1.1μm, at different ratios. The results show that the highest green density was obtained for the condition that weakens the aggregation force between particles by adding the most dispersant.

## INTRODUCTION

The condition of a ceramic slurry has an effect / influence on the properties of the gel-cast body (dried gel cast body). This is one of the most important factors determining the green properties, sintering behavior, and other characteristics. For example, as is well known, the aggregation structure contributes to fabricating voids in a green body that lower green density and strength. It is thought that aggregated particles can be dispersed by shear stress of flowing in the ceramic slurry process. To control green properties by controlling the slurry particle dispersion state, particle size distribution, and other slurry properties, it is necessary to determine the particle dispersion state of the slurry during flowing and molding.

The gel-casting process has the advantage of fabricating uniform fine shape ceramics products[1-3]. In this method, the ceramic slurry is solidified by an in situ- polymerization reaction. The dispersion state of the particles in a gel-cast slurry decides the green microstructure. To investigate particle aggregation behavior in a ceramic slurry during processing, the amount of dispersant added was controlled, and the rheological behavior of the gel-cast slurry and the properties of the resultant green body were measured.

## EXPERIMENTAL

### Materials and Method

Figure1 shows the experimental procedure of the gel-cast method. The alumina slurry for gel-casting was composed of powder, dispersant, an organic monomer, a cross-linker, and distilled water[1-3]. The concentration of alumina powder was constant at 55vol%.

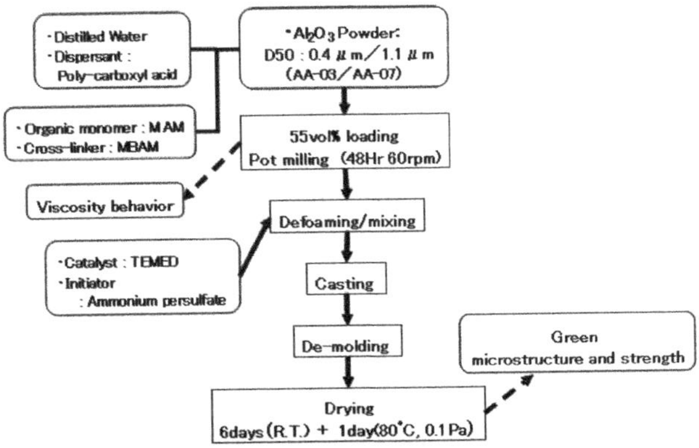


Figure 1. Procedure for preparing a green body by gel-cast method.

Two kinds of commercial alumina powder with different particle size; "A" powder (AA-03, D50: 0.4μm, Sumitomo Chemical Co., Ltd.), and "B" powder (AA-07, D50: 1.1μm, Sumitomo Chemical Co., Ltd.) were used, blending at different ratios. To control the dispersion state of the particles in the slurry, poly carboxyl acid (KAO-CER2100, Kao Corp.), was adjusted for the specific surface area to become 1.2mg/m$^2$, 2mg/m$^2$ and 7.8mg/m$^2$. An organic monomer, methacryl-amide (Wako Pure Cemicas Industries, Ltd.), and a cross-linker, N,N'-methylenebisacrylamide (Wako Pure Cemicas Industries, Ltd.) were added to the alumina powder amounts of 3.9wt%, and 1.3wt% respectively. The slurry was ball milled for 48 h, and polymerized was initiated by adding the initiator, ammonium persulfate (Wako Pure Cemicas Industries, Ltd.). The polymerization was accelerated by a catalyst, N, N, N', N' –tetramethylethylenediamine (Wako Pure Cemicas Industries, Ltd.). The polymerized gel-cast green body was de-molded and first dried in a humidity controlled chamber (Model FX206, Kusmoto Cemicals Co., Ltd.) for 6 days, and then in an oven (VO-320, Toyo Seisakusho Kaisha Ltd.) at 80°C for 3 h.

Analysis

The particle size distribution of the powders and in the ceramic slurry was measured with commercial equipment (Microtrac MT3300EX, Nikkiso Co., Ltd.).

The rheological properties of the ceramic material were analyzed via a static viscosity measurement by increasing and decreasing the rotational speed of rotor (Rheostress 600, EKO Instruments Co., Ltd.). The static viscosity, η, was measured at a rotational speed of the rotor 15 sec$^{-1}$ and plotted. The green density was analyzed by measuring the size of the dried green body and its weight.

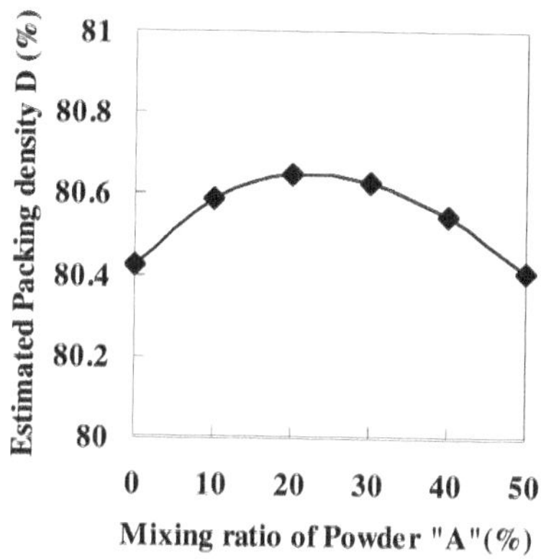


Figure 2. Estimated Particle Packing Density D with a Mixing Ratio of

RESULTS AND DISCUSSION

Estimating particle packing density with particle size distribution

Figure 2 shows the estimated packing density calculated with particle size distribution for each percentage mixing ratio of "A". The percentage value refers to the mixing ratio of "A" to the amount of "A+B". The model of Suzuki et al.[4-5] was used to calculate the relative void space, ε. It is assumed that monosize spherical particles will pack to a density of 74% in a hexagonal close-packed structure. Packing density, D, was estimated from the relative void space, ε, (D=1- ε). In Figure 3, the highest packing density was estimated with the mixing ratio "A" of 20-30%. In this paper, packing density, D, was controlled by blending "A" to "B", and the effects of particle size distribution on rheological and green properties were investigated.

Optimizing the amount of dispersant

In this investigation, the dispersion state of the particles was controlled by controlling the dispersant condition. The "B" was used to investigate the most optimize dispersant condition in the slurry, adjusted to 44vol%.

The static viscosity was lowest with the amount of dispersant fixed at 2.0mg/m$^2$. Thus, in this investigation, the optimized dispersant condition was fixed as 2.0mg/m$^2$, with the shortage dispersant condition as 1.2mg/m$^2$, and the excessive dispersant condition as 7.8mg/m$^2$ to determine the effects of dispersion state on rheological and green properties.

The effect of dispersion state on particle size distribution in slurry

Figure.3 shows the particle size distribution in the slurry with the: (a) shortage dispersant

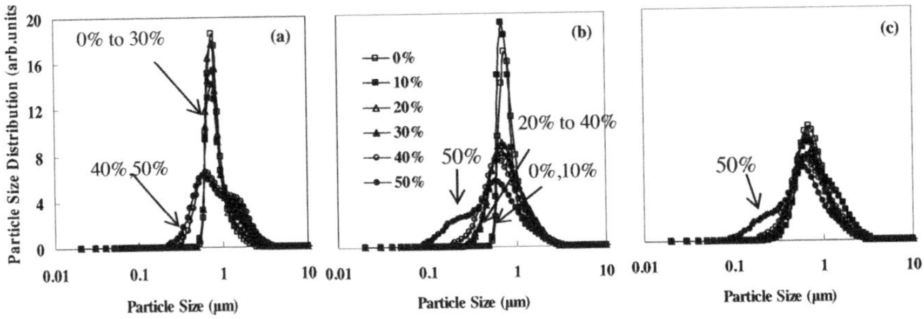


Figure 3. (a) shortage dispersant condition,1.2mg/m$^2$, (b) optimized dispersant condition, 2.0mg/m$^2$, and (c) excessive dispersant condition 7.8mg/m$^2$.

condition, (b) optimized dispersant condition, and (c) excessive dispersant condition. The powder “A” was blended to powder “B” with ratio of 0% to 50%.

In Figure 3, D50 in each distribution was about 0.7μm, which is smaller than powder “B”, D50: 1.1μm. It is supposed that the particles smaller than 0.7μm in the slurry are forming aggregates. Especially for the mixing ratio of 0% to 30% in the shortage dispersant condition and 0% to 10% in the optimized dispersant condition, the particles sized under 0.5μm are not measured. Almost all of these particles are aggregating. For the mixing ratio of 40% to 50% in the shortage dispersant condition, despite the increasing amount of smaller particles from powder ”A”, the larger particles, about 1.7 μm are increasing. This means the aggregation is increasing. Additionally, there is smaller peak around 0.6 μm. This is due to the aggregation of particle smaller than 0.5 μm increasing for other mixing ratios, 0-30%.

With an excessive amount of dispersant, Figure 3(c), for each distribution other than the mixing ratio 50%, the shape of the distributions is almost the same. This suggests that the same size aggregates, bridged by the dispersant molecular, are forming with mixing ratio of “A”.

Rheological behavior with particle size distribution

The rheological behavior was analyzed with mixing ratio of “A” to powder “B” for each dispersant condition. Figure 4 shows the static viscosity, η, with the mixing ratio “A”. The static viscosity, η, is lowest for the range of mixing ratio of 20-30%, in the optimized and excessive dispersant condition, where the highest packing density was estimated.

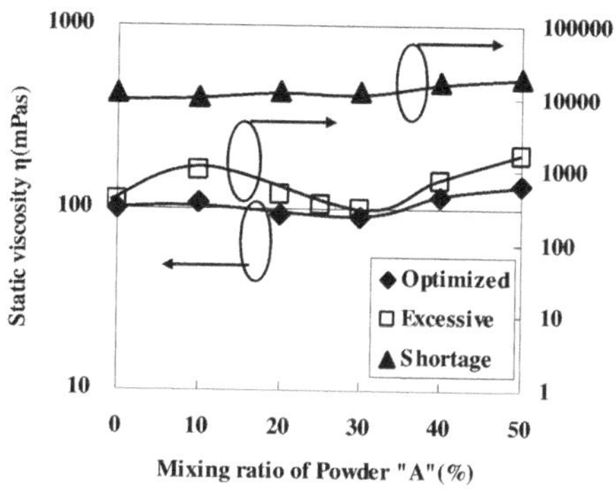


Figure 4. The static viscosity, η, with the percentage mixing ratio of "A", for each dispersant condition.

The static viscosity, η, and estimated packing density in Figure 2 are in inverse correlation. When there were no aggregates in the slurry, and the packing density is increasing, the amount of solvent necessary for the slurry to flow decreased. For the same concentration of powder, more solvent is necessary for flow with a lower packing density[6,7]. But the particle size distribution of these slurries is different from those calculated due the aggregation of particles sized under 0.5 μm. These results suggest that, for the optimized and excessive dispersant condition, almost all of the aggregates could be dispersed with by the shear stress of flowing. And, in the 20% to 30% range, the static viscosity, η, is lowered as the aggregation of particles is prevented.

For the shortage dispersant condition, the static viscosity, η, is almost constant and increases a little with increasing mixing ratio of "A". But in Figure 3(a), the particle size distributions in the slurry are different from each other, with most aggregation in 40% to 50%, with increasing mixing ratio of "A", but the static viscosity, η, is little increasing. These results indicate aggregation, and that these are harder to disperse with the shear stress of flowing compared to the optimized and excessive dispersant conditions,.

Green density and drying shrinkage with particle size distribution

The gel-cast green body was fabricated with different dispersant condition and particle size distribution. The green density and drying shrinkage were measured. Figure.5(a) shows green density.

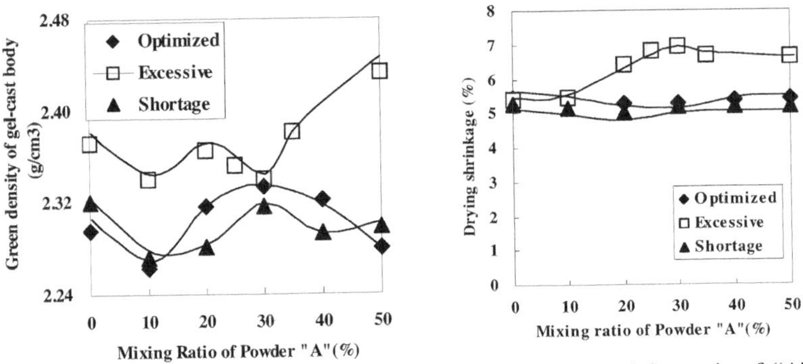


Figure 5 (a)Green density and (b) Drying shrinkage with mixing ratio of "A", in Optimized, 2,0mg/m$^3$,Excessive, 7.8mg/m$^3$, and Shortage, 1.2mg/m$^2$, dispersant condition.

Figure 5(b) shows drying shrinkage with mixing ratio of "A" to "B" for each dispersant condition. For the shortage and optimized dispersant condition, with mixing ratio of "A" 30%, the highest measured density of the green body is, 2.32g/cm$^3$ and 2.33g/cm$^3$, respectively, while the shrinkage value is almost constant. This result suggests that with increasing packing property of particle size distribution of the powder, the less aggregates the voids and higher the green density.

For the excessive dispersant condition, the green density is increasing for the mixing ratio of "A", 30% to 50%, and the highest measurement is 2.43g/cm$^3$ for 50%. The particles in the slurry are bridged by the dispersant molecular and aggregated. The aggregates have weaker attractive force than the directly aggregated ones[8]. It is supposed that the aggregating force is weaker than the drying shrinkage force, and enough to transform and achieve higher packing. By increasing the amount of small particles with the mixing ratio of "A", the contact point with weaker aggregation force is increasing. Therefore, for the excessive dispersant condition, the green density becomes higher with mixing ratio of "A".

## CONCLUSIONS

The following conclusions were obtained.

(1) Because of the different states of aggregation of the particles especially smaller than 0.5μm, the particle size distribution in the slurry is different from in the raw powder, and in the particle dispersion state.

(2) The aggregated particles are dispersed in optimized and excessive dispersant condition by

the stress of flowing, and the static viscosity correlates with the estimated packing density.
(3) The green density has highest with the excessive dispersant condition and a 50% mixing ratio of "A", which is caused by weaker aggregation, but it has the most shrinkage, which is inconvenient for fabricating a large sized green body.

REFERENCES


[1] C. G. Ha, Y.G. Jung, J.W. Kim, C.Y. Jo, U. Paik, "Effect of Particle size on gelcasting process and green properties in alumina", Mater. Sci. Eng., **A377**, 212-221(2002)

[2] Z. Zheng, D. Zhou, S. Gong, "Study of drying and sintering characteristics of gelcast BaTiO3-based ceramic parts" Ceram. Int., **34**, 551-555(2008)

[3] K. Prabohakaran, C. Pavithran, "Gelcasting of alumina from acidic aqueous medium using acrylic acid", J. Eur. Ceram. Soc., **20**, 1115-1119(2000)

[4] M. Suzuki, H. Kada, M. Hiroata, "Effect of size distribution on relation between coordination number and void fraction of spheres in a randomly packed bed", Adv. Powder. Technol., **10**, 353-365(1999)

[5] M. Suzuki, H. Sato, M. Hasegawa, M. Hirota, "Effect of size distribution on tapping properties of fine powder", Powder Technol., **118**, 53-57(2001)

[6] J.V. Robinson, "The Viscosity of Suspensions of Spheres", J. Phys. Colloid Chem. **53** (1949) 1042-1056.

[7] J.V. Robinson, "The Viscosity of Suspensions of Spheres. II", J. Phys. Colloid Chem. **53** (1950) 455-464.[8] M.Takahashi, M.Oya, M.Fuji, Transparent observation of particle dispersion in alumina slurry using in situ solidification, J.Soc.Powder Technol., Japan, **40**, 410-417(2003).

# EFFECT OF THE POLYMER STRUCTURE IN A GELCAST GREEN BODY ON GRAPHITIZATION

Tomoaki Kato, Takashi Shirai, Hideo Watanabe, Masayoshi Fuji* and Minoru Takahashi
Ceramics Research Laboratory, Nagoya Institute of Technology
Honmachi 3-101-1, Tajimi, Gifu, Japan *E-mail fuji@nitech.ac.jp

## ABSTRACT

We suggest a new method to fabricate alumina / carbon composites using the gelcasting method followed by sintering in an inert atmosphere. In this method, the carbon component has a graphite structure that gives the composite material good electrical conductivity. With the gelcasting method, the gel polymer structure was varied to optimize it for the desired shape. In this work, the effect of the polymer structure on graphitization is discussed. Varying the ratio of cross linker to initiator changed the structure. The structure change was confirmed by differences in the gelation behavior or the bending strength of the wet green body. However, the degree of graphitization and electrical conductivity does not vary significantly. It was confirmed that our method produces a composite with good electrical conductivity, even if the gel polymer structure is varied to optimize it for casting a complicated shape.

## INTRODUCTION

One of the problems with ceramics is the difficulty in machining them. To avoid machining, a near net shape process is required. The gelcasting method[1,2] is suitable for molding a complicated geometry[3,4]. For the gelcasting method, gellants have been studied to optimize the solidification time and the strength of the green body, depending on the desired shape. Vinyl monomer[2,3,5], epoxy resin[6,7], and polysaccharide[8-10] are typical gellants. In particular, a vinyl type monomer is often used because it is easy to change the gel structure by controlling the amount of initiator or cross linker. The gel structure significantly affects the solidification time and the strength of the green body.

Recently, we succeeded in fabricating alumina / carbon composites from gelcast green bodies sintered in an inert atmosphere[11,12]. This material has a good and homogeneous electrical conductivity, with a carbon content less than 1 mass%. Various methods have been proposed for fabricating electrically conductive ceramics. A common method for improving the electrical properties of an insulating ceramic is to incorporate electrically conductive fillers such as metals[13], metallic oxides[14] or carbon[15]. However, in order to achieve the desired level of electrical conductivity, a high percentage of filler is usually added to the composite material. Unfortunately, with high loadings, the mechanical properties can degrade, and/or there can be undesirable effects on the physico chemical properties (e.g., hardness and stiffness) of the composite. In our method, the polymer network in a gelcast green body transforms to carbon during the sintering process. The polymer network is a source of carbon that is forms between the alumina particles in the slurry. Thus, the carbon exists homogeneously around each alumina particle, and this results in good electrical conductivity. The carbon in the composite material has a graphite structure as confirmed by electron beam diffraction and Raman spectroscopy.

It is thought that our new method is suitable for fabricating electrically conductive ceramics with a complicated geometry. However, the influence of the gel structure on the carbon structure and the electrical conductivity in the composite has not been investigated. In this study, the effect of the polymer structure on graphitization in a gelcast body is examined and discussed.

Varying the ratio of initiator to cross – linker, changes the gel structure. After sintering, the electrical conductivity and the degree of graphitization are measured to determine the relationship with the gel structure.

## EXPERIMENTAL PROCEDURE

### Sample preparation

The slurry compositions studied are shown in Table 1. Each composition was decided based on the composition (A). The ratio of monomer and cross - linker was changed on the composition (B) and (C). And the amount of initiator system was changed on the composition (D) and (E). An alumina powder of mean particle diameter 0.5 μm (AL160-SG4, Showa Denko Co.) was used the source of the ceramics matrix. Methacryamide and N,N'-Methylene-bis-acrylamide were dissolved in distilled water as gellants. The alumina powder was dispersed in the solvent with a dispersant (Seruna D-305, an ammonium salt of polycarboxylate solution, Chukyo Yushi) by ball milling for 24 H. The well-dispersed slurry was agitated under a vacuum condition for 15 minutes to degas. Then, ammonium persulfate solution of 10 mass % as initiator, and N,N,N',N'-Tetramethylethylenediamine as a catalyst were added to the slurry in order to initiate a polymerization reaction. The slurry was then poured into the forming mold and solidified for 24 H. The solidified wet green bodies were demolded and dried at 25°C for 5 days under controlled humidity from 90% to 60%. The dried green-body was sintered at 1700°C for 2 H in an argon atmosphere at 150 kPa.

### Characterization

The viscoelasticity of the slurry during the gelation reaction was measured using a rheometer (Rheostress 600, Haake) with 35 mm serrated parallel plates (for 2 – $10^6$ mPa). First, the initiator was added to the slurry and stirred for 3 minutes. Then, the catalyst was added and stirred for 3 minutes. The starting time of the gelation reaction was assumed to begin with the catalyst addition. The slurry was then filled between the rheometer plates and measured at the frequency of 1 Hz and at a shear stress of 0.5 Pa. This measurement was started after 5 minutes (First 3 minutes was for mixing the catalyst and second 2 minutes was for the filling of the slurry) from the starting of the gelation reaction. The change in the viscoelastic property of gelling slurry was recorded every ten seconds.

The strength and flexibility of wet green body, 24 H after molding, was measured using a three – point bending test at a crosshead speed of 1 mm/min (AGS, Shimazu).

The electrical conductivity of sintered body was measured using the 4-probe method using

Table 1. Slurry Compositions

| Slurry Composition | Alumina (mass%) | Distilled Water (mass%) | Dispersant (mass%) | Monomer (mass%) | Cross-linker (mass%) | Initiator (μL/g) | Catalyst (μL/g) |
|---|---|---|---|---|---|---|---|
| A | 80.0 | 15.1 | 0.720 | 3.15 | 1.05 | 1.03 | 0.170 |
| B | | | | 3.92 | 0.240 | 1.03 | 0.170 |
| C | | | | 1.52 | 2.75 | 1.03 | 0.167 |
| D | | | | 3.15 | 1.05 | 0.515 | 0.0880 |
| E | | | | 3.15 | 1.05 | 5.15 | 0.850 |

DC voltage current source / monitor (R6243, Advantest). In this measurement, the probe current was fixed to 0.1 mA.

Raman spectroscopy for analysis of the carbon in the sintered body was recorded over the frequency ranging from 1000 to 2000 $cm^{-1}$ at an interval of 1.86 $cm^{-1}$ with a collection time of 10 s (Laser Raman Spectrometer NRS-3100, JASCO). A 11 mW, 532 nm laser beam was used as the excitation source.

## RESULTS AND DISCUSSION

Figure 1 shows the evolution of the complex modulus during the gelation of the slurry. The increase in the complex modulus can be divided into two phases[16]. One is the physical entanglement area based on the growth of polymer strand that shows a gradual slope initially upon gelation. The other is the chemical cross linking area with a steep slope later in the gelation reaction. Figure 2 shows tan(δ), where δ is the phase angle between loss and storage modulus. A value of tan(δ) = 1 suggests a gel point on the gelation reaction[17]. The gel point of each composition is shown in Table 2. The gelation behavior is clearly changed by the slurry composition as shown in these figures. In comparing compositions (A), (B) and (C), the effect of increasing the cross linker density is seen on increasing the slope in the chemical cross linking area, and decreasing the time for the gel point. The amount of initiator also significantly affected the gel point. The gel point of (E), already passed at the beginning of measurement, was expected to be about 5.5 minutes from the slope of the tan(δ) shown in Fig.2.

The correlation plot between bending strength and strain is shown in Figure 3. The green body becomes fragile with increasing cross linker density as shown in Fig.3 (A), (B) and (C). The cause of this is thought to be that the polymer strand is fixed at the cross linking point and loses flexibility. Both (D) and (E) show a little lower bending strength than (A). However, the maximum strain of them was almost same. It is thought that the amount of initiator on (D) was so small that non reacted monomers might have remained in the green body. In the case of (E), too much initiator produced a lot of short polymer strands. Thus, the polymers could not

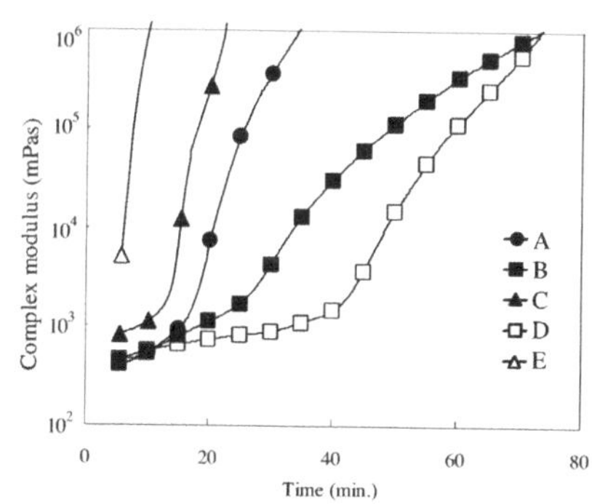


Fig. 1 Complex modulus with gelation time for each slurry composition in an oscillatory shear experiment at constant frequency and stress.

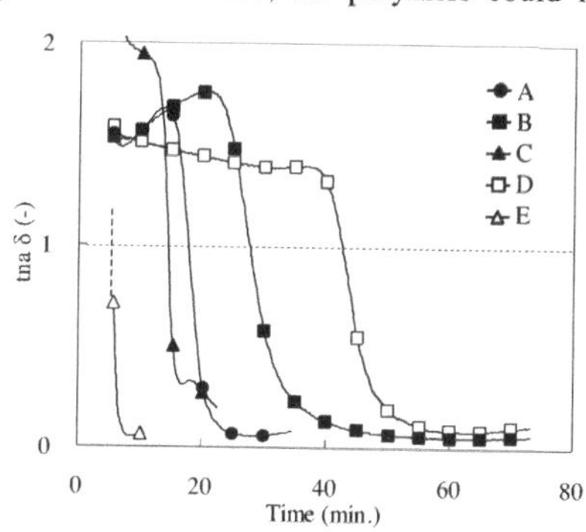


Fig. 2 The change of tan(δ) with gelation time in oscillated viscoelasticity measurement of each slurry composition, where δ is the phase angle between loss and storage modulus.

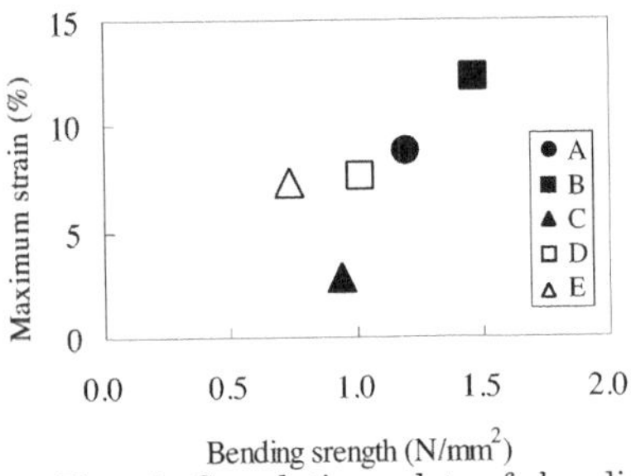


Fig. 3 Correlation plot of bending strength and maximum strain of a wet green gelcast body fabricated from each slurry composition.

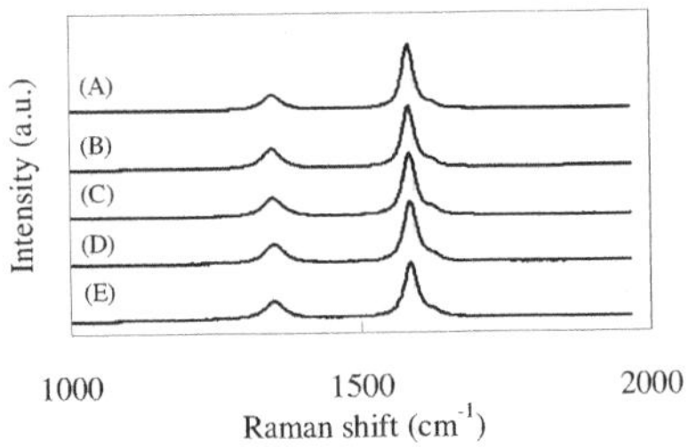


Fig. 4 Raman spectrum of the D-band and G-band for a gelcast body fabricated from each slurry after sintering at 1700°C.

entangle with each other, resulting in a decrease the bending strength. From these results, each gel polymer structure can be compared with the sample (A). Composition (B) gives soft gel structure based on the low cross linker density. On the other hand, the gel of (C), which has a high cross linker density, is hard. The main chain of (D) is the longest, with the lowest initiator system concentration and longest reaction time. Conversely, (E) has a short main chain. The different gel polymer structures can be estimated.

The Raman spectrum for each composition is shown in Figure 4, and the analyses of the results are in Table 2. An intensity ratio of D-band and G -band ($I_{1360}/I_{1580}$) indicates the degree of graphitization. The half band width of the G-band ($\Delta\nu_{1580}$) shows the crystallinity and turbostraticity of the graphite planes. Differences in the graphite structure depending on the gel structure are not seen in these results.

The electrical conductivity of (D) was lower than others as shown in Table 2. The unreacted monomer, suggested from the results in Fig.3, was thought as the reason for the low conductivity of (D). But, this difference seems small compared with the clear difference in the gel structure. The results indicate that the difference in gel structure seen in the gelation behavior or bending test do not significantly affect electrical conductivity. In our method, it is confirmed that the carbon fabricated from an optimized gel for a desire structure has a high degree of graphitization and that produces good electrical conductivity in the resultant alumina / carbon composite material.

The graphitization mechanism of the gel polymers in the gelcasted green body is interpreted as being similar to the graphitization mechanism of pitch. Major fabrication methods to fabricate carbon or graphite materials are pyrolysis of polymer strands or film, graphitize of pitch, and vapor deposition of hydrocarbon gas. If the graphitization mechanism of the gel polymer is similar to the pyrolysis of polyacrylonitrile, it is thought that the resultant graphite structure is affected by the gel polymer structure because the graphite plane should be formed by the cooperation of oriented adjoining chains. Also, higher temperature than 2000°C is needed to fabricate oriented graphite by vapor deposition or pyrolysis of polymer films. In our expectation, the gel polymer was decomposed into low molecular weight segments at low temperature on sintering. Then, these segments become pitch with the elimination of function groups and elements. The sintering of matrix and graphitization at the sintering temperature will reorient this pitch.

## CONCLUSION

The effect of the polymer structure on graphitization in a near net shape formed gelcast body was discussed. Varying the slurry composition changed the polymer structure, but it did not affect to the degree of graphitization and the electrical conductivity. It is confirmed that our method can produce a composite with good electrical conductivity, even when the gel polymer structure has been changed to optimize for casting complicated shape.

## ACKNOWLEDGEMENT

This research was carried out by the Ministry of Education, Culture, Sports, Science, and Technology (MEXT) financed project "Cooperation for Innovation Technology and Advanced Research in Evolution Area".

## REFERENCES

[1]O. O. Omatete, M, A, Janney and R. A. Strehlow, "Gelcasting – a new ceramic forming process", Am. Ceram. Soc., Bull., 70 (10) 1991 1641 – 1649

[2]A. C. Young, O. O. Omatete, M. A. Janney and P. A. Menchhofer, "Gelcasting of alumina", J. Am. Ceram. Soc., 74 (3) (1991) 612-618

[3]M. A. Janney, O. O. Omatete, C. A. Walls, S. D. Numm, R. J. Ogle and G. Westmoreland, "Development of low-toxicity gelcasting systems", J. Am. Ceram Soc., 81 (3) (1998) 581 – 591

[4]K. Niihara, B. Kim, T. Nakayama, T. Kusunose, T. Nomoto, A Hikasa and T. Sekino, "Fabrication of complex – shape alumina / nickel nanocomposites by gelcasting process", J. Euro. Ceram. Soc., 24 (2004) 3419 – 3425

[5]F. S. Ortega, P. Sepulveda and V. C. Pandlfelli, "Monomer systems for the gelcasting system", J. Euro. Ceram. Soc., 22 (2002) 1395 – 1401

[6]X. Mao, S. Shimai and S. Wang, "Gelcasting of alumina foams consolidated by epoxy resin", J. Euro, Ceram. Soc., 28(1) (2008) 217-222

[7]X. Mao, S. Shimai, S. Wang, M. Dong and L. Jin, "Rheological characterization of a gelcasting system based on epoxy resin", Ceram. International, 35 (1) (2009), 415-420

[8]L. Santacruz, M. Isabel and R. Moreno, "Rheological characterization of synergistic mixtures of carrageenan and locust gum for aqueous gelcasting of alumina", J. Am. Ceram Soc.,85 (10) (2002) 2432 – 2436

[9]A. J. Millan, M. I. Nieto, R. Moreno, C. Baudin, "Thermogelling polysaccharides for aqueous gelcasting – part 3 : mechanical and microstructural characterization of green bodies", J. Euro. Ceram. Soc., 22 (2002) 2223-2230

[10]K. Jono, M. Fuji and M. Takahashi, "Porous ceramics for Building materials Fabricated by in situ solidification method using natural polymers and west resources", J. Ceram. Soc. Japan, 112 (5) (2004) S138-S143

[11]R. L. Menchavez, K. Adachi, M. Fuji and M. Takahashi, "Semi-Conductivity of gelcast alumina sintered under nitrogen atmosphere", Adv. Mater. Research, 11-12 (2006) 493-496

[12] R. L. Menchavez, M. Fuji and M. Takahashi, "Electrically conductive dense and porous alumina with in-situ synthesized nanoscale carbon networks", Adv. Mater., 20 (2008) 2345 – 2351

[13] S. Hussain I. Barbariol, S. Roitti and O. Sbaizero, "Electrical conductivity of an insulator matrix (alumina) and conductor particle (molybdenum) composites", J. Eur. Ceram. Soc., 23, 315-321 (2002)

[14] C. Laurent, A. Peigney, O. Dumortier, and A. Rousset, "Carbon nanotubes – Fe – Alumina nanocomposites. Part II: microstructure and mechanical properties of the hot-pressed composites", J. Eur. Ceram. Soc., 18, 2005–2013 (1998).

[15] J. Fan, D. Zhao, M. Wu, Z. Xu and J. Song, "Preparation and microstructure of multi-wall carbon nanotubes-toughened $Al_2O_3$ composite", J. Am. Ceram. Soc., 89, 750–753 (2006)

[16] H. Winter and F. Chanbon, "Analysis of liner viscoelasticity of a crosslinking polymer at the gel point", J. Rheol. 30 (2) (1986) 367 – 382

[17] A. B. Rodd, J. J. Cooper-White D. E. Dunstan and D. V. Boger, "Polymer concentration dependence of the gel point for chemically modified biopolymer networks using small amplitude oscillatory rheometry", Polymer, 42 (8) (2001) 3923-3928

# FABRICATION OF ALUMINA DIAMOND PHOTONIC CRYSTALS WITH TWINNED INTERFACES TO DEVELOP TERAHERTZ WAVE RESONATORS

Toshiki Niki and Soshu Kirihara
Smart Processing Research Center, Joining and Welding Research Institute, Osaka University
11-1 Mihogaoka, Ibaraki, Osaka 567-0047, Japan

## ABSTRACT

Micrometer order alumina diamond lattice photonic crystals were fabricated using micro stereolithography, structural joining, and powder sintering. Twinned interfaces were formed intentionally as plane defects between mirror symmetric diamond lattices to realize efficient resonation and directional emission of terahertz waves. The electromagnetic wave properties were measured using a terahertz time domain spectroscopy. A complete photonic band gap was observed through the diamond lattices in the terahertz frequency range, which that showed good agreement with calculations based on a plane wave expansion method. A localized mode sharp emission peak formed in the perfect band gap through the optimized twinned lattice structure. In an electric field intensity distribution profile simulated using transmission line modeling, incident waves resonated and localized strongly through multiple reflections between the twinned diffraction lattices. The amplified waves were emitted from the opposite side as plane waves with directional propagation.

## INTRODUCTION

Photonic crystals with periodic arrangements of dielectric materials form photonic band gaps to totally reflect electromagnetic waves by Bragg diffraction[1-2]. When a structural defect is inserted in the periodic arrangement, a localized mode appears in the band gap. Structural defects can localize electromagnetic waves and permit wave transmission at resonation frequencies of the specific wavelength corresponding to the size and dielectric constant of the defect region. These features will be applied to various electromagnetic devices such as resonators, wavelength filters, and plane wave emitters in wide frequency ranges from the microwave to the optical wave regions[3-5]. Three dimensional photonic crystals with a diamond structure can form the complete photonic band gap to reflect electromagnetic waves in all directions and control the wave transmission effectively. In our research group, micrometer order photonic crystals with the diamond structures composed of ceramic lattices were fabricated using micro stereolithography of a computer aided design and manufacturing (CAD/CAM) system[6-7]. These crystals exhibited perfect photonic band gaps of micrometer order wavelength in the terahertz frequency range. In the near future, terahertz waves can be expected to be applied to various types of novel sensors to detect micro defects on material surfaces, cancer cells in human skin, and micro bacteria in vegetables[8-9]. In previous research, micrometer order photonic crystals with alumina diamond lattices were fabricated using micro stereolithography, structural joining, and powder sintering, and an air cavity as a point defect was introduced to form localized modes in the band gap[7]. In this study, twinned interfaces were formed intentionally as plane defects between two photonic crystals with mirror symmetric diamond lattices to realize efficient resonation and directional emission of terahertz waves. Adequate period numbers of the twinned lattices were investigated to obtain a sharp emission peak of the localized mode in the band gap. The transmission spectra of the terahertz waves were measured using terahertz time domain spectroscopy. The electric field intensity in the optimized twinned lattices was simulated at the localized frequency through electromagnetic simulation with the finite element method.

## EXPERIMENTAL PROCEDURE

Diamond lattices were designed by using a 3D-CAD software (Toyota Caelum Co. Ltd., Japan, Think-Design Ver. 9.0). The lattice constant of the diamond structure and the aspect ratio of the dielectric lattice were 500 μm and 1.5, respectively. The whole structure was 4x4x2 mm and consisting of 8x8x4 unit cells. In addition, the twinned diamond structures with plane defects between the mirror symmetric lattices for the Γ-L<111> crystal direction were designed. The twinned interfaces were sandwiched between the mirror symmetric lattice blocks composed of two, three, and four periods. The designed models were converted into stereolithography (STL) file format for rapid prototyping and sliced into a series of two dimensional cross sectional data 10 μm in layer thickness. These data were transferred to micro stereolithographic equipment (D-MEC Co. Ltd., Japan, SI-C 1000). Figure 1 shows an illustration of the micro stereolithography. Photosensitive acrylic resins including alumina particles of 170 nm in average diameter at 40 volume % were supplied on a glass substrate from a dispenser nozzle using air pressure. This paste was spread using a mechanically controlled knife edge. The thickness of each layer was set at 10 μm. A visible laser beam 405 nm in wavelength was introduced into a digital micro-mirror device (DMD), and a 2 μm resolution micro pattern was exposed on the resin surface through an objective lens. In this optical device, 1024x768 aluminum micro mirrors of 14 μm in edge length were assembled. Each mirror can be tilted independently using piezoelectric actuators. Through the layer by layer process under computer control, the acrylic resin components with the alumina particles dispersion were obtained. These composite precursors were dewaxed at 600 °C for 2 h and sintered at 1500 °C for 2 h in air. The relative density of the sintered samples was measured by Archimedes method. Their appearance was observed using a digital optical microscope (Keyence Co. Ltd., Japan, VH-Z 100). The transmission properties of incident terahertz waves were analyzed by using a terahertz time

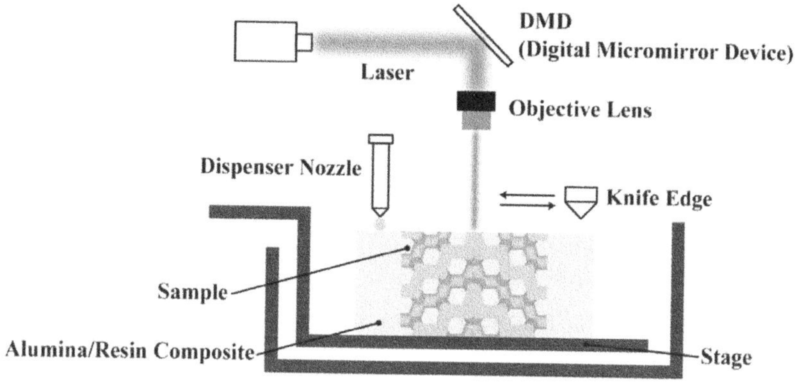


Figure 1. An illustration of the micro stereolithography used to form the photonic crystals.

domain spectroscopy (Advanced Infrared Spectroscopy Co. Ltd., Japan, J-spec 2001 spc/ou). An electromagnetic band diagram of the diamond lattice structure was graphed theoretically using a plane wave expansion (PWE) method. The electric field intensity in the optimized twinned diamond lattice was simulated at the localized frequency by using a transmission line modeling (TLM) simulator (Flomerics Co. Ltd., UK, Micro-Stripes Ver. 7.0).

## RESULTS AND DISCUSSION

Figure 2 shows the fabricated alumina diamond photonic crystals with or without the twinned lattices. The twinned interfaces parallel to the (111) crystal plane were formed as plane defects between the mirror symmetric lattice blocks composed of two, three and four periods.

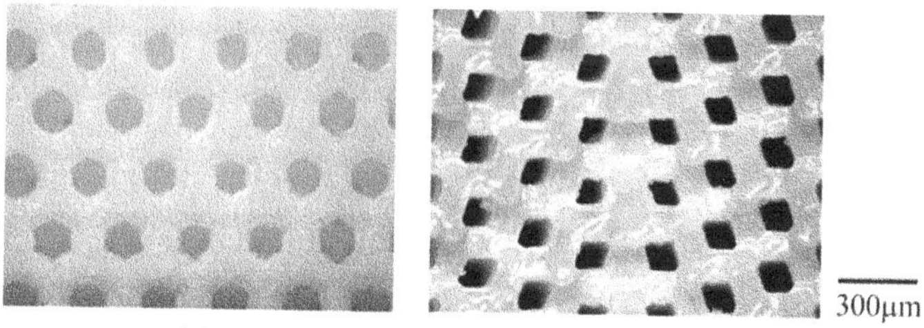


Figure 2. Alumina diamond photonic crystals: (a) without twinned lattices and (b) with twinned lattices.

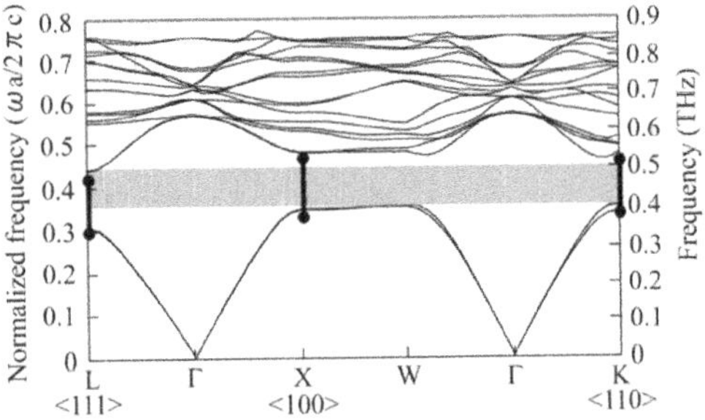


Figure 3. A photonic band diagram calculated using a plane wave expansion method. Closed circles and black bars are the measured photonic band gap frequencies. The dielectric constant of the lattice is 9.8. The gray color frequency range indicates a complete band gap is common for all directions.

The spatial resolution was within 10 μm. The average linear shrinkage was 25 %. The lattice constant of the sintered diamond structure was 375 μm. The relative density reached 97.5 %. Dense alumina lattice structures were obtained without cracking and deformation. The measured band gap frequencies through the diamond lattices were compared with the calculated ones by the PWE method as shown in Figure 3. The measured results have good agreement with the calculated regions. According to the photonic band diagram, it was demonstrated that a perfect photonic band gap opened between 0.40 and 0.47 THz. Figure 4 shows the measured transmission spectra for the Γ-L <111> crystal directions of the twinned lattice structures composed of two, three and four period. The clear photonic band gap region and the sharp emission peak of the localized mode can be obtained through the twinned lattices with three periods. The peak frequency was 0.409 THz and it was included in the complete band gap. The twinned lattices composed of three periods are the ideal structure for the sharp emission peak. The measured band gap range and the peak frequency were compared with the calculations by the TLM method. They were in good agreement. The electric field intensity distribution in the twinned diamond lattices composed of three periods was simulated at the localized frequency. Figure 5 shows a cross sectional image of the distribution. The incident electromagnetic waves propagated form the left to the right direction. The black and white color areas show the electric field intensity is high and low, respectively. The incident waves were resonated and localized strongly by the multiple reflections in the twinned defect interface between the mirror symmetric diffraction lattices. The amplified waves were propagated directionally for the opposite side as plane waves. At this frequency, the emission peak can be formed in the band gap.

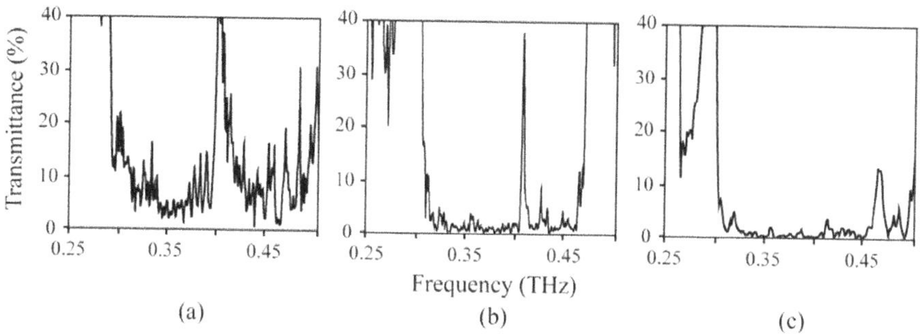


Figure 4. Measured transmission spectra for the Γ-L <111> crystal directions of the twinned diamond lattice structures composed of (a) two, (b) three and (c) four periods.

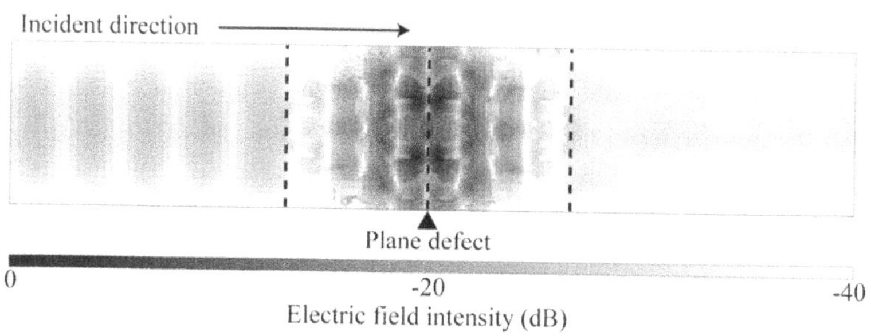


Figure 5. A cross sectional image of simulated electric field intensity at the localized frequency in the twinned diamond lattices.

## CONCLUSIONS

Alumina diamond photonic crystals with or without twinned interfaces were fabricated using micro stereolithography, structural joining, and powder sintering. A complete band gap was obtained through the diamond lattices in the terahertz frequency range from 0.40 to 0.47 THz. A localized mode sharp emission peak was observed through the optimized twinned lattice structure at 0.409 THz in the perfect band gap. In the electric field intensity distribution profile simulated using a TLM method, the strong localized mode was formed through multiple reflections in the twinned defect interface between the mirror symmetric diffraction lattices. The amplified waves were emitted from the opposite side as plane waves with directional propagation. The alumina diamond photonic crystal with the twinned interface can be applied to make efficient terahertz wave resonators and directional emitters.

ACKNOWLEDGEMENTS

This study was supported by Priority Assistance for the Formation of Worldwide Renowned Centers of Research - The Global COE Program (Project: Center of Excellence for Advanced Structural and Functional Materials Design) from the Ministry of Education, Culture, Sports, Science and Technology (MEXT), Japan.

REFERENCES

[1] E. Yablonovitch, "Inhibited Spontaneous Emission in Solid-State Physics and Electronics", Phys. Rev. Lett., **58**, 2059-2062 (1987).

[2] S. John, "Strong Localization of Photons in Certain Disordered Dielectric Superlattices", Phys. Rev. Lett., **58**, 2486-2489 (1987).

[3] B. Temelkuran, M. Bayindir, E. Ozbay, R. Biswas, M. M. Sigalas, G. Tuttle, and K. M. Ho, "Photonic Crystal-Based Resonant Antenna with a Very High Directivity", J. Appl. Phys., **87**, 603-605 (2000).

[4] H. Takano, B. S. Song, T. Asano, and S. Noda, "Highly Efficient in-Plane Channel Drop Filter in a Two-Dimensional Heterophotonic Crystal", Appl. Phys. Lett., **86**, 241101 (2005).

[5] B. Kuhlow, G. Przyrembel, S. Schlüter, W. Fürst, R. Steingrüber, and C. Weimann, "Photonic Crystal Microcavities in SOI Photonic Wires for WDM Filter Applications", J. Lightwave Technol., **25**, 421-431 (2007).

[6] W. Chen, S. Kirihara, and Y. Miyamoto, "Three-Dimensional Microphotonic Crystals of $ZrO_2$ Toughened $Al_2O_3$ for Terahertz Wave Applications", Appl. Phys. Lett., **91**, 153507 (2007).

[7] H. Kanaoka, S. Kirihara, and Y. Miyamoto, "Terahertz Wave Properties of Alumina Microphotonic Crystals with a Diamond Structure", J. Mater. Res., **23**, 1036-1041 (2008).

[8] Y. Oyama, L. Zhen, T. Tanabe, and M. Kagaya, "Sub-Terahertz Imaging of Defects in Building Blocks", NDT&E International, **42**, 28-33 (2008).

[9] V. P. Wallace, A. J. Fitzgerald, S. Shankar, N. Flanagan, R. Pye, J. Cluff, and A. J. Fitzgeraldet, "Terahertz Pulsed Imaging of Basal Cell Carcinoma ex Vivo and in Vivo", Br. J. Dermatol., **151**, 424–432 (2004).

# FABRICATION OF ALUMINA DENTAL CROWNS USING STEREOLITHOGRAPHY

Mitsuyori Suwa[a], Soshu Kirihara[a] and Taiji Sohmura[b]

[a] Joining and Welding Research Institute, Osaka University, 11-1, Mihogaoka, Ibaraki, Osaka, Japan, 567-0047

[b] School of Dentistry, Osaka University, 8-1, Yamadaoka, Suita, Osaka, Japan, 565-0871

## ABSTRACT

A dental-crown composite composed of alumina and glass was successfully fabricated using stereolithogaraphy. A photo sensitive acrylic resin containing alumina particles 1.8 μm in diameter at 70 vol. % was spread on a metal substrate in a 60 μm thick layer. A 355 nm wave length ultra violet laser was scanned on the surface to create cross-sectional planes. After a layer stacking process, the desired shape body was formed. The body was dewaxed and sintered at 600°C and 1600°C in air, respectively. Flexural test specimens 1.2×4×20 mm in dimensions were fabricated using the same materials and processes. The average bending strength after sintering was about 123 MPa. $La_2O_3$-$B_2O_3$-$Al_2O_3$-$SiO_2$ glass was coated on the alumina specimens to close micro cracks on the surface. The bulk ceramics coated with glass powder were heated at 1100°C for 2 h in air, after which the excess glasses was removed by sand blasting. The average flexural strength improved to 462 MPa after glass infiltration.

## INTRODUCTION

In recent years, all ceramic dental crowns have been focused in the area of dental restoration for dentures. They have desirable aesthetics and superior biocompatibility compared with traditional metallic ones. A variety of methods have been contrived to manufacture the ceramic structure[1-3]. In particular, cutting work is currently a predominant method of forming ceramic crowns. However, the high hardness of ceramics makes cutting work difficult. Besides, they can fabricate only a single crown at a time by this method. By contrast, stereolithography of layer lamination makes it possible to fabricate a complicated configuration. Our research group has succeeded in forming complex three dimensional ceramic structures by using stereolithography[4-5]. It is important for dental-crowns to have sufficient mechanical property for clinical use, and a precise degree of accuracy to fit into an individual shape. In this study, we estimated the flexural strength and dimensional accuracy of fabrication, and the linear shrinkage after thermal treatment.

## EXPERIMENTAL PROCEDURE

Initially, flexural test specimens of 4.0×20×1.2 mm in dimensions were manufactured. Figure 1 illustrates a stereolithography CAD/CAM system. The three

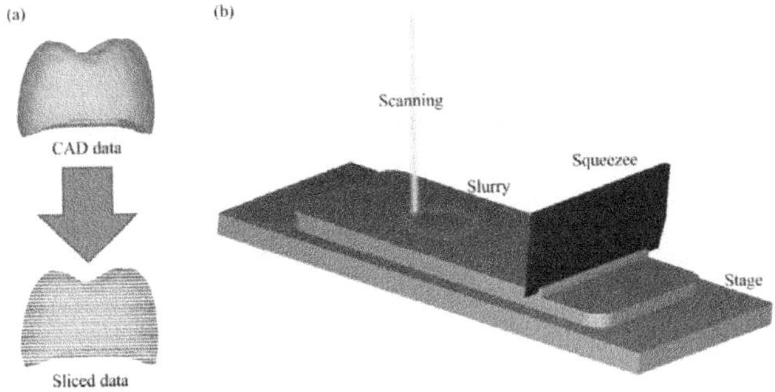


Figure 1. Illustrations of stereolithography using a CAD/CAM system: (a) Slicing CAD data, and (b) Stereolithography process

dimensional graphic model was converted into the STL file format for stereolithography, and sliced into a series of two dimensional data with uniform thickness using the CAD application. These numerical data were transferred into the stereolithography equipment (SCS-300P, D-MEC, Japan). A photo sensitive acrylic resin containing $Al_2O_3$ particles 1.8 μm in diameter at 70 vol. % was spread on a metal substrate and smoothed to form a 60 μm thick layer with a moving mechanical knife edge. A 355 nm wave length ultra violet laser was scanned over the deposited layer to create cross-sectional planes. The 100 mW power laser beam was adjusted to have a 100 μm spot size. After a layer by layer stacking process, rectangle flexural test specimens were fabricated. These precursors were dewaxed at 600°C for 2 h using a heating rate of 0.5°C /min, and sintered at 1400, 1500 and 1600°C for 2 h with using a heating rate of 8°C/min in air. Twelve test specimens were fabricated at each sintering temperature, and six of them were coated with $La_2O_3$-$B_2O_3$-$Al_2O_3$-$SiO_2$ glass ceramic powder (In-Ceram-Alumina, Zahnfabrik, Germany) and heated at 1100°C for 2 h in air. This glass powder is widely used for ceramic coatings on metal teeth in dentistry. The excess glass powder on the surface was eliminated by sand blasting. The microstructures of the sintered

Figure 2. Illustrations of a hollow cylindrical column: (a) Bottom view, (b) Side view, and (c) Cross-sectional view.

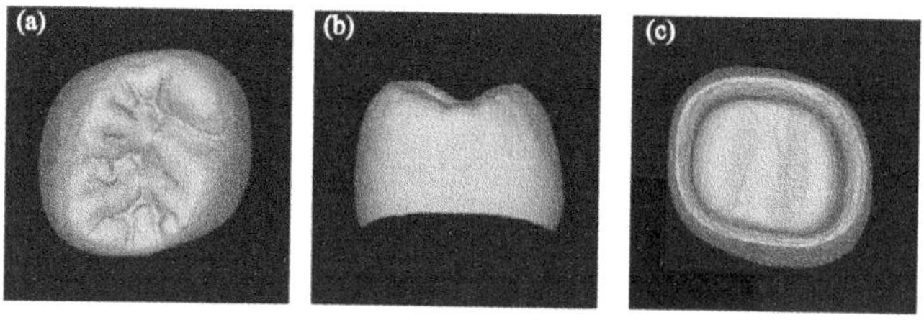


Figure 3. Different directional views of a CAD model of an upper molar tooth.
(a) Top view (b) Side view (c) Bottom view.

and glass infiltrated specimens were observed by scanning electron microscope (SEM). Flexural strengths and relative densities were measured for these test bars.

Subsequently, the hollow cylindrical column shown in Figure 2 was fabricated using the same materials and methods to measure the linear shrinkage after the sintering. This hollow structure was dewaxed for 2 h at 600°C with the heating rate of 0.5 °C/min, and sintered for 2 h at 1600°C with a heating rate of 8°C /min in air. The measured values were fed back for computer designing to fit the objective shape.

A dental-crown model of an upper molar tooth (Figure 3) produced using the CT scanning system was manufactured using the ideal heat treatment pattern to obtain the maximum value of flexural strength.

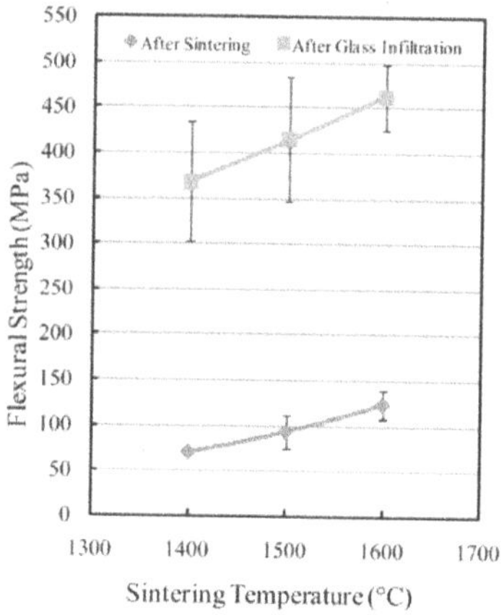


Figure 4. Relationship between sintering temperatures and flexural strengths before and after glass infiltration

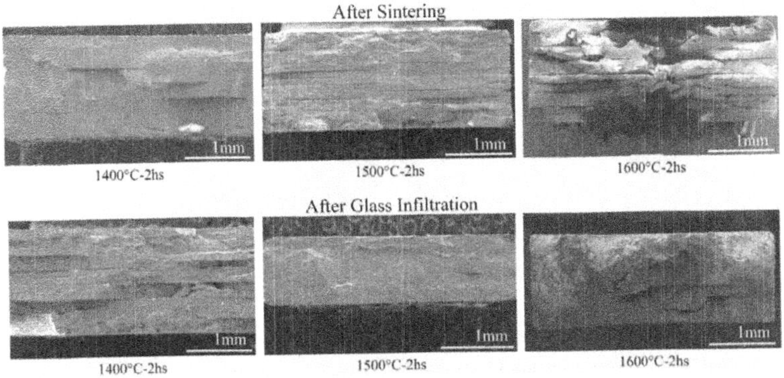


Figure 5. $Al_2O_3$ flexural test specimens with or without glass infiltrations

RESULTS AND DISCUSSION

Figure 4 shows the relationship between sintering temperature and average flexural strength of the test specimens. The average flexural strength of sintered and glass infiltrated test bars increased with sintering temperature. The average relative density of sintered test specimens was approximately 96 %. The maximum flexural strength of the grass infiltrated sample is 462±36 MPa. This value exceeds the required level for clinical use of 400 MPa.

Figure 5 shows the cross sectional surfaces of the sintered samples. The macro cracks were formed parallel to the stacked layers in the stereolithography process. Compared with these results, the macro cracks disappear after glass ceramics infiltration (Figure 5).

Figure 6 shows the detail of the cylindrical hollow column model with its dimensional accuracy and linear shrinkage after sintering. The dimensional accuracy of fabrication was approximately within 200 μm and the linear shrinkage ratios in the horizontal and vertical axes are about 7.24 % and 12.7 %, respectively. The larger shrinkage ratio in the vertical axis is considered to be a gravity effect. The different directional views of the fabricated and sintered cylindrical hollow model are also shown in Figure 6. There is no large deformation and the cylindrical form is precisely shaped. As shown in Figure 7 which shows different directional views of the fabricated and sintered dental-crown model, there is no large deformation or cracking, and the complicated grooves in the top surface of the green body are precisely shaped.

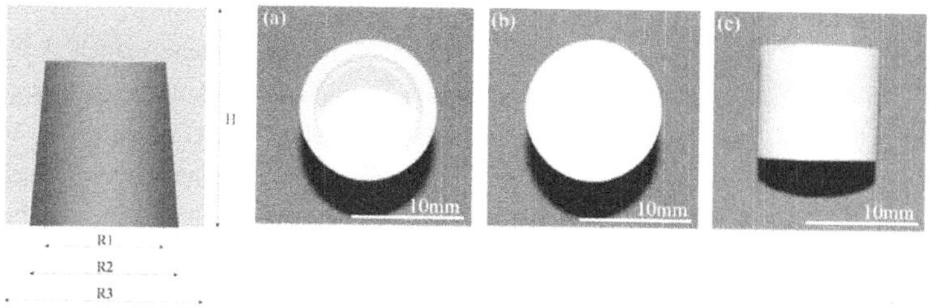


| | R1 | R2 | R3 | H |
|---|---|---|---|---|
| CAD Model (mm) | 13.82 | 11.11 | 10.03 | 13.70 |
| Green Body (mm) | 14.08 | 10.89 | 9.800 | 13.68 |
| After Sintering (mm) | 13.06 | 10.21 | 9.14 | 11.94 |
| Linear Shrinkage Ratio (%) | 7.24 | 6.24 | 6.73 | 12.72 |

Figure 6. The sintered $Al_2O_3$ cylindrical hollow structures and measured linear shrinkages for vertical and horizontal axes.

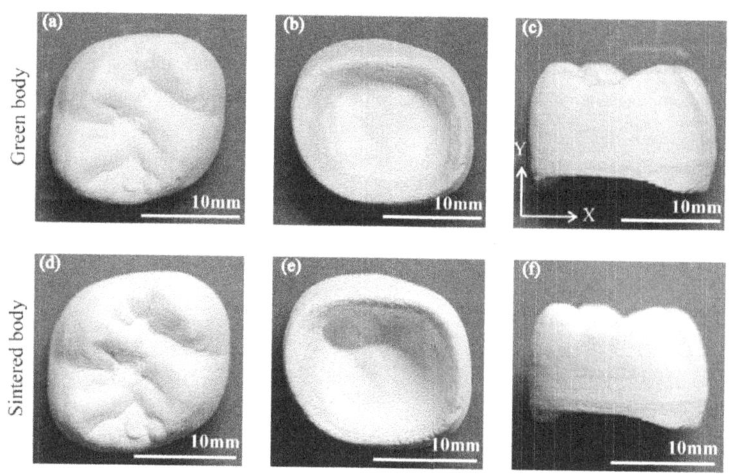


Figure 7. Appearance in different directions of the $Al_2O_3$ / acryl crowns: (a), (d) Top view; (b), (e) Bottom view: (c), (f) Side view.

## CONCLUSION

A dental crown model composed of acrylic resin containing alumina particle 1.8 μm in diameter at a 70 % in volume ratio was fabricated using stereolithography. A dense alumina ceramic dental-crown was successfully obtained after dewaxing at 600°C and sintering at 1600°C. $La_2O_3$-$B_2O_3$-$Al_2O_3$-$SiO_2$ glass ceramic was coated on the fabricated alumina crown. Mechanical properties were measured for plate specimens of sintered alumina with or without grass infiltrations. The maximum flexural strength of the glass coated sample is 462±36 MPa and exceeds the required level of 400 MPa for clinical use. Linear shrinkage ratios in the horizontal and vertical axes were measured for cylindrical hollow column models. Through optimization, a dental crown model with high dimensional accuracy was obtained.

## REFERENCES

[1] J. Stampfl, H. C. Liu, S. W. Nam, K. Sakamoto, H. Tsuru, S. Kang, A. G. Cooper, A. Nikel, F. B. Prinz, "Rapid prototyping and manufacturing by gelcasting of metallic and ceramic slurries", *Materials Science and Engineering*, **A334**, 187-192 (2002).

[2] H. H. Tnag, F. H. Liu, "Ceramic laser gelling", *Journal of the European Ceramic Society*, **25**, 627-632 (2005).

[3] R. Zauner, "Micro powder injection molding", *Microelectronic Engineering*, **83**, 1442-1444 (2006).

[4] S. Kirihara, Y. Miyamoto, K. Takenaga, M. Takeda, K. Kajiyama, "Fabrication of electromagnetic crystals with a complete diamond structure by stereolithography", *Solid State Communication*, **121**, 435-39 (2002).

[5] M. Ishikawa, S. Kirihara, Y. Miyamoto, T. Sohmura, "Freeform fabrication of alumina dental-crown model by using stereolithography", *Advanced Processing and manufacturing tecnnologies for Structural and Multifunctional Materials II, Ceramic Engineering and Science Proceeding* **Volume 29**, Issue 9 (2008).

STRUCTURE AND PHOTOCHEMICAL PROPERTIES OF A $TiO_2$–$In_2O_3$–$SnO_2$–$Al_2O_3$ FILM PREPARED BY ELECTROLYSIS OF AN AQUEOUS SOLUTION

Naoki Matsunaga[1], Keigo Oshiro[1], Yoshihiro Hirata[1], Yousuke Fukuda[1], Soichiro Sameshima[1], Toshifumi Yoshidome[1] and Takamasa Ishigaki[2]

[1]Department of Chemistry, Biotechnology, and Chemical Engineering, Kagoshima University, 1-21-40 Korimoto, Kagoshima 890-0065, Japan

[2]National Institute for Materials Science, Nano Ceramics Center, 1-1 Namiki, Tsukuba, Ibaraki 305-0044, Japan

ABSTRACT

Electrolysis of $TiCl_4$ aqueous solutions of 0–0.1 mol/l was carried out at 10 V/cm of electric filed strength between a dense alumina/tin-doped indium oxide (ITO) cathode and a carbon anode. The composite cathode dissolved in the solution and was precipitated with $TiO_2$ on the surface of the electrode. SEM, XRD, EPMA, Raman spectra, UV-vis spectra and XPS were used to analyze the structure of the deposited film. The structure of the $TiO_2$ formed on the surface of the film changed from amorphous to rutile with an increase in heating temperature. The deposited $TiO_2$ film efficiently absorbed photon energy in the wide wavelength range from 200 to 800 nm. The photocurrent flowed between the deposited $TiO_2$ film and a Pt wire in oxalic acid solutions at pH 5.8, and accelerated the decomposition of oxalic acid at both the electrodes ($C_2O_4^{2-} + 2h^{\bullet} \rightarrow 2CO_2$, $2H^+ + 2e^- \rightarrow H_2$).

INTRODUCTION

Titanium oxide ($TiO_2$) is well known as an excellent photocatalyst in the ultraviolet wavelength range with applications as a cosmetic, solar cell, or coating material for windows. Titania shows a phase transition from anatase (band gap, 3.2 eV) to rutile (band gap, 3.0 eV) at about 800°C. Anatase has higher catalytic activity than rutile. Thus, low temperature processing is needed to deposit anatase on a substrate for many applications. Zhou et al.[1] prepared a $TiO_2$ film on F-doped $SnO_2$ (FTO)-coated glass using a $TiO_2$ aqueous suspension and electrophoretic deposition (EPD). The 480 nm thick $TiO_2$ film was composed of aggregated $TiO_2$ particles with sizes above 200 nm. Hernández et al.[2] deposited a $TiO_2$ and a $TiO_2$-carbon composite film on ITO-coated glass. These two papers report significant absorption of UV light shorter than 400 nm of wavelength. Yoon et al.[3] fabricated Au/$TiO_2$ or Pt/$TiO_2$ films on quartz glass or ITO glass substrates by co-sputtering. Heat treatment of these films of 200–300 nm thickness at 600°C improved the absorption of the light at 650–700 nm of wavelength.

Recently we prepared a thick alumina-ITO film with a thin $TiO_2$ layer on a dense $Al_2O_3$-ITO composite electrode using a dissolution-reprecipitation process by the application of external voltage[4]. This paper reports the analysis of the structure of the deposited film by SEM, XRD, EPMA, Raman spectra, UV-vis spectra and XPS. The photocurrent of $Al_2O_3$-ITO composite with a $TiO_2$ layer was also measured at 300–700 nm of wavelength in oxalic acid solutions.

EXPERIMENTAL

The detailed method for preparing the $Al_2O_3$-ITO electrode and the electrolysis apparatus were described in our previous paper[4]. The bulk density of 60 vol% $Al_2O_3$-40 vol% ITO (90 mass% $In_2O_3$-10 mass% $SnO_2$) electrode, measured by the Archimedes method with double distilled water, was 98.7 % theoretical density (5.25 g/cm$^3$). The sintered electrode (diameter 10 mm, height 4 mm) and a cylindrical carbon (internal diameter 30 mm, external diameter 50 mm, height 35 mm) were used as the working and counter electrodes, respectively. Ammonium sulfate ($(NH_4)_2SO_4$) and titanium (IV) chloride solution (Ti: 16.0–17.0 mass%) were dissolved in double distilled water to control the $TiCl_4$ concentration (0–0.10 mol/l) and $(NH_4)_2SO_4$ concentration (0.2 mol/l) at pH 0.5–2.0. The dense $Al_2O_3$-ITO electrode was set at the center of the cylindrical carbon electrode with an Ag/AgCl reference electrode. The electrolysis was carried out in the current density range from 50 to 390 mA/cm$^2$ for 24 h. The corresponding electric field strength was 10 V/cm. After the electrolysis, the $Al_2O_3$-ITO composite was dried at 100°C for 24 h in air to observe the Al, In, Sn and Ti distribution by electron probe X-ray microanalyzer (EPMA, JXA-8600S/M, Japan Electron Optics Lab. Co., Ltd., Japan) and the microstructure by scanning electron microscope (SM300, Topcon Technologies, Inc., Tokyo, Japan). Phase identification of the surface layers of $Al_2O_3$-ITO composite heated at 700–900°C were conducted using X-ray diffraction (RINT 2200PCH/KG, Rigaku Co., Japan), Raman spectra (JASCO, NR-1800, Japan) and XPS (ESCA-1000, Shimazu Co., Japan). The absorption of photon by the $Al_2O_3$-ITO-$TiO_2$ film heated at 200–900°C was measured by UV-visible spectrophotometer (UV-2450, Shimadzu Co., Japan). The photocurrent between the $Al_2O_3$-ITO-$TiO_2$ composite electrode and a Pt wire was also measured using a fluorescence spectrophotometer (F-3000, Hitachi Ltd., Japan) with a 150 W Xe lamp in the wavelength rang from 300 to 700 nm in an oxalic acid ($H_2C_2O_4$) solution at pH 4 or 6.

RESUTLS AND DISCUSSION

Figure 1 shows the cross section of $Al_2O_3$-ITO composite (A) and the elemental distribution (B) of Al, In, Sn and Ti in the deposited layer after the electrolysis in a solution containing 0.05 M $TiCl_4$ for 24 h. The deposited layer of 1.12 mm thickness contained Al, In and Sn components. The concentration of Al or In of deposited layer was comparable to that of Al or In inside the electrode. In the deposited layer a higher concentration of Sn was measured as compared with the inside concentration. On the other hand, Ti component was rarely observed inside of the electrode. The Ti component was highly measured on the surface of deposited layer. That is, Ti did not co-precipitate inside the deposited layer.

Figure 2 shows the X-ray diffraction patterns for the surface of the $Al_2O_3$-ITO

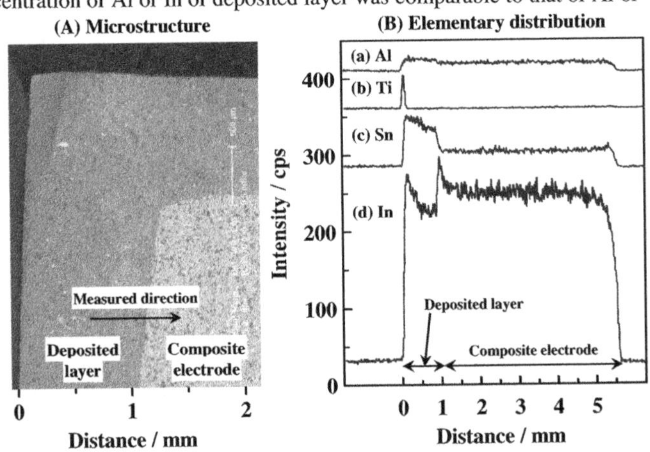


Figure 1. Cross-section of $Al_2O_3$-ITO composite (A) after the electrolysis in the solution containing 0.05 M $TiCl_4$ for 24 h and elementary distribution (B) of Al, In, Sn and Ti in the deposited layer.

composite heated at 700–900°C after the electrolysis in 0.05 M $TiCl_4$ solution. Alpha-$Al_2O_3$, $In_2O_3$ solid solution (ITO) and $SnO_2$ were recognized in the starting powder and heated composite. Rutile was also included in the composite heated at 800–900°C. In addition, a perovskite structure compound $InAlO_3$ was formed in the heated composite by the reaction between $Al_2O_3$ and ITO. The diffraction peaks of $InAlO_3$ could be assigned based on the structure of $InFeO_3$ (JCPDS card No. 41-0038). The detailed electrochemical reaction was discussed in our previous paper[4]. $Ti(OH)_4$ deposited on the surface of composite changed to amorphous phase below 700°C and crystallized to rutile at a higher temperature.

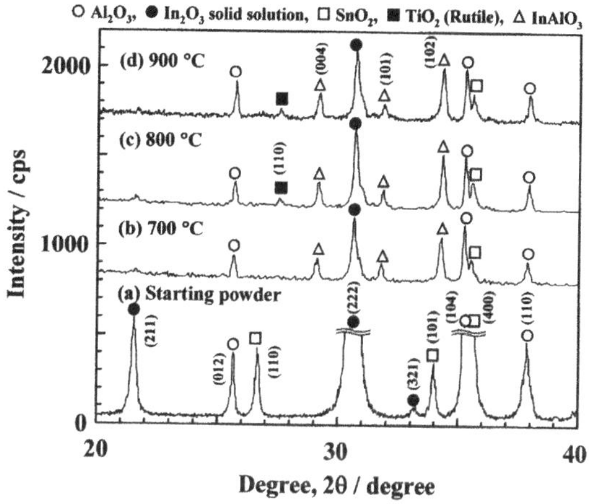


Figure 2. X-ray diffraction patterns of the starting mixed powder and $Al_2O_3$-ITO composite heated at 700-900°C after the electrolysis in 0.05 M $TiCl_4$ solution.

Figure 3 shows the Raman spectra of $Al_2O_3$-ITO composite before and after the electrolysis in $TiCl_4$ solutions. After the electrolysis, samples were heated at 900°C in air. The absorption peaks at 131, 305, 364, 495 and 630 $cm^{-1}$ correspond to $\delta(InO_6)$ or $\nu(InO_6)$ mode of $In_2O_3$[5,6]. The absorption peaks at 426, 498 and 632 $cm^{-1}$ represent $A_{2g}$, TO-$A_{2u}$ and $A_{1g}$ modes[7-9] of $SnO_2$, respectively. On the other hand, the peaks at 422, 448 and 613 $cm^{-1}$ correspond to $E_g$ or $A_{1g}$ mode[10,11] of rutile $TiO_2$. The $E_g(\nu_6)$ and $E_g(\nu_1)$ modes of anatase $TiO_2$ appear at 146 and 641 $cm^{-1}$, respectively[12]. The observed peaks were interpreted by the above information. In the $Al_2O_3$-ITO electrode before the electrolysis (sample a), $In_2O_3$-related peaks were observed. No peaks related to $Al_2O_3$ were identified at 100–900 $cm^{-1}$ of the wavenumber. After the electrolysis (sample b), the intensity of $In_2O_3$-related peaks decreased and the intensity of $SnO_2$-related peaks at 436 $cm^{-1}$ became stronger. This result agreed with the concentration profile measured in Fig. 1. The increased intensity of peak at 630 $cm^{-1}$ in sample

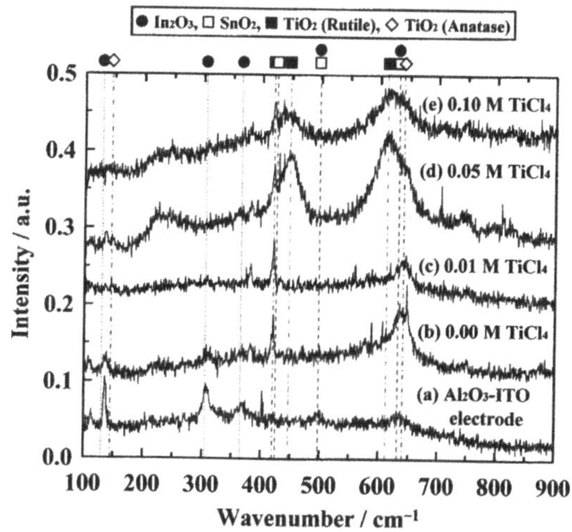


Figure 3. Raman spectra of (a) $Al_2O_3$-ITO composite before electrolysis, and (b-e) $Al_2O_3$-ITO-$TiO_2$ film heated at 900°C after the electrolysis in $TiCl_4$-containing solutions.

(b) reflects the deposition of $SnO_2$ on the $Al_2O_3$-ITO electrode (Fig. 1). In samples (d) and (e), two strong peaks at 450 and 610 $cm^{-1}$ appeared and were related to the formation of rutile $TiO_2$. Since no peak appeared at 146 $cm^{-1}$, the formation of anatase $TiO_2$ was denied. In sample (c), a weak absorption peak at 630 $cm^{-1}$ was measured, suggesting little formation of $TiO_2$ in the dilute $TiCl_4$ solution.

Figure 4 shows XPS spectra at 450–470 eV of binding energy for the $Al_2O_3$-ITO composite heated at 900°C after the electrolysis in $TiCl_4$ solutions. The peaks for In $3d_{3/2}$, Ti $2p_{3/2}$ and Ti $2p_{1/2}$ appear at 451.2, 458.5 and 464.2 eV, respectively[13]. In Figure 4, the corresponding peaks appeared at 451–455 eV (In $3d_{3/2}$), 458–462 eV (Ti $2p_{3/2}$) and 463–467 eV (Ti $2p_{1/2}$), respectively. No peaks of Ti $2p_{3/2}$ and Ti $2p_{1/2}$ were observed in the as-prepared electrode. In samples (d) and (e), Ti-related peaks appeared apparently. However, little $TiO_2$ were formed in sample (c) and this result agreed with the observation by Raman spectra in Figure 3. The slight detection of Ti in sample (b) may be the contamination from the carbon electrode just after the electrolysis experiment of sample (e). Similarly, the XPS peaks related to Al, Sn and O were measured in samples (a)-(e).

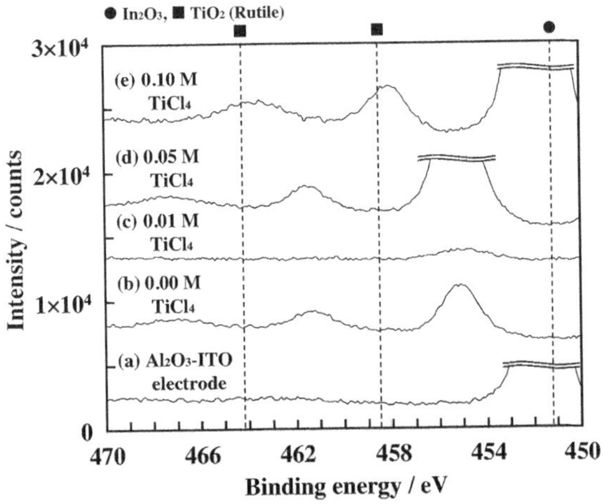


Figure 4. XPS spectra of $Al_2O_3$-ITO-$TiO_2$ composite heated at 900°C after the electrolysis in $TiCl_4$ solutions.

Table I summarizes the results from XRD, EPMA, Raman and XPS for $Al_2O_3$-ITO-$TiO_2$ composite heated at 900°C after the electrolysis. The as-prepared $Al_2O_3$-ITO electrode dissolves by the electrolysis at DC 10 V to form a $SnO_2$-rich $Al_2O_3$-$In_2O_3$ film on the electrode. When $TiCl_4$ concentration is increased to 0.05–0.10 mol/l, the surface of the $Al_2O_3$-$In_2O_3$-$SnO_2$ film is covered with a $Ti(OH)_4$ or amorphous $TiO_2$ layer. The conductivity of the electrode is

Table I. Microstructure information for the $Al_2O_3$-ITO-$TiO_2$ composite heated at 900°C after the electrolysis in $TiCl_4$ solutions.

| Method (detection depth) | Sample (a) As-sintered electrode | Sample (b) 0.00 M $TiCl_4$ | Sample (c) 0.01 M $TiCl_4$ | Sample (d) 0.05 M $TiCl_4$ | Sample (e) 0.10 M $TiCl_4$ |
|---|---|---|---|---|---|
| XRD (< 0.2 mm) | $Al_2O_3$<br>$In_2O_3$ s.s.<br>$SnO_2$<br>$InAlO_3$ | $Al_2O_3$<br>$In_2O_3$ s.s.<br>$SnO_2$<br>$InAlO_3$ | $Al_2O_3$<br>$In_2O_3$ s.s.<br>$SnO_2$<br>$InAlO_3$ | $Al_2O_3$<br>$In_2O_3$ s.s.<br>$SnO_2$<br>$InAlO_3$<br>$TiO_2$ (rutile) | $Al_2O_3$<br>$In_2O_3$ s.s.<br>$SnO_2$<br>$InAlO_3$<br>$TiO_2$ (rutile) |
| EPMA (0.1 μm – 1 mm) | | | | Al ($Al_2O_3$)<br>In ($In_2O_3$ s.s.)<br>Sn ($SnO_2$)<br>Ti ($TiO_2$) | |
| Raman (100 -1000 nm) | $In_2O_3$ | $In_2O_3$<br>$SnO_2$ | $In_2O_3$<br>$SnO_2$ | $In_2O_3$<br>$SnO_2$<br>$TiO_2$ (rutile) | $In_2O_3$<br>$SnO_2$<br>$TiO_2$ (rutile) |
| XPS (1-10 nm) | Al<br>In<br>Sn<br>O | Al<br>In<br>Sn<br>Ti<br>O | Al<br>Sn<br>O | Al<br>In<br>Sn<br>Ti<br>O | Al<br>In<br>Sn<br>Ti<br>O |

maintained by the continuous connection of thee $In_2O_3$-$SnO_2$ components. The $TiO_2$ layer formed crystallized to rutile by heat-treatment above 800°C.

Figure 5 shows the reflectivity of $Al_2O_3$-ITO-$TiO_2$ composite at 600 nm of wavelength as a function of calcination temperature. The reflectivity of the composite with the deposited $TiO_2$ film was very low, below 300°C, and increased above 400°C. The reflectivity of the $Al_2O_3$ powder was almost 100% in the wavelength of 200–800 nm. The reflectivity of ITO powder was small in the wavelength range shorter than 350 nm, and increased in the longer wavelength range. However, the $Al_2O_3$-ITO composite sintered at 1500°C absorbed the photon energy in the wavelengths from 200–800 nm. After the electrolysis in 0.1 mol/l $TiCl_4$, the formed $Al_2O_3$-ITO-$TiO_2$ films showed a low reflectivity of the visible light.

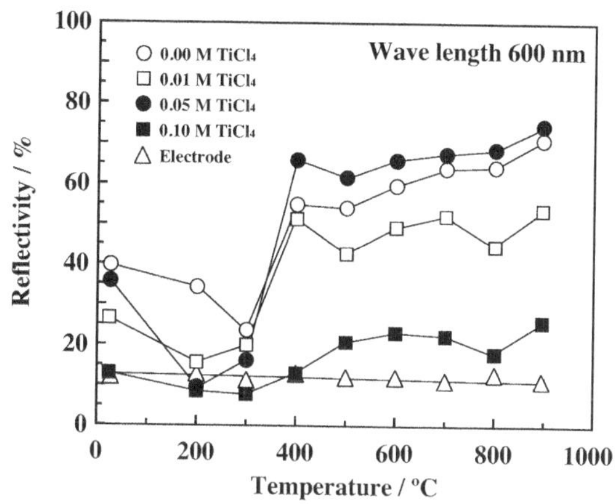


Figure 5. Reflectivity of $Al_2O_3$-ITO-$TiO_2$ composite after electrolysis at 600 nm of wavelength as a function of calcination temperature.

Figure 6 shows the photocurrent of $Al_2O_3$-ITO-$TiO_2$ film in an oxalic acid solution as a function of irradiation time. The photocurrent was represented as a ratio to the current ($I_0$) in a dark solution. The pH of starting buffer solution (potassium hydrogen phthalate solution) and oxalic acid solutions were 4.4 and 5.8, respectively. The $Al_2O_3$-ITO-$TiO_2$ film was deposited on the electrode in a 0.1 mol/l $TiCl_4$ solution and heated at 200°C in air. The interaction between the $Al_2O_3$-ITO-$TiO_2$ film and photon may produce electrons (conduction band) and holes (valence band) in $TiO_2$. The increased number of electric carrier is detected as an increased photocurrent. In an oxalic acid solution, the formed holes and electrons react with $C_2O_4^{2-}$ and $H^+$ ions of oxalic acid ($H_2C_2O_4$), respectively: Anode ($Al_2O_3$-ITO-$TiO_2$ film) $C_2O_4^{2-} + 2h^{\bullet} \rightarrow 2CO_2$, Cathode (Pt wire) $2H^+ + 2e^- \rightarrow H_2$. The electrons formed in the anode film by the interaction with photon are carried out to the external cell. As a result, the pH of the solution

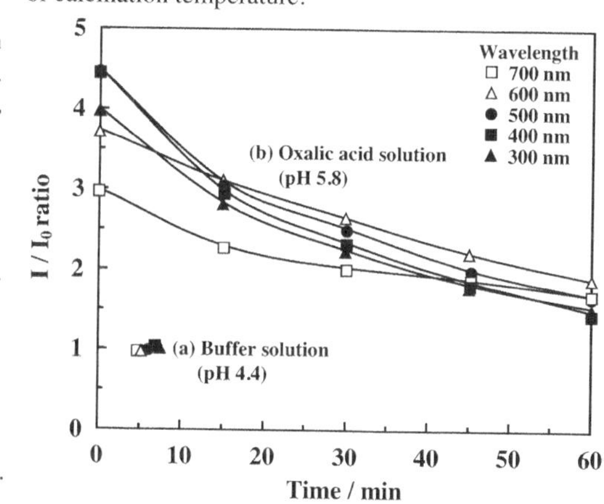


Figure 6. Photocurrent between $Al_2O_3$-ITO-$TiO_2$ film and Pt wire as a function of irradiation time. I and $I_0$ represent the current in the dark and irradiated solutions, respectively.

increases because of the formation of $H_2$. Figure 6 shows no change of photocurrent in the buffer solution for the irradiation of UV-visible light and pH (4.4) of the solution was constant after 50 minutes of the continuous experiment (700 nm → 300 nm). In the buffer solution, the cathodic reaction is suppressed by the action of buffer solution, giving no change of the current as compared with the dark solution. As seen in Figure 6, the electric current in the oxalic acid solutions at pH 5.8 increased when UV-visible light at 300–700 nm wavelength was irradiated and decreased gradually with increasing time of irradiation. The result is explained by the interaction of $Al_2O_3$-ITO-$TiO_2$ films with photon. The pH (5.8) of the starting solutions increased to pH 7–8 after the irradiation of 60–120 minutes. In addition to the reaction with oxalic acid, $OH^-$ ions in the solution may react with holes to form $O_2$ : $2OH^- + 2h^\bullet \rightarrow H_2O + 1/2O_2$.

## CONCLUSIONS

Electrolysis of aqueous solutions containing $TiCl_4$ of 0–0.10 mol/l was carried out at 10 V/cm of electric field strength between a dense $Al_2O_3$ (60 vol%)-ITO (40 vol%) cathode and a carbon anode. The composite cathode dissolved in the solution and then an $Al_2O_3$-$In_2O_3$-$SnO_2$-$TiO_2$ film deposited on the surface of the electrode. The $Al_2O_3$-$In_2O_3$-$SnO_2$-$TiO_2$ film formed efficiently absorbed photon energy in the wavelengths of 200–800 nm. The photocurrent flowed between the formed film and a Pt wire in an oxalic acid solution at pH 5.8. The increased current was used to accelerate the decomposition of oxalic acid.

## REFERENCES


[1]M. Zhou, J. Yu, S. Liu, P. Zhai, and L. Jiang, "Effects of Calcination Temperatures on Photocatalytic Activity of $SnO_2/TiO_2$ Composite Films Prepared by an EPD Method", *J. Hazard. Mater.*, **154**, 1141-1148 (2008).

[2]J.M.P. Hernández, J. Manríquez, Y.M. Vong, F.J. Rodríguez, T.W. Chapman, M.I. Maldonado, and L.A. Godínez, "Photocatalytic Properties of Nano-Structured $TiO_2$-Carbon Films Obtained by Means of Electrophoretic Deposition", *J. Hazard. Mater.*, **147**, 588-593 (2007).

[3]J.W. Yoon, T. Sasaki, and N. Koshizaki, "Dispersion of Nanosized Noble Metals in $TiO_2$ Matrix and Their Photoelectrode Properties", *Thin Solid Films*, **483**, 276-282 (2005).

[4]N. Matsunaga, Y. Hirata, Y. Fukuda, and S. Sameshima, "Preparation of Titania-Coated Alumina Film by Electrolysis of $TiCl_4$-Containing Solution", *Ceram. Inter.*, **35**, 693-698 (2009).

[5]Y. Chen, X. Zhou, X. Zhao, X. He, and X. Gu, "Crystallite Structure, Surface Morphology and Optical Properties of $In_2O_3$-$TiO_2$ Composite Thin Films by Sol-Gel Method", *Mater. Sci. Eng. B*, **151**, 179-186 (2008).

[6]M. Kaur, N. Jain, K. Sharma, S. Bhattacharya, M. Roy, A.K. Tyagi, S.K. Gupta, and J.V. Yakhmi, "Room-Temperature $H_2S$ Gas Sensing at ppb Level by Single Crystal $In_2O_3$ Whiskers", *Sens. Actuators B*, **133**, 456-461 (2008).

[7]K.N. Yu, Y. Xiong, Y. Liu, and C. Xiong, "Microstructure Change of Nano-$SnO_2$ Grain Assemblages with the Annealing Temperature", *Phys. Rev.* B, **55**, 2666-2671 (1997).

[8]J.F. Scott, "Raman Spectrum of $SnO_2$", *J. Chem. Phys.*, **53**, 852-853 (1970).

[9]A. Diéguez, A.R. Rodríguez, A. Vilà, and J.R. Morante, The Complete Raman Spectrum of Nanometric $SnO_2$ Particles, *J. Appl. Phys.*, **90**, 1550-1557 (2001).

[10]R.J. Betsch, H.L. Park, and W.B. White, "Raman Spectra of Stoichometric and Defect Rutile", *Mater. Res. Bull.*, **26**, 613-622 (1991).

[11]T. Ohsaka, F. Izumi, and Y. Fujiki, "Raman Spectrum of Anatase, $TiO_2$", *J. Raman Spectrosc.*, **7**,

321-324 (1978).
[12]S.P.S. Porto, P.A. Fleury, and T.C. Damen, "Raman Spectra of $TiO_2$, $MgF_2$, $ZnF_2$, $FeF_2$, and $MnF_2$", *Phys. Rev.*, **154**, 522-526 (1967).
[13]NIST X-ray Photoelectron Spectroscopy Database 20, Version 3.5.

# Materials Design

# PREDICTING THE RELIABILITY OF BRITTLE MATERIALS

S.W. Freiman
Freiman Consulting
Potomac, MD

Jeffrey Fong
Department of Mechanical Eng. And Mechanics
Drexel University
Philadelphia, PA

## ABSTRACT

All ceramics as well as most semiconducting materials are brittle. In addition, most of these materials are susceptible to environmentally enhanced crack growth, leading to possible delayed failure over time. This paper describes the fracture mechanics background, testing procedures, and statistical analysis techniques that can be used to confidently predict safe lifetimes of such brittle materials under stress in crack growth enhancing environments, usually water.

## INTRODUCTION

Designing ceramic components for safe operation over long periods of time involves issues of both the statistical nature of their fracture and the fact that most ceramics undergo slow crack growth in moisture containing environments. As a further complication, stresses can arise in components, not just due to mechanical loads, but also because of thermal gradients, phase transformations, or the presence of electric fields.

This paper outlines the fracture mechanics expressions, experimental procedures, and statistical techniques needed to determine the mechanical reliability of brittle materials. As an example, data taken from a previous publication[1] describing design procedures for all-glass aircraft windows is presented, but now makes use of more modern statistical techniques to determine the overall uncertainty in lifetime prediction for the windows. A more detailed description of the statistical procedures is being prepared for publication[2].

## EXPRESSION FOR PREDICTED LIFETIME

A power law expression (Equation 1) is typically used to relate the crack growth rate, $V$ to the stress intensity factor, $K_I$.

$$V = V_O \left[ \frac{K_I}{K_{IC}} \right]^N \quad (1)$$

$V_O$ is a material-dependent constant, and $N$ is the environmentally enhanced crack growth parameter, ranging from around 15 for glasses to 50 or greater for ceramics such as silicon nitride and silicon carbide. $K_{IC}$ is the critical stress intensity factor, used here as a scaling parameter.

We also use the fact that:

$$K_I = Y\sigma a^{\frac{1}{2}} \tag{2}$$

where σ is the far-field stress on the flaw, $a$ is the flaw depth in the surface, and $Y$ is a geometric constant dependent on flaw geometry and location.

Failure in brittle materials typically begins from surface flaws caused by grinding, machining, or some form of handling. It is also reasonable to assume that there will be a local residual stress associated with each flaw generated by the way they were introduced[3], Therefore Equation (1) is modified to the following:

$$V = V_O\left[\frac{K_I}{K_{IC}}\right]^{N'} \tag{3}$$

$N' = (3N + 2) / 4$ for point-type flaws[3] which will be the most prevalent.

Given these expressions plus knowledge of the local stress field, an expression for the time-to-failure, $t_f$, in a part can be derived[3]:

$$t_f = \frac{\lambda}{N'+1}\left(\frac{S}{S_v}\right)^{N'-2}\sigma^{-N'} \tag{4}$$

where $S$ *is* the strength of the as-finished material; $S_v$ is the strength of an indented specimen of the material. $\lambda$, and $N'$, are constants determined through stressing rate, i.e. "dynamic fatigue" tests (Equation 5), and σ is the stress on the part:

$$\sigma_f^{N'-1} = \lambda\frac{d\sigma}{dt} \tag{5}$$

It is important to carry out all of the test procedures used to determine the parameters in Equation 4 in such a manner so as to minimize errors and uncertainties, as described in the following section.

## TESTING PROCEDURES

The initial strength of a part, S, is based on the flaw distribution just before it is put into service. In carrying out a reliability analysis, a key assumption is that no, more severe, flaws than those in the initial flaw distribution are introduced during operation, and that flaws grow only due to environmentally enhanced crack growth.

While it would be beneficial to measure the strength of the actual part, this requirement often poses a difficult engineering problem as well as an economic constraint. Consequently, tests are conducted on small pieces of the same material. It is essential that the processing history and surface treatment of these specimens be identical to that in the part if valid predictions are to be made.

Because of the propensity of brittle materials to fail from surface flaws, flexural tests (either uniaxial or biaxial) are used. The decision of whether to conduct uniaxial or biaxial tests is usually based on whether edges in the actual part will be subject to significant stress. To minimize edge failures in uniaxial flexural specimens, edges should be rounded or chamfered (as should the edges of the part). In either case, measurements must be conducted at a high loading rate and in an inert environment, e.g., a dry gaseous nitrogen environment, in order to eliminate slow crack growth effects. If a component is likely to contain internal flaws such as pores or inclusions, tensile tests may be required. A minimum of 30 tests is required to provide adequate statistics.

The dynamic fatigue tests used to obtain the crack growth parameters make use of indented rather than as-finished specimens because as noted earlier the indentations provide a more realistic and conservative value of the environmental exponent for the actual flaws. The tests should be carried out in the harshest environment that will be seen by the part, usually liquid water. In order to obtain accurate values for N', it is necessary to take the dynamic fatigue data over as wide a stressing rate as possible, but certainly not less than three orders of magnitude.

Because indented specimens were used to determine N′, measurements of the strength of such specimens are needed to calculate $S_V$. Again, a minimum of 30 tests must be carried out at a high loading rate in an inert environment.

The stress state in the component can be obtained using finite element analysis.

Even when tests are carried out in an optimal manner, there will be scatter in the data obtained from each of the procedures. These uncertainties must be combined to develop an overall confidence in the calculated time to failure.

## COBINING MEASUREMENT UNCERTAINTIES

From Equation (4), we note that the time to failure, $t_f$ , is a function of five parameters, namely, $\lambda$ , $N'$, $S$ , $S_v$ , and $\sigma$

A formula for the variance of the time to failure, *var* ($t_f$), can be derived using established error propagation formulas[4] as a function of the variances of those five parameters, namely, *var* ($\lambda$), *var* ($N'$), *var* ($S$), *var* ($S_v$), and *var* ($\sigma$).

$$s.d.(t_f) = t_f\sqrt{\frac{\text{var}(\lambda)}{\lambda^2}+\frac{\text{var}(N')}{(N'+1)^2}+(N'-2)^2*\left[\frac{\text{var}(S)}{S^2}+\frac{\text{var}(S_v)}{(S_v)^2}\right]+(N')^2*\frac{\text{var}(\sigma)}{\sigma^2}} \tag{6}$$

Equation 6 can be used to ascertain the uncertainty in the lifetime prediction of both the specimens as well as larger parts manufactured from the same material.

## CASE STUDY

The experimental data in Reference 1 consisted of three sets: 1) 31 as-manufactured borosilicate (BK7) glass discs tested rapidly in biaxial flexure in dry $N_2$ gas, 2) 35 indented biaxial discs tested in water as a function of stressing rate, and 3) 33 indented glass discs tested the same as Set 1.

In determining the initial strength distribution, S, we made use of a publicly available software program called "dataplot"[5]. The 31 data points in the initial strength set were entered into the program. Dataplot fits the data to a Weibull-type expression, allowing the exponent, location, and scale to vary. The difference between the various curves and the actual data is called the "probability plot correlation coefficient" (ppcc)[6-8]. The ppcc for each of 50 different exponents (here called gamma) is shown in Figure 1.

A PPCC plot as described in the Dataplot program is a graphical data analysis technique for determining the "best" distributional fit to the data, and is an alternative to the maximum likelihood method. The PPCC plot can be used to estimate the shape parameter of a Weibull distribution. After finding the best value of the shape parameter, the probability plot can be used

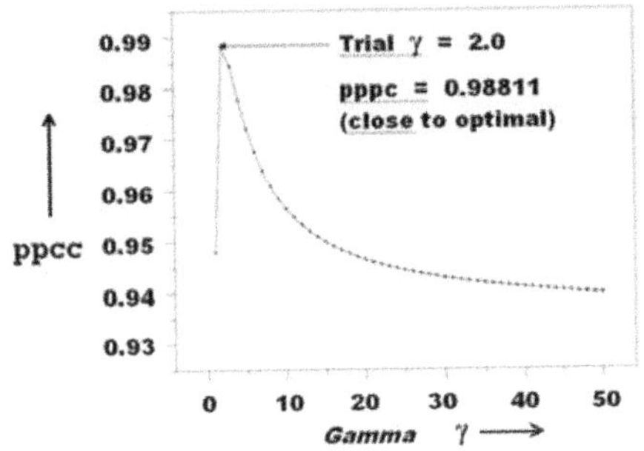


Figure 1: Plot of ppcc (probability plot correlation coefficient) versus $\gamma$ of a 3-parameter Weibull plot with $\gamma = 2.0$, which is close to optimal.

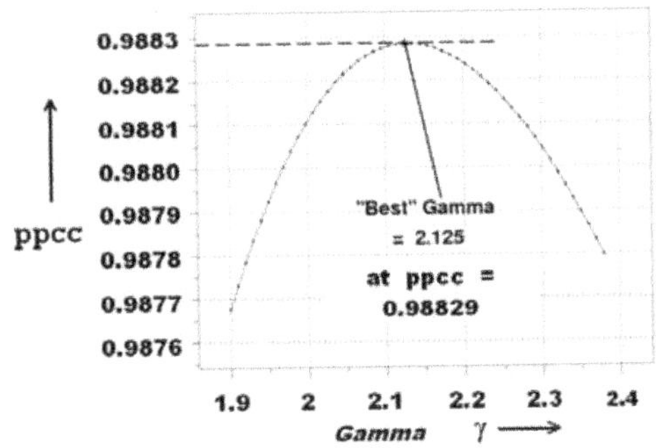


Figure 2: Plot of ppcc at its optimum $\gamma$ of 2.125.

to estimate the location and scale parameters of a distribution. The resulting PPCC plot thus consists of: vertical axis which is the probability plot correlation coefficient value and a horizontal axis, the distributional family parameter value. The value of the distributional parameter (on the horizontal axis) which corresponds to the maximum of the PPCC plot curve (on the vertical axis) indicates the best-fit member of the family.

The process can be refined to obtain the optimal $\gamma$ of 2.125 as shown in Figure 2.

Figure 3 shows the selected 3-paramenter Weibull distribution. The lowest strength in the distribution, seen in the plot as ppa0, is given as 15.95 ksi, with a standard deviation (s.d.) = 0.472. The 95% confidence limits are then twice the s.d., or ±.0.94.

The dynamic fatigue data are plotted in Figure 4. The crack growth parameters determined from this fit are $N' = 1/C = 15.382$ (s.d. = 0.286), and $\lambda = 10^{A} = 47.461$ (s.d. = 0.076).

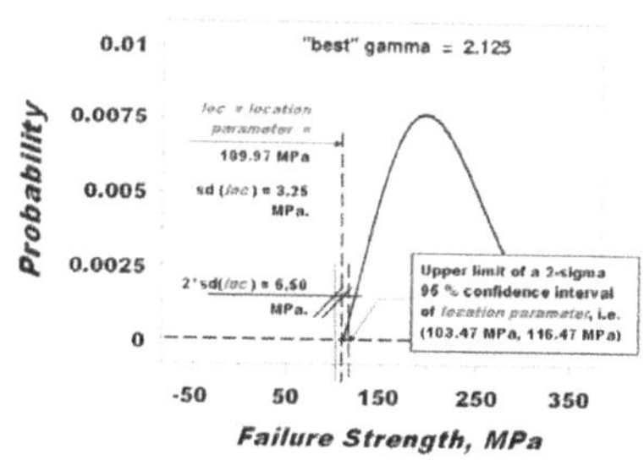


Figure 3: Probability vs. Failure Strength Plot for a 3-parameter Weibull distribution fit of a 31-sample data set at the optimal $\gamma$.

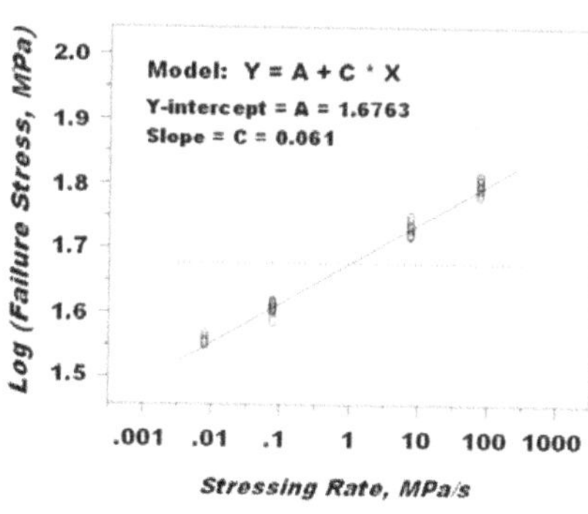


Figure 4: Dynamic fatigue data for indented glass discs

Figure 5 shows the plot of the reference data set. The ppa0 of this set is 14.339 (s.d. = 0.119).

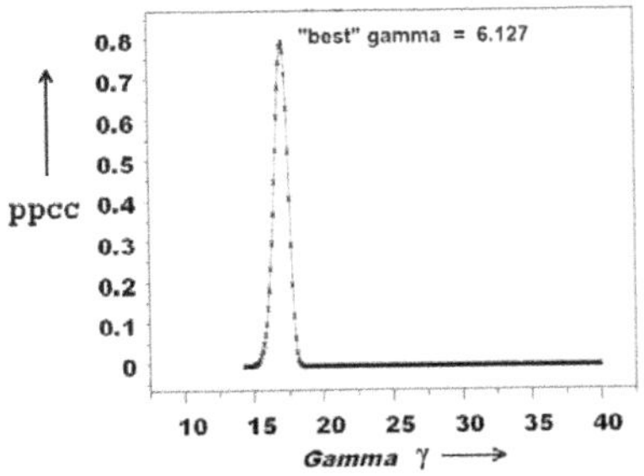


Figure 5: Weibull strength distribution for indented glass discs

From Equation 6 we obtain:

$$s.d.(\ t_f) = t_f\sqrt{0.000003 + 0.000346 + 0.156827 + 0.012333 + 236.6 * \frac{\mathrm{var}(\sigma)}{\sigma^2}} \tag{7}$$

We can derive two simplified formulas:

Case 1: Since $var(\lambda)$, $var(N')$, and $var(S_v)$ are negligible compared to the $var(S)$, and assuming that the $var(\sigma)$ is also negligibly small, we obtain:

$$s.d.(\ t_f)\ =\ (\ N' - 2\ )\ *\ (\ t_f\ /\ S\ )\ *\ s.d.(\ S\ ). \tag{8}$$

Case 2: However, if there is uncertainty regarding the finite element analysis, then:

$$s.d.(\ t_f)\ =\ t_f\ \sqrt{(N'-2)^2\frac{\mathrm{var}(S)}{S^2} + (N')^2\frac{\mathrm{var}(\sigma)}{\sigma^2}} \tag{9}$$

Depending on the uncertainty in $\sigma$, the s.d. in $t_f$ will be larger.

We must now also recognize that this predicted time to failure, $t_f$, applies to an object whose stressed area corresponds to that of a flexural disc. Since the stressed area of a window is roughly 50 times this, we must account for the greater probability of failure in the window itself. This is accomplished through the statistical concept known as "coverage"[9], which allows us to calculated tolerance limits of an entire population from knowledge of that applicable to a particular data set. Tolerance factors can be obtained from established tables of such data.

When the possibility of failure at very short times is unacceptably large, there are some alternatives. The parameters $N'$ and $\lambda$ are fixed by the choice of material and the environment; and $S_v$ is an artificial data set, so only changes in S will affect the result. If S can be increased or the s.d.(S) can be reduced, the confidence limits can be improved. However, there is another possibility for reducing the failure probability, namely by increasing the size of the data set. We have shown[2] that if four artificial data points are added to the Weibull distribution in Figure 3, the tolerance limits can be increased. Even more conservative values could be obtained with further enlargements in the data set.

## SUMMARY AND CONCLUSIONS

In this paper we have tried to outline how one can confidently predict the reliability of brittle materials through the use of proper testing and statistical procedures. Designs must ultimately be based on making sure that the part serves its function in an efficient manner while still maintaining structural integrity.

## REFERENCES

[1] E. R. Fuller, Jr., S. W. Freiman, J. B. Quinn, G. D. Quinn, and W. C. Carter, "Fracture Mechanics Approach to the Design of Glass Aircraft Windows: A Case Study", *Proc. Conf., SPIE - The International Society for Optical Engineering*, 26-28 July 1994, San Diego, CA, **2286**, 419-430 (1994).

[2] J.T. Fong, J. J. Filliben, N. A. Heckert, and S.W. Freiman, "Uncertainty Analysis of Time-to-Failure of Brittle Materials Based on Experimental Data and Error Propagation Formulas", in preparation for publication

[3] E. R, Fuller, B. R. Lawn, and R. F. Cook, "Theory of Fatigue for Brittle Flaws Originating from Residual Stress Concentrations", J. Am. Ceram. Soc., **66**, 314-21 (1983).

[4] H. H. Ku, "Notes on the Use of Propagation of Error Formulas", *NBS Journal of Research*, **70C**, No. 4,. 263-273 (1966).

[5] J. J. Filliben, N. A. Heckert, N. A., *Dataplot: A Statistical Data Analysis Software System,* A Public Domain Software by National Institute of Standards and Technology (NIST), Gaithersburg, MD 20899, http://www.itl.nist.gov/div898/software/dataplot.html (2002).

[6] J. J. Filliben, "The Probability Plot Correlation Coefficient Test for Normality", *Technometrics*, **17**, No. 1, 111-17 (1975).

[7] Fong, J. T., and Filliben, J. J., "A Data Analysis Methodology as Applied to the PVRC Round Robin NDE Test Program", ASME Special Publication NDE-1, pp. 183-222 (1986).

[8] Fong, J. T., Filliben, J. J., Heckert, N. A., and Freiman, S. W., "An Engineering Reliability Mehodology with Uncertainty Estimation of Time-to-Failure of Brittle Materials under Stress in Crack Growth Enhancing Environments", manuscript in preparation for NIST Journal of Research, Jan. 2010 (contact fong@nist.gov for .pdf file).

[9] P.R. Nelson, M. Coffin, and K.A. Copeland, *Introductory Statistics for Engineering Experimentation*. Elsevier (2003)

# EFFECTIVE METHODS TO DETERMINE HARDNESS AND TOUGHNESS OF STRUCTURAL CERAMICS

R Morrell, National Physical Laboratory, Teddington, UK, TW11 0LW

## ABSTRACT

In the accepted senses of the descriptors 'hardness' and 'toughness', ceramic materials are hard, but not tough. However, their applications in the modern engineering world derive from both hardness and toughness, but not in ways that are fully understood. This paper reviews the current measurement good practice for these properties and how these methods relate to performance in applications. Four particular cases are considered – orthopaedic components, metal-forming tools, bearing elements, and ballistic armour – representing key markets for advanced technical ceramics. In orthopaedic components, lifetime is generally governed by toughness, while hardness tends to have more influence on counterfaces than on the ceramic itself. In metal forming tools and in bearing elements, toughness plays a dominant role in resisting damage, rather than hardness. In contrast, in ballistic armour there is as yet no clear correlation between any conventional measures of hardness and toughness and resistance to projectile penetration. Prediction of potential performance of new materials or in new applications remains difficult and uncertain.

## INTRODUCTION

Compared with most metallic materials, ceramics offer high hardness but poor toughness, so the advantageous aspect for many mechanical applications is counterbalanced by a characteristic that has to be handled carefully through material optimisation and component design. While neither hardness nor toughness can be used as a design parameter, there remains a strong driver to ensure that the material delivers the expected performance, and hence there is a strong need to characterise hardness and toughness for quality assurance purposes.

## HARDNESS MEASUREMENT

The classical method of hardness measurement is to create an indentation in a material using a diamond indenter and a known indentation force. This has long been standardised for metals using a variety of indenter types: Brinel (spherical), Rockwell (spherically ended cone), Vickers (pyramidal), and Knoop (elongated pyramid). In more recent times the advent of nanoscale indentation has seen the use of Berkovitch (triangular pyramid) and cube-corner indenters. It has therefore been tempting to employ such methods to ceramic materials, but the principal issue arising is that unlike in metals the deformation under the indenter is seldom purely plastic, is inhomogeneous, and can lead to fragmentation and the development of an associated crack pattern. In order to reduce the extent of cracking it is necessary to operate at rather lower force levels than is conventionally used for metals. For example, Vickers indentation at a force derived from a mass of 30 kg (HV30) typically used for metals would produce an unreadable crater. In some materials, the indentation diagonal becomes clearly measurable only at HV10 or HV5, but there is still a strong risk of obtaining unacceptable indentations. Clearly, a much lower force (microhardness) could be employed, but as is well-known the scatter of measurement rises as the indentation covers less and less representative microstructure, and the relative measurement uncertainties become larger because of the limits of optical resolution. Consequently, a compromise has to be sought, and HV1 is the recommended level for most ceramic materials (at 1500 HV1 an indentation diagonal length of typically 20 μm usually covers representative microstructure), or even HV0.5 and HV0.2 if the material is particularly brittle, such as boron carbide or some silicon carbides.

Cracking is reduced if a Knoop indenter with its much shallower penetration is used, and the maximum recommended force is derived from a mass of 2 kg (HK2). In practice it has been found that HV and HK values have similar uncertainties.

These force levels were first embodied in standardised methods for ceramics in ASTM C1327 (HV) and C1326 (HK) in 1996, and at about the same time in CEN EN843-4, both of which are now in line with ISO 18756. It should be noted that these standards recommend a departure from metals practice with the use preferred units of GPa, rather than a hardness number with kgf/mm$^2$ implied.

A further feature of hardness testing of ceramics is an indentation-force dependence of hardness number which is usually much more marked than in metals, especially at the microhardness level. Although the existence of this as a real effect has been questioned, recent work using glass has shown that while cracking contributes to apparent reducing hardness, there is still an effect when no cracking occurs[1-2]. It is therefore absolutely essential that material comparisons are made using the same indenter type and indentation force.

A further issue for the reproducibility of measurement centres on the eyesight of the individual making the measurement. The recommended levels of force result in HV indentations which may be only a 20-30 μm diagonal length, and in HK indentations which may be ~100 μm long diagonal length. Errors of a few micrometres simply due to visual acuity and the criterion used to position microscope cross-wires at the correct place can lead to significant deviations between operators. The latter point has had to be explicitly covered in the standards. Further, it is recommended that a reference block of similar hardness is used in order to ensure that the eye is fully attuned to correct practice[*]. It can therefore be appreciated that although it is a superficially easy measurement to make, conventional hardness is not a good discriminator of materials. Attempts to use the approach of depth-sensing hardness to eliminate operator bias have also been tried, but with similar issues of scatter[3].

## TOUGHNESS MEASUREMENT

The fracture toughness of advanced ceramic materials is low in general materials terms, but important for material ranking as well as fracture diagnosis. Historically, there has been a variety of approaches based on test methods for metals, but it is only within the past decade that methods based on flexural strength test-pieces have been fully evaluated and converted into formally standardised methods[4]. The general approach has been the determination of a 'fast-fracture' toughness, i.e. under conditions when the crack is running quickly but in a controlled way from a sharp pre-crack. This condition arises in the single-edge pre-cracked beam (SEPB) method (ASTM C1421, EN ISO 15732, starting crack of length ~1.5 mm), and in the surface crack in flexure (SCF) method (ASTM C1421, EN ISO 18756, starting crack semielliptical ~ 100 μm half-axis). In the chevron notch beam (CNB) method (ASTM C1421, EN 14425-3, ISO 24370), the crack should develop controllably from the sharp tip of the remaining material chevron after notching, and grow to a length of ~1 mm at the point of peak force application. More recently, the single-edge vee-notched (SEVNB) beam method has been introduced, in which a sharpened tip to the notch is developed using a razor blade polishing technique (ISO 23146), considered to approximate to a sharp crack in may materials.

In fine-grained materials that do not suffer from subcritical crack growth or so-called R-curve (rising crack resistance with increasing crack length) effects, these tests have been demonstrated to give closely equivalent results[5]. However, differences can arise if the material shows R-curve behaviour, since different propagation distances are involved. It should also be noted that for various reasons not all test types work effectively on all material types, e.g. the SCF method is ineffective when the grain size approaches the pre-crack size.

Methods that have been explicitly excluded from formal standardisation are the notched beam (the notch tip is usually too blunt), double torsion (mixed mode stress intensity), indentation fracture (quasi-static crack lengths, uncertain fracture mechanics)[6], and indentation strength (residual stresses, uncertain fracture mechanics).

One of the challenging factors for engineers and materials scientists is understanding the relevance of long-crack toughness to materials performance issues, such as contact damage, machining damage and subcritical crack growth. The scale of such fracture initiating flaws may be a factor 10 or more smaller than even the SCF precracks. It is often noted that fracture initiates from microstructural abnormalities which may have lower fracture toughness associated with them.

## SECTORAL RELEVANCE

### Orthopaedic and dental applications

There are strict controls on the compositional types of materials that are appropriate for orthopaedic use which means that in general the properties are fairly well defined by meeting compositional, density and grain size criteria (ISO 6474-1, ISO DIS 6474-2, ISO 13354). Regular measurement of hardness or toughness can be used for quality assurance, and have been outside the outside the scope of material specifications, but are being introduced. While the toughness of high-purity fine-grained alumina can readily be measured by a variety of methods achieving equivalent numerical results (Figure 1), the same cannot be said for Y-TZP. This material continues to pose problems for its correct characterisation. Principally these problems arise because of incorrect application of standardised procedures, or of the use of non- standardised methods. The academic community is the main guilty party by continuing to cite the discredited indentation fracture method. There are numerous instances in the literature where

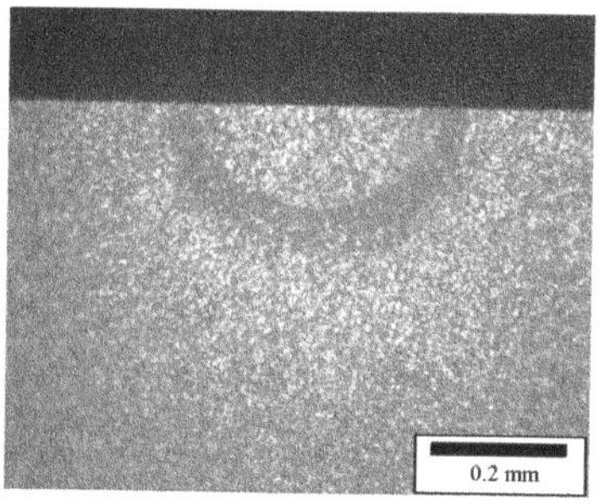


Figure 1: SCF method HK10 pre-crack in a high-purity fine-grained alumina test-piece after fracture. The pop-in crack undergoes subcritical crack growth when stressed, giving the dark halo, which has to be taken into account in measurement of pre-crack dimensions.

unrealistically high values are claimed even for baseline 3Y-TZP. Only the SEPB and modified SCF methods seem to give good results for sharp pre-cracks. The increasing use of zirconia strengthened alumina for orthopaedic applications is based on supposed improvements in toughness and resistance to subcritical crack growth compared with alumina alone, and initial indications suggest that SEVNB and SCF are appropriate test methods for this material class[7].

Certainly, by comparison with other materials, small improvements in fast fracture toughness can be viewed as somewhat insignificant compared with the suggested benefits of reducing subcritical crack growth in use, and potentially premature failure. More important for the manufacturer and orthopaedic supplier is control of the quality of machining and achievement of the design geometry than the intrinsic fast fracture toughness[8].

Dental materials based on glassy porcelains, either in bulk or as veneers on stronger materials such as 3Y-TZP, generally have poor toughness and are prone to damage by chipping. Conventional fracture toughness measurement is not particularly useful, but some efforts have been made to study the chipping process as a method of assessing relative behaviour, and can be viewed as a way of improving understanding the role of edge design[9]. The relationship between edge chipping resistance and bulk toughness with which it is correlated still requires research, different authors finding different relationships with different types of indenter.

Cutting and forming tools

Ceramic cutting tools find niche applications where the properties of coated cemented carbides are inadequate, such as in machining cast iron vehicle parts. Although many suitable ceramics are harder than cemented carbides, they are clearly less tough, and have to be used with care in continuous cutting operations. Under interrupted cutting their behaviour is generally unreliable. Used under conditions which avoid fracture, the life of a cutting tool is principally determined by the chemical wear process that takes place of the tip[10], and that is controlled by the chemical species present, rather than by any mechanical property or microstructural detail[11]. Interactions between the work-piece, the cutting tool surface and the atmosphere lead to erosion of the tool edge, requiring greater input power to continue machining, and in bad cases leading to notching and fracture.

Key requirements are therefore high bulk strength to withstand the clamping and cutting forces, good resistance to deformation at high cutting tip temperatures, and high fracture toughness to minimise the risks of chipping. (The study of edge flaking was initiated by issues of cutting and forming tool damage.) A high toughness also makes the strength less sensitive to process defect size.

Conventionally measured strength and fast fracture toughness are thus relevant comparative parameters for cutting tool materials, although as far as is known there are no standards or specifications in existence for ceramic cutting tools. As with dental materials, characterising the risk of damage by chipping is useful, and simple edge indentation tests have been developed[12-13].

In forming tools, contact forces are generally compressive, but high tensile stress concentrations can occur particularly at contact peripheries. Fracture toughness plays a key role in restricting the initiation and propagation of pop-in cracks, but under severe conditions the design of the system plays a dominant role. Application of compressive stresses across likely pop-in crack paths seems to be a crucial aspect for achieving reliable behaviour, as exemplified by a recent study of silicon nitride wire-forming rolls[14].

Bearing balls and rollers

The performance of rolling bearing elements, balls and rollers, is crucially controlled by the nature of surface defects (Figure 2) and their resistance to propagation under cyclic Hertzian loading. Specifications for bearing materials have evolved in ASTM and ISO for the most commonly used material, silicon nitride, in which fracture toughness is a key quality control parameter. Optimal rolling contact fatigue performance is obtained for certain microstructures which exhibit elongated grains, but with an overall fine grain size, and toughness measurement can be used to screen potentially suitable materials. However, standardised methods of toughness measurement require flexural test-pieces which are difficult to machine from balls, and in any event would reflect the interior of the ball, rather than the near-surface microstructure, which may be different as a consequence of atmosphere effects during material firing or hot isostatic pressing. The alternative option that is currently permitted according to the material specification standards (ASTM F2094, ISO 26602) is to make the test on test-pieces cut from specially prepared plates, but these may not be representative of the process route for balls.

To overcome this problem, the bearing community has resorted to the indentation fracture method to provide a measure described as 'indentation fracture resistance' to avoid any unrealistic comparisons with a fast fracture toughness. Although it is recognised that this is a subjective test, it currently provides the only viable method for independently determining a measure of toughness on small balls. At present there is no formalised method, but one is planned in ISO. Specifications on surface preparation, indentation force and the method for observing cracks may well be required to ensure reproducibility, and a round-robin on representative products would be a valuable exercise to ensure this.

It may in the future be possible to adapt the SCF method to the recently developed 'C-ball'[15] or 'notched-ball'[16] type of compression test, but the technique needs some development before it can be considered to be robust.

Ceramic armour

The properties needed for good performance of ceramic armour remain unclear. While hardness and toughness are thought to play a key role in disrupting the incoming projectile, the situation is much more complicated. The only meaningful correlation appears to be that the tougher the ceramic, the less fragmentation there is away from the impact site, and the better chance the panel has of surviving a second hit. Furthermore, such conventional

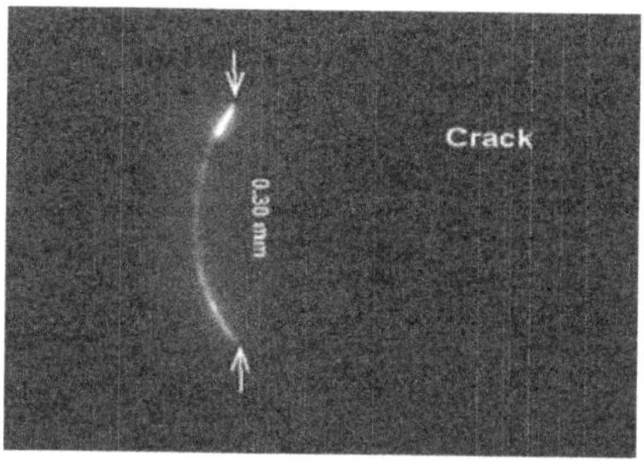


Figure 2: Partial Hertzian ring crack in a production ball bearing observed using dye penetrant.

Determinations of hardness and toughness at room temperature seem to bear little resemblance to the varied conditions of ballistic impact where high temperatures and pressures can be developed and elastic limits exceeded, but remain the only simple methods available to link performance with composition and microstructure.

Some recent unpublished work in the UK studied the effects of using very high force indentation to see if materials with similar room-temperature behaviour but different ballistic behaviour could be distinguished. By using the depth-sensing method with a polycrystalline diamond Vickers penetrator it could be seen that some materials showed continuous penetration, while others, notably $B_4C$ and TiC, showed discontinuous behaviour consequent upon multiple localised fracture (Figure 3). The extension of this approach to high temperatures and high loading rates may provide a fruitful route for the future as an alternative to the split Hopkinson bar method for generating the high impact pressures.

CONCLUDING REMARKS

Numerical values of hardness and fracture toughness determined under carefully defined and controlled conditions are of little direct value in many of the applications of ceramics, but are used mainly for material selection and quality assessment, with increasing use being made of such tests in material minimum specifications. Fracture toughness values can also be used with caution in fractographic assessment of failures. However, more generally it appears that more complex sets of conditions affect material suitability and component survival in applications in which ceramics can be used advantageously. This situation continues to spur research to find and fully understand alternative economical testing methods to rank materials for these more complex conditions, such as an edge chipping test, or a ball strength test, or when there is a need to account for temperature, time and environmental effects. There is also a need to understand better crack initiation and growth under service conditions to better understand delayed failure and fatigue effects. But over and above such testing there is no doubt that control of microstructural defects and minimizing machining damage are the keys to reliability for a majority of successful applications, since these define fracture strength and hence the forces that a component will withstand.

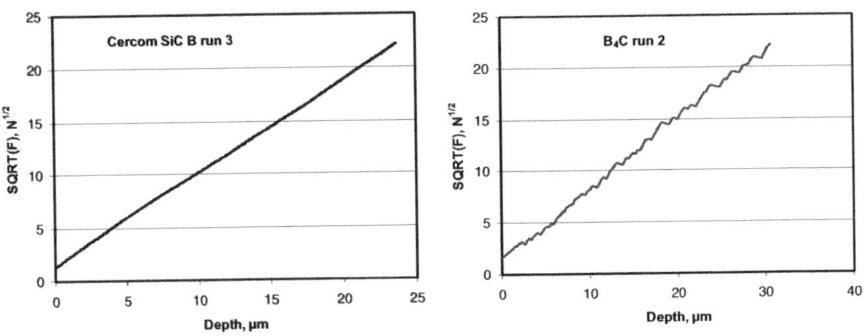


Figure 3: High-force depth-sensing hardness: plots of $\sqrt{F}$ *vs. h* for (left) Cercom SiC B with smooth behaviour, and (right) Ceradyne $B_4C$ with discontinuous behaviour.

ACKNOWLEDGMENTS

Thanks are due to numerous collaborations with organisations in the four above-cited fields, especially noting Morgan Advanced Ceramics, Metalloceramica Vanzetti, Montan-universität Leoben, SKF, University of Wolverhampton and Advanced Defence Materials Ltd.

REFERENCES

1. J.B. Quinn and G.D. Quinn, 'On the hardness and brittleness of ceramics', *Key Engg. Mater.*, 1997, 132-6, 460-463.
2. G.D Quinn, 'Indentation size effect for glasses – Yes, there is a fracture contribution', *Fracture Mechanics of Ceramics - Active Materials, Nanoscale Materials, Composites, Glass and Fundamentals*, edited by Bradt, R.C., White, K.W., Munz, D., Sakai M., Springer, 2005, pp.149-171.

3. C. Ullner, J. Beckmann, and R. Morrell, 'Instrumented indentation test on advanced technical ceramics', *J. Eur. Ceram. Soc.*, 2002, 22(8), 1183-1189.
4. R. Morrell, 'Fracture toughness testing for advanced technical ceramics – internationally agreed good practice', *Adv. in Applied Ceram.*, 2006, 105(2), 1-11.
5. G.D. Quinn, R. Gettings, and K Xu, 'Standard Reference Material 2100: Ceramic fracture toughness', *Cer. Eng. Sci. Proc.*, 1999, 20(3), 513-23.
6. R. Morrell, 'Fractography and fracture toughness measurement', *Key Engg. Mater.* 2009, 409, 17-27.
7. G.D. Quinn and R.C. Bradt, 'On the Vickers indentation toughness test', *J. Amer. Ceram. Soc.*, 2007, 90(3), 673-80.
8. R. Morrell and S. Hughes, 'Factors influencing the reliability of ceramic femoral components', *J. Eur. Ceram Soc.*, 2007, 27, 1535-41.
9. J.B. Quinn and I.K. Lloyd, 'Flake and scratch size ratios in ceramics', *Ceram. Trans.*, 2001, 112, 57-72.
10. R. Edwards, 'Cutting tools', Institute of Materials, London, 1993.
11 A. Ferrari, G. Schwier, R. Morrell, R. Calzavarini, and S. Martelli, 'Development and testing of nanocomposite cutting tool materials based on silicon nitride', *Silicates Industrielles*, 2004, 69(7-8), 261-7.
12. N.J. McCormick and E.A. Almond, 'Edge flaking of brittle materials', *J. Hard Mater.*, 1990, 1(1) 25–51.
13. R. Morrell, and A Gant, 'Edge chipping – what does it tell us?', *Ceram. Trans.*, 2001, 122, 23-42.
14. M. Lengauer and R. Danzer, 'Silicon nitride tools for the hot rolling of high-alloyed steel and superalloy wires – Crack growth and lifetime prediction', *J. Eur. Ceram. Soc.*, 2008, 28, 2289-98.
15. A.A. Wereszczak, T.P. Kirkland, and O.M. Jadaan, 'Strength measurement of ceramic spheres using a diametrally compressed 'C-sphere' specimen', *J. Am. Ceram. Soc.*, 2007, 90, 1843-1849.
16. P. Supancic, R. Danzer, S. Witschnig, E. Polaczek, and R. Morrell, 'A new test method to determine tensile strength of brittle balls – the notched ball test', *J. Eur. Ceram. Soc.*, 2009, 29(12), 2447-59.

* SRM 2831, silicon nitride Knoop hardness reference material is available from NIST, Gaithersburg, USA

# DESIGN AND RELIABILITY OF LAYERED CERAMICS

Y.N. Kim[1], H.G. Shin[1], J.K. Song[1], M.C. Shin[2], B.K. Shin[2], T.M. Noh[2], and H.S. Lee[2]
[1]Machinery & Material Testing Center, Korea Testing Laboratory, Seoul 152-718, KOREA
[2]School of Materials Science & Engineering, Pusan National University, Busan 609-735, KOREA

## ABSTRACT

In the present paper, two case studies are discussed on the reliability assessment and improvement of layered ceramics for catalytic ceramic filters and conductive ceramic coatings. The reliability assessment and improvement cover characterization, durability tests, failure analysis (failure mode and mechanism), lifetime prediction, feedback research, robust design, etc. The lifetime predictions were carried out through accelerated tests, statistical modeling, and a finite element analysis (FEA) simulation based on the physics of failure. As the final step, feedback research was applied to develop more robust designs.

## INTRODUCTION

Recently, new concept hybrid materials, generally comprised of different materials exhibiting strong chemical interactions between the components, have been introduced[1]. They have received increasing attention as materials of the future because of their hybrid functions retaining enhanced performance and hybrid microstructure, namely, chemical bonding between different classes of materials, such as ceramics, metals, and polymers. This non-uniform microstructure is obtained by hybridization of different crystallographic states (crystalline/amorphous/pore), size scales (nano/micro/macro), and various processing techniques. A number of hybrid materials have already been applied to numerous fields. Layer-structured ceramic, e.g. multilayer ceramic chip for display devices is one of the widespread applications of ceramic-based hybrid materials. As parts of a study on the hybrid layer-structured materials, we have investigated the design and reliability of catalytic ceramic filters and conductive ceramic coatings since the early 2000s, for simultaneous removal of dust and hazardous gas and for electronic devices, respectively.

Advanced ceramics are classified into three major categories: monolith ceramics, ceramic matrix composites (CMCs), and ceramic coating systems. They are wear-resistant, corrosion-resistant, light weight, and superior to many materials in high temperature stability. The combination of these properties provides great potential to resolve a wide range of today's material challenges in the field of processing industries, power generation, aerospace, transportation, and military applications. Despite this advantage, they have been used only in a small percentage of applications that could benefit from their capabilities. Many end-users are reluctant to utilize them because of concerns about insufficient reliability, higher costs than competing materials, insufficient design, and in some cases, inadequate properties to meet the needs of demanding applications[2].

Reliability is defined as the ability of materials, devices, or systems to perform the required functions under stated conditions for a specified period of time. It can elucidate the performance decay below an acceptable level and the failure modes that can lead to catastrophic failure. There have been recent attempts to assess the reliability of the materials that make up devices and systems because most failures generally originate from defects in the materials, mismatches between materials, etc. An important process to improve the performance of materials and to prolong their lifetime is a thorough understanding of their failure modes and mechanisms. A considerable amount of research has been conducted in recent years to develop analytical design and lifetime prediction methods for monolithic ceramics and ceramic matrix composites[3-4], whereas little has been done for layered ceramics as hybrid materials. Accordingly, as the applications of layered ceramics expand to meet the demands of increasingly harsh conditions and to extend lifetimes, efforts to establish better design and lifetime prediction capabilities must be enhanced.

In order to develop methods to evaluate and to enhance the durability and reliability of layered ceramics, existing reliability assessment tools have been modified to be suitable for catalytic ceramic filters and conductive ceramic coatings. The failure modes and mechanisms of layered ceramics were analyzed, and their lifetimes were predicted via durability tests and accelerated tests. The accelerated tests suitable for each layered ceramic have been designed in consideration of their characteristics, fabrication processes, service conditions, and field failures. In addition, the failure or degradation behavior of layered ceramics was simulated on the basis of the physics of failure.

The reliability assessment and improvement include characterization, durability test, failure analysis (failure mode and mechanism), lifetime prediction, feedback, robust design, etc. Specifically, this paper focuses on the lifetime prediction and the feedback research, in an attempt to improve the reliability of layered ceramics.

## EXPERIMENTAL PROCEDURE

A reliability assessment of layered ceramics was conducted under consecutive processes of durability tests, failure analysis, accelerated tests, and lifetime prediction. A reliability assessment and improvement of catalytic ceramic filters and conductive ceramic coatings (indium tin oxide, ITO) were carried out in the present study. Durability tests were performed to find the most influential factor degrading the performances of each of the materials under various stresses, such as temperature, thermal cycling, humidity, applied voltage, etc. Accelerated tests were carried out for the materials under their own major stress that causes a critical influence on their performance decay. Their failures were judged by the performance decay below acceptable levels of each material, and the failure data were acquired via the accelerated tests. The failure mechanisms were analyzed in each case, and the lifetimes were finally predicted using statistical analysis and an acceleration model. The lifetime of ITO coatings was derived by analyzing the diffusion behavior of oxygen in ITO films with FEA. In an attempt to improve the durability and reliability of ceramic filters and ITO thin films, feedback research and robust design as the final step were carried out with the results of failure analysis. In order to prolong their lifetime and to improve their reliability, inorganic compounds were added to ceramic filters during manufacturing processes, and ITO thin films were post-annealed in a reducing atmosphere, respectively.[5-8)]

## RESULTS AND DISCUSSION

### Catalytic ceramic filters

Catalytic ceramic filters for simultaneous removal of dust and hazardous gas (NOx) were studied[9]. They were manufactured with an apparent porosity of 58%, a compressive strength of 10 MPa, and a pressure drop of 1200 Pa at the face velocity of 5 cm/sec at 400°C. The effect of thermal shock and high temperature on the filtration layer and catalyst was observed through acceleration tests, which may induce a reduction in filter durability and catalyst efficiency, respectively.

It was found that the failure mode for ceramic filters was the decrease of pressure drop, and their failure mechanism was the generation and growth of cracks originated by thermal shock[8]. These results were similar to those from Eriksson's group[10]. To predict their lifetime, the variation in pressure drop was measured as a function of thermal shock temperatures and cycles (Figure 1). The lifetime was defined as the time for their pressure drop to decrease as much as 15% relative to that of the as-prepared samples. The Weibull distribution could be obtained from a linear regression analysis using the failure data of thermal shock test at $\Delta T$ = 500°C and $\Delta T$ = 600°C, respectively (Figure 2). The Weibull function is expressed as:

$$F(t) = 1 - \exp(-(t/\eta)\beta) \quad (1)$$

where, F(t) is the failure percentage, t is the time, $\eta$ is the characteristic lifetime and $\beta$ is the shape parameter. It was confirmed that there was accelerating relation between each stress level due to the similar shape parameter at each stress level. The shape parameter was about 1.74 at $\Delta T$ = 500°C and 1.86 at $\Delta T$ = 600°C, respectively. Mean time to failure of the filters calculated from the characteristic lifetimes was 161 cycles at $\Delta T$ = 500°C and 58 cycles at $\Delta T$ = 600°C, respectively. The acceleration factor (AF) between two stress levels could be calculated by Eq. 2 on the basis of the Coffin-Manson relationship[11-12].

$$AF = N_{f\ at\ \Delta T1}/N_{f\ at\ \Delta T2} = (C_0/\Delta T_1^{q})/(C_0/\Delta T_2^{q}) = (\Delta T_2/\Delta T_1)^{q} \qquad (2)$$

$C_0$ and q are material constants, $N_f$ is the mean number of thermal cycling to failure, $\Delta T$ is thermal shock temperature. The AF between $\Delta T$ = 300°C and $\Delta T$ = 500°C was calculated to be 18. This means that the effect of 18 cycles at $\Delta T$ = 300°C equals that of 1 cycle at $\Delta T$ = 500°C. If thermal shock temperature in normal operating condition is $\Delta T$ = 300°C, the lifetime of ceramic filters under normal operating conditions can be predicted to be 2913 cycles (about 1.5 years).

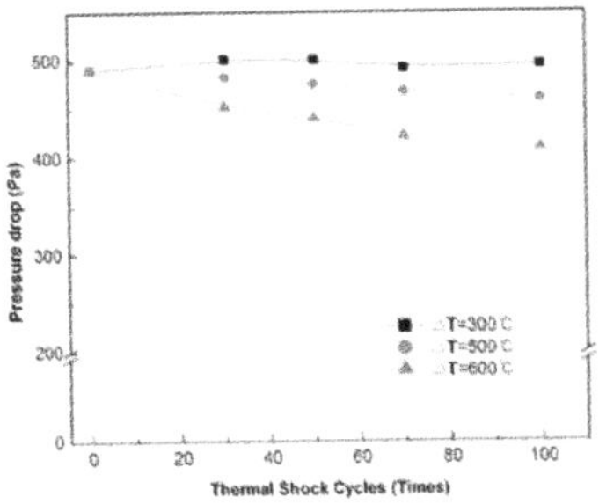


Figure 1. Pressure drop of the ceramic filters for dust removal after thermal shock cycle test.

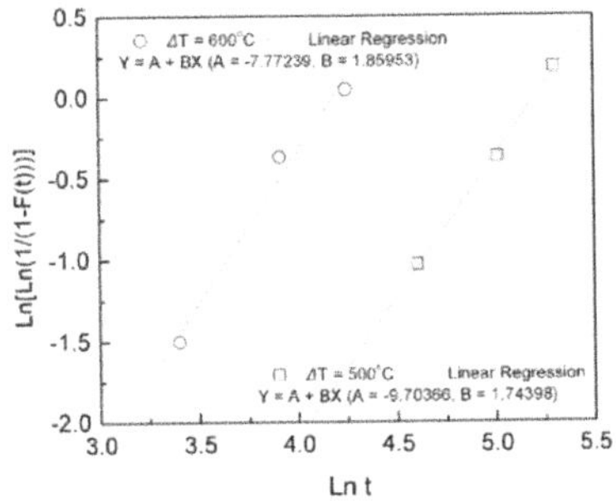


Figure 2. Weibull plot of thermal shock-accelerated lifetime test for ceramic filters.

To analyze the chemical degradation behavior of the catalyst, the specimens were heated at temperatures ranging from 450 to 600°C for 200 h in air and in SOx gas of 600 ppm, and then the active sites of catalytic filters were examined using FT-IR spectroscopy in order to find out the main factor for catalyst deactivation. Although the catalytic filters were heated up to high temperatures, their crystal structure and microstructure did not change. However, the active sites of the catalyst decreased as the aging temperature increased.

Figure 3 shows the variation of the IR-spectra of catalytic filters with adsorbed ammonia under ambient conditions for 1 h after durability tests. The chemisorption of ammonia with the V=O double bond is known to create Lewis acid sites and Bronsted acid sites at 1605 and 1425 $cm^{-1}$, respectively[13]. These sites are the active sites of the $V_2O_5$ catalyst. From the results in Figure 3, we could estimate that the intensities of active sites decreased because the bond strength of V=O double bond was weakened by high temperature. It was found that the failure mode of catalytic filters was the decrease of NOx conversion and their failure mechanism was the decline of active sites due to the high temperature. It was confirmed that there was thermal durability of catalytic filters between each stress level due to the Arrehenius model (Eq.3). The failure of the catalytic filters was defined as the time to decrease to 10% below the initial efficiency.

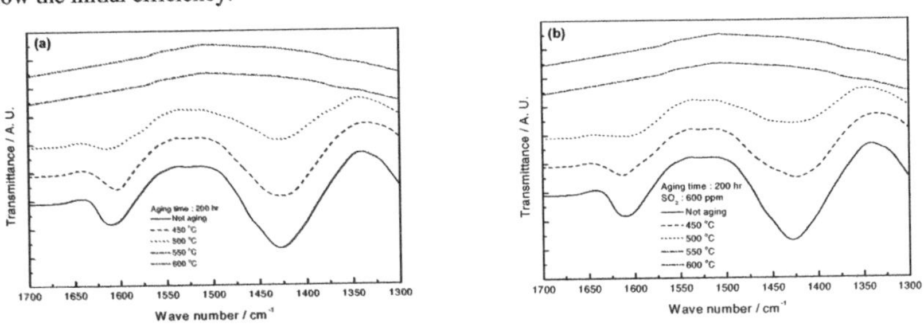


Figure 3. FT-IR spectra of catalytic filters before & after durability test:
(a) thermal durability and (b) thermal & chemical durability

The lifetimes of catalytic filter were $8.678x10^5$ sec at 450°C, $8.843x10^4$ sec at 500°C and $2.820x10^4$ sec at 550°C. In order to present their lifetime at normal operating temperature, the

Arrhenius model was used, expressed as:

$$L = A \cdot \exp(E_a / kT) \quad (3)$$

$E_a$ is Activation energy (1.767 eV), A and k are the characteristic constant, T is temperature. The AF between 350°C (normal) and 450°C (accelerated) was calculated to be 94.835 by Eq 4. This means repeatedly that the effect of 1 cycle at 450°C equals that of 94 cycles at 350°C. If thermal durability at normal operating conditions is 350°C, the lifetime of catalytic filters under normal operating condition can be predicted to be about 2 years.

$$AF = \exp[E_a / k (1/T_1 - 1/T_2)] \quad (4)$$

In order to improve the durability and reliability of catalytic ceramic filters for simultaneous removal of dust and hazardous gas cleaning, feedback research and a robust design were developed. Failure analysis indicated that the primary failure was determined to be the decrease of pressure drop in the filtration layer.

To extend their lifetime and to improve their reliability, inorganic compounds were added to the ceramic filters during the manufacturing processes. They had an apparent porosity of 56%, a compressive strength of 14 MPa, and hence exhibited higher strength. If thermal shock temperature in normal operating condition is $\Delta T = 300°C$, the lifetime of the cordierite filters under normal operating condition can be predicted to be 3700 cycles (about 2 years). As a result, the higher strength of ceramic filters, the better their durability against thermal shock, and therefore their lifetime can be extended.

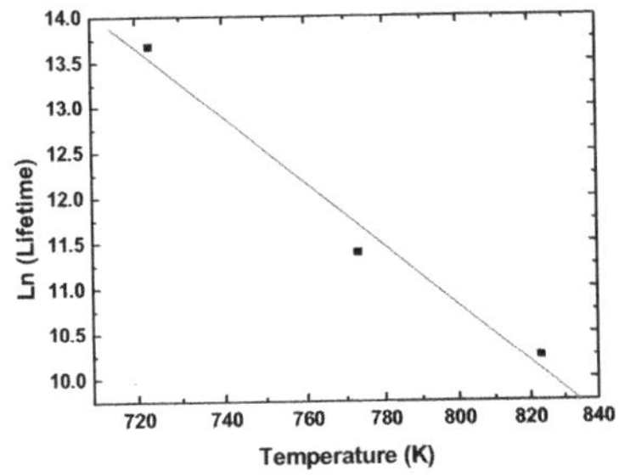


Figure 4. Performance degradation of catalytic filters as a function of time.

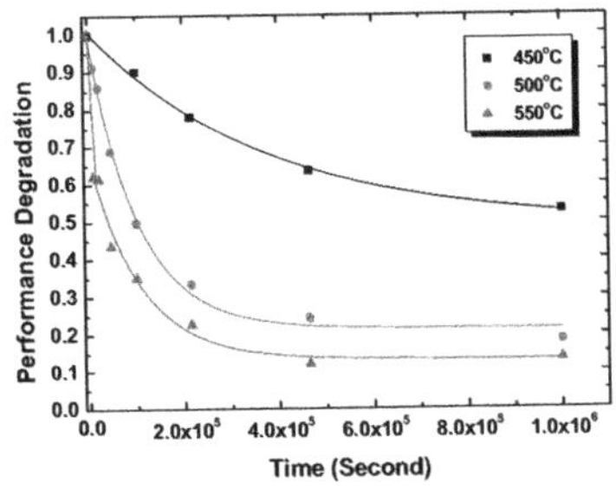


Figure 5. Arrhenius plot between lifetime and temperature.

Conductive ceramic coatings

ITO thin films were deposited on glass substrates by RF magnetron sputtering and then post-annealed in a reducing atmosphere. The deposited ITO thin films had the thickness of 550 nm, the resistivity of $8.164\times10^{-4}$ Ω·cm, and the optical transmittance of 86.0%[5]. In previous research, the potential failure mode of ITO films, which were thermally degraded at temperatures below 250°C, turned out to be an increase in resistivity, and their failure mechanism was the diffusion of oxygen into the ITO films[6]. With the stress factor of temperature, the ITO thin films were forced to be thermally degraded at the stress levels of 200°C, 225°C and 250°C. Figure 6 shows the variation of resistivity as a function of heat treatment time. In order to analyze the variation extent of resistivity as a function of time, the generalized chemical reaction model (Eq. 5) was used, expressed as:

$$(\Delta\rho)_a = (K \cdot t)^\beta \tag{5}$$

where, $\Delta\rho$ is the variation extent of resistivity of thermally-degraded ITO film, $(\Delta\rho)_a$ is the average value of distributed variation extents, K is the reaction rate, t is the time, and $\beta$ is a constant. The linear relation between $Ln((\Delta\rho)_a)$ and Ln(t), shown in Figure 6, was acquired by logarithm of Eq. 5, and K and $\beta$ could be calculated using a linear regression. K and $\beta$ were 0.5004 and 0.5150 at 200°C, 1.3301 and 0.8388 at 225°C and 28.3503 and 0.7063 at 250°C, respectively.

If the lifetime of ITO films is defined as the time for their resistivity to increase as much as 10% compared to their initial value, the lifetime can be calculated to be about 175 h, 12 h and 1 h at 200°C, 225°C and 250°C, respectively. Therefore, thermal degradation of ITO coatings is a temperature-activated process, and thus their lifetimes under temperature stress can be analyzed using the Arrhenius model[15]. The activation energy was calculated to be about 2.24 eV from a linear fitting of the Arrhenius plot between the lifetimes of ITO films and the absolute temperatures. The acceleration factor (AF) between two stress levels was 656 between 150°C and 200°C, calculated by the reported relationship[15]. This means that the effect of 656 h at 150°C on degradation equals to that of 1 h at 200°C. If a normal operating temperature is 150°C, and thermal degradation mechanism of oxygen diffusion is effective at 150°C, the lifetime of the ITO films under the normal operation condition can be predicted to be $1.148\times10^5$ h (175 h×656).

According to XPS results of our previous study, the oxygen concentrations in the ITO films and at the surface were assumed as $6.24\times10^4$ and $6.60\times10^4$ mol/m$^3$, respectively[6]. If the resistivity of ITO films increases as much as 10% compared to their initial value, the carrier concentration must decrease as much as 9.1% on the assumption that the mobility remains invariable. The carrier concentration is proportional to the concentration of oxygen vacancies. Therefore, when the resistivity increases as much as 10%, the oxygen concentration needs to be increased up to $6.37\times10^4$ mol/m$^3$. Using a finite element analysis, FEA, simulation, the lifetime of the ITO films can be evaluated with above assumptions. According to the theory of mass transfer, the product of diffusivity and diffusion time shows a constant value[16]. Based on this theory, the diffusivities of oxygen in ITO films were $4.480\times10^{-20}$ m$^2$/s, $6.53\times10^{-19}$ m$^2$/s, and $7.83\times10^{-18}$ m$^2$/s at 200°C, 225°C and 250°C, respectively, as derived from the experimental results.

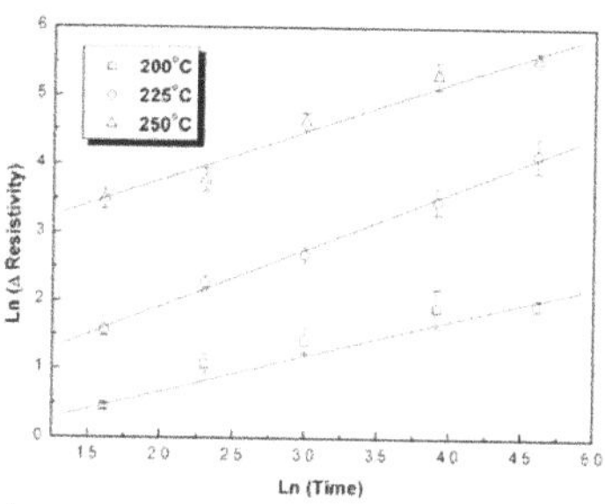

Figure 6. Variation in resistivity of ITO thin films as a function of thermal degradation time.

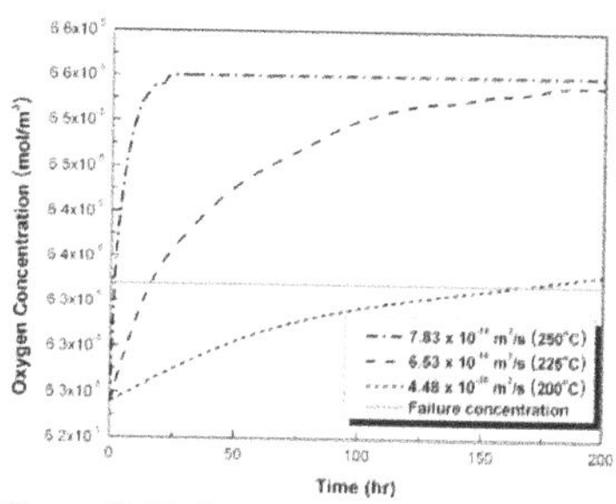

Figure 7. Variation of oxygen concentration in ITO thin films as a function of thermal degradation time, simulated through FEA.

Figure 7 shows the variation of oxygen concentration in the ITO films as a function of heat treatment time obtained by FEA using the experimental results. Diffusivity at 150°C was $7.50\times10^{-23}$ m$^2$/s, as derived using the Arrhenius relation. Using this value, the diffusion of oxygen in ITO films was simulated, and their lifetime could be estimated to be $9.722\times10^4$ h.

In an attempt to improve the durability and reliability of ITO thin films, feedback research was applied to develop a robust design. Failure analysis indicated that the variation in resistivity of ITO films because of thermal degradation was largely dependent on carrier concentration rather than mobility[6]. In order to prolong their lifetime and to improve their reliability, ITO thin films were post-annealed in a reducing atmosphere to increase the oxygen vacancies so that they have a higher carrier concentration. Figure 8 (a) and (b) show the variation in carrier concentration and in resistivity of ITO films as a function of post-annealing temperature, respectively, and Figure 8 (c) shows the variation in resistivity of ITO films with different carrier concentrations as a function of thermal degradation time. As shown in Figure 8 (c), the higher the carrier concentration of ITO films, the better their durability against thermal degradation, and the longer the lifetime that can be achieved. When the carrier concentration of ITO films increased from $1.716\times10^{20}$ cm$^{-3}$ to $6.865\times10^{20}$ cm$^{-3}$, their lifetime was prolonged from $1.148\times10^5$ h to $7.352\times10^5$ h.

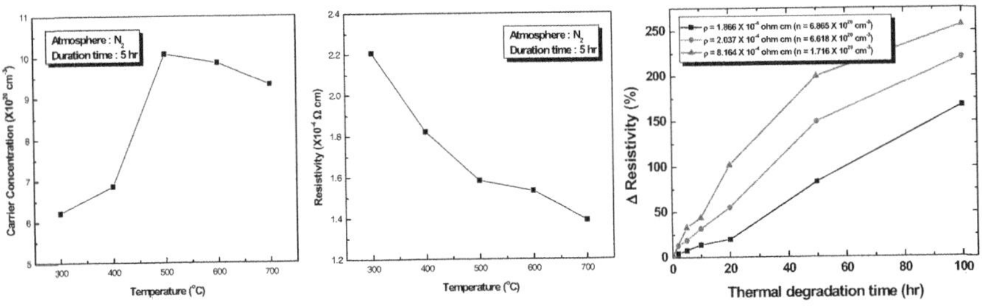

(a) (b) (c)

Figure 8. Variation in (a) carrier concentration and (b) resistivity of ITO films as a function of post-annealing temperature and (c) variation in resistivity of ITO films degraded at 200°C as a function of thermal degradation time.

The procedures for the reliability assessment and improvement for ITO coatings can be summarized as: (1) durability tests with various stresses, (2) identification of the failure mode and mechanism via failure analysis, (3) lifetime prediction using accelerated degradation tests and statistical modeling, (4) FEA simulation based on the physics of failure, and (5) reliability improvement through feedback research to develop a more robust design.

## CONCLUSIONS

There is a great deal of interest in improving the reliability of layered ceramics under harsh application conditions. This requires the establishment of systematic reliability assessment tools. A methodology, which allows the determination of the reliability of layered ceramics based on the physics of failure and accelerated degradation tests, was discussed. A reliability assessment system was successfully designed for catalytic ceramic filters and conductive ceramic coatings. Their degradation behavior, performance decay, and failure mechanisms were investigated through durability tests, failure analyses, and accelerated tests.

These reliability assessment tools can be effectively applied to other layered ceramics in different applications. As the layered ceramics expand their applications, substituting for conventional materials, data acquisition and evaluation methods need to be further emphasized and optimized on commercial scales. Further studies are being conducted to develop more elaborate degradation models for multiple stresses, and to apply them to various hybrid materials, such as catalytic porous ceramics, multi scale coatings, and membrane electrode assembly (MEA) of alkaline anion exchange membrane fuel cells (AAEMFCs).

## ACKNOWLEDGEMENTS

This research was supported by a grant from the Fundamental R&D Program for Core Technology of Materials funded by the Ministry of Knowledge Economy (M-2009-01-0028) and partially supported by grants-in-aid for the National Core Research Center Program from MEST/KOSEF (No. R15-2006-022-01001-0), Republic of Korea.

REFERENCES

[1]G. Kickelbick, Hybrid Materials: Synthesis, Characterization, and Applications (Wiley Interscience, (2007), p4

[2]Energetics, Incorporated and Richerson and Associates, Advanced Ceramics Technology Roadmap–charting our course, December 2000

[3]D.K. Shetty, "Mixed-mode fracture criteria for reliability analysis and design with structural ceramics", J. Eng. Gas Turbines Power, 109, 282 -289 (1987).

[4]H. Fischer, M. Weber and R. Marx, "Lifetime Prediction of All-ceramic Bridges by Computational Methods", J. Dent. Res., 82, 238-242 (2003).

[5]Y.N. Kim, J.H. Park, H.G. Shin, J.K. Song and H.S. Lee, "Fabrication of Thin Film Using Nanosized Indium Tin Oxide (ITO) Powder and Its Degradation", Mater. Sci. Forum, 449-452, 285 (2004).

[6]Y.N. Kim, H.G. Shin, J.K. Song, D.H. Cho, Y.G. Jung and H.S. Lee, "Thermal degradation behavior of indium tin oxide thin films deposited by radio frequency magnetron sputtering", J. Mater. Res., 20, 1574-1579 (2005).

[7]M.S. Jeon, J.K. Song, E.J. Lee, T.H. No, D.H. Kim I.G. Park and H.S. Lee, "Reliability during Thermal Shock Testing of NiCr Films on Mn-Ni-Co Oxide Substrates", Key Eng. Mater., 553 317-318 (2006).

[8]M. C. Shin, J. S. Cha, J. H. Lee, S. H. Lee and H. S. Lee, "Simultaneous Removal of SOx and NOx by Catalytic Filter", Key Eng. Mater., 713, 317-318, (2006).

[9]J.S. Lee, K.H. Shin, M.C. Shin, D.H. Cho and H.S. Lee, "Properties of Catalytic Filter for the Hot Gas Cleaning", Mater. Sci. Forum, 1181, 449-452 (2004).

[10]T. Eriksson, J. Isaksson, P.K. Stahlberg, E. Kurkela and V. Helanti, "Durability of ceramic filters in hot gas filtration", Bioresource Technol., 46, 103-112 (1993).

[11]L.F. Coffin, "A study of the Effects of Cyclic Thermal Stresses on a Ductile Metal, *ASME Transactions*, 76, 931 (1954).

[12]S.S. Manson, "Fatigue: A Complex Subject-Some Simple Approximations", Experimental Mechanics, 5, 193 (1965).

[13]N. Economidis, R. F. Coil, P. G. Sirmiots, "Catalytic performance of $Al_2O_3/SiO_2/TiO_2$ loaded with $V_2O_5$ for the selective catalytic reduction of NOx with ammonia", Catal. Today, 40, 27-37 (1998).

[14]S. M. Goodnick, J. F. Wager, and C. W. Wilmsen, "Thermal degradation of indium-tin-oxide/*p*-silicon solar cells", J. Appl. Phys., 51, 527 (1980).

[15]J. F. Lawless, Statistical Models and Methods for Lifetime Data, Wiley Interscience, 271 (2003).

[16]Y.A. Cengel, Heat Transfer – A Practical Approach (WCB McGraw-Hill), 635 (1999).

# METALLIC BONDING WITH THE USE OF COPPER COMPOUND POWDERS

Kazuaki Ihara[1], Y. Kobayashi[1], N. Shibamiya[1], M. Minato[1], Y. Yasuda[2], T. Morita[2] and S. Yamada[2]
[1] Department of Biomolecular Functional Engineering, College of Engineering, Ibaraki University, 4-12-1 Naka-narusawa-cho, Hitachi, Ibaraki 316-8511, Japan
[2] Materials Research Laboratory, Hitachi Ltd., 7-1-1 Omika-cho, Hitachi, Ibaraki 319-1292, Japan

## ABSTRACT

This paper describes a method for preparing CuI nanoparticles in aqueous solution, and demonstrates the feasibility of metallic bonding using CuI nanoparticles. A colloidal solution of CuI nanoparticles with an average size of 128 nm was prepared from $CuCl_2$ (0.01 M), KI (0.01 M), and a reducing reagent, $Na_2SO_3$ (0.05 M), in water at room temperature. A metallic Cu stage and plate were successfully bonded by annealing at 400°C and pressurizing at 1.2 MPa for 5 min in $H_2$ gas with the help of CuO obtained by pre-annealing the CuI particles in air at 400°C. The shear strength required to separate the bonded Cu substrates was 14.8 MPa.

## INTRODUCTION

A bonding technique is required to connect solid state materials in various fields such as civil engineering, the construction industry, and electronics[1]. In the bonding of metallic materials, a process of annealing and pressurizing is usually used[2]. This process provides diffusion of the components on their interface, and consequently the metallic materials are bonded. Since the metallic powders melt and diffuse into the metallic materials under heating at high pressure, bonding can be helped by inserting metallic powders between the metallic materials. Among metallic powders, Cu particles are low cost promising materials for the metallic bonding. However, they face at a problem of easy oxidation in air[3-4], which spoils bonding properties.

Cu in an oxidative state, such as a Cu salt, may be available as a precursor to metallic Cu, since it can be reduced to metallic Cu with a reducing agent or under a reducing atmosphere. Therefore, Cu salts may also be useful as powders for bonding metallic materials. During the bonding of metallic materials in a reducing atmosphere, it is expected that ions of Cu and iodine will diffuse into the metallic materials intensively due to concentration gradient. Simultaneously to the diffusion, the Cu ions are reduced to metallic Cu in the reducing atmosphere. Consequently, the metallic materials will be bonded.

Cuprous iodide (CuI) is a candidate as the insertion powder, since CuI is chemically stable, and can be easily prepared in aqueous solution. Preparation of CuI in aqueous solution has been reported previously. Zhou et al. prepared from $CuCl_2$ and $KI/Na_2SO_3$[5]. Precipitate that was obtained by mixing the solutions of $CuCl_2$ and $KI/Na_2SO_3$ was dissolved in KI/ethanol/water solution. After dissolving, precipitate of CuI was obtained with addition of water to the solution. Yang et al. reported preparation of porous spherical CuI nanoparticles from $Cu(CH_3COO)_2$, KI, NaOH and $NH_2$-OH·HCl[6]. A suspension of CuO was prepared by aging a mixture of $Cu(CH_3COO)_2$ and NaOH for 2 days. Then, $NH_2$-OH·HCl and KI aqueous solutions were added to the CuO suspension, which produced CuI nanoparticles. These methods successfully fabricated CuI crystallites. However, they required a long time and many steps.

A present work proposes an alternative method for preparing CuI particles in aqueous solution. In our method, CuI colloid particles were prepared by mixing $CuCl_2$, KI and $Na_2SO_3$ in water at room temperature, in which their concentrations were adjusted to a concentration that produces a colloid solution of particles instead of precipitates. Metallic bonding with the use of the CuI particles was also studied in the present work.

## EXPERIMENTAL

### Chemicals

Copper chloride dihydrate ($CuCl_2 \cdot 2H_2O$) (Kanto Chemical Co., Inc., >99 %) and potassium iodide (KI) (Kanto Chemical Co., Inc., >99.5 %) were used as CuI precursors. Sodium sulfite (Kanto Chemical Co., Inc., >97.0 %) was used as a reducing reagent. All chemicals were used as received. Water that was ion-exchanged and distilled with Shimadzu SWAC-500 was used in all the preparations.

### Preparation

Colloid solutions of CuI nanoparticles were synthesized by redox reaction. Freshly prepared 5 ml of 0.2 M $Na_2SO_3$ aqueous solution containing 0.033 g KI was added to 15 ml of 0.013 M $CuCl_2$ aqueous solution under vigorous stirring at room temperature, which resulted in a CuI concentration of 0.01 M in the final solution. Color of the mixture turned yellow-green immediately after the addition of $KI/Na_2SO_3$ aqueous solution to $CuCl_2$ aqueous solution. The yellow-green products were CuOH, since the addition of $Na_2SO_3$ brought about an increase in pH. After the color turning, the mixture became opaque gradually, which implied production of a colloid solution of CuI particles. The reaction time was 1 day.

### Characterization

The CuI particles were characterized by transmission electron microscopy (TEM) and X-ray diffractiometry (XRD). TEM photographs were taken with a JEOL JEM-2000FX II microscope operating at 200 kV. Samples for TEM were prepared by dropping and evaporating the particle colloid on a colloid ion-coated copper grid. XRD measurements were carried out with a Rigaku RAD-B X-ray diffractometer at 50 kV and 150 mA with $CuK\alpha_1$ radiation. For preparing samples for the XRD measurements, supernatant of the particle colloid was removed with decantation, and then residue of the colloid was dried at room temperature for 24 h in vacuum.

Metallic bonding property was investigated by sandwiching powder samples between a Cu stage (diameter: 10 mm, thickness: 5 mm) and a Cu plate (diameter: 5 mm, thickness: 2.5 mm), pressing the powder samples under annealing at 1.2 MPa in $H_2$ for 5 min with an Ultex flip chip bonder, and measuring shear strength required for separating the bonded stage and plate with a Seishin SS-100KP bond tester. Cross section of the bonded part was observed with a Hitachi S-4800 scanning electron microscope (SEM).

## RESULTS AND DISCUSSION

### CuI nanoparticles

Figure 1 shows TEM image of as-prepared particles. The particles were quasi spherical. They appeared to form micron-sized aggregates. Since the particles were colloidally stable, the aggregates were probably produced during preparation of TEM sample. The particle size was 128±34 nm.

Figure 2 (a) shows XRD pattern of the as-prepared particles. The pattern showed peaks at 25.6, 29.5, 42.3, 50.0, 52.4, 61.2, 67.4, 69.4 and 77.2 degree, which were attributed to those of γ-CuI (JCPDS card No. 6-246), and no peak besides the peaks was detected. Accordingly, this result indicated that the present method successfully prepared crystalline CuI nanoparticles.

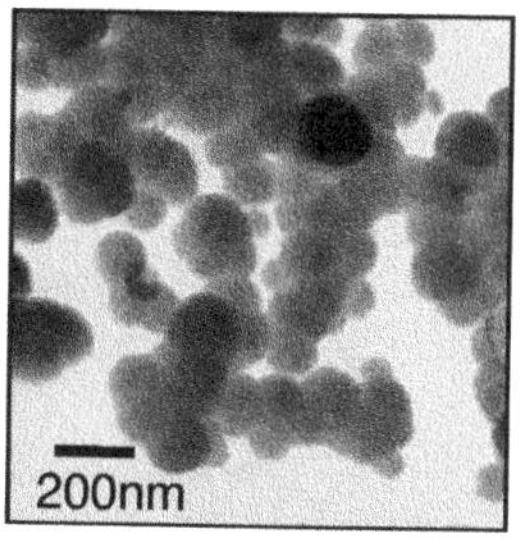


Figure 1. TEM image of as-prepared particles.

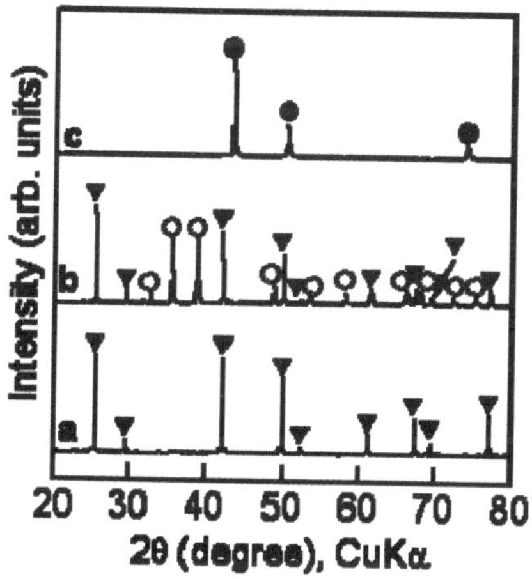


Figure 2. XRD patterns of particles. Sample (a) was as-prepared particles. Sample (b) was prepared by annealing the sample (a) at 400oC in air. Sample (c) was prepared by annealing the sample (a) at 400°C in air and follwingly at 350°C in $H_2$ gas. ●: metallic Cu, ○: CuO, ▼: γ-CuI.

Bonding Property

In a preliminary examination, bonding with the use of as-prepared CuI particles was performed in $H_2$ gas at 350°C. The Cu stage and plate could not be bonded. Figure 3 shows photograph of the bonded area.

Yellowish CuI powder was left on the stage after the examination, which indicated that the CuI was not reduced to metallic Cu with such examination conditions. Consequently, Cu bonding could not be made. Since it was speculated that the existence of iodine probably prevented formation of metallic Cu, an attempt to remove iodine from the CuI powder partially was performed, or, the as-prepared CuI particle powder was pre-annealed in air prior

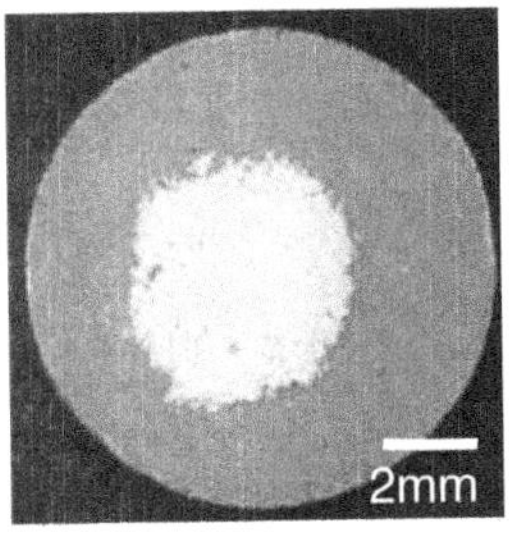


Figure 3. Photograph of bonded area. Bonding was performed in $H_2$ gas at 350°C for as-prepared CuI nanoparticles.

to bonding in $H_2$ gas. Figure 2 (b) shows XRD pattern of the particles annealed at 400°C in air. Besides peaks due to γ-CuI, several peaks attributed to CuO (JCPDS card No. 5-661) were detected at 32.5, 35.5, 38.7, 48.8, 53.5, 58.3, 61.5, 65.8, 66.2, 68.1, 72.4 and 75.2 degree, which indicated that partial removal of iodine from the CuI particles was successfully attained, and consequently copper oxide was produced.

The pre-annealed CuI particles were annealed in $H_2$ gas, and their XRD was measured. Figure 2 (c) shows XRD pattern of particles pre-annealed in air at 400°C followed by in $H_2$ gas at 350°C. The pattern showed peaks at 43.3, 50.4 and 74.0 degree, which were attributed to those of metallic Cu (JCPDS card No. 4-836), and no peak besides the peaks was detected. This result implied that the pre-annealed particles could be used as insertion powder for metallic bonding.

With the use of the pre-annealed powder, bonding examination was performed in $H_2$ gas at 400°C. As a result, the coppers were successfully bonded, and the bond had a shear strength of 14.8 MPa. Figure 4 shows photograph of the bonded area. Glossy red product that was obviously metallic Cu was observed in widespread area on the stage. This result indicated that the pre-annealing in air was effective for the formation of metallic Cu, and consequently the successful bonding could be performed.

Figure 5 shows SEM images of cross-sections of the bonded region. Sintering of particles took place and micron-sized domains were produced, though some voids were also observed and no large crack was formed. This observation confirmed the successful bonding.

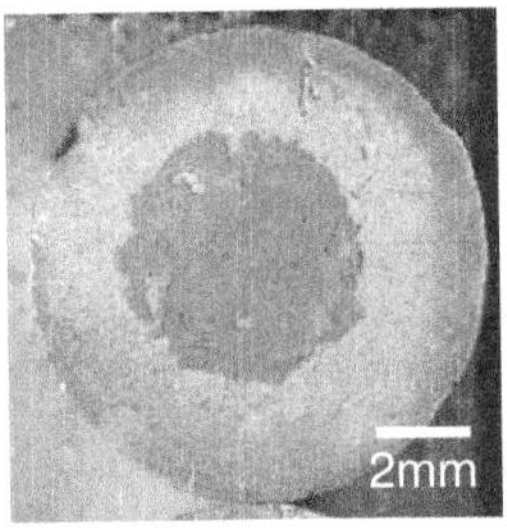


Figure 4. Photograph of bonded area. Bonding was performed in $H_2$ gas at 350°C for CuI particles pre-annealed in air at 400°C.

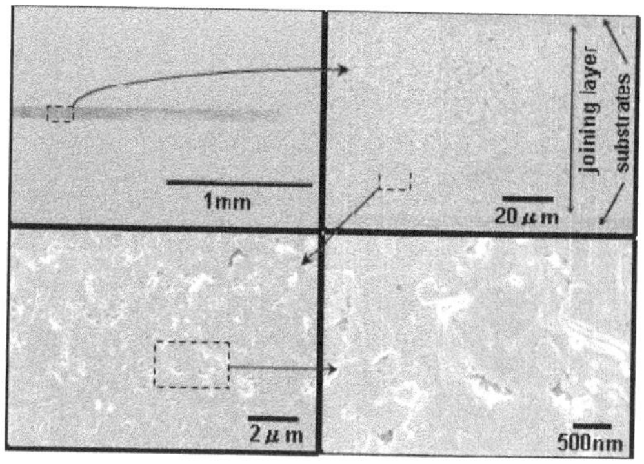


Figure 5. SEM images of cross-section of bonded region. Bonding conditions were the same as in Fig. 3. An area surrounded with a rectangle was magnified to an image indicated with an arrow.

## CONCLUSIONS

A method for preparing CuI nanoparticles in aqueous solution was proposed. CuI nanoparticles were prepared by reacting $CuCl_2$ with $Na_2SO_3$ in the presence of KI in aqueous solution at room temperature. Metallic Cu materials were bonded in $H_2$ gas with the use of the CuI particles pre-annealed in air at 400°C. A shear strength of 14.8 MPa was required to separate the Cu substrates bonded at an annealing temperature of 400°C and a pressure of 1.2 MPa. The present work demonstrates the feasibility of metallic bonding with the use of a metal salt. Precise optimization of the experimental conditions, and detailed characterization of particles are in progress.

## ACKNOWLEDGEMENTS

This work was partially supported by Hitachi, Ltd. We express our thanks to Dr. T. Noguchi and Mr. M. Hayashi in College of Science of Ibaraki University, Japan for their help for TEM observation, and to Prof. O. Abe in College of Engineering of Ibaraki University, Japan for his help for XRD measurements.

## REFERENCES

[1] S. Matsuoka and H. Imai, "Direct welding of different metals used ultrasonic vibration": *J. Mater. Process. Technol.*, **209**, 954-960(2009).

[2] C. Weisman, ed.: 'Welding Handbook' 7th edn, Vol. 1, 18-19; 1976, Miami, American Welding Society.

[3] P. Kanninen, C. Johans, J. Merta, and K. Kontturi, "Influence of ligand structure on the stability and oxidation of copper nanoparticles", *J. Colloid Interface Sci.*, **318**, 88–95(2008).

[4] B. K. Park, D. Kim, S. Jeong, J. Moon and J. S. Kim, "Direct writing of copper conductive patterns by ink-jet printing", *Thin Solid Films*, **515**, 7706–7711(2007).

[5] Y. Zhou, M. Lü, G. Zhou, S. Wang and S. Wang, "Preparation and photoluminescence of γ-CuI nanoparticles", *Mater. Lett.*, **60**, 2184–2186(2006).

[6] M. Yang, J.-Z. Xu, S. Xu, J.-J. Zhu and H.-Y. Chen, "Preparation of porous spherical CuI nanoparticles", *Inorg. Chem. Commun.*,**7**, 628-630(2004).

# EFFECT OF PLASMA SPRAY COATING PARAMETERS ON THE MICROSTRUCTURE AND PERFORMANCE OF AN ALUMINA/COPPER COMPOSITE

Kuan Hong Lin[a], Zi Hao Xu[b] and Shun Tian Lin[b]
[a]Department of Mechanical Engineering, Tungnan University, Taipei, Taiwan
[b]Department of Mechanical Engineering, National Taiwan University of Science and Technology, Taipei, Taiwan

## ABSTRACT

A coating of alumina powder was plasma sprayed onto a Cu substrate to fabricate a composite with dielectric performance. The effect of the plasma spray coating parameters on the microstructure, thermal resistance, and dielectric properties of the alumina layer are discussed in this study. Experimental results indicate that higher power influences the fine powder process by forming a thicker layer with a deteriorated surface roughness for the alumina/copper composite. The thermal resistance of the alumina layer increases with surface roughness. The layer's microstructure, thickness, and absorbed moisture in the porosity affect the breakdown voltage of the alumina layer. The optimized parameters for plasma spray coating included an alumina powder with a particle size ranging from 22 to 45μm, a plasma power of 40 kW, a spraying velocity of 750 m/s, an argon flow rate of 45 L/min, a spraying distance of 140mm, and a spraying angle of 90°. With a coated layer thickness of 20μm, a relatively low thermal resistance (0.03 K/W), a low surface roughness (3.13 μm), and a highly reliable breakdown voltage (40 V/μm) was attained in this work.

## INTRODUCTION

Plasma spray coating technology has been widely applied in the field of aerospace, chemistry, petrifaction, steelmaking and mechanical industries for thermal dissipation, thermal barriers, insulation, wear-resistant, and corrosion protection of the parts [1-3]. The advantages of plasma spray coating include spraying materials are unrestricted, high deposition rates of 4 to 9 kg/hr, and the layered surface roughness can be controlled within 3 μm. Nevertheless, some drawbacks are also found, such as porosity of layers in the range of 5%~30%, no sustainability of bending or impact stress, and relatively high cost equipment. [4].

Most electronic components have a dielectric thin-film coated onto a high thermal conductivity spreading substrate, such as CuW, CuMo, Al, or Cu to fabricate a composite possessing a high thermal dissipation. Spherical ceramic powders such as AlN, SiC and $Al_2O_3$ with particles sizes ranging from 10 to 100 μm are preferred for the dielectric layer. However, the porosities within the coated-layer of AlN will easily absorb moisture, resulting in hydrolysis and a decrease in the thermal dissipation performance [5]. The higher cost, low electric resistivity, and leakage current of SiC also make it unsuitable for a dielectric layer. The thermal conductivity of $Al_2O_3$ powder (k= 30 W/m·K) is lower than that of AlN and SiC; however, its easy production and lower cost make it suitable for use as the dielectric layers of high-power electric devices.

In this work, alumina powders were coated onto a Cu substrate by plasma spray coating. A dielectric layer with high thermal dissipation, low thermal resistance, and high electric insulation was expected. The objective of this work was to determine the optimal working parameters to produce a plasma spray coating with improved heat dissipation performance.

## EXPERIMENTAL PROCEDURES

The surface of the copper substrate was grit-blasted before it was plasma spray-coated. The constant working parameters for plasma spray coating included a spraying velocity of 750 m/s, an argon flow rate of 45 L/min, a spraying distance of 140 mm, and a spraying angle of 90°. The parameters varied and the specimen designation numbers are shown in Table 1.

The cross-section images of the plasma spray-coated alumina/copper composites were characterized using a field emission scanning electron microscope (FESEM, JEOL, JSM-6500F). The surface roughness of the alumina layer was measured using a surface roughness tester (HOMMELWERKE, LV-50E). The thermal resistance perpendicular to the interface of the alumina layer and copper substrate was measured using a thermal interface material tester (LONG WIN SCIENCE, LW-9091), over a test area 0.0254 m x 0.0254 m, at a clamp pressure of 8 kgf, with an input power of 80 W. The electrical resistivity ($\rho$) of the layered alumina was measured using a four-point probe resistivity measurement tester (KEITHLEY, 2400); its maximum value was $10^{14}$ Ω·cm. The mean value among five measuring points was recorded for each specimen. A puncture and insulation tester (TOPWARD, TPT-500) was used to measure the breakdown voltage of the alumina layer. The effect of ambient moisture on the breakdown voltage of the alumina layer was also determined.

Table 1. Parameters of plasma spray coating and designation number of specimens in this test.

| Sample Designation | A1 | A2 | A3 | A4 | A5 | A6 |
|---|---|---|---|---|---|---|
| | B1 | B2 | B3 | B4 | B5 | B6 |
| Plasma Power (kW) | 35.2 | 40 | 45 | 35.2 | 40 | 45 |
| Layer Thickness (μm) | 20 | | | 40 | | |
| Alumina Particle Size (μm) | 5 to 22 (A1 to A6) | | | | | |
| | 22 to 45 (B1 to B6) | | | | | |

RESULTS AND DISCUSSION

Figure 1 shows the secondary electron images (SEI) of cross-sectional of plasma spray-coated alumina/copper composites. The region between the Cu and epoxy as indicated by arrows is the alumina layer, which is bonded with the Cu substrate excellently in all specimens. However, the alumina layers in all specimens are porous. The porosity of the alumina layers increased with the thickness of the layer, which can be revealed by the comparison between Fig. 1(b) and Fig. 1(d). On the other hand, the porosity decreased with an increase of particle size, which can be revealed as compared with the Fig. 1(b) and Fig. 1(f).

The mean thickness of the alumina layers in the specimen A1 was about 20 μm, which was very close to the original scheme. However, the layer thickness of specimen A3 was approximately 40 μm, exceeding the intended value by around 100%. A similar result was observed in specimen A6 as displayed in Fig. 1(d). The average thickness of the alumina layers in specimen B group was very close to the intended values, with the exception of specimen B3 and B6, in which the thickness was about 20% higher, due to a higher spraying power (45 kW) used. Accordingly, the spraying power had a greater influence on the thickness of fine powder (5 to 22 μm) than that of the coarse powder (22 to 45 μm).

Figure 2 shows the surface roughness (Ra) of all of the alumina layers formed by plasma spray coating. The average roughness of layer in the specimens A1 to A3 (particle size of 5 to 22 μm) is slightly higher than that of B1 to B3 (particle size of 22 to 45 μm) by 0.09 μm. Similarly, the mean roughness of the specimens A4 to A6 is also slightly higher than that of group B4 to B6

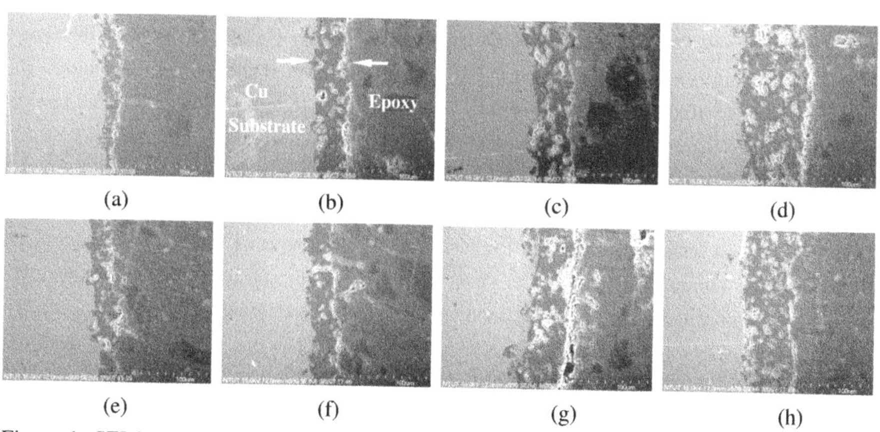


Figure 1. SEI image of cross-section of plasma spray coating alumina/copper composites, (a) A1-20μm, (b) A3-20μm, (c) A4-40μm, (d) A6-40μm, (e) B1-20μm, (f) B3-20μm, (g) B4-40μm, (h) B6-40μm.

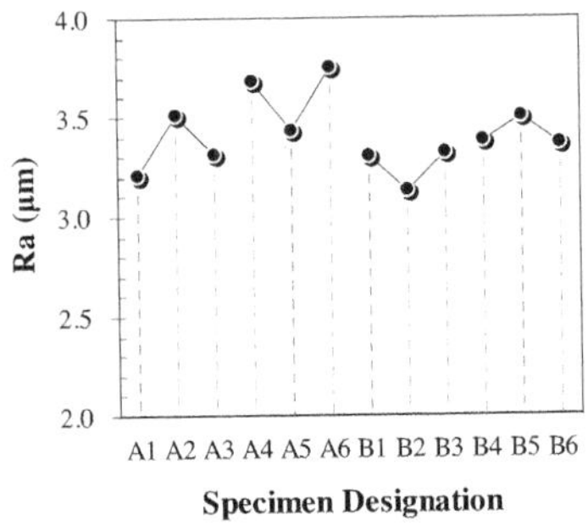


Figure 2. Surface roughness of all alumina layers, over a test area of 0.0254 m × 0.0254 m.

by 0.21 μm. This result reveals that the surface roughness is not only related to the alumina layer thickness, but also the mass effect of the powder. The lower mass of the smaller alumina powder exhibited a relatively lower kinetic energy and enthalpy during plasma spray coating. When the alumina particles impacted the Cu substrate during plasma spray coating, the smaller alumina particles cooled down before completely embedding into the coated layer. Consequently, many pores were formed within the alumina layer, and the surface roughness of the coated layer was thus increased.

The thermal resistance of all alumina layers formed by plasma spray coating is shown in Fig. 3. The thermal resistance slightly increases with the layer thickness. However, the average thermal resistance of B4 to B6 (0.045 K/W) is almost the same as B1 to B3 (0.042 K/W). Such a result is ascribes to the similar mean surface roughness values of these two groups. Besides, a linear relationship is observed between the surface roughness and the thermal resistance. The thermal resistance of alumina layer formed by plasma spray coating is only 0.03 K/W at a power density of 0.125 W/mm$^2$, which is equivalent to a thermal conductivity (K) of 3.1 W/m·K. These values are superior to those of a printed circuit board, which has a thermal conductivity of 0.3 W/m·K [6]. They are also better than those of an insulated metal substrate (IMS), such as a Hitachi–Bergquist substrate, on which the insulating layer composed of the epoxy and $Al_2O_3$ powder with a thermal resistance of 0.35 K/W. The Eurolam substrate, the insulating layer consisting of epoxy and glass fibers was reported a thermal resistance of 0.8 K/W [7]. Apparently, the thermal resistance of the plasma spray-coated alumina layer in this work is four to ten times lower than those of the aforementioned substrates.

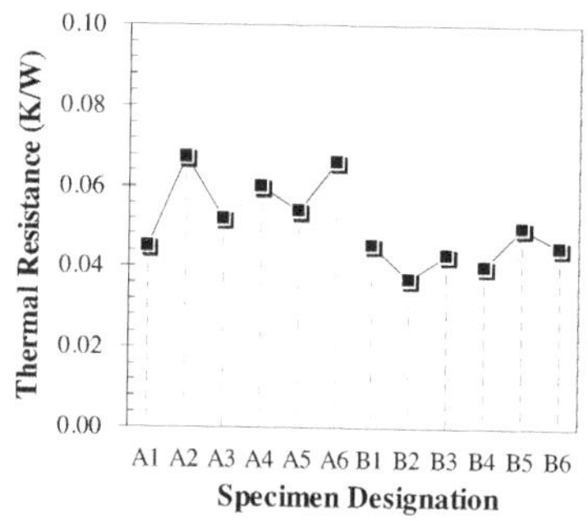


Figure 3. Thermal resistance of all alumina layers, over a test area of 0.0254 m × 0.0254 m.

All of the alumina dielectric layers formed by plasma spray coating exhibited excellent insulation properties, and the electric resistivity was more than $10^{14}$ Ω·cm, due to the presence of alumina. In this work, a continuous and uniform layer around 20 μm thick was formed using a plasma spray power of 35.2 to 45 kW, and alumina particles sizes ranging from and 5 to 45 μm. The leakage current of the substrate was effectively prevented in this work.

It is reported that the plasma spray coatings tend to be very porous. A porosity around 5 to 30 vol.% was found by researchers [4]. The pores will absorb moisture in a wet and hot environment, and thus reduce the breakdown voltage of the alumina layer. In this work, all specimens were baked at 100°C for 1 to 2 H to compare the breakdown voltages, which are plotted in Fig. 4. The breakdown voltage of specimen increased with the plasma spraying power. This phenomenon is caused not only by the improved microstructure, but also by the slightly increased thickness of the layer, which is confirmed by the SEM images shown in Figs. 1(a), 1(b) and Figs. 1(c), 1(d). Furthermore, increasing the thickness of the alumina dielectric layer also increases the breakdown voltage. Such a result is very consistent with other investigations [4].

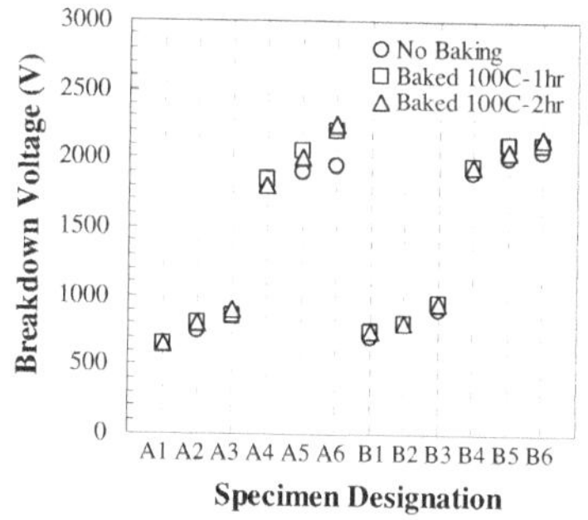


Figure 4. Breakdown voltage of alumina layer, no baking or baked at 100 °C for 1 to 2 hours.

## CONCLUSION

Smaller alumina particles (5 to 22μm) and higher spraying power (45 kW) with the plasma spray coating process forms a thicker layer with a higher porosity and higher surface roughness. The surface roughness increases with the thickness of alumina layer, and thus increases the thermal resistance. The electric insulation of the alumina dielectric layer is excellent, and the electric resistivities of all specimens are higher than $10^{14}$ Ω·cm. The breakdown voltage of the composite is determined by the layer thickness, and increases slightly by baking. In this work, the optimal parameters for plasma spray coating are a plasma power of 40 kW, alumina powder with particle sizes ranging from 22 to 45μm, a coated alumina layer thickness of 20μm, a spraying velocity of 750 m/s, an argon flow rate of 45 L/min, a spraying distance of 140mm, and a spraying angle of 90°.

## REFERENCES

[1] Y. Zeng, S. W. Lee, C. X. Ding, "Plasma Spray Coatings in Different Nano-size Alumina", *Materials Letters*, **57**, 495~501 (2002).

[2] A. Koutsomichalis, C. Panagopoulos, H. Badekas, "Wear Behaviors of Alumina Plasma-Sprayed Coating on an Al-Cu Alloy", *Materials Letters*, **18**, 19~24 (1993).

[3] D. W. Richerson, "Modern Ceramic Engineering-Properties, Processing, and Use in Design", Marcel Dekker, Inc., New York, 222~224 (1990).

[4] C. W. Mossor, "Electrical Breakdown of Thermal Spray Alumina Ceramic Applied to AlSiC Baseplates Used in Power Module Packaging", Virginia Polytechnic Institute and State University, M.S. Thesis, (1999).

[5] F. Miyashiro, N. Iwase, A. Tsuge, F. Ueno, "High Thermal Conductivity Aluminum Nitride Ceramic Substrates and Packages", *IEEE Transactions Components, Hybrids and Manufacturing Technology*, **13(2)**, 313~319 (1990).

[6] T. F. Lemczyk, B. Mack, J. R. Culham, M. M. Yovanovich, "PCB Trace Thermal Analysis and Effective Conductivity", Proceedings - IEEE Semiconductor Thermal and Temperature Measurement Symposium, Phoenix, 15~22 (1991).

[7] S. D. Pascoli, P. E. Bagnoli, C. Casarosa, "Thermal Analysis of Insulated Metal Substrates for Automotive Electronic Assemblies", *Microelectronics Journal*, **30**, 1129 ~1135 (1999).

# FABRICATION OF NANO-$TiO_2$ COATED MORPHGENETIC CARBON AND ITS PHOTOCATALYTIC PERFORMANCE

D. Huang[1], Y.Miyamoto[2], T.Matsumoto[2], T.Tojo[2], T. Fan[1], J. Ding[1], Q. Guo[1] and D.Zhang[1]

[1] State Key Lab of Metal Matrix Composites, Shanghai Jiaotong University, Shanghai 200030, PR China

[2] Toyotanso Joint Research Chair, Joining and Welding Research Institute, Osaka University, Ibaraki, Osaka 567-0047, Japan

## ABSTRACT

Morphgenetic carbons (MCs) made from biomass are expected to be used as catalyst supports due to the muti-hierarchal structure of their pores. Hydrolytic precipitation was used to deposit a tetra-n-butyl titanate monomer onto morphgenetic carbons made from water bamboo leaves using the sol-gel method, and then the precursors were heated at 400°C, resulting in formation of $TiO_2$-coated morphgenetic carbon composites ($TiO_2$/MCs). The $TiO_2$/MC composites were characterized by SEM, XRD, and optical absorption spectroscopy. Fine $TiO_2$ particles were coated on the walls of the micrometer-sized pores. Methylene blue in an aqueous suspension of the $TiO_2$/MC powders was degraded efficiently by ultraviolet light irradiation.

## INTRODUCTION

Nano-crystalline $TiO_2$ is used in a wide range of applications, especially in photocatalysis[1-2]. The photocatalytic activity is due to the $h^+/e^-$ couples generated on $TiO_2$ surface in the presence of UV light[3]. Study of properties and applications of this material under light illumination started with the discovery of the photocatalytic splitting of water on a $TiO_2$ electrode in 1972 by A. Fujishima and K. Honda[4]. Subsequently, in 1977, S. N. Frank and his co-workers reported the application of photocatalytic purification for the first time[5]. Since then, $TiO_2$ has been regarded as an efficient photocatalyst due to its stability, low expense, and good photocatalytic performance.

$TiO_2$ particles tend to coagulate during the degradation reaction. This process can hamper the light incidence on active centers of the $TiO_2$ and consequently reduce the catalytic activity.[6] Additionally, small-sized $TiO_2$ particles with most photocatalytical activity have the highest filtration costs[7]. TGo circumvent this, researchers started to load $TiO_2$ onto porous supports, such as activated carbon, silica, alumina, and zeolite[8-11].

Morphgenetic carbons (MCs) made from biomass are expected to be an effective support for catalysts due to their environmental friendliness and the multi-hierarchical structure of pores including replicated natural pores (micrometer size) and newly processed nanometer-sized pores.

This kind of structure can disperse the $TiO_2$ to maintain the active surface which is expected to develop a synergistic effect between $TiO_2$ and the supports. In this report, the fabrication and photocatalytic performance of $TiO_2$-coated morphgenetic carbons ($TiO_2$/MCs) are described. Hydrolytic precipitation of titanium tetraethoxide (TTED) was deposited onto morphgenetic carbon made from water bamboo leaves, followed by heat treatment to fabricate $TiO_2$/MC. The fabricated composites showed high catalytic performance.

## EXPERIMENTAL

### MCs preparation

The water bamboo leaves were cut into 70 mm pieces. The pieces were washed with deionized water, and then dried at 80°C. The dried pieces were heated (the heating rate: 10°C/min) to 850°C under an air flow (55Pa in the oven). Samples were held at 850°C for 3h before cooling down under vacuum. Nanometer-sized pores were newly formed by this activation process. The products were washed sequentially with 0.5M HCl, and distilled water to remove residual mineral matters, and then dried at 110 °C.

### Photocatalysts preparation

$TiO_2$/MCs were fabricated through the sol-gel method. TTED was dissolved into an absolute methanol (20ml). MCs (0.5g) were added to the solution. After 15 min, 0.48ml nitric acid of 0.28M was added. The resulting suspension was stirred for 3 h at room temperature, followed by the addition of distilled water under vigorous stirring. And then, the gel was heated at about 60°C. The resulted solid product was washed with distilled water and methanol, dried in vacuum at 60°C, and calcined at 400°C for 3 h. The $TiO_2$/MCs samples made from precursors containing 3.0g, 1.5g and 0.5g of TTED were denoted as TC30, TC15 and TC05, respectively.

### Characterization of products

Morphologies of $TiO_2$/MCs were observed by FE-SEM. The crystalline phases and sizes were identified at 40 kV, 40mA (Cu $K_\alpha$ = 1.542 Å) by XRD. The scan speed was 1°/min. The strongest diffraction peak of $TiO_2$ corresponding to anatase (101) was selected to evaluate the crystallite size. The mean crystallite size was determined by the Scherer equation. UV–vis spectrophotometer was used to obtain the absorption spectra of the catalysts over a range of 200–550 nm.

### Photocatalyic capability test

The photocatalytic activities of $TiO_2$/MCs composites were investigated by photocatalytic degradation of Methylene Blue (MB, $3\times10^{-4}$mol/L). The removal rates (R) of MB were measured and calculated using the following equation:

$$R = \left(1 - \frac{C}{C_0}\right) \times 100\% \tag{1}$$

where $C_0$ is the original concentration of MB, and C is the remained concentration of MB after photocatalytic activity. In all tests, the $TiO_2$ concentration was 4g/L, and the concentration of MCs was 5g/L. The amount of $TiO_2$ and MCs was equal to that of TC15. The wavelength of UV light used for photodegradation was 260nm.

## RESULTS AND DISCUSSION

### Morphologies of MCs, $TiO_2$, and $TiO_2$/MCs

In Figure 1 (a) and (b), the morphologies of $TiO_2$ and $TiO_2$ particles coated on MCs are shown, respectively. It can be seen in Figure 1 (a) that nanometer-sized $TiO_2$ agglomerates. The primary particle size is about 200 to 300 nm. Whereas, in $TiO_2$/MCs composites, $TiO_2$ is well dispersed onto the wall of MCs' micrometer-sized pores and it formed a thin porous film. The size of well-dispersed $TiO_2$ particles is much smaller, because the substrate of MCs kept $TiO_2$ from agglomerating.

### XRD analysis

To characterize the crystalline phase of the samples, the XRD patterns of MCs and $TiO_2$/MCs composites were obtained (Figure 2). In the XRD pattern of MCs, the peaks at about 24.3° and 43.8° could be assigned to the diffractions of (002) and (100) planes of graphite, respectively[12]. In the XRD pattern of $TiO_2$/MCs, five diffraction peaks appear which are consistent with the anatase phase of $TiO_2$. The strongest peak of $TiO_2$ corresponding to anatase (101) was selected to evaluate the crystallite size of $TiO_2$. That was 7.8(2) nm using Scherer equation.

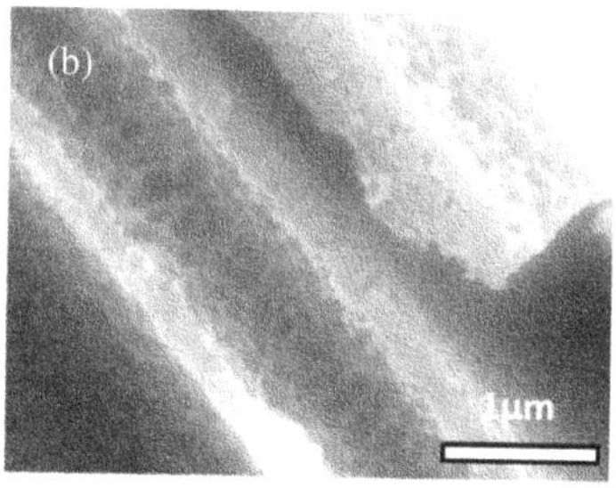


Figure 1. SEM images of (a) pure $TiO_2$, (b) $TiO_2$/MCs

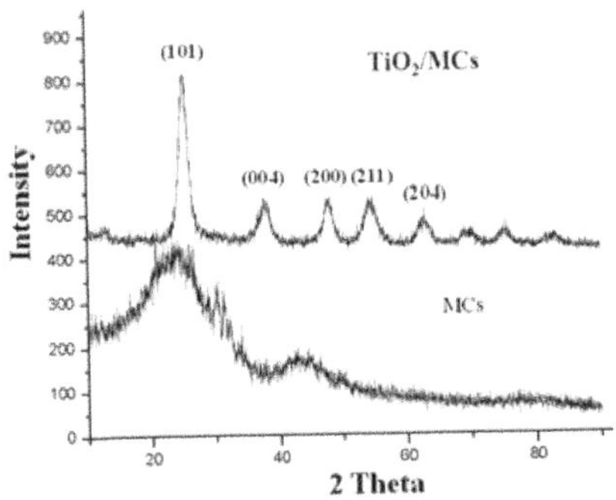


Figure 2. XRD patterns of MCs and $TiO_2$/MCs

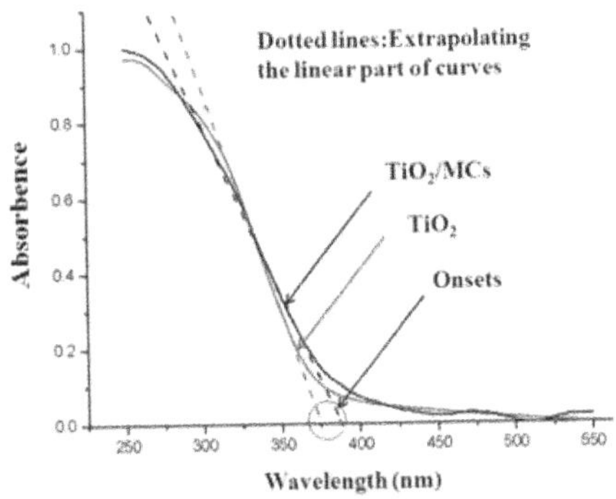


Figure 3. UV/vis absorption spectra of $TiO_2$ and $TiO_2$/MCs

UV/vis absorption spectra scanning

UV/vis spectroscopy was used to characterize the optical absorbance of $TiO_2$ and $TiO_2$/MCs, which were dispersed in methanol. Figure 3 shows the UV/vis absorption spectra of pure $TiO_2$and as-synthesized composites. Both spectra show main absorption peaks at about 270nm. However, MCs still affected the optical absorption of $TiO_2$. In the region of visible light zone, the absorption spectrum of $TiO_2$/MCs shows some small absorption peaks, which do not exist in that of bare $TiO_2$. Additionally, the absorption onset of $TiO_2$ in composites is 390nm, larger than that of pure $TiO_2$ (375nm, seen from dotted lines in Figure 3). According to Bedja[13], absorption peaks in a visible light region and the red shift of absorption onset were caused by trapped electrons in $TiO_2$. The band gap excitation of $TiO_2$ leads to charge separation followed by trapping of electrons. It might become the trap state centers of excitons, resulting in absorption peaks in the visible light region and the red shift of absorption onset.

Removal capability of $TiO_2$/MCs in MB solution

The photocatalytic performance of $TiO_2$/MCs was determined by using the oxidation of MB. The removal rates of MB with $TiO_2$/MCs and the reference are plotted as a function of irradiation time, which is shown in Fig.4. MB degradation without catalysts was less than 3%, much smaller than removal rates with $TiO_2$/MCs and $TiO_2$. The observed photocatalytic removal of MB comes predominantly from the photocatalytic degradation by $TiO_2$. It can be seen that $TiO_2$/MCs composites show a good photocatalytic performance and their removal rate is higher than the sum of removal rates of pure $TiO_2$ and MCs.

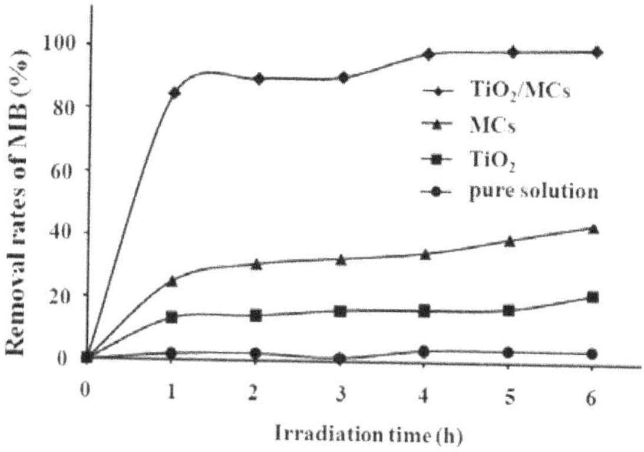


Figure 4. Removal rates of MB with $TiO_2$ , $TiO_2$/MCs, MCs, and pure solution by bare UV light

The results indicate the enhancement of photocatalytic performance of $TiO_2$ with MCs substrates. In the system of $TiO_2$/MCs composites, MCs not only acted as supports for $TiO_2$, but also as a physical absorbent. MB absorbed by MCs was transmitted to the surface of $TiO_2$ by diffusion, and was concentrated around $TiO_2$ particles for photocatalytic reaction.

## SUMMARY

New porous composites consisting of morphgenetic carbon derived from biomass and coated with nanometer sized $TiO_2$ prepared using the sol-gel method to get higher performance in photocatalytic degradation of organic pollutants in water. The results can be summarized as follows:

(1) The walls of the MC's micro-sized pores provided sites for the $TiO_2$ coating particles.

(2) There is a synergistic effect between the MCs and the $TiO_2$ for the degradation of MB, in which MB can be absorbed by MCs and photo-degraded by $TiO_2$ efficiently.

## ACKNOWLEDGEMENTS

The degradation rate of MB was measured successfully by using UV-VIS Spectrometer. We thank to Prof. S. Ohara and Prof. K. Sato in JWRI, Osaka University for their kind help.

## REFERENCES

[1] J. Muscat, V. Swamy, and N.M. Harrison, "First-principles calculations of the phase stability of $TiO_2$", *Phys. Rev. B*, **65,** 2241121-5 (2002).

[2] H. Zhang, J.F. Banfield, "Thermodynamic analysis of phase stability of nanocrystalline titania", *J. Mater. Chem.*, **8**, 2073-76 (1998).

[3] R. Vogel, K. Pohl, and H. Weller, "Sensitization of highly porous, polycrystalline $TiO_2$ electrodes by quantum sized CdS", *Chem. Phys. Lett.*, **174**, 241-246 (1990).

[4] A. Fujishima and K. Honda, "Electrochemical photolysis of water at a semiconductor electrode", *Nature*, **238**, 37-38 (1972).

[5] Steven N. Frank, and Allen J. Bard, "Heterogeneous photocatalytic oxidation of cyanide and sulfite in aqueous solutions at semiconductor powders", *J. Phys. Chem.*, **81**, 1484-88 (1977).

[6] K.E. O'Shea, E. Pernas and J. Saiers, "Influence of mineralization products on the coagulation of $TiO_2$ photocatalyst", *Langmuir,* **15,** 2071-76 (1999).

[7] B.J. Aronson, C.F. Blandford, A. Stein, "Solution-Phase Grafting of Titanium Dioxide onto the Pore Surface of Mesoporous Silicates: Synthesis and Structural Characterization", *Chem. Mater.*, **9**, 2842-51 (1997).

[8] Z. Ding, X. Hu, P.L. Yue, G.Q. Lu, and P.F. Greenfield, "Synthesis of anatase $TiO_2$ supported on porous solids by chemical vapor deposition", *Catalysis Today*, **68**, 173-182 (2001).

[9] R. Van Grieken, J. Aguado, M.J. López-Muoz, and J. Marugán, "Synthesis of size-controlled silica-supported $TiO_2$ photocatalysts", *J. Photochem. and Photobio. A: Chemistry*, **148**, 315-322 (2002).

[10] M. Nozawa, K. Tanigawa, M. Hosomi, T. Chikusa, and E. Kawada, "Removal and decomposition of malodorants by using titanium dioxide photocatalyst supported on fiber activated carbon", *Water Sci. Technol.*, **44**, 127-133 (2001).

[11] M. I. El-Barghouthi, A. H. El-Sheikh, Y.S.Al-Degs, and G. M. Walker, "Adsorption behavior of anionic reactive dyes on H-type activated carbon: Competitive adsorption and desorption studies", *Sep. Sci. Technol.*, **42 (10)**, 2195-2220 (2007).

[12] Z. Zhu, A. Li, S. Zhong, F. Liu, and Q. Zhang, "Preparation and characterization of polymer-based spherical activated carbons with tailored pore structure", *J. Appl. Polym. Sci.*, **109,** 1692-98 (2008).

[13] I. Bedja, S. Hotchandani, and P.V. Kamat, "Photoelectrochemistry of quantized $WO_3$ colloids. Electron storage, electrochromic, and photoelectrochromic effects", *J. Chem. Phys.*, **97**, 11064-70 (1993).

# MICROSTRUCTURE AND MECHANICAL PROPERTIES OF FRICTION STIR WELDED INCONEL 718 ALLOY

K.H. Song* and K. Nakata
Joining and Welding Research Institute, Osaka University,
11-1 Mihogaoka, Ibaraki, Osaka 567-0047, Japan

## ABSTRACT

This study was carried out to investigate the microstructure and mechanical properties of friction stir welded Inconel 718 alloy. Friction stir welding (FSW) was performed at a rotation speed of 200 rpm and a welding speed of 150 mm/min. As a result, the grain size was significantly refined from 5-20 μm in the base material to 1-3 μm in the stir zone, accompanied by dynamic recrystallization. This has an effect on the mechanical properties such that the microhardness and tensile strength in the stir zone are increased approximately 30 % and 25 %, respectively, compared to the base material.

## INTRODUCTION

Inconel 718, age-hardenable alloy, is widely used in aircraft engines, power plants, and gas turbines due to its good properties such as creep and corrosion resistance, and high strength at high temperature[1]. In general, to manufacture these parts, fusion welding such as with a $CO_2$ laser, a Nd-YAG laser or an electron beam has been applied lately[2-4]. However, for the case of applying the fusion welding on Inconel 718 alloy, there are some problems such as Boron / Niobium segregation, Laves phase, and liquation cracking in the weld zone or heat affected zone (HAZ) due to its higher heat input[5-6]. To avoid these weak points, the application of FSW is being investigated. FSW makes it possible to weld in the solid state, resulting in lower heat input than for fusion welding. However, FSW research on Inconel 718 has not been reported until now. Therefore, the present study was carried out to evaluate the application of FSW and to investigate the microstructure and mechanical properties of friction stir welded Inconel 718 alloy.

## EXPERIMENTAL PROCEDURES

The material used in this study was Inconel 718 alloy, and its chemical composition is shown in Table I. To carry out the FSW, 150 mm × 32.5 mm × 2 mm sheets samples were prepared. FSW was conducted using a WC-Co tool with a 15 mm diameter shoulder, and a 6 mm diameter probe, 1.8 mm in length. Also, to obtain sound welds, the tool was tilted at 3° forward

*Corresponding author. Tel.: 81-6-6879-8668; Fax.: 81-6-6879-8668
*E-mail address*: song@jwri.osaka-u.ac.jp (K.H. Song)

from the vertical, and argon gas was utilized to prevent surface oxidation during the welding. FSW was performed at a tool rotation speed of 200 rpm, a tool down-force of $39.2 \times 10^3$ N, and a traveling speed of 150 mm/min.

Table I. Chemical composition of the Inconel 718 alloy.

| Element | Ni | Cr | Fe | Mo | Nb | C | Mn | Si | S | Al | Ti |
|---|---|---|---|---|---|---|---|---|---|---|---|
| Mass % | 52.60 | 18.60 | 18.44 | 3.06 | 5.30 | 0.053 | 0.23 | 0.06 | 0.003 | 0.53 | 1.01 |

In order to observe the macrostructures of the welds, 15 ml HCl, 10 ml $HNO_3$ and 10 ml $CH_3COOH$ solution was prepared, and samples were surface etched after polishing with abrasive paper. Also, to evaluate the distribution of MC system carbides in the microstructure, scanning electron microscopy (SEM) and energy dispersive spectroscopy (EDS) analysis were employed. To evaluate the mechanical properties, Vickers microhardness and tensile tests were employed. Vickers hardness was carried out on the cross section of the weld zone with a load of 9.8 N and a dwell time of 15 s. For the tensile tests, two types of tensile test specimens were used to evaluate the transverse tensile strength of friction stir welds and the longitudinal tensile strength of the stir zone, respectively.

## RESULTS AND DISCUSSION

The top view and macrostructure of friction stir welded Inconel 718 alloy is shown in Figure 1. The specimen prepared at a welding speed of 150 mm/min was welded without any defects on the surface, as seen in Figure 1(a). Also, this specimen was penetrated 1.7 mm from the top without any defects in the weld zone; however, a band structure was observed in the center of the stir zone, as seen in Figure 1(b). A similar result was previously obtained in a research conducted on materials with a high melting point[7]. The temperature hysteresis during FSW is shown in Figure 2. The temperature was measured on the back side of plate at the center of the stir zone. The maximum temperature in the stir zone was approximately 740°C, and it was perfectly cooled from 740°C to 100°C for 30 seconds, as seen in Figure 2.

Results of SEM and EDS analyses of the welded material are shown in Figure 3. In the initial state, the base material consisted of grains ranging between 5 and 20 μm in size, with annealing twins in the grains and MC system carbides (arrows in the microstructure) in the grains and grain boundaries, as shown in Figure 3(a). On the other hand, the stir zone had more refined grains than those of the base material, ranging between 1 and 3 μm, as shown in Figure 3(b). Also, MC system carbides were observed similar to the base material; however,

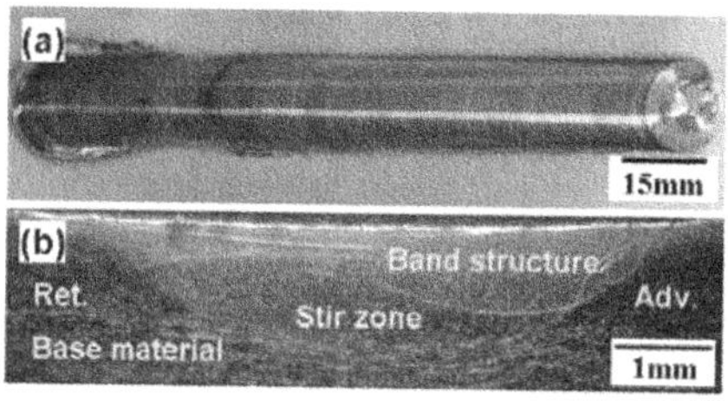


Figure 1. (a) An external shape and (b) macrostructure of the weld zone. Adv. and Ret. indicate advancing side and retreating side, respectively.

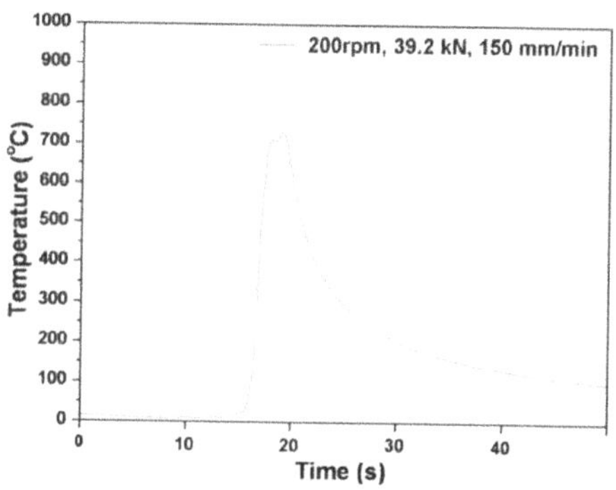


Figure 2. The temperature hysteresis in the stir zone during the friction stir welding.

more refined than those of the base material. Results of EDS on the MC system carbides observed in the microstructures are simply shown in Figure 3(c).

The grain refinement can be explained in terms of the dynamic recrystallization from the severe deformation and heat input during the FSW. FSW can be accompanied by a high strain rate (stored energy) and friction heat between material and tool[8]. Furthermore, the peak temperature measured in this work was 740°C in the center of the stir zone, enough to recrystallize Inconel 718 alloy, as seen in Figure 2. Also, the cooling time from the peak temperature (740°C) to low temperature (100°C) was approximately 30 seconds, which is enough to protect against the grain growth after the recrystallization, as seen in Figure 2. Therefore, accompanying the high strain rate, enough heat input and fast cooling time during the FSW can lead to the grain refinement more easily.

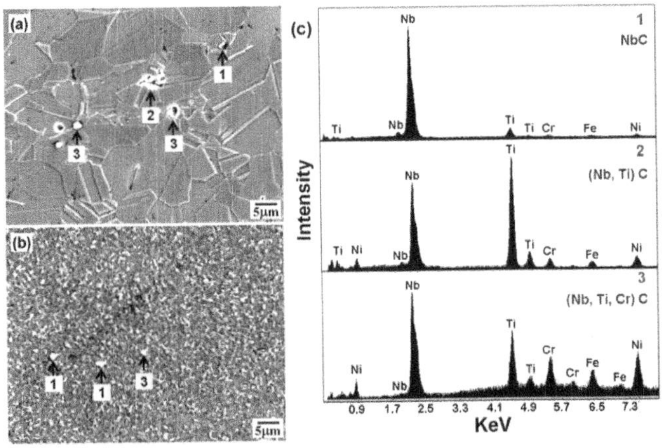


Figure 3. (a-b) SEM images and (c) EDS spectrums of the base material and the stir zone.

The material used in this study was Inconel 718 alloy, which has the low stacking fault energy of FCC metals[9]. In general, these materials can easily undergo dynamic recrystallization, compared to the material with high stacking fault energy such as Al alloys, while it is difficult to rearrange the dislocation by the dynamic recovery[10]. In other words, the materials with low stacking fault energy can make recrystallization nuclei more easily, compared to the materials with high stacking fault energy. Also, accompanying the higher dislocation density by severe deformation during the FSW can notably increase the density of recrystallization nuclei. Therefore, the recrystallization nuclei can be coincidently created at the grains and grain boundaries, which have a higher density of dislocations. As a result, grain refinement can be obtained by FSW.

The microhardness of the welded material is shown in Figure 4. The Vickers hardness measured for the base material was distributed between 265 Hv and 285 Hv, with an average hardness of 273 Hv. On the other hand, the microhardness of the stir zone was notably increased relative to that of the base material, ranging between 335 Hv and 370 Hv, with an average microhardness of 352 Hv. These increases in microhardness was caused by the grain refinement in the stir zone, with grains ranging between 1 and 3 μm, compared to the base material (between 5 and 20 μm), as seen in Figure 3. Generally, the grain size affects the mechanical properties such as hardness and strength. Therefore, this work showed clearly the effect of the grain refinement.

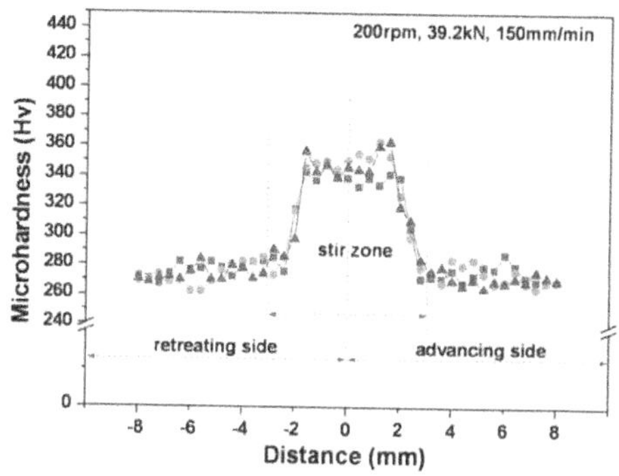


Figure 4. Distributions of Vickers microhardness in friction stir welded Inconel 718 alloy.

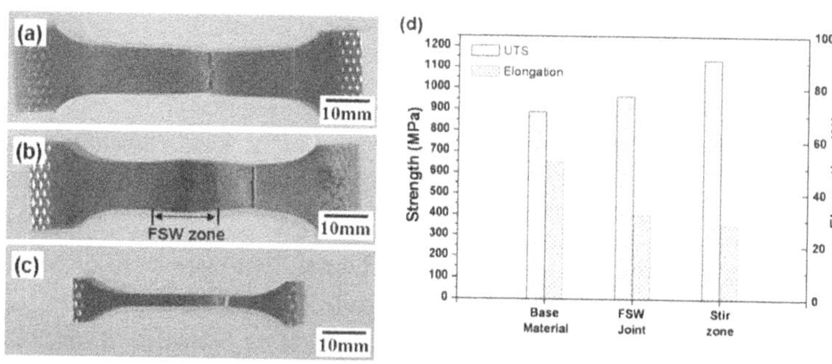


Figure 5. Transverse (weld joint) and longitudinal (stir zone) tensile properties of friction stir welded Inconel 718 alloy.

The results of the tensile tests on the friction stir welded Inconel 718 alloy are shown in Figure 5. The base material was fractured in the center of tensile specimen, with an elongated shape in the specimen, as seen in Figure 5(a), which exhibited 886 MPa in ultimate tensile strength (UTS) and 52% in elongation, as seen in Figure 5(d). However, the FSW jointed specimen was fractured at the base material, as seen in Figure 5(b). The stir zone specimen also exhibited an elongated shape and fractured, as seen in Figure 5(c). The application of the welding led to a slight increase in UTS, while the elongation was decreased, compared to the base material. As a result, the UTS in the transverse direction of the welded material was 961 MPa, and the elongation was 32%. In case of the stir zone specimen, the UTS was 1135 MPa, significantly higher than that of the base material, while the elongation was 28%. In the FSW jointed specimen, the base material was first deformed and fractured due to its coarsened grain size, compared to the stir zone, which showed correctly the increase of strength by the grain refinement in the stir zone.

## CONCLUSIONS

FSW of Inconel 718 alloy can be successfully performed at 150 mm/min in welding speed without any defects in the welds. As a result, the application of FSW led to the grain refinement from 5-20 μm in the base material to 1-3 μm in the stir zone, accompanied by the dynamic recrystallization. This grain refinement was effective to achieve an improvement in mechanical properties. The microhardness notably increased from 273 Hv in the base material to 352 Hv in the stir zone. Tensile strength also increased from 886 MPa in the base material to 1135 MPa in the stir zone. Therefore, the application of FSW is effective to achieve the increase of grain refinement and mechanical properties of Inconel 718 alloy.

## REFERENCES

[1]E.A. Loria, “Status and prospects of alloy 718”, *J. Met.*, 40, 36-41 (1988).

[2]J.K. Hong, J.H. Park, N.K. Park, I.S. Eom, M.B. Kim and C.Y. Kang, “Microstructures and mechanical properties of Inconel 718 welds by $CO_2$ laser welding”, *J. Mater. Process. Tech.*, 201, 515-520 (2008).

[3]S. Gobbi, L. Zhang, J. Norris, K.H. Richter and J.H. Loreau, “High powder $CO_2$ and Nd-YAG laser welding of wrought Inconel 718” *J. Mater. Process. Tech.*, 56, 333-345 (1996).

[4]C.A. Huang, T.H. Wang, C.H. Lee and W.C. Han, “A study of the heat-affected zone (HAZ) of an Inconel 718 sheet welded with electron-beam welding (EBW)”, *Mater. Sci. Eng. A*, 398, 275-281 (2005).

[5]W. Chen, M.C. Chaturvedi and N.L. Richards, “Effect of boron segregation at grain boundaries on heat affected zone cracking in wrought Inconel 718”, *Metall. Trans. A*, 32, 931-939 (2001).

[6]T.J. Kelly, “Elemental effects on the cast 718 weldability”, *Weld. J.*, 68, 44-51 (1989).

[7]K.H. Song, H. Fujii, K. Nakata, “Evaluation of grain refinement and mechanical property on friction stir welded Inconel 600”, *Mater. Trans.*, 50, 832-836 (2009).

[8]R.S. Mishra and M.W. Mahoney, “Friction stir welding and Processing”, ASM Int, (2007)

[9]J.M. Howe, Interfaces in Materials, Wiley-Interscience, (1997).

[10]F.J. Humphreys and M. Hatherly, Recrystallization and Related Annealing Phenomena. 2nd ed.: Elsevier, (2004).

# IN SITU OBSERVATION OF PHASE TRANSFORMATIONS IN Fe-0.08C ALLOY

Di Zhang, Hidenori Terasaki and Yuichi Komizo
Joining and Welding Research Institute, Osaka University
11-1 Mihogaoka, Ibaraki, Osaka, 562-0032, Japan

Corresponding Author: Di Zhang
Tel & Fax: +81-6-6879-8671, E-mail: zhangdi@jwri.osaka-u.ac.jp

## ABSTRACT

In situ observations were carried out using laser scanning confocal microscopy to track the phase transformations in Fe-0.08C alloy. The nucleation and growth processes of bainite during isothermal holding, and martensite during cooling were clearly observed.

## INTRODUCTION

When the austenite (γ) phase with a face-centered cubic (fcc) structure in steels is cooled to low temperature, new phases are formed with a body-centered cubic (bcc) or a body-centered tetragonal structure. Rapid quenching to the room temperature often results in the formation of martensite, a very hard structure in which the carbon, formerly in solid solution in the austenite, goes into solution in the new phase[1]. At a relatively lower cooling rate or in the medium temperature range, the bainite transformation occurs in which lath-shaped fine aggregates of ferrite and cementite are formed that possess some of the properties of the high temperature reactions involving ferrite and pearlite as well as some of the characteristics of the martensite transformation[2]. The rate of cooling in the bainite transformation must be appropriate to suppress the higher temperature diffusion-controlled ferrite and pearlite reactions, as well as low temperature shear-controlled martensite reactions.

In this paper, we studied morphological development and phase transformations during isothermal and continuous cooling processes. The nucleation and growth of bainite and martensite were directly observed in real time by laser scanning confocal microscopy.

## EXPERIMENTAL PROCEDURE

The materials used in the present study were Fe-0.08C alloys. The chemical compositions are listed in Table I. The sample disc was machined into 5 mm in diameter and 2 mm in height, and the observed plane was mirror polished. In situ observation was carried using a high-temperature laser scanning confocal microscopy (LSCM) combined with an infrared imaging furnace (IIF), schematically shown in Figure 1. An elaborate description was shown in the previous literature[3-5].

The sample was directly austenitized at 1400°C for 60 s, isothermally held at 480°C for 900 s and continuously cooled to the ambient temperature. The input and output thermal cycles are shown in Figure 2.

Table I. Chemical compositions of the material

| C | Si | Mn | P | S | Cr | Ni | Mo | N | O |
|---|---|---|---|---|---|---|---|---|---|
| 0.08 | 0.5 | 1.8 | 0.003 | 0.001 | 1.0 | 3.0 | 0.65 | 0.002 | 0.002 |

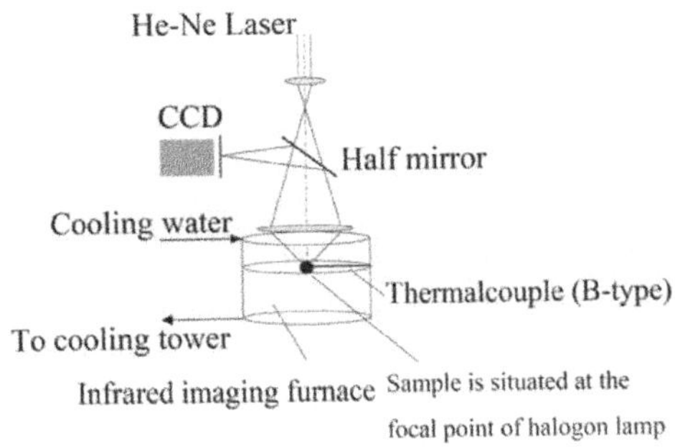


Figure 1. Illustration of laser scanning confocal microscopy.

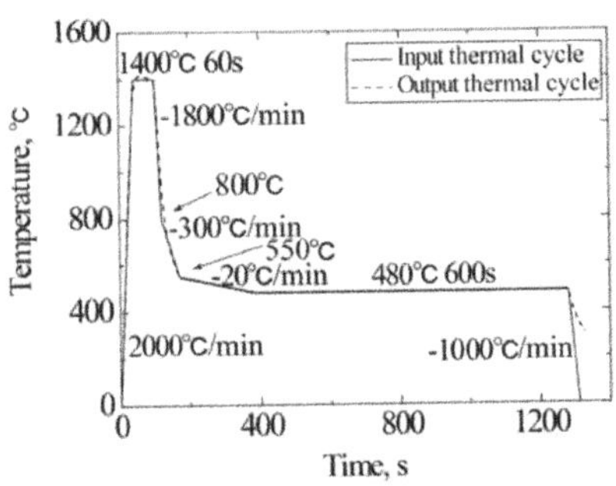


Figure 2. In situ thermal cycles.

## RESULTS AND DISCUSSION

Figure 3 shows the results of in situ morphological evolution of bainite transformation. It is shown in Figure 3(b) that bainite transformation takes place both at the prior austenite grain boundary and in the austenite grain. Subsequently, the bainitic ferrite nucleate sympathetically at the broad side of the primary bainitic ferrite, which is called sympathetic nucleation, typically shown in Figure 3(c) and (d). During isothermal holding, the laths, blocks, and packets gradually appear within the austenite grain by sympathetic nucleation and cause impingement between the bainitic ferrites, shown in Figure 3(e) and (f). It is interesting to note that the transformation rate of bainitic ferrite decreases with increasing isothermal holding time. The untransformed austenite phase in each austenite grain is consumed by the lower temperature product during continuous cooling process.

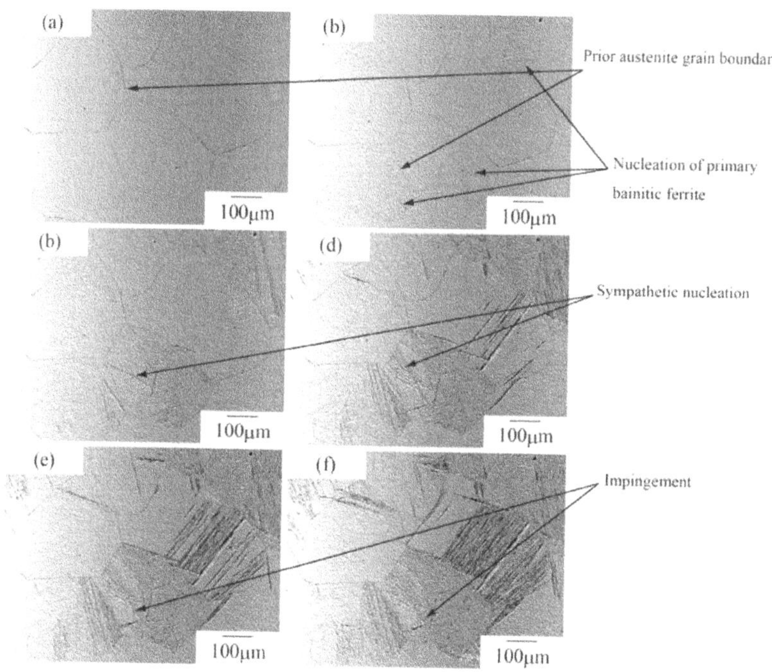


Figure 3. Morphological evolution of bainite transformation. (a) 642°C, 0s; (b) 498°C, 182.3s; (c) 483°C, 225.3s; (d) 480°C, 259.5s; (e) 480°C, 294.1s; (f) 480°C, 374.3s.

Figure 4 shows the results of in situ morphological evolution of martensite transformation during cooling process. The martensite reaction begins at a martensite start temperature, $M_s$, as shown in Figure 4(b), until the reaction ceases at the martensite finish temperature, $M_f$, shown in Figure 4(f). The martensite transformation is readily distinguished from bainite transformation by its appearance under the in situ observation. The retained austenite after bainite transformation, transforms by the sudden formation of thin plates or laths, blocks, packets of martensite with striking crystallographic character. These are the surface relief to the polished surface (observed plane)[6-8].

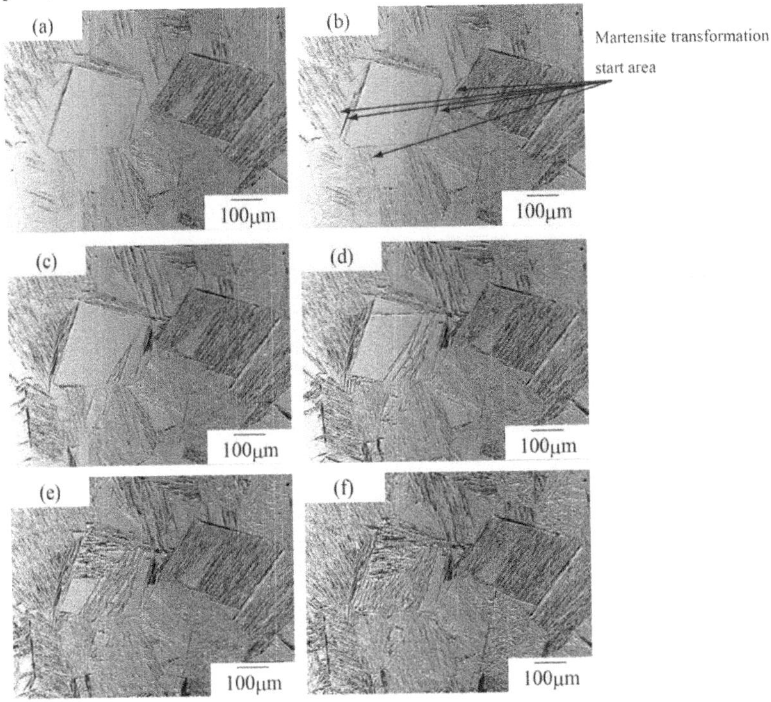


Figure 4. Morphological evolution of martensite transformation (a) 480°C, 0s; (b) 404°C, 47.7.3s; (c) 387°C, 51.6s; (d) 380°C, 53.3s; (e) 375°C, 54.6s; (f) 363°C, 57.9s.

The experimental results of the transformation process can be summarized as shown in Figure 5, with time and temperature. The $\gamma$ to $\alpha$ transformation takes place at 498°C. The martensite transformation initiates at 419°C and finishes at 363°C.

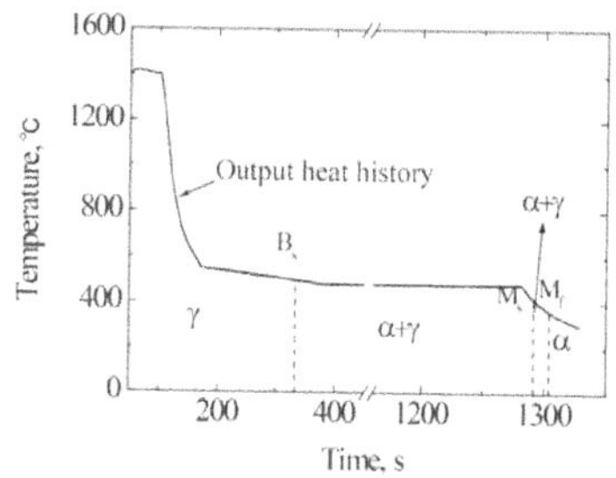


Figure 5. Summarized phase transformations during the cooling process. The symbols of α, γ, $B_s$, $M_s$ and $M_f$ represent α-ferrite and γ-austenite, bainite start temperature, martensite start temperature and martensite finish temperature, respectively.

## CONCLUSIONS

This paper investigated the phase transformation process in Fe-0.08C alloy. Nucleation at the grain boundary, sympathetic nucleation, and impingement of bainitic ferrite were directly shown in real time. Subsequently, laths, blocks and packets of martensite suddenly appeared in each grain of retained austenite. Additional experiments and modeling are needed for more in-depth investigation of the bainite and martensite transformations in steels.

## REFERENCES

[1]R.W.H. Honeycombe, Steels Microstructure and Properties, Edward Arnold (Publishers) Ltd., 1981 pp77.

[2]R.W.H. Honeycombe, Steels Microstructure and Properties, Edward Arnold (Publishers) Ltd., 1981 pp106.

[3]H. Terasaki, Y. Komizo, "In situ observation of morphological development for acicular ferrite in weld metal", Sci. Technol. Weld. Join., **11**, 561-566 (2006).

[4]E. Schmidt, Y. Wang, S. Sridhar, "A study of nonisothermal austenite formation and decomposition in Fe-C-Mn alloys", Metall. Mater. Trans., **37A**, 1799-1809 (2006).

[5]H. Chikama, H. Shibata, T. Emi, M. Suzuki, "In situ real time observation of planar to celluar and celluar to dendritic transition of crystals growing in Fe-C alloy melts", ISIJ Int., **37**, 620-626 (1996).

[6]D. Zhang, H. Terasaki, Y. Komizo, "In situ observation of phase transformation in Fe-0.15C binary alloy", J. Alloy. Compd, **484**, 929-933 (2009).

[7]D. Zhang, H. Terasaki, Y. Komizo, "Morphological and crystallographic evolution of bainite transformation in Fe-0.15C binary alloy", Microsc. Res. Tech, in press.

[8]S. Zhang, H. Terasaki, Y. Komizo, "In situ observation of martensite transformation of low carbon high alloy steel using high temperature laser scanning confocal microscopy and X-ray diffraction by synchrotron radiation", Tetsu-to-hagane, **96**, 8-13 (2010).

# Author Index

Printed and bound by CPI Group (UK) Ltd, Croydon, CR0 4YY

16/09/2026

14963457-0005